Marine Cosmeceuticals

Trends and Prospects

Marine Cosmeceuticals

Trends and Prospects

Edited by

Se-Kwon Kim

CRC Press
Taylor & Francis Group
Boca Raton London New York

CRC Press is an imprint of the
Taylor & Francis Group, an **informa** business

CRC Press
Taylor & Francis Group
6000 Broken Sound Parkway NW, Suite 300
Boca Raton, FL 33487-2742

Version Date: 20111012

International Standard Book Number-13: 978-1-4398-6029-8 (Hardcover)

Visit the Taylor & Francis Web site at
http://www.taylorandfrancis.com

and the CRC Press Web site at
http://www.crcpress.com

Contents

Preface

Science is the intellectual activity carried on by humans that is formulated to discover newer and novel information about the natural world in which we thrive. Humans have developed great skills to organize and use this basic information and translate it into meaningful output for their well-being through logical approaches. Cosmeceutical science is one branch of this never-ending curiosity of humans to design the best out of the existing framework of resources in the natural environment. Although a very promising field, it is yet to achieve its bloom. A vast exploration of various animal and plant sources in search of cosmeceuticals has led to some efficient products. However, most of them are now being associated with secondary effects of unknown magnitude of complications. The idea of this book is to investigate oceanic sources as an alternative to these complications and survey the possibilities of bringing together this vast treasure of marine environment toward safer cosmeceutic development. As an editor of this book, I have aimed to accumulate the latest research in this field from across the globe and highlight the potential of marine micro- and macroflora and fauna as effective agents for the development of novel cosmeceuticals. The concept of this book had been developing since I started researching hair growth factors and anti-inflammatory compounds of marine origin. I always used to marvel at the gems of the aquatic milieu for their possession of substances par excellence and wonder why the exploration and exploitation of marine cosmeceuticals is a long-time task. The question of why, despite this great potential, just a few cosmeceutical products from marine sources have entered the cosmetic market has pushed me to commit to the complex task of compiling a book on this topic of utmost interest. I looked into the research groups working in the field of aquatic sciences, marine environment in particular, and started reviewing their concepts of developing marine-derived cosmeceuticals. After thorough research, I decided on a collective contribution of researchers from different countries for their input, and finally wrote this book to present a consolidated overview of the marine environment as a productive source of novel cosmeceuticals, and the future prospects of seeing marine sources as cosmeceutics factories.

With contributions from Korea, Japan, China, India, Spain, France, Chile, Malaysia, Canada, Italy, Sri Lanka, and Taiwan, this book holds great insight for cosmetologists and marine researchers in general. My sincere thanks to all the contributors, who have brought the best themes and concepts to the book. The layout of the book has been kept simple and understandable with explanations wherever possible. Part I starts with a general introduction to cosmeceuticals and marine cosmeceutical sources. It covers an overview of the biological activities of different marine sources and the class of cosmeceutical agents that they may deliver. This is followed by the chemical properties of marine cosmeceuticals, with special emphasis on chitosan as a cosmeceutical agent, in Part II. The most extended part of this book, Part III, covers the range of biological properties and bioactive promise that the oceanic environment holds in the field of cosmetology. The information covered in this part is useful for cosmetologists and marine researchers looking for marine alternatives to existing cosmeceuticals in terms of skin whitening, antiwrinkling, photo protection, tissue repair, and antiacne, as well as hair growth. Collagen is an important protein in our skin and its degradation results have been associated with aging. Part III also covers information on marine fish and sponge-derived collagens as effective skin treatment agents to be utilized in antiaging formulations. Biotechnology is an emerging science that in conjunction with marine biology and microbiology may lead to wonders in the field of cosmetology. I have explored the role of biotechnology in marine cosmeceutical science, and Part IV of this book deals with this, where we have presented different strategies for enhanced cosmeceutical production. Part V wraps up the book with the industrial applications of marine cosmeceuticals. Apart from other applications, this part covers information on aquatic and marine bioactive antimicrobial peptides for the cosmeceutical industry, which is an important aspect of marine protein chemistry. This book offers comprehensive knowledge on novel marine cosmeceutical agents, their biological and chemical properties, as well as industrial applications.

I owe deep gratitude to the people who helped me and encouraged me in the process of writing this book. My students, who are the budding researchers, have extended their helping hands for this effort of mine. I would also like to thank Taylor & Francis for their willingness in this project and publishing this book with their in-depth professionalism and sense of excellence. I hope reading this book is a great experience for you.

Se-Kwon Kim

Editor

Se-Kwon Kim, PhD, is a professor of marine biochemistry in the Department of Chemistry, Pukyong National University (PKNU), Busan, South Korea. Dr. Kim received his MSc and PhD degrees from PKNU and became a faculty member there. He conducted his postdoctoral study at the Bioprocess Laboratory, Department of Food Science and Technology, University of Illinois, Urbana-Champaign, Illinois (1988–1989). He became a visiting scientist at the Memorial University of Newfoundland in Canada (1999–2000).

In 2004, Dr. Kim became the director for the Marine Bioprocess Research Center (MBPRC) at Pukyong National University. He served as president of the Korean Society of Chitin and Chitosan (1986–1990) and the Korean Society of Marine Biotechnology (2006–2007). Dr. Kim was also the chairman for 7th Asia-Pacific Chitin and Chitosan Symposium, which was held in South Korea in 2006. He is one of the board members of the International Society of Marine Biotechnology (IMB) and the International Society for Nutraceuticals and Functional Foods (ISNFF).

He was the editor-in-chief of the *Korean Journal of Life Sciences* (1995–1997), the *Korean Journal of Fisheries Science and Technology* (2006–2007), and the *Korean Journal of Marine Bioscience and Biotechnology* (2006–present). To the credit of his research, he won Best Paper awards from the American Oil Chemists' Society (AOCS) and the Korean Society of Fisheries Science and Technology (KSFST) in 2002.

His major research interests are investigation and development of bioactive substances derived from marine organisms and their application as cosmeceuticals and nutraceuticals via marine bio-processing and mass production technologies. He expanded his research fields into the industrialization of antiphotoaging and photoprotective substances from marine organisms. To date, Dr. Kim has authored over 450 research papers and holds 72 patents. In addition, he has written or edited more than 30 books.

Contributors

Masayuki Abe
Faculty of Fisheries Sciences
Hokkaido University
Hokkaido, Japan
and
Kaneka Co.
Osaka, Japan

Niuris Acosta Contreras
Instituto de Estudios Biofuncionales
Departamento de Química Física II
Facultad de Farmacia
Universidad Complutense
Madrid, Spain

Chang-Bum Ahn
School of Food Technology and Nutrition
Chonnam National University
Busan, Republic of Korea

Abdul Bakrudeen Ali Ahmed
Institute of Biological Sciences
Faculty of Science
University of Malaya
Kuala Lumpur, Malaysia

Yasantha Athukorala
School of Nutrition
Ryerson University
Toronto, Ontario, Canada

Sumathy Babitha
Department of Biological Engineering
National Research Lab of Skin Bioactive Materials
Inha University
Incheon, Republic of Korea

Soon-Sun Bak
Marine Bioprocess Research Center
Pukyong National University
Busan, Republic of Korea

Ira Bhatnagar
Marine Biochemistry Laboratory
Department of Chemistry
Pukyong National University
Busan, Republic of Korea
and
Infectious Diseases
Centre for Cellular and Molecular Biology
Hyderabad, India

Laura Calderón Muñoz
Instituto de Estudios Biofuncionales
Departamento de Química Física II
Facultad de Farmacia
Universidad Complutense
Madrid, Spain

Pamela Chávez-Crooker
Biotechnologies Aguamarina Research Center (CR97)
Antofagasta, Chile

Paola del Ciotto
R&D Lab
MAVI Sud
Aprilia (LT), Italy

Ruth Expósito Harris
Instituto de Estudios Biofuncionales
Departamento de Química Física II
Facultad de Farmacia
Universidad Complutense
Madrid, Spain

Véronique Fabien-Soulé
Regulatory Affairs
Puteaux, France

Dinanath B. Fulse
Synthetic Cellular Chemistry Laboratory
Advanced Science Institute RIKEN
Saitama, Japan

Yoshihiko Hayashi
Department of Cariology
Graduate School of Biomedical Sciences
Nagasaki University
Nagasaki, Japan

Ángeles María Heras Caballero
Instituto de Estudios Biofuncionales
Departamento de Química Física II
Facultad de Farmacia
Universidad Complutense
Madrid, Spain

Masashi Hosokawa
Faculty of Fisheries Sciences
Hokkaido University
Hokkaido, Japan

Takeshi Ikeda
Department of Cariology
Nagasaki University Hospital
Nagasaki, Japan

Jae-Young Je
School of Food Technology and Nutrition
Chonnam National University
Yeosu, Republic of Korea

You-Jin Jeon
Faculty of Marine Biomedical Science
Jeju National University
Jeju, Republic of Korea

Fatih Karadeniz
Department of Chemistry
Pukyong National University
Busan, Republic of Korea

Eun-Ki Kim
Department of Biological Engineering
National Research Lab of Skin Bioactive Materials
Inha University
Incheon, Republic of Korea

Se-Kwon Kim
Department of Chemistry, and
Marine Bioprocess Research Center
Pukyong National University
Busan, Republic of Korea

Spandana R. Kopalli
Department of Pharmacology
College of Medicine
Seoul National University
Seoul, Republic of Korea

Sushruta Koppula
College of Biomedical and Health Sciences
Konkuk University
Chungbuk-Do, Chungju, Republic of Korea

Maheshika S. Kurukulasuriya
Department of Animal Science
Faculty of Agriculture
University of Peradeniya
Peradeniya, Sri Lanka

Elena Lecumberri Lima
Instituto de Estudios Biofuncionales
Departamento de Química Física II
Facultad de Farmacia
Universidad Complutense
Madrid, Spain

Xiuping Lin
Key Laboratory of Marine Bioresources Sustainable Utilization
Guangdong Key Laboratory of Marine Materia Medica
Research Center for Marine Microbes
South China Sea Institute of Oceanology
Chinese Academy of Sciences
Guangzhou, China

Yonghong Liu
Key Laboratory of Marine Bioresources Sustainable Utilization
Guangdong Key Laboratory of Marine Materia Medica
Research Center for Marine Microbes
South China Sea Institute of Oceanology
Chinese Academy of Sciences
Guangzhou, China

María Ángeles López Mengíbar
Instituto de Estudios Biofuncionales
Departamento de Química Física II
Facultad de Farmacia
Universidad Complutense
Madrid, Spain

Jenn-Kan Lu
Department of Aquaculture
National Taiwan Ocean University
Keelung, Taiwan

Rosna Mat Taha
Institute of Biological Sciences
Faculty of Science
University of Malaya
Kuala Lumpur, Malaysia

Nana Mikami
Faculty of Fisheries Sciences
Hokkaido University
Hokkaido, Japan
and
Faculty of Nursing and Human Nutrition
Yamaguchi Prefectural University
Yamaguchi, Japan

Kazuo Miyashita
Bio-Functional Material Chemistry
Faculty of Fisheries Sciences
Hokkaido University
Hakodate, Japan

Gianluca Morganti
R&D Lab
MAVI Sud
Aprilia (LT), Italy

Pierfrancesco Morganti
Applied Cosmetic Dermatology
II University of Naples
and
China Medical University
Shenyang, China
and
I.S.C.D.
Rome, Italy
and
Nanoscience Centre
Mavi Sud S.r.l Via dell'Industria
Aprilia, Italy

Karla Moya
Biotechnologies Aguamarina Research Center (CR97)
Antofagasta, Chile

Johanna Obreque
Biotechnologies Aguamarina Research Center (CR97)
Antofagasta, Chile

Ramjee Pallela
Marine Bioprocess Research Center
Pukyong National University
Busan, Republic of Korea

Ratih Pangestuti
Marine Biochemistry Laboratory
Department of Chemistry
Pukyong National University
Busan, Republic of Korea

G.V.S. Bhanu Prakash
Toxicology Unit
Biology Division
Indian Institute of Chemical Technology
Hyderabad, India

Kenji Sato
Division of Applied Life Sciences
Graduate School of Life and Environmental Sciences
Kyoto Prefectural University
Shimogamo Kyoto, Japan

Mahinda Senevirathne
Marine Bioprocess Research Center
Pukyong National University
Busan, Republic of Korea

Yasutaka Shigemura
Division of Applied Life Sciences
Graduate School of Life and Environmental Sciences
Kyoto Prefectural University
Shimogamo Kyoto, Japan
and
Osaka Yuhigaoka Gakuen Junior College
Osaka, Japan

P.N. Sudha
Department of Chemistry
DKM College
Thiruvalluvar University
Tamilnadu, India

Jayachandran Venkatesan
Department of Chemistry
Pukyong National University
Busan, Republic of Korea

Janapala Venkateswara Rao
Toxicology Unit
Biology Division
Indian Institute of Chemical Technology
Hyderabad, India

Vazhiyil Venugopal
Seafood Technology Section
Food Technology Division
Bhabha Atomic Research Center
Mumbai, India

Jeanette Vera
Biotechnologies Aguamarina Research Center (CR97)
Antofagasta, Chile

Janak K. Vidanarachchi
Department of Animal Science
Faculty of Agriculture
University of Peradeniya
Peradeniya, Sri Lanka

Noel Vinay Thomas
Marine Biochemistry Laboratory
Department of Chemistry
Pukyong National University
Busan, Republic of Korea

M. Airanthi K. Widjaja-Adhi
Bio-functional Material Chemistry
Faculty of Fisheries Sciences
Hokkaido University
Hakodate, Japan

Isuru Wijesekara
Marine Biochemistry Laboratory
Department of Chemistry
Pukyong National University
Busan, Republic of Korea

W.A.J.P. Wijesinghe
School of Marine Biomedical Science
Jeju National University
Jeju, Republic of Korea

Jen-Leih Wu
Institute of Cellular and Organismic Biology
Academia Sinica
Taipei, Taiwan

Shizuka Yamada
Department of Cariology
Graduate School of Biomedical Sciences
Nagasaki University
Nagasaki, Japan

Kajiro Yanagiguchi
Department of Cariology
Nagasaki University Hospital
Nagasaki, Japan

Bin Yang
Key Laboratory of Marine Bioresources Sustainable Utilization
Guangdong Key Laboratory of Marine Materia Medica
Research Center for Marine Microbes
South China Sea Institute of Oceanology
Chinese Academy of Sciences
Guangzhou, China

Xian-Wen Yang
Key Laboratory of Marine Bioresources Sustainable Utilization
Guangdong Key Laboratory of Marine Materia Medica
Research Center for Marine Microbes
South China Sea Institute of Oceanology
Chinese Academy of Sciences
Guangzhou, China

Yvonne V. Yuan
School of Nutrition
Ryerson University
Toronto, Ontario, Canada

Xue-Feng Zhou
Key Laboratory of Marine Bioresources Sustainable Utilization
Guangdong Key Laboratory of Marine Materia Medica
Research Center for Marine Microbes
South China Sea Institute of Oceanology
Chinese Academy of Sciences
Guangzhou, China

1 Cosmeceuticals from Marine Resources

Prospects and Commercial Trends

Se-Kwon Kim and Isuru Wijesekara

CONTENTS

1.1 INTRODUCTION

Cosmetics are defined as articles intended to be applied to the human body for cleansing, beautifying, promoting attractiveness, or altering the appearance without affecting body structure or functions (Nelson and Rumsfield, 1988). Cosmetic preparations that are also intended to prevent disease or affect the function or structure of the human skin are considered drugs, for example, sunscreen or antidandruff shampoos. Cosmeceuticals contain active ingredients as creams, lotions, and ointments. In addition, ingestible beauty products or nutricosmetics are offered as liquids, pills, or functional foods. These products are formulated with ingredients or nutrients believed to promote healthy skin, hair, and nails at cellular levels, and their key ingredients often include vitamins, minerals, botanical extracts, and antioxidants. The world's cosmeceuticals market in 2007 was estimated at US$60 billion, with a growth rate of 8 to 12%, but it increased

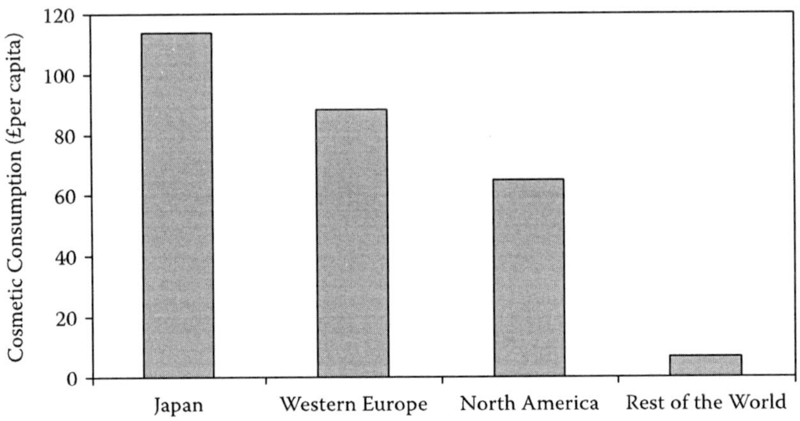

FIGURE 1.1 Cosmetic products consumption by various nations, excluding soap, toothpaste, and razors. 2008 provisional estimates. (From 2008 L'Oreal estimates, manufacturer selling prices. With permission.)

to £110.3 billion in 2008. In terms of the geographical market for cosmeceuticals, Japan is by far the biggest market, valued at $6 to $8 billion. The United States is estimated at around $5 to $6 billion, and the EU at $3 to $5 billion. China became the fourth largest market in 2008, and the South Korean cosmetic market is the fourth biggest among Asian cosmetic markets. Over the last few years, the cosmetic market has witnessed a huge change that has resulted in an entire shift in its value chain, with cosmetics production being shifted to low-cost countries like India and China. At the same time new channels of distribution are emerging, such as online retailing. The cosmetic products consumption in the world is depicted in Figure 1.1. Recently, a great deal of interest has been developed by consumers toward novel bioactive compounds as ingredients in cosmeceuticals from natural resources because of their numerous beneficial effects against synthetic ingredients. Marine resources have been demonstrated as rich sources of structurally diverse biologically active compounds with great cosmeceutical potential (Table 1.1). As more than 70% of the world's surface is covered by oceans, the wide diversity of marine organisms offers a rich source of natural products. The marine environment contains a source of functional materials, including polyunsaturated fatty acids (PUFAs), polysaccharides, essential minerals and

TABLE 1.1
Some Potential Cosmeceutical Ingredients from Marine Resources

Cosmetic Ingredient	Marine Resources
Potassium alginate	Brown algae
Squid ink powder	*Sepia subaculeate*
Mussel glycogen	*Nattallia japonica*
Aluminum silicate	Sea mud
Chitin/chitosan	Crustaceans
Shell powder	Oysters
Turtle oil	*Chelonia mydas*
Squalene	Shark species
Whale wax	*Physeter macrocephalus*
Dried sea water substance	Sea water
Carrageenan	Red algae
Fucoidan	Brown algae

vitamins, antioxidants, enzymes, and bioactive peptides (Kim and Wijesekara, 2010; Pomponi, 1999). This chapter focuses on prospects and novel trends of potential cosmeceutical ingredients from marine resources and presents a brief overview of their cosmeceutical activities with health benefits.

1.2 PROSPECTS AND TRENDS OF MARINE-DERIVED COSMECEUTICAL INGREDIENTS

Marine resources have been well recognized for their biologically active substances with a great potential to be used as cosmetics (Kim et al., 2008). Capitalizing on the growing need to look younger through natural and organic means, the need of marine biotechnology is the best way to help aging and inflammation, minimize the formation of free radicals, and slow the degradation of the skin. Moreover, much attention has been paid recently by the young generation to beauty as well as maintaining a young appearance with novel cosmetics containing natural bioactive ingredients. Therefore, the search for safe and inexpensive natural bioactive ingredients from marine resources as cosmetic ingredients is promising, and a few potential examples follow.

Phycoerythrin pigment protein in red algae can be used as a pigment in cosmetics. Incorporating marine proteoglycans as a nutricosmetic in skin capsules is said to boost collagen and help strengthen cell cohesion and improve skin density, creating firmer skin. New technologies are also on the horizon for incorporating popular nutritional ingredients into cosmetic formulas, including marine-derived ingredients such as omega-3 fatty acids. Among the reported effects of omega-3 fatty acids on skin is the stimulation of tissue repair and enhancement of collagen production. Bioactive peptides derived from Hawaiian red seaweeds are incorporated some cosmeceuticals to enhance collagen formation and firmer skin appearance. *Laminaria japonica*, a brown seaweed, stores marine minerals and trace elements in a highly concentrated form and uses them to generate special algae-based active ingredients that are used as a protective shield against all harmful influences, for example, UV radiation. *Laminaria* produces extremely effective moisture binding agents that prevent the alga from drying out at low tide. Therefore, *Laminaria* extract could be used as a potential cosmetic ingredient to make the skin firm and smooth. Marine phytoplankton is a skin "super food" that contains more than 65 amino acids, essential fats, vitamins, key minerals, trace elements, antioxidants, electrolytes, nucleic acids, and enzymes—virtually all the raw materials necessary for complete cellular regeneration. Able to deeply penetrate skin, it increases circulation and oxygenates tissues to repair, rebuild, and replenish healthy skin. In addition, marine-derived proteins such as collagen have potential as functional ingredients in cosmetics such as sunscreen lotions, shampoo, conditioners, hair gel, nail polish, and lipstick. Recently, the cosmetic application of collagen and gelatin, derived from terrestrial animals like cow and pig, have been declining because of animal diseases and some ethnic or religious barriers. New trends are pointing toward the use of other food sources such as marine fish-derived collagen and gelatin. These are excellent functional ingredients for the cosmetic industry. Collagen and gelatin have high moisturizing properties and can be produced as novel cosmetic creams and gels. With the invention of UV radiation protection compounds, antiwrinkling agents, and antiaging compounds in the cosmetic industry, new trends have focused on the manufacture of sunscreen lotions, creams, and other cosmetics. In this regard, marine-derived bioactive functional ingredients are promising functional ingredients for novel cosmeceuticals with health benefits.

1.3 POTENTIAL COSMECEUTICAL INGREDIENTS FROM MARINE RESOURCES

1.3.1 Seaweed Extract

Among marine organisms, seaweeds have been identified as an underexploited plant resource and a source of cosmeceutical ingredients. In addition, they have long been used in food diets as well

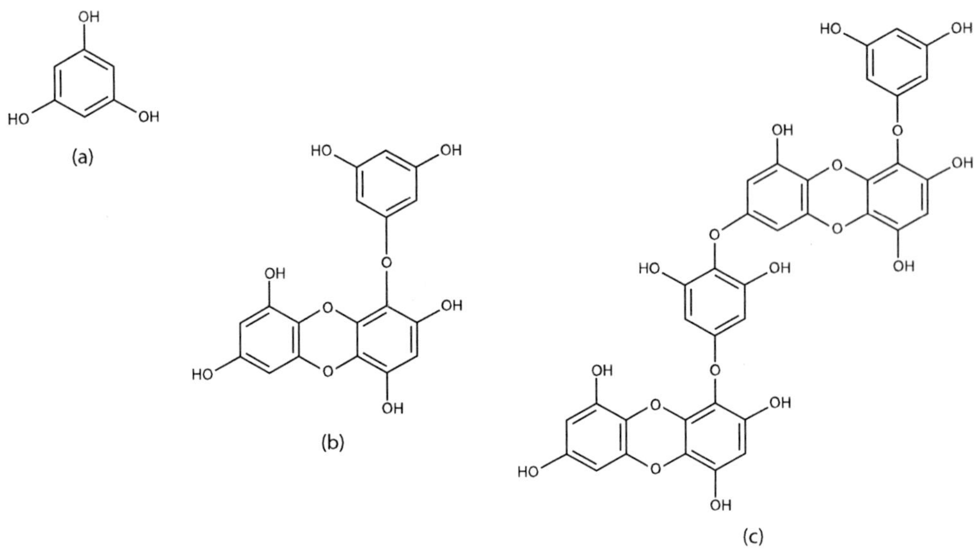

FIGURE 1.2 Some phlorotannins derived from marine algae (a) phloroglucinol, (b) eckol, and (c) dieckol.

as traditional remedies in Asian countries, mainly China, Japan, and Korea. Said to be rich in vitamins and minerals, and well known to tolerate desiccation, temperature variations, and high saline water, the components make up a product that could possibly head up the growing trend—enticing manufacturers with its multifunctional skin care properties. Recently, their importance as a source of novel bioactive substances to be used in the cosmetic industry has grown rapidly and researchers have revealed that marine algal originated compounds exhibit various biological activities (Wijesekara et al., 2010, 2011). In general, marine brown and red seaweeds are the common types used in cosmeceutical products. Seaweed extracts are rich in cosmeceutical ingredients such as phlorotannins, polysaccharides, and carotenoid pigments.

1.3.1.1 Phlorotannins

Phlorotannins are formed by the polymerization of phloroglucinol (1,3,5-tryhydroxybenzene) monomer units and biosynthesized through the acetate-malonate pathway, also known as the polyketide pathway. Phlorotannins are highly hydrophilic components with a wide range of molecular sizes ranging between 126 Da and 650 kDa. Marine brown algae such as *Ecklonia cava* accumulate a variety of phloroglucinol-based polyphenols as phlorotannins of low, intermediate, and high molecular weight containing both phenyl and phenoxy units. Furthermore, pholorotannins consist of phloroglucinol units linked to each other in various ways, and are of wide occurrence among marine organisms, especially brown and red algae (Wijesekara et al., 2011). Many researchers have shown that phlorotannins (Figure 1.2) derived from marine brown algae have potential effects on human skin health and beauty.

1.3.1.2 Polysaccharides

The cell walls of marine algae are rich in various bioactive polysaccharides such as fucoidans in brown algae, carrageenans in red algae, and ulvans in green algae. These polysaccharides exhibit many beneficial cosmeceutical effects. Compared to other polysaccharides, fucoidans (Figure 1.3) are commercially important ingredients that are widely available in various brown seaweeds such as *Laminaria japonica*, *Fucus vesiculosus*, *Undaria pinnatifida*, and *Hizikia fusiformis*; hence more and more fucoidans have been investigated in recent years to develop novel cosmetic products. Fucoidans have a high fucose polymer backbone and are usually highly branched, sulfated, and acetylated. Throughout the past few decades, there has been a growing interest in *in vitro* research

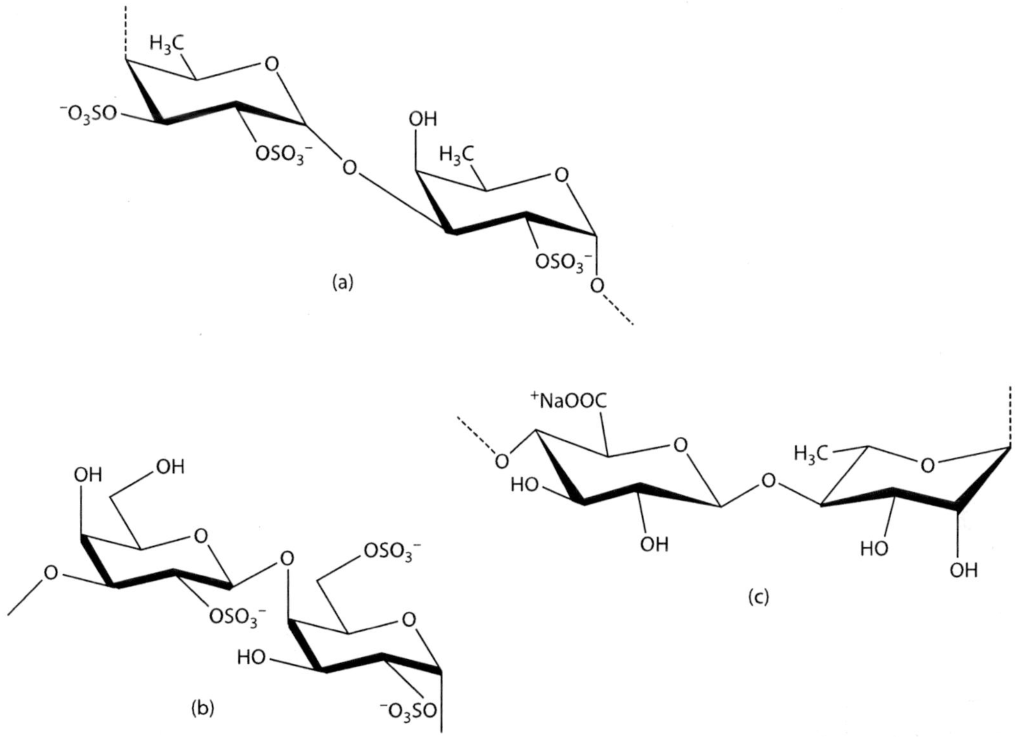

FIGURE 1.3 Sulfated polysaccharides derived from marine algae (a) fucoidan, (b) carrageenan, and (c) ulvan.

about bioactive properties of fucoidans. According to most of these studies, fucoidans have a promising potential in cosmetic use.

Alginates also are cell wall ingredients of mainly brown seaweeds. They are chain-forming heteropolysaccharides made up of blocks of mannuronic acid and guluronic acid. Alginates have the potential to be used as cosmetic additives to improve, modify, and stabilize the texture of cosmeceuticals. Moreover, alginates are highly functional hydrocolloids to thicken and stabilize emulsions at low pH values.

Carrageenans are sulfated linear polysaccharides with galactose and are extracted from red seaweeds. The most common source of carrageenan is *Gigartina* from southern Europe. This sulfated polysaccharide has been used as a viscosity-altering and thickening agent in cosmeceutical products.

1.3.1.3 Carotenoid Pigments

Carotenoids are a family of pigmented compounds that are synthesized by plants, algae, fungi, and microorganisms, but not animals. The cosmeceutical effects of carotenoid pigments depend on the number of conjugated double bonds of the molecule and carotenoid end groups or the nature of substituents in carotenoids containing cyclic end groups. Numerous seaweed-derived carotenoids are considered to be potential cosmetic ingredients. In this regard, fucoxanthin and astaxanthin derived from marine algae have been known to be major ingredients (Figure 1.4) of marine algal carotenoids with beneficial cosmeceutical effects.

1.3.2 Microalgae Extract

Microalgae extracts have been mainly used in the formulation of cosmetic and skin care products. There is a growing trend in marine microalgae, which offer a unique opportunity to deliver

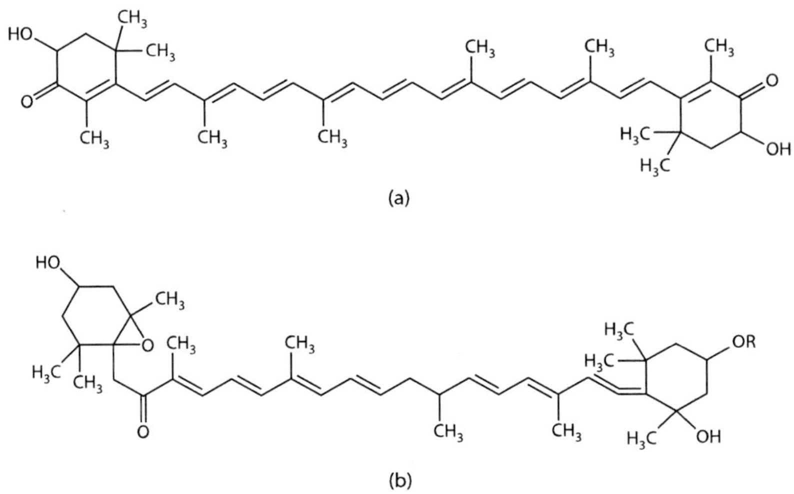

FIGURE 1.4 Caratenoids derived from marine algae (a) fucoxanthin and (b) astaxanthin.

functional ingredients for the cosmetic market with the advantages of being plant derived, natural, sustainable, and highly novel. Existing microalgal products include carotenoids and fatty acids, but it is certain that a vast number of new microalgal-based functional ingredients will appear over the coming years. Among these are novel polysaccharide products that could replace existing seaweed-derived polysaccharides and provide a wide range of novel functional saccharides for the development of functional cosmeceutical products. Some cosmetic companies maintain their own microalgal production systems. Extracts from marine microalgae are a rich source of bioactive proteins, vitamins, minerals, and carotenoid pigments such as astaxanthin. The extracts from *Arthrospira* and *Chlorella*, marine microalgal species, have been used in the cosmeceutical industry to make skin care products. *Chorella*, a unicellular green alga, contains various valuable proteins and has great potential to be used in novel cosmeceuticals.

1.3.3 Collagen

Collagen is the most abundant animal protein polymer, representing nearly 30% of the total protein in the animal body. It is the main structural element of bones, cartilage, skin, tendons, ligaments, blood vessels, teeth, cornea, and all other organs of vertebrates. The molecular structure of collagen contains three polypeptide α chains wound together in a tight triple helix. Each polypeptide α chain consists of a repeated sequence of triplets $(Gly\text{-}X\text{-}Y)_n$, where X and Y are often proline (Pro) and hydroxyproline (Hyp). For industrial purposes, collagen is extracted mainly from skin and bones of cattle and pigs. However, due to highly infectious and contagious animal diseases and some religious barriers, industrial use of collagen from these sources is becoming limited. Therefore, marine-derived collagen is an alternative with safe and economical advantages (Senaratne et al., 2006). Additionally, collagen improves the elasticity of skin by stimulating the production of collagen by the skin cells themselves, thus maintaining youthful and vibrant skin and neutralizing the skin's continuous aging deterioration by redensifying the dermis.

1.3.4 Bioactive Peptides

Marine-derived bioactive peptides are inactive within the sequence of their parent protein and can be released by enzymatic hydrolysis of marine organisms and seafood by-products. Bioactive peptides usually contain 3–20 amino acid residues, and their activities are based on their amino acid

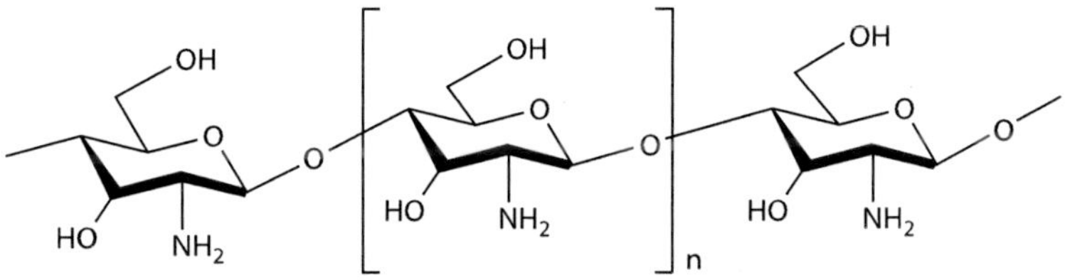

FIGURE 1.5 Structure of chitosan.

composition and sequence. Biologically active marine peptides are food-derived peptides that function beyond their nutritional value, having physiological, hormone-like properties with possible roles in improving skin health.

1.3.5 Chitooligosaccharide (COS) Derivatives

Chitin is the second most abundant biopolymer on the earth after cellulose and one of the most abundant polysaccharides. It is a glycan of β (1→4)-linked N-acetylglucosamine units, and it is widely distributed in crustaceans and insects as the protective exo-skeleton and cell walls of most fungi. Chitin is usually prepared from the shells of crabs and shrimp. Chitosan (Figure 1.5), a partially deacetylated polymer of N-acetylglucosamine, is prepared by alkaline deacetylation of chitin (S. K. Kim et al., 2006). COSs are chitosan derivatives (polycationic polymers comprised principally of glucosamine units) and can be generated via either chemical or enzymatic hydrolysis of chitosan (Jeon and Kim, 2000a, 2000b). Recently, COSs have been the subject of increased attention in terms of their cosmeceutical and pharmaceutical applications (Kim and Rajapakse, 2005), due to their lack of toxicity and high solubility as well as their positive physiological effects on skin health.

1.3.6 Enzymes

Marine bacteria isolated from the deep sea produce specific enzymes to catalyze removal reactions of free radicals caused by UV damage at high temperatures. Specific enzymes from *Thermus thermophilus* have been used for developing cosmeceuticals for skin protection (Kim et al., 2008).

1.3.7 Sea Mud

Sea mud contains various nutrients and minerals, and even antibacterial components. Hence sea mud has been used for skin care and cosmeceutical products. Dead Sea mud is especially recognized for its therapeutic properties for psoriasis and other skin-related disorders. Dead Sea mud and waters rich in minerals can be applied to either the entire body or just parts. Dead Sea mud masques stimulate, tighten, clean, and invigorate the skin with a refreshing glow (Kim et al., 2008).

1.4 BIOLOGICAL ACTIVITIES OF MARINE COSMETIC INGREDIENTS ON SKIN HEALTH

1.4.1 Antioxidant Activity

Antioxidants may have a positive effect on human skin health as they can protect human skin against damage by reactive oxygen species (ROS), which attack macromolecules such as membrane lipids, proteins, and DNA, leading to skin disorders. Lipid oxidation by ROS such as superoxide anion, hydroxyl radicals, and H_2O_2 also causes a decrease in the youthful appearance of skin. Therefore, in the cosmeceutical as well as pharmaceutical industries, many synthetic commercial antioxidants such as butylated hydroxytoluene (BHT), butylated hydroxyanisole (BHA), tert-butylhydroquinone

(TBHQ), and propyl gallate (PG) have been used to retard the oxidation and peroxidation processes. However, the use of these synthetic antioxidants must be under strict regulation due to potential health hazards (Hettiarachchy et al., 1996; Park et al., 2001). Hence the search for natural antioxidants as safe alternatives is important in the cosmeceutical industry. Recently, there has been considerable interest in the cosmeceutical industry the development of antioxidants from natural sources, such as marine flora and fauna.

Among marine resources, seaweeds represent one of the richest sources of natural antioxidants such as phlorotannins, sulfated polysaccharides, and fucoxanthins (Mayer and Hamann, 2002; Ruperez, 2001). Marine-derived chitooligosaccharide derivatives and bioactive peptides recognized as potential antioxidants with free radical scavenging effects have the potential to be used as novel cosmeceutical ingredients.

1.4.2 Antiwrinkling Activity

Matrix metalloproteinase enzymes (MMPs) play an important role in the digestion of extracellular matrix components and are closely associated with wrinkle formation. Several types of MMPs, such as MMP-2 and MMP-9, have been reported. The inhibition of MMP activity by marine-derived chitooligosaccharides, flavonoides, polyphenols, and fatty acids has been previously reported (Zhang and Kim, 2009). M. M. Kim et al. (2006) for the first time reported a detailed study on the *in vitro* inhibitory effects of phlorotannins derived from *E. cava* on MMP activities. A novel gelatin digestion assay could visualize complete inhibition of bacterial collagenase-1 activity at 20 μg/ml of *E. cava* extract during preliminary screening assays. A sensitive fluorometric assay revealed that *E. cava* phlorotannins can specifically inhibit both MMP-2 and MMP-9 activities significantly ($p < 0.001$) at 10 μg/ml. In addition, artificially induced activities of MMP-2 and MMP-9 in human dermal fibroblasts and HT 1080 cells have been inhibited by *E. cava* extract in a more or less similar manner to the positive control doxycycline.

1.4.3 Antityrosinase Activity

Melanin contributes to the skin color and plays a crucial protective role against skin photocarcinogenesis. However, the production of abnormal melanin pigment is a serious skin problem. Melanin biosynthesis can be inhibited by avoiding UV exposure, the inhibition of tyrosinase enzyme, the inhibition of melanocytes metabolism and proliferation, or the removal of melanin with corneal ablation. Therefore, in an effort to inhibit the activity of tyrosinase, many different types of tyrosinase inhibitors have been developed from marine resources. These compounds are applicable for cosmetics as skin-whitening agents and also as drugs for use in the treatment of pigmentation disturbances. For example, a phlorotannin, 7-phloroeckol from marine brown seaweed, *Ecklonia cava* has been reported as a potential skin-whitening agent with antityrosinase activity via reducing melanogenesis (Yoon et al., 2009).

1.4.4 Antiacne Activity

Propionibacterium acnes is a bacterium, a major inhabitant of adult human skin, where it resides within sebaceous follicles, and has been implicated in acne vulgaris (pimples) formation. *P. acnes* bacteria are Gram-positive, non-spore-forming, anaerobic rods. They release lipases to digest a surplus of the skin oil, sebum, that has been produced, which stimulates an intense local inflammation that bursts from the hair follicle. This causes acne formation, a common cosmetic issue among youth close to adolescence. The inhibition of the growth of *P. acnes* has been recognized as a treatment for acne in the cosmeceutical industry. Sargafuran from marine brown alga *Sargassum macrocarpum* has been shown to have potent antiacne activity against *P. acnes* (Kamei et al., 2009). Chitooligosaccharide derivatives are well-known antibacterial compounds from marine resources.

Therefore, marine-derived bioactive compounds are potential cosmetic ingredients to be used as antiacne cosmetics.

1.5 CONCLUDING REMARKS

Natural cosmetics and personal care products that incorporate marine-based extracts are being increasingly sought after in the industry, with more and more consumers demanding products that are of natural origin, with the latest move into marine-based skin care and cosmeceuticals boosting their profile within the competitive market. Since the ocean is where life first started, it is no surprise, perhaps, that the biggest potential source for new, bioactive ingredients for the health and beauty market may originate from the same place. With so many new species of marine resources still to be discovered, the potential for new marine-derived compounds and extracts is immense, and the cosmeceutical industry is poised for accelerated development in the near future.

REFERENCES

Hettiarachchy, N. S., Glenn, K. C., Gnanasambandan, R., and Johnson, M. G. (1996). Natural antioxidant extract from fenugreek (*Trigonella foenumgraecum*) for ground beef patties. *Journal of Food Science*, 61, 516–519.

Jeon, Y. J., and Kim, S. K. (2000a). Continuous production of chitooligosaccharides using a dual reactor system. *Process Biochemistry*, 35, 623–632.

Jeon, Y. J., and Kim, S. K. (2000b). Production of chitooligosaccharides using ultrafiltration membrane reactor and their antibacterial activity. *Carbohydrate Polymers*, 41, 133–141.

Kamei, Y., Sueyoshi, M., Hayashi, K., Terada, R., and Nozaki, H. (2009). The novel anti-*Propionibacterium acnes* compound, Sargafuran, found in the marine brown alga *Sargassum macrocarpum*. *Journal of Antibiotics*, 62, 259–263.

Kim, M. M., Ta, Q. V., Mendis, E., Rajapakse, N., Jung, W. K., Byun, H. G., Jeon, Y. J., and Kim, S. K. (2006). Phlorotannins in *Ecklonia cava* extract inhibit matrix metalloproteinase activity. *Life Sciences*, 79, 1436–1443.

Kim, S. K., Nghiep, N. D., and Rajapakse, N. (2006). Therapeutic prospectives of chitin, chitosan and their derivatives. *Journal of Chitin Chitosan*, 11 (1), 1–10.

Kim, S. K., and Rajapakse, N. (2005). Enzymatic production and biological activities of chitosan oligosaccharides (COS): A review. *Carbohydrate Polymers*, 62, 357–368.

Kim, S. K., Ravichandran, Y. D., Khan, S. B., and Kim, Y. T. (2008). Prospective of the cosmeceuticals derived from marine organisms. *Biotechnology and Bioprocess Engineering*, 13, 511–523.

Kim, S. K., and Wijesekara, I. (2010). Development and biological activities of marine-derived bioactive peptides: A review. *Journal of Functional Foods*, 2, 1–9.

Mayer, A. M. S., and Hamann, M. T. (2002). Marine pharmacology in 1999: Compounds with antibacterial, anticoagulant, antifungal, antihelmintic, anti-inflammatory, antiplatelet, antiprotozoal and antiviral activities affecting the cardiovascular, endocrine, immune and nervous systems, and other miscellaneous mechanisms of action. *Comparative Biochemistry and Physiology Part C*, 132, 315–339.

Nelson, F. P., and Rumsfield, J. (1988). Cosmetics: Content and function. *International Journal of Dermatology*, 27, 665–672.

Park, P. J., Jung, W. K., Nam, K. D., Shahidi, F., and Kim, S. K. (2001). Purification and characterization of antioxidative peptides from protein hydrolysate of lecithin-free egg yolk. *Journal of American Oil Chemists Society*, 78, 651–656.

Pomponi, S.A. (1999). The bioprocess-technological potential of the sea. *Journal of Biotechnology*, 70, 5–13.

Ruperez, P. (2001). Antioxidant activity of sulphated polysaccharides from the Spanish marine seaweed Nori. In *Proceedings of the COST 916 European Conference on bioactive compounds in plants foods* (p. 114). Tenerife, Canary Islands, Spain: Health Effects and Perspectives for the Food Industry.

Senaratne, L. S., Park, P. J., and Kim, S. K. (2006). Isolation and characterization of collagen from brown backed toadfish (*Lagocephalus gloveri*) skin. *Bioresource Technology*, 97, 191–197.

Wijesekara, I., Pangestuti, R., and Kim, S. K. (2011). Biological activities and potential health benefits of sulfated polysaccharides derived from marine algae. *Carbohydrate Polymers*, 84, 14–21.

Wijesekara, I., Yoon, N. Y., and Kim, S. K. (2010). Phlorotannins from *Ecklonia cava* (Phaeophyceae): Biological activities and potential health benefits. *Biofactors*, 36, 408–414.

Yoon, N. Y., Eom, T. K., Kim, M. M., and Kim, S. K. (2009). Inhibitory effect of phlorotannins isolated from *Ecklonia cava* on mushroom tyrosinase activity and melanin formation in mouse B16F10 melanoma cells. *Journal of Agricultural and Food Chemistry*, 57, 4124–4129.

Zhang, C., and Kim, S. K. (2009). Matrix metalloproteinase inhibitors (MMPIs) from marine natural products: The current situation and future prospects. *Marine Drugs*, 7, 71–84.

2 Chemical and Biological Aspects of Marine Cosmeceuticals

Bin Yang, Xiuping Lin, Xue-Feng Zhou, Xian-Wen Yang, and Yonghong Liu

CONTENTS

2.1 INTRODUCTION

Cosmeceuticals are cosmetics formulated for topical application to enhance skin transformation and regeneration, wound healing, and antiaging of the skin to prevent skin wrinkles. The markets for cosmeceuticals have grown exponentially over the years and have become a feasible substitute for cosmetic or nonsurgical procedures and surgeries. Suffice to say that as long as there is a desire to maintain a young-looking appearance, there will be a business surge in the cosmetics market.

The world's cosmetic industry is worth tens of billions of U.S. dollars, and the industry is constantly seeking new, especially natural, products. This has resulted in numerous natural cosmetic products incorporating extracts of coconut, jasmine, lemon grass, longan, algae, and many other sources commonplace on the shelves of the world's cosmetic shops.

Marine organisms form a prominent component of the oceanic population which significantly contribute in the production of biologically efficient moieties with cosmeceutical and pharmaceutical properties. More than 12,000 novel chemicals from marine plants, animals, and microbes have been explored, and a large number of pharmaceutical products, enzymes, and biomaterials have been developed to date. Various compounds exhibit diverse degrees of biological activities like antibacterial, anticancerous, antiphotoaging, anti-inflammatory, antioxidative, etc.

This chapter depicts various compounds from marine sources for further elaborative research and their probable use in the cosmeceutical and pharmaceutical industries.

2.2 ANTIPHOTOAGING AND PHOTOPROTECTIVE COMPOUNDS

For convenience ultraviolet (UV) radiation is split into four major bands. Vacuum UV is radiation with a wavelength less than 200 nm. UV-C radiation occupies the region between 200 and 280 nm. Neither vacuum UV nor UV-C reaches the surface of the present-day earth because of atmospheric Rayleigh scattering and ozone absorption, although these regions are important parts of the extraterrestrial spectrum. UV-B radiation is energetically less damaging than UV-C. In the scientific literature, UV-B radiation is often defined as 280–320 nm. However, the legal definition provided by the International Commission on Illumination sets the UV-B radiation range as 280–315 nm. On earth, most of the UV-B radiation is attenuated by the ozone column that absorbs strongly in the Hartley region (200–300 nm) and weakly in the Huggins band (300–360 nm). Finally, UV-A radiation (315–400 nm) reaches the surface of the earth relatively unattenuated and is still less energetic than UV-B radiation.[1]

UV radiation, and particularly the higher-energy wavelengths, has a range of effects. They include DNA damage in most organisms, inhibition of photosynthetic primary productivity in microorganisms and higher plants, inhibition of nitrogenase activity, inhibition of heterocyst formation in some cyanobacteria, reduction in microbial motility, and a diversity of other responses that have been reviewed elsewhere.[2–5] The shorter the wavelength, the more energetic and potentially harmful is the radiation. UV-B radiation was recognized long ago as the cause of skin erythema (sunburn), and accumulated exposure results in DNA damage and immunosuppression, eventually leading to skin cancer.[1]

In response to intense solar radiation, organisms have evolved certain mechanisms such as avoidance, repair, and protection by synthesizing or accumulating a series of photoprotective compounds to counteract the toxicity of UV (particularly UV-B) radiation.[1] A number of photoprotective compounds, such as melanins, mycosporines, mycosporine-like amino acids (MAAs), scytonemin, parietin, usnic acid, carotenoids, phycobiliproteins, phenylpropanoids, flavonoids, and several other UV-absorbing substances of unknown chemical structure, have anti-inflammatory and antioxidant properties with possible applications in the cosmetic and pharmaceutical industries.[6–9]

2.2.1 Mycosporine-Like Amino Acids

2.2.1.1 Structure of MAAs

As a class of compounds, MAAs are polar (water soluble) with low molecular weights (<400 Da) and have a chemical structure based on either a cyclohexenone or cyclohexenimine ring structure with amino acid substituents. There are more than 20 different MAAs found in various organisms, some of which are shown in Figure 2.1. Generally, the ring system contains a glycine subunit at the third carbon atom. Some MAAs also contain sulfate esters or glycosidic linkages through the imine substituents. The difference in the absorption spectra of MAAs is due to the variations in the attached side groups and nitrogen substituents.[10,11] The conjugated double bonds within the molecule result in broad-band absorptions with wavelength maxima (λ_{max}) ranging from 310 nm in the ultraviolet-B for cyclohexenone-based structures (UV-B, 280–320 nm) to 360 nm in the ultraviolet-A (UV-A,

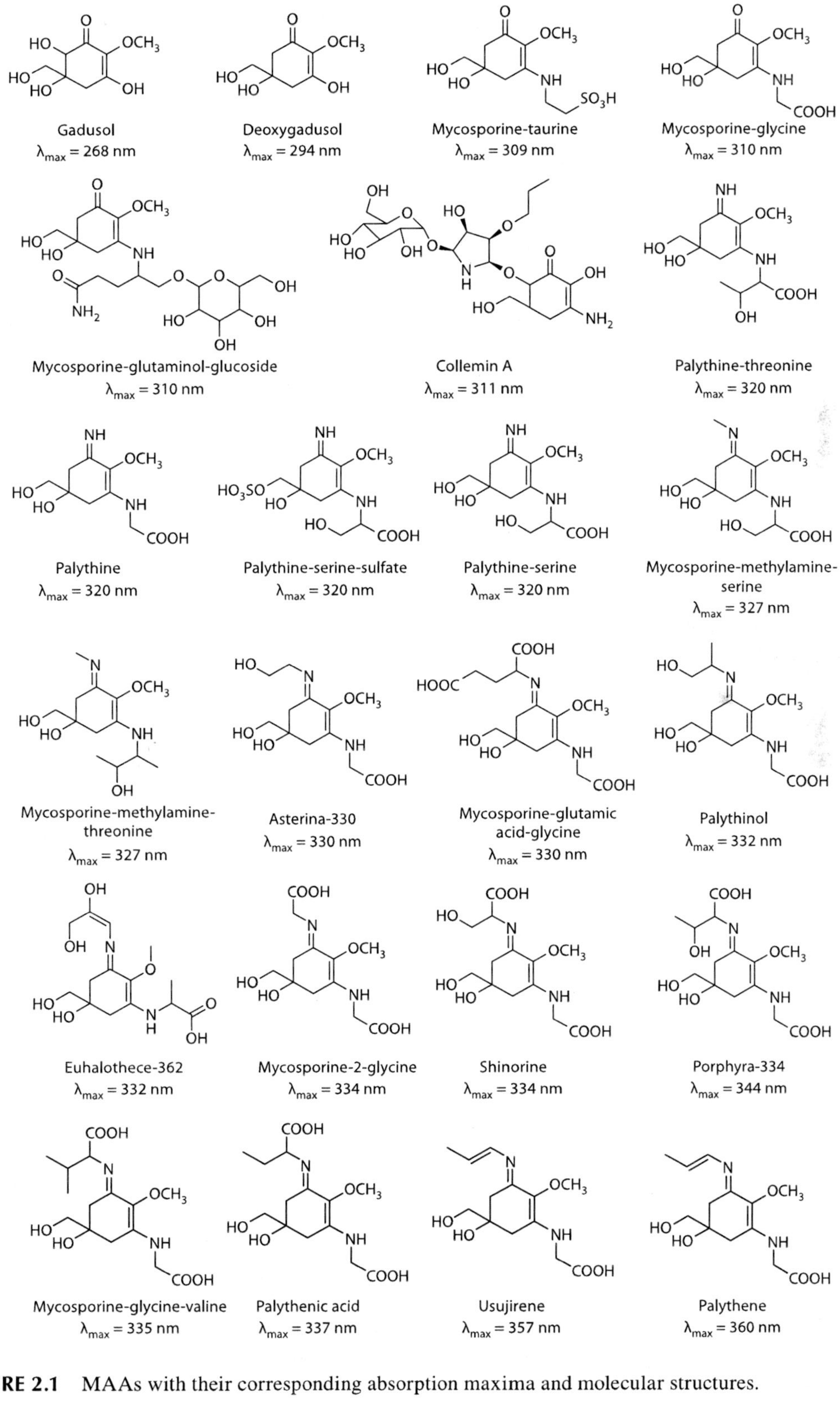

FIGURE 2.1 MAAs with their corresponding absorption maxima and molecular structures.

320–400 nm) for cyclohexenimine-based structures. Changes in the λ_{max} are caused by substituent effects, with organisms generally containing a variety of MAAs protecting across a wide UV band range.[1,12]

MAAs are favored as photoprotective compounds because they have maximum UV absorption between 310 and 362 nm, high molar extinction coefficients (e = 28, 100–50,000 M^{-1} cm^{-1}), the capability to dissipate absorbed radiation efficiently as heat without producing reactive oxygen species (ROS), and photostability and resistance to several abiotic stressors.[8,13,14] It was found that the examined MAAs protected cells from UV-induced cell death and had a protective effect on human cells. It is further expected that these compounds may have potential applications in cosmetics as antiphotoaging/photoprotective agents.[9]

2.2.1.2 MAAs in Marine Sources

Several species of macroalgae, belonging to Rhodophyceae (red algae), Phaeophyceae (brown algae), and Chlorophyceae (green algae), are commonly exposed to elevated solar radiation and synthesize and accumulate high concentrations of MAAs as UV-sunscreen compounds. Tsujino and Saito first reported the photoprotective role of UV-absorbing substances in macroalgal species in 1961.[15] Since then several scientists reported MAAs in all major macroalgae.

The MAA content in macroalgae varies between classes and with increasing depth and latitudes. Many macroalgae produce one or several UV-absorbing compounds. Most MAA-producing macroalgae belong to Rhodophyceae, followed by Phaeophyceae, and only a few green algae produce MAAs.[16] Several species of red algae, such as *Acanthophora*, *Bangia*, *Bostrychia*, *Caloglossa*, *Catenella*, *Devaleraea*, *Ceramium*, *Chondrus*, *Corallina*, *Devaleraea*, *Gelidiella*, *Gelidium*, *Gracilaria*, *Iridea*, *Palmaria*, *Phyllophora*, *Polysiphonia*, *Porphyra*, *Stictosiphonia*, etc., have been reported to produce high concentrations of different MAAs.[8,17,18] A high degree of MAA diversity has been found in the majority of Rhodophyceae; for instance, *Gracilaria changii* contains up to seven different UV-absorbing compounds.[19] Porphyra-334 is an MMA widely distributed among marine algae. de la Coba et al. documented the cutaneous photoprotective ability of the high UV-absorbing MAAs porphyra-334 and shinorine (P-334+SH), isolated from the red alga *Porphyra rosengurttii*, by *in vivo* procedures on mouse skin.[20] They also reported the presence of the MAAs asterina-330 and palythine from *Gelidium corneum*, as well as shinorine from *Ahnfeltiopsis devoniensis*.[21]

In contrast to Rhodophyceae, the occurrence of MAAs in Chlorophyceae and Phaeophycea is limited. Within the Chlorophyceae, most species do not contain MAAs or only trace concentrations of these compounds. In an investigation of the occurrence of MAAs in 13 macroalgal Chlorophyceae collected from the intertidal zone of tropical Hainan Island, Karsten et al. found a significant concentration of photoprotective compounds, such as mycosporine-glycine and porphyra-334, in only two green algae, *Boodlea composita* and *Caulerpa racemosa*, respectively.[19]

The synthesis of MAAs also occurs in cyanobacteria, bacteria, phytoplankton, and lichens. Cyanobacteria are phylogenetically a primitive group that possess higher-plant-type oxygenic photosynthesis. MAA-producing cyanobacteria are abundant in marine environments. Shibata first reported the accumulation of large amounts of MAAs in cyanobacteria from the Great Barrier Reef.[22] Kedar et al. isolated and characterized mycosporine-2-glycine in *Euhalothece* sp.[23] Three MAAs, M-tau, dehydroxylusujirene, and M-343, have been found to play a photoprotective role in *Synechocystic* sp. PCC 6803, and dehydroxylusujirene and M-343 induction was mainly UV-A dependent, but M-tau could be induced by both UV-A and UV-B.[24] As in macroalgae and cyanobacteria, several MAAs have also been isolated and identified from a number of marine phytoplankton species.[18] MAAs have been reported to occur predominantly in members of the Dinophyceae, Bacillariophyceae, and Haptophyceae (or Prymnesiophyceae). Some species of dinoflagellates such as *Alexandrium excavatum*[25] and the prymnesiophyte *Phaeocystis pouchetii*[26] are known to produce MAAs in high concentrations. Mycosporine-glutaminol-glucoside was the major UV-absorbing compound found in the basidiomycetous yeasts *Rhodotorula minuta* and

R. slooffiae isolated from lakes. This is the first identification of a mycosporine in yeasts.[27] A novel photoprotective mycosporine was isolated from the lichenized ascomycete *Collema cristatum*. Biological activity was measured in terms of protection against UV-B-induced membrane destruction and pyrimidine dimer formation in cultured human keratinocytes, and prevention of UV-B-induced erythema.[28]

In certain cases, the debate is still on to determine photoprotective compounds and their sure role in photoprotection in marine organisms. The synthesis of MAAs doesn't occur in animals, because they lack the shikimate pathway. Even so, certain marine animals such as arthropods (copeposize and accumulate krill, sea spiders, isopods, and amphipods), mollusks, chordates (fish), cnidarians (sea anemones, jellyfish, corals), protochordates (ascdians), eubacteriobionts, poriferans (sponges), nemertineans (ribbon worms), echinodermates (sea urchins, sea stars, brittle stars, sea cucumbers, sea hares), crinoids, platyhelminthes (planarians), annelids (polychaetes), bryozoans, and protozoans have also been reported to have MAAs derived from their diet or from symbiotic partnerships.[8,18,29–31] Thus, MAAs provide protection from UV radiation not only in their producers, but also in primary and secondary consumers. MAAs have also been shown to be highly resistant to abiotic stressors such as temperature, UV radiation, various solvents, and pH.[14]

2.2.1.3 Biosynthesis of MAAs

There is still a great deal of controversy concerning the precise mechanisms of MAA biosynthesis. However, it is evident that 3-dehydroquinate, which is formed during the early stages of the shikimate pathway (Figure 2.2), serves as a precursor for the synthesis of fungal mycosporines and MAAs via gadusols.[7,8,10] Mycosporine-glycine is considered a prime MAA synthesized in the shikimate pathway that further goes through chemical or biochemical conversions to produce secondary MAAs.

The inhibition of MAA synthesis in *Stylophora pistillata*, by the application of N-phosphonomethlyglycine (glyphosate), which is a specific shikimate pathway inhibitor, has provided the first direct evidence for MAA synthesis through the shikimate pathway in marine organisms.[32] Singh et al. revealed the probable role of certain genes involved in MAA biosynthesis in cyanobacteria and proposed that the YP_324358 and YP_324357 gene products are involved in the biosynthesis of the common core (deoxygadusol) of all MAAs. The YP_324879 gene product is exclusively involved in the shikimate pathway catalyzing the formation of dehydroquinate, while the YP_324358 gene product in conjunction with the YP_324357 gene product (O-methyltransferase) catalyzes the formation of deoxygadusol, which is the core of all MAAs.[33]

Several environmental factors have been reported to affect the production of photoprotective compounds in various organisms. MAA was found to increase when induced by different wavelength bands of UVR. Elevated levels of photosynthetically active radiation (PAR) (400–700 nm), blue light (400–500 nm), UV-A and UV-B wavelengths have been implicated in the MAA induction response.[34–36] The dinoflagellate *Alexandrium excavatum*, isolated from the continental shelf near Buenos Aires, shows changes in MAA composition and an overall increase in UV absorption under low to high PAR.[33] Photoinduction of MAA synthesis was also found in the dinoflagellate *Gyrodinium dorsum* in the presence of PAR and only UV-A and UV-B radiation.[37,38] It shows that changes in the level of UVR also play an important role in the interconversions of different MAAs.[8] Recently, Singh et al. also observed the conversion of mycosporine-glycine into porphyra-334 and shinorine, under the influence of PAR and UVR in the cyanobacterium *Anabaena doliolum*.[39] Certain other factors such as osmotic stress, nutrients, salt stress, and temperature have been found to affect the biosynthesis of MAAs.[40–42]

Overall, some aspects of the biosynthesis and in particular biotransformations of different MAAs in marine organisms are still unsolved and need more investigation.

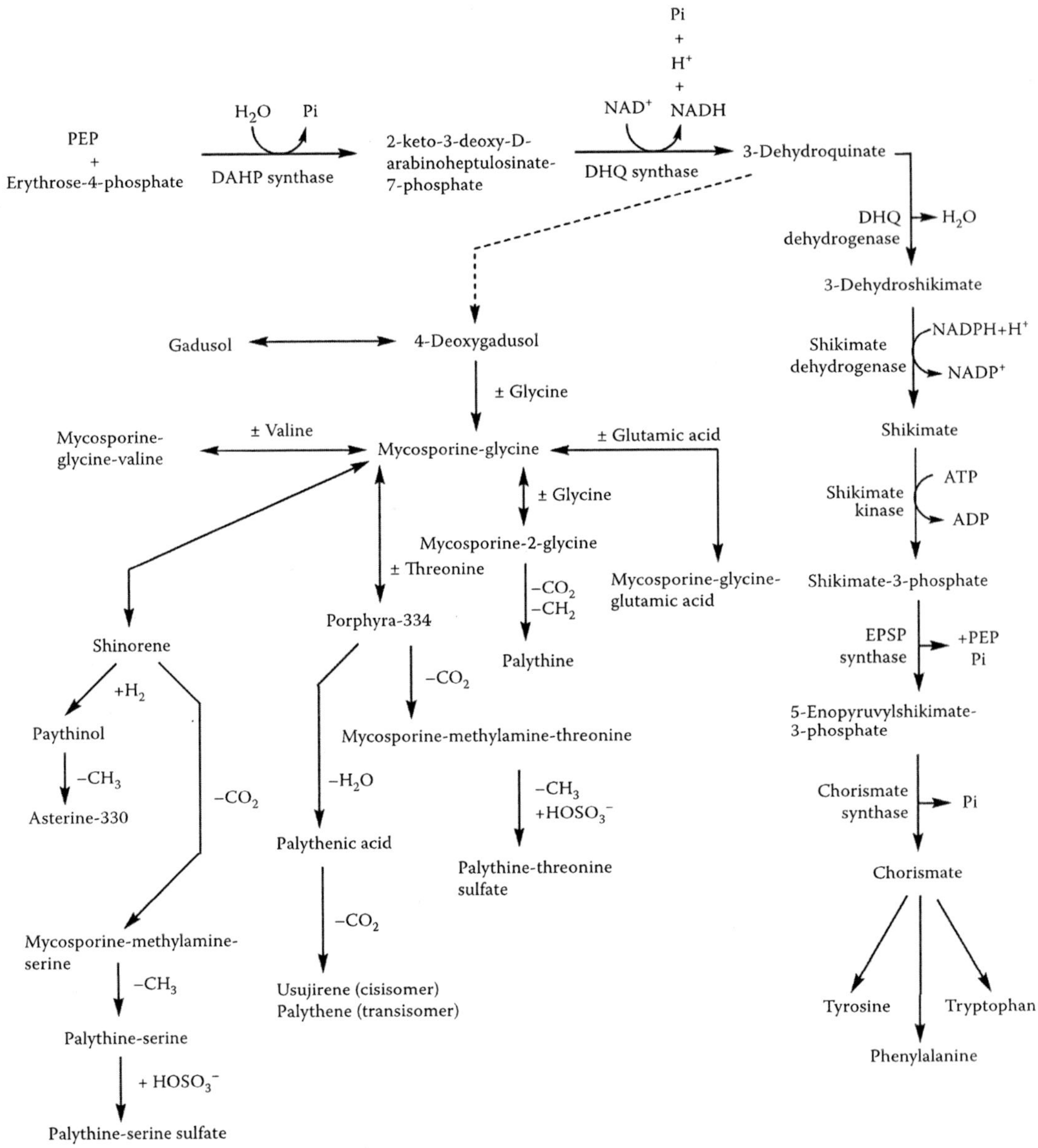

FIGURE 2.2 Biosynthetic routes for the formation of MAAs via the shikimate pathway and their feasible chemical or biochemical conversion. Broken line represents the putative biosynthetic connection between dehydroquinate, gadusols, and MAAs. PEP, phosphoenolpyruvate; DAHP, 3-deoxy-D-arabinoheptulosinate-7-phosphate; DHQ, dehydroquinate; EPSP, 5-enolpyruvylshikimate-3-phosphate.

2.2.2 Carotenoids

2.2.2.1 Structure of Carotenoids

Carotenoids are multifunctional, taxonomically widespread, and biotechnologically important pigments. These are derived from five carbon isoprene units that are polymerized enzymatically to form regular, highly conjugated 40-carbon structures with up to 15 conjugated double bonds. One or both ends of the carbon skeleton may go through cyclization to form β-ionone ring end groups, which additionally may be substituted by oxo, hydroxy, or epoxy groups at dissimilar positions to form the different xanthophylls.[43] Depending on the length of their conjugated double-bond chain and the nature of its substituents, carotenoids most often absorb light in the 300–600 nm range to appear yellow, orange, or red. Carotenoids are structurally divided into two classes: carotenes, which are exclusively hydrocarbons, and xanthophylls, which are oxygenated.[44]

Carotenoids are important components of photosynthetic light-harvesting complexes, where carotenoids dissipate excess energy and radicals from excited oxygen and (bacterio) chlorophyll molecules, physically structure the photosynthetic reaction center, and act as accessory light-harvesting pigments.[45–47] In all organisms carotenoids may function as antioxidants and promote oxidative stress resistance,[48] and even act as a virulence factor in *Staphylococcus aureus* by promoting resistance to neutrophil oxidative burst.[49] Membrane fluidity and proton permeability may also be modulated by carotenoids in all organisms, depending on carotenoid structure and concentration[50–51]; these latter functions remain poorly studied, especially *in vivo*. Carotenoids can also be cleaved to form apocarotenoids. These include retinal (vitamin A), rhodopsin (xanthorhodopsin), plant hormones, fungal pheromones, and antifungal compounds.[52–55]

Carotenoids are biotechnologically high-value compounds with an annual market estimated to exceed US$1 billion by 2010.[56] Several commercial applications include natural food coloring agents, food additives, drugs, enhancers of the color of egg yolks, improvement of the health and fertility of cattle, and use in the cosmetic industry.[57,58] Carotenoids may directly provide photoprotection against UVR-induced photooxidation in the skin and also provide protection against inflammation, ulcers due to *Helicobacter pylori* infection, and age-related diseases.[59–61] It has been shown that β-carotene can modulate UV-A-induced gene expression in human keratinocytes.[62] At low oxygen tension, carotenoids also scavenge peroxyl radicals, inhibiting the process of lipid peroxidation. Some carotenoids offer provitamin A activity,[63] and high consumption of carotenoids may diminish threats of certain pathologies.[64]

2.2.2.2 Biosysthesis of Carotenoids

Carotenoid biosynthesis serves as a model system for understanding the evolution of secondary metabolism. Regarding the biosynthetic pathway of carotenoids in diverse organisms, several studies have recently been published.[65–68] Figure 2.3 describes representative carotenoid biosynthetic pathways and known chemical structures. The carotenoid pathways proceeding from the C5 isoprenoid precursors isopentenyl diphosphate and dimethylallyl diphosphate all the way to lycopene are similar in amino acid sequence to the corresponding enzymes in green algae, land plants, and certain species of cyanobacteria.[67,69] Plant and cyanobacterial carotenoid pathways begin to diverge with the cyclization of lycopene to yield β-carotene, and subsequent pathway reactions, including the hydroxylation of each β-ring of β-carotene to produce zeaxanthin (3,3′-dihydroxy-β-carotene), a near omnipresent photoprotective pigment in cyanobacteria,[70] algae, and plants,[71] are catalyzed by distinctly different enzymes in cyanobacteria and plants.

Some species of bacteria, yeasts, and fungi, as well as algae and higher plants, can synthesize a large number of carotenoids, and some marine bacteria can produce dicyclic, monocyclic, or acyclic carotenoids.[58] More than 750 carotenoids with different molecular structures have been isolated from natural sources.[72] Shindo et al. isolated and characterized a novel red pigment glyco-carotenoic acid ester, methyl glucosyl-3,4-dehydro-apo-8-lycopenoate, from the marine bacterium *Planococcus maritimus*. The compound showed potent antioxidative activity.[73] In general, animals do not synthesize carotenoids *de novo*, and those found in animals are either directly accumulated from food or partly modified through metabolic reactions. Maoka et al. have identified and given the structure of a new C37 skeletal carotenoid, named hydratopyrrhoxanthinol 3′-ester, from the clam *Paphia amabillis*.[72]

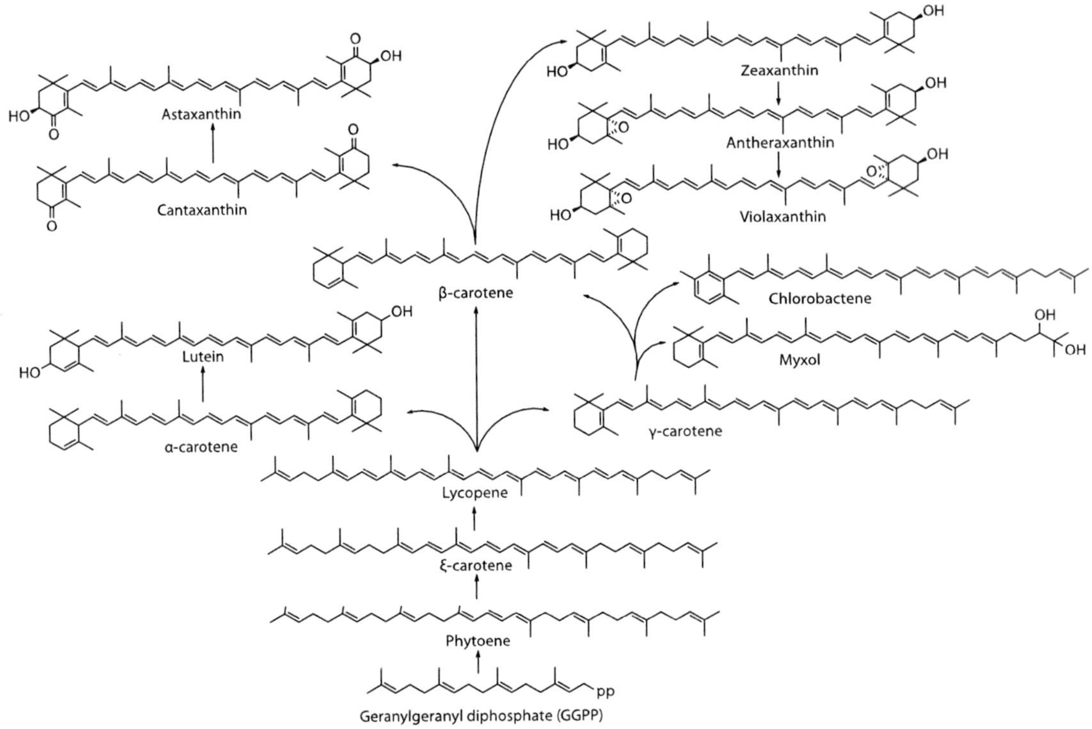

FIGURE 2.3 Biosynthetic pathway of some carotenoids.

2.2.3 Scytonemin

2.2.3.1 Structure of Scytonemin

This UV-absorbing/screening pigment was first reported in some terrestrial cyanobacteria by Nägeli[74] and later termed scytonemin (**1**).[75] Scytonemin has been proposed to serve as a UV-A sunscreen. The occurrence of scytonemin is restricted to cyanobacteria, but it is widespread among this diverse group, and there are more than 300 species.[6] It is a yellow-brown lipid-soluble pigment located in the extracellular polysaccharide sheath.[76,77] Scytonemin has an *in vivo* absorption maximum at 370 nm. Purified scytonemin has an absorption maximum at 386 nm, but it also absorbs significantly at 252, 278, and 300 nm.[6] It is a dimer composed of indolic and phenolic subunits having a molecular mass of 544 Da. Scytonemin exists in oxidized (yellow) and reduced (red) forms, which were named fuscochlorin and fuscorhodin (Figure 2.4), respectively.[78,79] The existence of two forms of scytonemin depends upon the redox and acid-base conditions during the process of extraction. The linkage between the two subunits in scytonemin is an olefinic carbon atom that is unique among natural products. Hence, scytonemin possesses a novel ring systemin nature for which the trivial name scytonemin skeleton has been proposed.[11,78] Three analogs of scytonemin, dimethoxyscytonemin (**2**), tetramethoxyscytonemin (**3**), and scytonin (**4**) (Figure 2.5), were isolated from *Scytonema* sp.[80]

2.2.3.2 Biosysthesis of Scytonemin

A number of papers about the biosynthesis and molecular genetics of scytonemin in cyanobacteria have been published.[81–87] The biosynthesis of scytonemin probably involves tryptophan and tyrosine derivatives, and can be induced by a high-photon fluence rate. A scytonemin-deficient *Nostoc punctiforme* ATCC 29133 mutant was created using random transposon mutagenesis. The authors found the specific region in the genome of *N. punctiforme* ATCC 29133 and revealed a cluster of 18 genes,

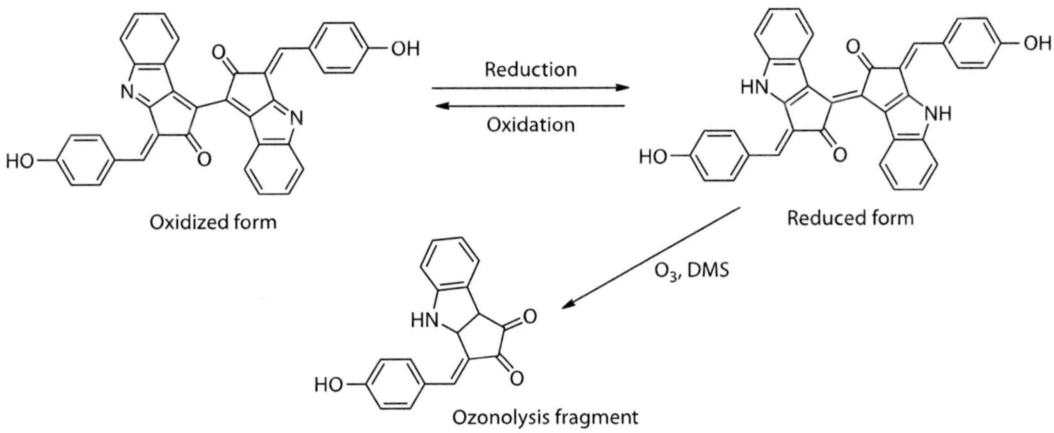

FIGURE 2.4 Scytonemin, its reduction product, and monomer formed by ozonolysis.

NpR1259–NpR1276, associated with scytonemin biosynthesis.[79] Balskus and Walsh proposed the probable biosynthetic route for scytonemin (Figure 2.6) and identified the acyloin reaction as a key step in constructing the carbon framework of this ecologically and evolutionary important pigment. They performed *in vitro* characterization of three enzymes encoded in the biosynthetic gene cluster: ScyA (NpR1276), ScyB (NpR1275), and ScyC (NpR1274). ScyB, a leucine dehydrogenase homolog, converts L-tryptophan to indole-3-pyruvic acid. ScyA is a thiamin (ThDP)-dependent enzyme responsible for the regioselective acyloin coupling of indole-3-pyruvic acid and *p*-hydroxyphenylpyruvic acid to afford a single, labile β-ketoacid product. ScyC, an enzyme encoded by the scytonemin cluster, catalyzes cyclization and decarboxylation of ScyA product β-ketoacid to form ketone, thereby constructing the indole-fused cyclopentane moiety of the natural product.[82,83] The products of NpR1263 and NpR1269 open reading frames (ORFs) are still to be functionally characterized. From a comparative study of four other cyanobacterial strains, Soule et al. suggested that two supplementary conserved clusters (NpF5232 to NpF5236) and a putative two-component

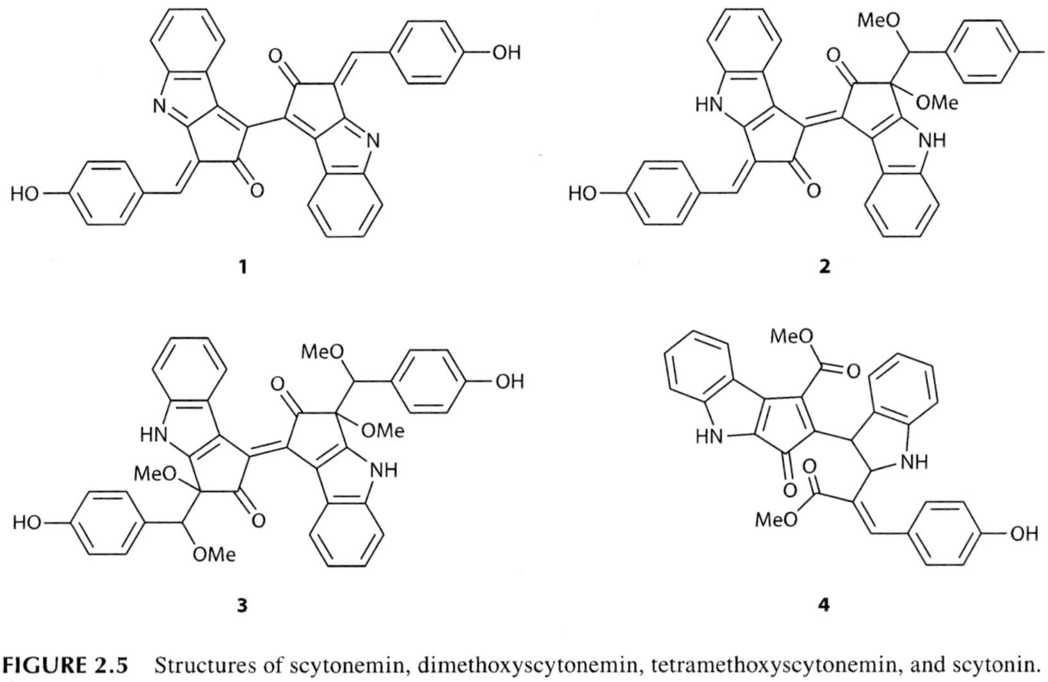

FIGURE 2.5 Structures of scytonemin, dimethoxyscytonemin, tetramethoxyscytonemin, and scytonin.

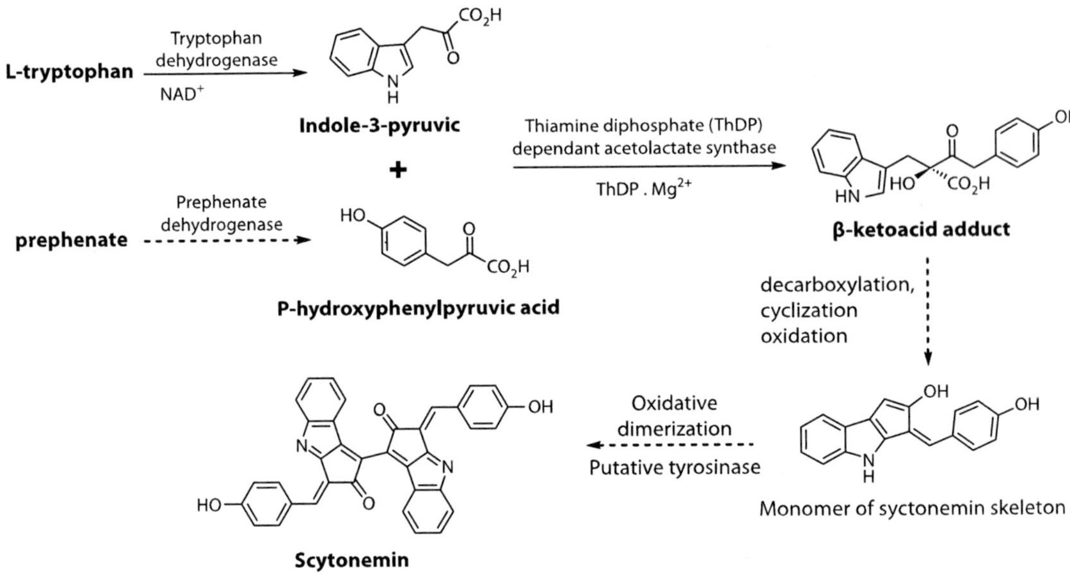

FIGURE 2.6 Proposed biosynthetic pathway for the photoprotective compound scytonemin.

regulatory system (NpF1277 and NpF1278) are possibly involved in scytonemin biosynthesis and regulation, respectively, on the basis of conservation and location.[8,86]

UV-A treatment is very efficient in inducing the synthesis of scytonemin, whereas blue, green, or red light at the same fluence rates does not cause any significant increase in scytonemin.[88] The effect of other correlated stress factors, including heat and osmotic and oxidative stress on the synthesis of scytonemin in the cyanobacterium *Chroococcidiopsis* sp., has been investigated in the presence and absence of UV-A irradiation.[89] Both an increase in temperature and oxidative stress in combination with UV-A have a synergistic effect on high production of scytonemin.[89] However, the osmotic stress decreased scytonemin synthesis in both the presence and absence of scytonemin-inducing irradiance. Thus, scytonemin induction may be regulated as a part of a complex stress response pathway in which multiple environmental signals affect its synthesis.[89]

2.2.4 Other Marine Photoprotective Compounds

Besides the occurrence of well-known photoprotective compounds in the marine organisms mentioned above, some other UVR-absorbing/screening compounds have been identified in marine organisms (Figure 2.7).

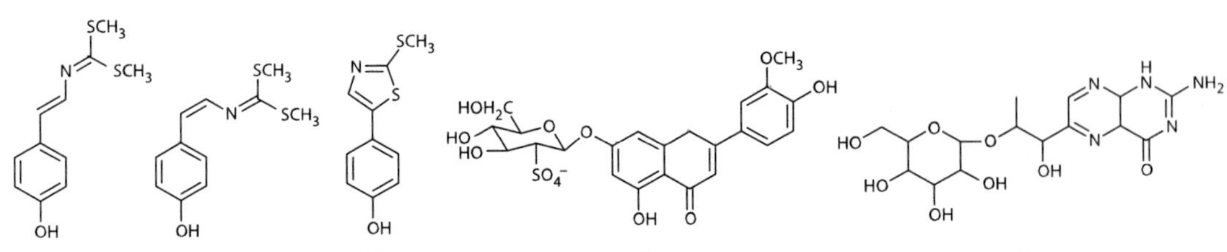

5 6 7 8 9

FIGURE 2.7 Other marine photoprotective compounds.

Three novel phenolic metabolites, tridentatols A-C (**5–7**), with an uncommon sulfur-containing functional group, have been isolated from the marine hydroid *Tridentata marginata.* These metabolites also may function to protect *T. marginata* from damaging solar ultraviolet radiation.[90] Thalassiolin B (**8**), extracted from the marine plant *Thalassia testudinum*, contributes to skin repair by allowing for increased synthesis of collagen fibers utilizing hyperactive fibroblasts.[91] Matsunaga purified a UV-A-absorbing pigment, biopterin glucoside (**9**), with absorption maxima at 256 and 362 nm, from a marine planktonic cyanobacterium *Oscillatoria* sp.[92] It has been found that the fluorescent pigment granules (FPGs) of corals may supplement the role of MAAs by blocking some of the UV-A.[93] Melanins are a ubiquitous class of biological pigments which play an important role in the photoprotection of skin. The various types of melanin show different physicochemical properties, suggesting that their photobiological properties are not unique.[94] A number of bacteria are reported to produce melanin from phenolic degradation products, such as 3,4-dihydroxyphenylacetate, that accumulate during the catabolism of tyrosine.[95] Synthesis of melanin has also been reported from a marine bacterium *Shewanella colwelliana*.[96,97] Several other UV-absorbing substances of unknown chemical structure from different organisms have been developed to counteract photodamage.

2.3 SKIN WHITENING COMPOUNDS

Skin whitening products are commercially available for cosmetic purposes in order to obtain a lighter skin appearance. They are also utilized for clinical treatment of pigmentary disorders such as melasma or postinflammatory hyperpigmentation. Many of them are known as competitive inhibitors of tyrosinase. Others inhibit the maturation of this enzyme or the transport of pigment granules (melanosomes) from melanocytes to surrounding keratinocytes.[98] Tyrosinase is a multifunctional copper-containing oxidase that catalyzes three distinct reactions of melanin synthesis, the hydroxylation of tyrosine to 3,4-dihydroxyphenylalanine (DOPA), the oxidation of DOPA to dopaquinone, and the conversion of 5,6-dihydroxyindole to melanochrome.[99–101] The enzymatic oxidation of L-tyrosine to melanin is of considerable importance because melanin has many functions, and alterations in melanin synthesis occur in many disease states. Recently, tyrosinase inhibitors have become increasingly important in cosmetic and medical products in relation to hyperpigmentation.[102,103]

In the skin, melanocytes are situated on the basal layer, which separates the dermis and epidermis. One melanocyte is surrounded by approximately 36 keratinocytes. Together, they form the so-called epidermal melanin unit. The melanin produced and stored inside the melanocyte in the melanosomal compartment is transported via dendrites to the overlaying keratinocytes.[98,104] The biosynthetic pathway for melanin formation in various life forms was first elucidated by Raper[105] and Mason,[106] and has recently been modified by Cooksey et al.[104] and Schallreuter et al.[107] (Figure 2.8). The melanin pigment is a polymer produced inside the melanosomes and synthesized from the amino acid L-tyrosine that is converted by the enzyme tyrosinase to dopaquinone. The subsequent dopaquinone is converted to dopa and dopachrome through autooxidation. Dopa is also the substrate of tyrosinase and oxidized to dopaquinone again by the enzyme. Finally, eumelanin is formed through a series of oxidation reactions from dihydroxyindole (DHI) and dihydroxyindole-2-carboxylic acid (DHICA), which are the reaction products from dopachrome. In the presence of cysteine or glutathione, dopaquinone is converted to cysteinyldopa or glutathionyldopa. Subsequently, pheomelanin is formed. In addition to eumelanin and pheomelanin, another "melanin" relying on phenolic monomers different from tyrosine is termed allomelanin.[108–110]

One of the most obvious cellular targets for depigmenting agents is the enzyme tyrosinase. Antityrosinase active compounds have been of great concern as functional personal care products for skin whitening effects and for preventive and therapeutic effects on local hyperpigmentation diseases. Tyrosinase inhibitors can be classified as competitive, uncompetitive, mixed type, and noncompetitive inhibitors.[108] A number of compounds have been identified as tyrosinase inhibitors from plants, microorganisms, and synthetic compounds.[109,110] In recent years, attention has

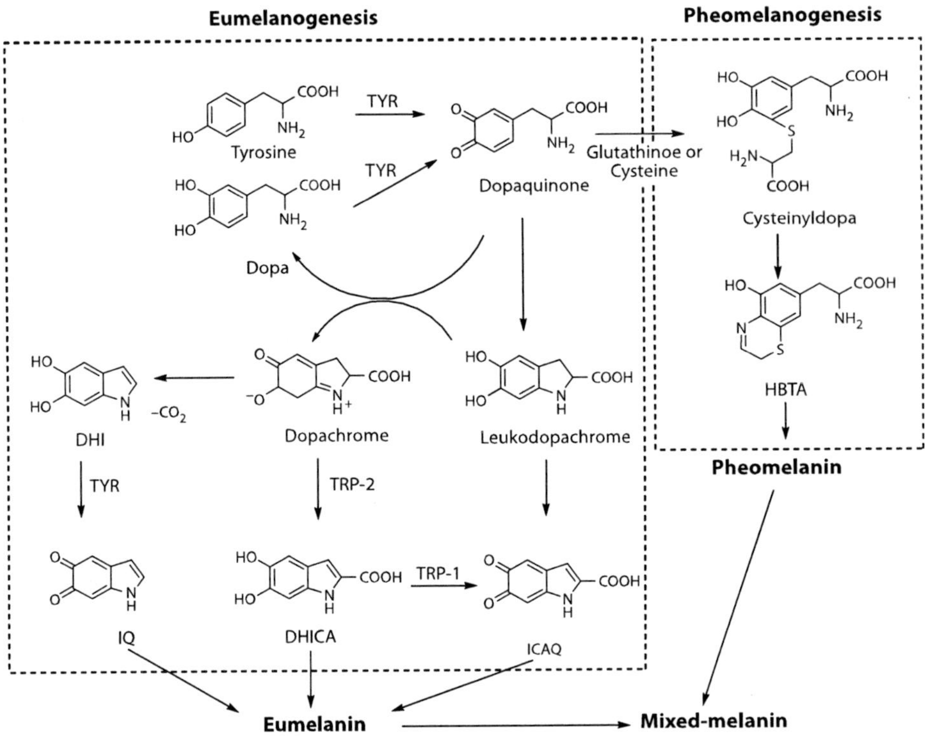

FIGURE 2.8 Biosynthetic pathway of melanin. TYR, tyrosinase; TRP, tyrosinase-related protein; dopa, 3,4-dihydroxyphenylalanine; DHICA, 5,6-dihydroxyindole-2-carboxylic acid; DHI, 5,6-dihydroxyindole; ICAQ, indole-2-carboxylic acid-5, 6-quinone; IQ, indole- 5,6-quinone; HBTA, 5-hydroxy-1,4-benzothiazinylalanine.

been focused on bioactive compound-producing marine organisms and a number of novel compounds have been isolated (Figure 2.9). Three strains that were isolated from sediment samples of Off-Izu Islands and Sagami Bay were found to produce tyrosinase inhibitors. Among the three, strain H1-7 (*Trichoderma* sp.) had significantly and constantly the strongest inhibitory activity (1,000–2,500 U/ml).[111] Tsuchiya et al. isolated and characterized a tyrosinase inhibitor, homothallin II, from *Trichoderma viride* strain H1-7.[112] Homothallin II (**10**) was also isolated as an antibiotic from *Trichoderma koningii* and *T. harzianum*.[113] Li et al. have found and given the structures of 6-*n*-pentyl-α-pyrone, trichodenone A, cyclonerodiol, and cyclopentenones designated as myrothenones A and B produced by the marine-derived fungus *Myrothecium* sp. 6-*n*-pentyl-α-pyrone (**11**) and myrothenone A (**12**) exhibited a tyrosinase inhibitory activity with ED_{50} values of 0.8 and 6.6 μM, respectively, which are more active than kojic acid (ED_{50} = 7.7 μM), currently being used as a functional personal care compound.[114–116] Zhang et al. reported a new α-pyrone derivative, 6-[(*E*)-hept-1-enyl]-α-pyrone (**13**), from a marine isolate of the fungus *Botrytis* sp., which exhibited a tyrosinase inhibitory activity with an IC_{50} value of 4.5 μM.[117]

Heo et al. investigated the tyrosinase inhibitory effect of 21 species of marine algae. It was observed that *Ishige okamurae* extract showed the highest levels of inhibitory effect (78.73%), and *Ishige sinicola* extract also evidenced a 42.87% inhibitory effect on tyrosinase. The active compound was identified as a kind of phlorotannin, diphlorethohydroxycarmalol (DPHC) (**14**), isolated from *I. okamurae*. The DPHC evidenced a stronger tyrosinase inhibitory effect (IC_{50} = 142.20 μM) than that observed with the commercial whitening agent arbutin (IC_{50} = 384.82 μM). The inhibitory effect of melanin synthesis was also evaluated for the determination of inhibiting melanogenesis in B16F10 mouse melanoma cell lines. DPHC possessed stronger inhibition (IC_{50} = 37.73 μM) than

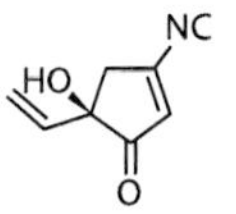

10

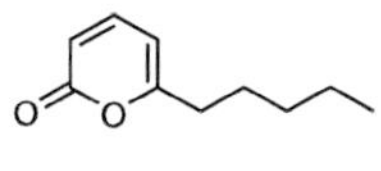

11

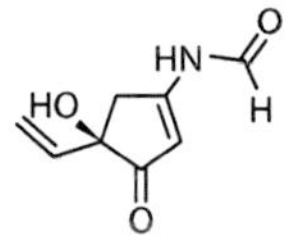

12

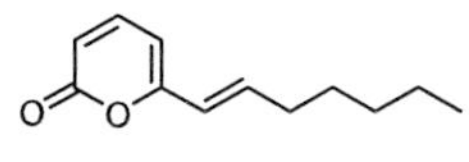

13

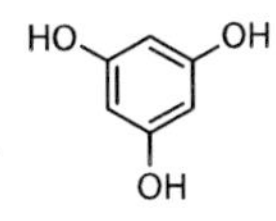

15

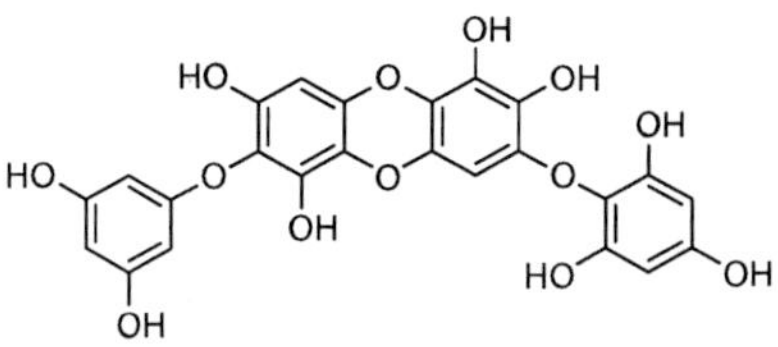

14

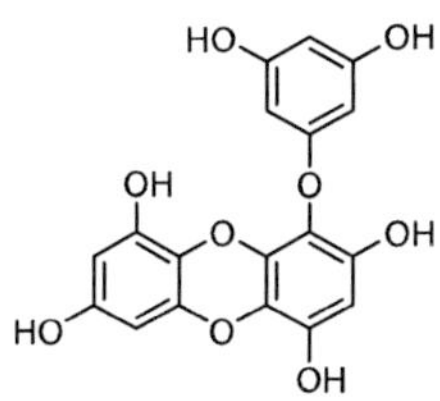

16

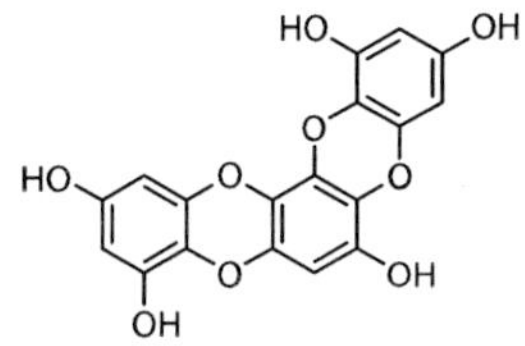

17

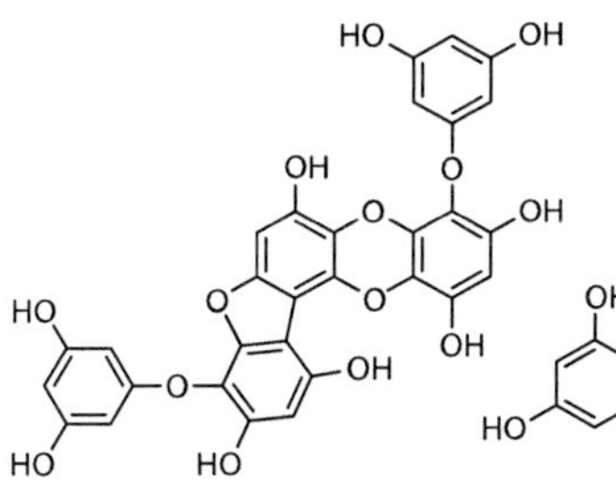

18

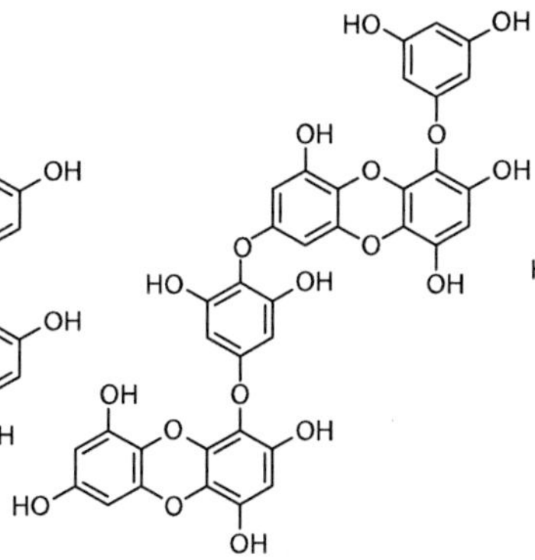

19

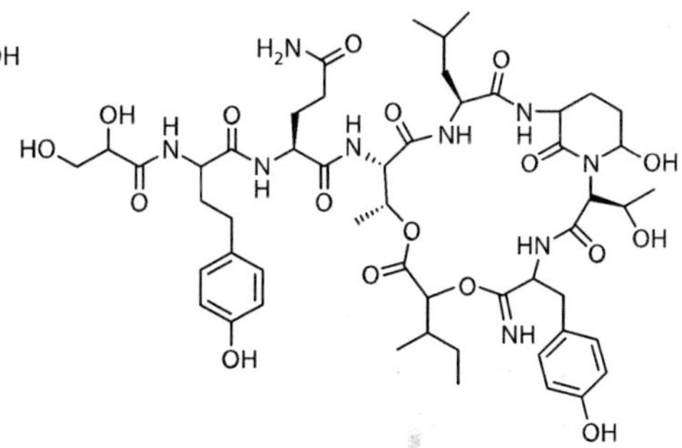

20

FIGURE 2.9 Structures of tyrosinase inhibitors from marine natural compounds.

retinol (IC_{50} = 50.25 μM), and was similar to phenylthiouren (PTU) (IC_{50} = 32.72 μM), as the positive controls.[118] Five phloroglucinol derivatives, phloroglucinol (**15**), eckstolonol (**16**), eckol (**17**), phlorofucofuroeckol A (**18**), and dieckol (**19**), were isolated from the methanolic extract of the edible brown alga *E. stolonifera*. All compounds were found to inhibit the oxidation of L-tyrosine catalyzed by mushroom tyrosinase with IC_{50} values of 92.8, 126, 33.2, 177, and 2.16 μg/ml, respectively. The IC_{50} values of the five compounds was compared with those of kojic acid and arbutin, well-known tyrosinase inhibitors, with IC_{50} values of 6.32 and 112 μg/ml, respectively.[101] It has also been reported that phlorotannins are antiplasmin inhibitors and exhibit antiallergic, antitumor, and antioxidant activities.[119–123] Sano and Kaya described the isolation and structure elucidation of oscillapeptin G (**20**), a novel tyrosinase-inhibiting cyclic depsipeptide, from toxic cyanobacterium *Oscillatoria agardhii*.[124]

2.4 ANTIOXIDANT COMPOUNDS

Antioxidants are able to hinder oxidation when only a small amount of them are used. Antioxidants, also called bioantioxidants, attract much attention since aging and degenerative diseases are related to the oxidation of biological components induced by reactive oxygen species (ROSs).[125] ROSs are implicated in or are an initiating factor in skin cancers, cutaneous aging, and many inflammatory disorders.[126] ROSs can not only be generated in the skin by oxidative metabolism, but also arise from exogenous sources such as UV radiation, ozone, and air pollution. Recently, aging, cancer, atherosclerosis, and some other serious diseases have been confirmed to correlate with low-density lipoprotein (LDL), cell membranes, and DNA exposed to oxidative stress.[126,127] Supplementations of antioxidants to maintain health and cure diseases are an important strategy in therapy, called antioxidant therapy. Antioxidants are composed of a large amount of enzymatic and nonenzymatic compounds. For example, superoxide dismutase (SOD), catalase, glutathione peroxidase, glutathione reductase, and glutathione transferase are enzymatic antioxidants in human plasma and erythrocytes.[128,129] Nonenzymatic antioxidants involve many small molecular organic compounds such as R-tocopherol (TOH), ascorbic acid, ubiquinol, β-carotene, etc.[130]

Currently, many kinds of synthetic antioxidants such as butylated hydroxyanisole (BHA), butylated hydroxytoluene (BHT), tertiary butyl hydroquinone (TBHQ), and propyl gallate have been used as materials or additives as an oxidation suppressant in food, cosmetic, and drug compositions. However, the use of these synthetic antioxidants for food or medicine components has been restricted by the toxicity and safety that can lead to the problems of the potential health in humans. Due to these reasons, for the past several years many researchers have tried to find more effective oxidation inhibitors that may be used as antioxidants for food or medicine compositions without side effects. So many researchers have paid attention to many kinds of natural antioxidants that can be used without toxicity in humans.[131,132]

The previously characterized marine antioxidant substances are mainly chemicals that are structurally related to plant-derived antioxidants.[133] These marine antioxidants include pigments such as chlorophylls and carotenoids and tocopherol derivatives such as vitamin E and related isoprenoids.[134–138] A new pheophorbide A-related compound named chlorophyllone A (**21**) was isolated as an antioxidative compound from the extract of the short-necked clam, *Ruditapes philippinarum*.[139] Several other chlorophyll-related compounds have been isolated from the short-necked clam, scallop, and juvenile abalone. These compounds showed stronger antioxidative activity than α-tocopherol in the dark.[140] A novel α-tocomonoenol (MTD) (**22**) was isolated from the chum salmon eggs. The antioxidant activity of MDT was found to be identical to that of α-tocopherol under the experimental conditions of measurement.[141] Mycosporine-like amino acids (MAAs) and carotenoids have the ability to act as antioxidants and thus protect cells and organisms against photooxidation. These metabolites are abundant in marine organizes and play diverse roles in photobiology, photochemistry, and photomedicine.[21,142]

Many kinds of antioxidative components have been isolated from different kinds of marine organisms to use as antioxidants (Figure 2.10).[132,143,144] Seaweed is known to have especially rich components, such as fucoxanthin, catalase, and polyphenol-containing phlorotannins, that have antioxidant effects. Phlorotannins, which is a class of compounds with polymerized phloroglucinol units existing in brown algae, have been greatly noteworthy. They have shown broad therapeutic perspectives, such as antidiabetes, radiation protection, anticancer, anti-HIV, as well as antiallergic activities.[132,145,146] Zhang et al. isolated a new hemiletal steroid, named cladiellin A (**23**), from the soft coral *Cladiella* sp.[147] Two new brominated phenolic compounds, subereaphenols B (**24**) and C (**25**), isolated from the Red Sea sponge *Suberea mollis*, displayed significant antioxidant activity.[148] Li et al. isolated and identified six bromophenols (**26–31**) from the marine red alga *Polysiphonia urceolata* with α,α-diphenyl-β-picrylhydrazyl (DPPH) radical scavenging activity. All were found to be potent, with IC_{50} values ranging from 6.1 to 35.8 μM, compared to the positive control, butylated hydroxytoluene (BHT), with an IC_{50} of 83.8 μM.[149] A new pyrroloiminoquinone alkaloid (**32**) belonging to the makaluvamine family has been isolated from the sponge *Zyzzya* cf. *fuliginosa*. When analyzed on ABTS radical cation decoloring assay, compound **32** demonstrated moderate activity expressed as Trolox equivalent antioxidant capacity (TEAC = 0.341).[150] Pure natural products isolated from marine sponges, algae, and cyanobacteria were examined for antioxidant activity using a 2,2-diphenyl-1-picrylhydrazyl radical (DPPH) solution-based chemical assay and a 2′,7′-dichlorodihydrofluorescein diacetate (DCFH-DA) cellular-based assay by Takamatsu's group. The algal metabolites cymopol (**33**), 7- hydroxycymopol (**34**), avrainvilleol (**35**), and fragilamide (**36**), and the invertebrate constituent puupehenone (**37**) showed strong antioxidant activity in both systems. The DCFH system detects only antioxidants that penetrate cellular membranes. Several

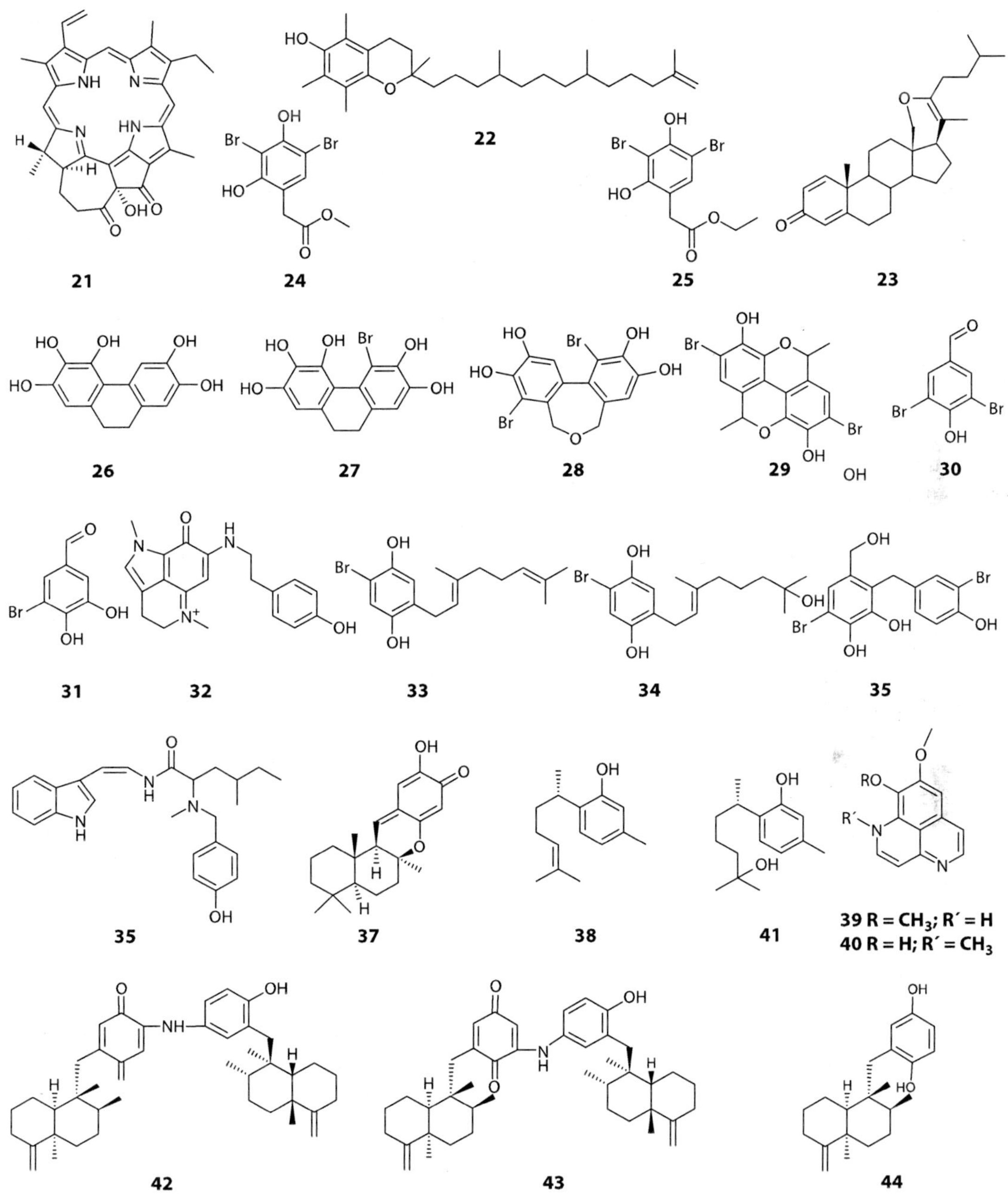

FIGURE 2.10 Structures of antioxidant compounds.

compounds were active in the DPPH assay but significantly less active in the DCFH system. The sponge metabolites (1*S*)-(+)-curcuphenol (**38**), aaptamine (**39**), isoaaptamine (**40**), and curcudiol (**41**) and the cyanobacterial pigment scytonemin (**1**) showed strong antioxidant activity in the DPPH assay, but were relatively inactive in the DCFH system.[151] Bioassay-guided fractionation of the $CHCl_3$ extract led to the isolation of the active DPPH scavenging compounds. Popolohuanones A (**42**) and F (**43**) and arenarol (**44**) showed DPPH radical scavenging activity with IC_{50} values of 35, 35, and 19 μM, respectively.[152]

2.5 ANTI-INFLAMMATORY COMPOUNDS

Inflammatory skin diseases are extremely common dermatological problems that present in a variety of forms, from occasional rashes accompanied by skin itching and redness to more chronic conditions such as atopic dermatitis, rosacea, seborrheic dermatitis, and psoriasis.[153] Cutaneous inflammation has now been linked to many diseases, including cancer, as well as visible skin aging. Visible skin aging can be reduced and prevented by the daily use of antioxidant or anti-inflammatory cosmeceuticals, coupled with a diet rich in anti-inflammatory and antioxidant foods.[154]

Inflammation is caused by a variety of stimuli including physical damage, ultraviolet irradiation, microbial invasion, and immune reactions. The classical key features of inflammation are redness, warmth, swelling, and pain. Inflammation cascades can lead to the development of diseases such as chronic asthma, rheumatoid arthritis, multiple sclerosis, inflammatory bowel disease, and psoriasis.[155] Modern therapeutic approaches are still not satisfactory, and shock, sepsis, and organ failure remain the great challenges in intensive care medicine. However, the treatment of inflammatory diseases also represents an unresolved problem.

Many factors contribute to the complex course of inflammatory reactions. Microbiological, immunological, and toxic agents can initiate the inflammatory response by activating a variety of humoral and cellular mediators. These mediators include prostaglandins (PGs), leukotrienes (LTs), nitric oxide (NO), tumor necrosis factor-α (TNF-α), and cytokines of the interleukin (IL) families.[155] Cytokines and eicosanoids play a crucial role in the maintenance of homeostasis and regulation of skin inflammation. Dermal and epidermal cell types constitutively produce various cytokines and eicosanoids, levels of which are regulated by physiological and pathophysiological events.[156–158] Arachidonic acid (AA), the mother substance of the pro-inflammatory eicosanoids, is released from membrane phospholipids in the course of inflammatory activation and is metabolized to PGs and LTs. Various strategies have been evaluated to control the excessive production of lipid mediators on different levels of biochemical pathways, such as inhibition of phospholipase A_2 (PLA_2), the trigger enzyme for release of AA, blockage of cyclooxygenase (COX) and lipoxygenase (LOX) pathways, and the development of receptor antagonists against platelet-activating factor (PAF) and LTs.[156,159]

Several reviews have been published recently on natural anti-inflammatory compounds.[155,156] A number of anti-inflammatory compounds have been identified from marine organisms and microorganisms (Table 2.1, Figure 2.11).

Heteronemin (**45**), a marine sesterterpene isolated from the sponge *Hyrtios* sp., inhibits TNF-α-induced NF-κB activation through proteasome inhibition and induces apoptotic cell death.[160] Tedanol (**46**), a new brominated and sulfated pimarane diterpene, isolated from the Caribbean sponge *Tedania ignis*, showed a potent anti-inflammatory activity at 1 mg/kg evaluated *in vivo* in a mouse model of inflammation. The anti-inflammatory activity was coupled with a strong inhibition of COX-2 expression, inhibition of cellular infiltration measured as mieloperoxidase (MPO) levels, and inhibition of iNOS expression.[161] Scalaradial (**47**), isolated from marine sponge *Cacospongia* sp., selectively inhibited $sPLA_2$. Scalaradial efficiently inhibited the bee venom PLA_2 (IC_{50} = 0.07 μM) and purified human recombinant type II $sPLA_2$ activity *in vitro* (IC_{50} = 5.4 μM) but displayed a weak inhibition of cytosol 85 kDa PLA_2 from U937 cells (IC_{50} = 28 μM).[162–165] Pseudopterosins A (**48**) and E (**49**), isolated from the marine gorgonian *Pseudopterogorgia elisabethae*, were both effective in reducing phorbol myristate acetate (PMA)-induced mouse ear edema when administered topically. *In vitro*, the pseudopterosins inhibited PGE_2 and LTC_4 production in zymosan-stimulated murine peritoneal macrophages.[166] Three new diterpenoid hexose-glycosides, calyculaglycosides A–C, were isolated from the Caribbean gorgonian *Eunicea* sp. Calyculaglycoside B (**50**) inhibits the synthesis of both PGE_2 and LTC_4, suggesting that it is a nonselective inhibitor of the 5-LOX and COX pathways.[167] Norcembrenolide (**51**) and sinuleptolide (**52**), isolated from the Okinawan soft coral *Sinularia* spp., inhibited TNF-α production in a dose-dependent manner, and showed a more potent effect than prednisolone at a concentration of 33 mg/ml. They also exhibited an inhibitory

TABLE 2.1
Anti-Inflammatory Compounds from Marine Sources

Compounds	Targets, Cells	Source	Reference
Heteronemin	TNF-α	*Hyrtios* sp.	160
Tedanol	COX-2	*Tedania ignis*	161
Scalaradial	PLA_2	*Cacospongia* sp.	162–165
Pseudopterosin E	PGE_2, LTC_4	*Pseudopterogorgia elisabethae*	166
Pseudopterosin A			
Calyculaglycoside B	5-LOX and COX	*Eunicea* spp.	167
Norcembrenolide	TNF-α	*Sinularia* spp.	168
Sinuleptolide			
Lobohedleolide		*Sinularia crassa*	169
Tolypodiol		*Tolypothrix nodosa*	170
Epitaondiol	$hsPLA_2$	*Stypopodium flabelliforme*	171
Cavernolide	PLA_2, TNF-α, NO and PGE_2	*Fasciospongia cavernosa*	172
Cacospongionolide B	Synovial PLA_2	*F. cavernosa*	173
Cacospongionolide E			
Dysidine	Synovial PLA_2	*Dysidea* sp.	174
Bolinaquinone	iNOS, secretory PLA_2	*Dysidea* spp.	175–176
Dysidotronic acid	TNF-α, IL-1	*Dysidea* spp.	177
Avarol		*Dysidea avara*	178
Avarone			
Manoalide	PLA_2	*Luffariella variabilis*	179–180
Secomanoalide	PLA_2	*Luffariella variabilis*	173
Cyclolinteinone	iNOS and COX-2	*Cacospongia linteiformis*	181
FMA	PLA_2		182
Ircinin	PLA_2		183
Ambrein	Elastase	*Physeter macrocephalus*	184
Didemnilactone A	5-LOX	*Didemnum moseleyi*	173
Didemnilactone B			
Spongidines A-D	PLA_2	*Spongia* sp.	185
Symbioimine	COX-2	*Symbiodinium* spp.	186
Neosymbioimine			

effect on NO production not influenced by cytotoxicity.[168] Lobohedleolide (**53**), isolated from the soft coral *Sinularia crassa*, produced the maximum effect at a dose of 10 mg/kg, and this is comparable to that of the reference compound, indomethacin.[169] Tolypodiol (**54**), isolated from the terrestrial cyanobacterium *Tolypothrix nodosa*, showed potent anti-inflammatory activity in the mouse ear edema assay.[170] Epitaondiol (**55**), isolated from *Stypopodium flabelliforme*, was tested *in vitro* on $sPLA_2$ in an ear mice edema assay and in other functions of a human neutrophils assay. The IC_{50} value measured for the decrease of TXB_2 production was 3.8 μM, whereas the effect on LTB_4 inhibition was lower. It blocked $hsPLA_2$ activity and showed topical anti-inflammatory effects on TPA-induced mouse ear edema and was more potent than indomethacin.[171] Cavernolide (**56**) is a novel C-21 terpene lactone isolated from the sponge *Fasciospongia cavernosa*. Cavernolide inhibited human synovial PLA_2 and decreased the myeloperoxidase degranulation process using different stimuli. Cavernolide also inhibited TNF-α, NO, and PGE_2 production in intact cell experiments.[172] Cacospongionolides B (**57**) and E (**58**), isolated from the Mediterranean sponge *Fasciospongia cavernosa*, exhibited specific inhibition against human and synovial PLA_2, respectively.[173] Dysidine (**49**), isolated from a sponge *Dysidea* sp., significantly inhibited the human synovial PLA_2 and

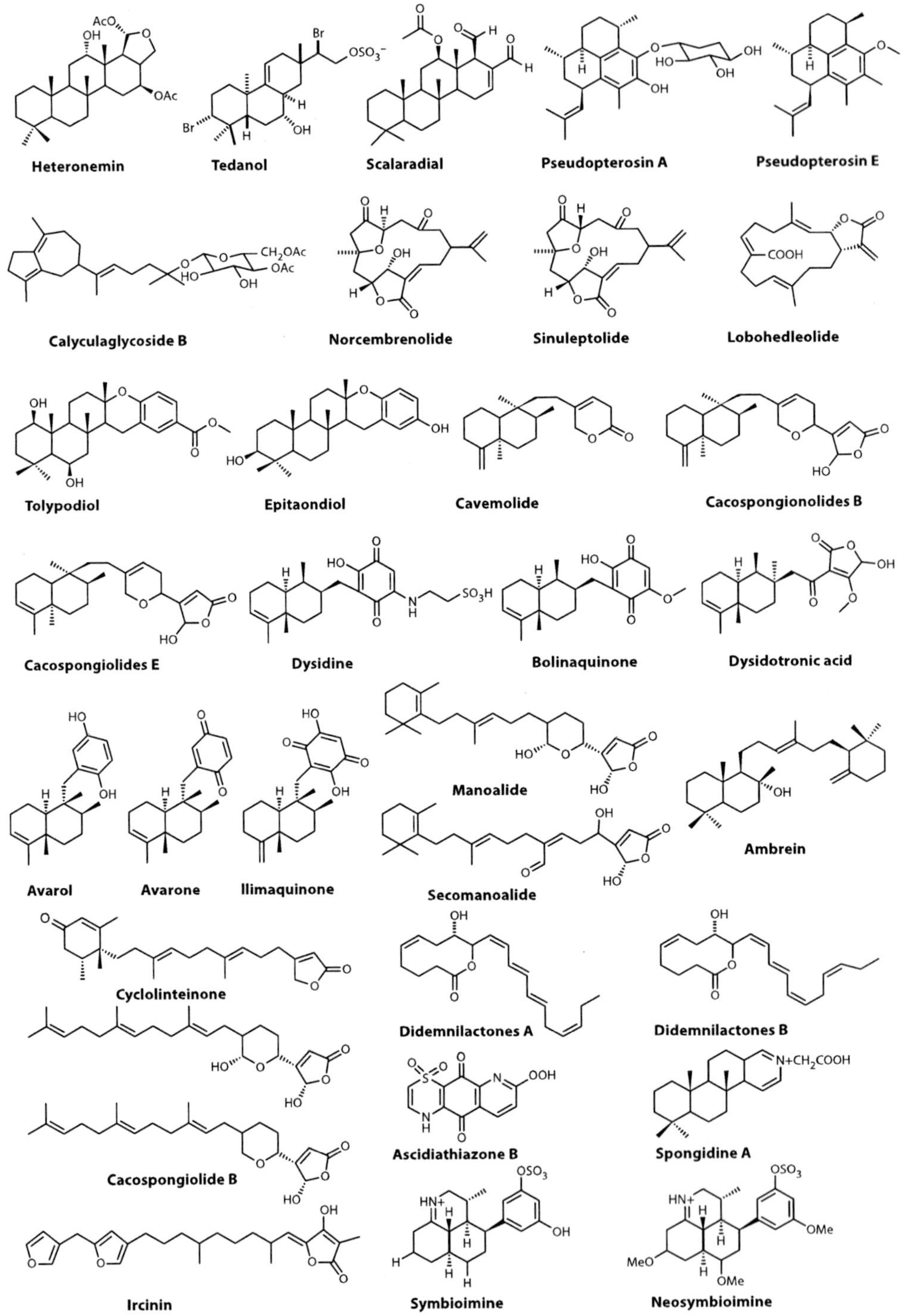

FIGURE 2.11 Chemical structures of anti-inflammatory compounds.

exerted a higher potency and selectivity toward this enzyme than manoalide. It also showed the inhibition of secretory PLA_2 belonging to *Naja naja* venom, porcine pancreatic enzyme, bee venom, and $cPLA_2$ from 264.7 cells. In addition, dysidine modulated the other human leukocyte functions such as degranulation and superoxide production.[174] Bolinaquinone (**60**), isolated from a *Dysidea* sp., significantly inhibited cytokine release, iNOS expression, and eicosanoid (LTB4, PGE2) generation *in vitro* and *in vivo* in several models of inflammation through secretory PLA_2 inhibition.[175,176] Dysidotronic acid (**61**), isolated from *Dysidea* spp., inhibited the production of TNF-α and IL-1 as well as the production of NO, PGE_2, and LTB_4. Decreased NO generation was the consequence of inhibition of the expression of iNOS, whereas PGE2 and LTB4 reduction was due to inhibition of AA bioavailability through a direct inhibitory effect of dysodotronic acid on secretory PLA_2.[177] Two sesquiterpenoids, avarol (**62**) and avarone (**63**), isolated from the Mediterranean sponge *Dysidea avara*, potently inhibited paw edema induced by carrageenan, as well as ear edema induced by PMA in mice. Avarol and avarone effectively controlled acute inflammation in experimental models after either oral or topical administration, and their anti-inflammatory activity may result from the inhibition of eicosanoid release and depression of superoxide generation in leukocytes.[178] Manoalide (**64**), isolated from the sponge *Luffariella variabilis*, was reported to irreversibly inhibit PLA_2 by modifying a specific number of its lysine residues.[179,180] Three congeners of manoalide (**65–67**) were also obtained from the same sponge, of which secomanoalide was more potent against bovine pancreatic PLA_2.[173] Cyclolinteinone (**68**), isolated from the Caribbean sponge *Cacospongia linteiformis*, downregulates iNOS and COX-2 protein expression by inhibiting NF-κB activation, and it has been suggested that it may represent a novel anti-inflammatory compound capable of controlling the excessive production of PGs and NO occurring in several inflammatory diseases.[181] De Rosa et al. synthesized some manoalide analogs, of which an analog containing a farnesyl residue linked to the pyranofuranone (FMA) was found to be a more potent PLA_2 inhibitor than manoalide. It inhibited human synovial and bee venom PLA_2 with IC_{50} values of 0.9 and 0.5 μM, respectively. FMA was tested in the mouse carrageenan paw edema at an oral dose of 10 mg/kg and showed an activity comparable to that of indomethacin.[182] Cholbi et al. investigated the marine diterpene ircinin for its effects on secretory and cytosolic PLA_2 activities *in vitro* as well as for inhibition of cellular functions in human neutrophils and inflammatory response in mice. Ircinin (**69**) is a new inhibitor of PLA_2 activity, and these results suggest a potential role for this marine product as an inhibitor of inflammatory processes.[183] Ambrein (**70**), isolated from *Physeter macrocephalus*, inhibited the elastase release from human neutrophils.[184] Didemnilactones A (**71**) and B (**72**), isolated from tunicate *Didemnum moseleyi*, showed inhibition against 5-LOX in human PMNL with IC_{50} values of 9.4 and 8.5 μM, respectively.[173] Four bioactive pyridinium alkaloids, spongidines A–D (**73–76**), from a Vanatua sponge of the genus *Spongia*, mainly inhibited human synovial PLA_2.[185] Two amphoteric iminium alkaloids, symbioimine (**77**) and neosymbioimine (**78**), were isolated from a cultivated symbiotic marine dinoflagellate *Symbiodinium* spp. These compounds have a characteristic 6,6,6-tricyclic iminium ring and an aryl sulphate moiety. Symbioimine significantly inhibited COX-2 activity.[186] All these compounds exhibited anti-inflammatory activity through different molecular mechanisms. Although these active compounds are not drugs per se, they deserve further investigation as potential candidates for anti-inflammatory drug development through preclinical and clinical studies.

2.6 OTHER COMPOUNDS

Many seaweeds produce hydrocolloids, associated with the cell wall and intercellular spaces.[187] The principal *phycocolloids* extracted from seaweeds are alginate, carrageenan, and agar; these *phycocolloids* have many applications and are used in the food, cosmetics, and pharmaceutical industries as emulsifiers and gelling agents.[188] These phycocolloids are all derived from different types of algae. Agar phycocolloids are derived from red seaweed genera, such as *Gelidium*, *Ahnfeltia*, *Pterocladia*, *Gelidiella*, *Acanthopeltis*, and *Gracilaria*. Carageenan is derived from the red algae groups *Chondrus crispus*, *Gigartina stellata*, *Eucheuma*, *Phyllophora*, *Iridea*, and *Gigartina*. Alginates are derived from brown algae groups, such as *Macrocystis*, *Nereocystis*, *Sargassum*, *Fucus*, *Ascophyllum*, *Laminaria*, and *Ecklonia*.[189]

REFERENCES

1. Cockell, C. S., and Knowland, J., Ultraviolet radiation screening compounds, *Biological Reviews* 74 (3), 311–345, 1999.
2. Sinha, R. P., Singh, N., Kumar, A., Kumar, H. D., Hader, M., and Hader, D. P., Effects of UV irradiation on certain physiological and biochemical processes in cyanobacteria, *Journal of Photochemistry and Photobiology B—Biology* 32 (1–2), 107–113, 1996.
3. Smith, R. C., Baker, K. S., Holmhansen, O., and Olson, R., Photoinhibition of photosynthesis in natural waters, *Photochemistry and Photobiology* 31 (6), 585–592, 1980.
4. Tevini, M., and Teramura, A. H., UV-B effects on terrestrial plants, *Photochemistry and Photobiology* 50 (4), 479–487, 1989.
5. Karentz, D., Cleaver, J. E., and Mitchell, D. L., Cell survival characteristics and molecular responses of *Antarctic phytoplankton* to ultraviolet-B radiation, *Journal of Phycology* 27 (3), 326–341, 1991.
6. Sinha, R. P., Klisch, M., Groniger, A., and Hader, D. P., Ultraviolet-absorbing/screening substances in cyanobacteria, phytoplankton and macroalgae, *Journal of Photochemistry and Photobiology B—Biology* 47, 83–94, 1998.
7. Singh, S. P., Kumari, S., Rastogi, R. P., Singh, K. L., and Sinha, R. P., Mycosporine-like amino acids (MAAS): Chemical structure, biosynthesis and significance as UV-absorbing/screening compounds, *Indian Journal of Experimental Biology* 46 (1), 7–17, 2008.
8. Rastogi, R. P., Richa, Sinha, R. P., Singh, S. P., and Haeder, D. P., Photoprotective compounds from marine organisms, *Journal of Industrial Microbiology and Biotechnology* 37 (6), 537–558, 2010.
9. Pallela, R., Na-Young, Y., and Kim, S. K., Anti-photoaging and photoprotective compounds derived from marine organisms, *Marine Drugs* 8 (4), 1189–1202, 2010.
10. Shick, J. M., and Dunlap, W. C., Mycosporine-like amino acids and related gadusols: Biosynthesis, accumulation, and UV-protective functions in aquatic organisms, *Annual Review of Physiology* 64, 223–262, 2002.
11. Sinha, R. P., and Hader, D. P., UV protectants in cyanobacteria, *Plant Science* 174 (3), 278–289, 2008.
12. Ehling-Schulz, M., and Scherer, S., UV protection in cyanobacteria, *European Journal of Phycology* 34 (4), 329–338, 1999.
13. Conde, F. R., Churio, M. S., and Previtali, C. M., The photoprotector mechanism of mycosporine-like amino acids. Excited-state properties and photostability of porphyra-334 in aqueous solution, *Journal of Photochemistry and Photobiology B—Biology* 56 (2–3), 139–144, 2000.
14. Groniger, A., and Hader, D. P., Stability of mycosporine-like amino acids, *Recent Research Developments in Photochemistry and Photobiology* 4, 247–252, 2000.
15. Tsujino, I., and Saito, T., Studies on the compounds specific for each group of marine algae. I. Presence of characteristic ultraviolet absorbing material in Rhodophyceae, *Bulletin of the Faculty of Fisheries Hokkaido University* 12, 49–58, 1962.
16. Sagan, C., Ultraviolet radiation selection pressure on the earliest organisms, *Journal of Theoretical Biology* 39 (1), 195–200, 1973.
17. Groniger, A., Sinha, R. P., Klisch, M., and Hader, D. P., Photoprotective compounds in cyanobacteria, phytoplankton and macroalgae—A database, *Journal of Photochemistry and Photobiology B—Biology* 58 (2–3), 115–122, 2000.
18. Sinha, R. P., Singh, S. P., and Hader, D. P., Database on mycosporines and mycosporine-like amino acids (MAAs) in fungi, cyanobacteria, macroalgae, phytoplankton and animals, *Journal of Photochemistry and Photobiology B—Biology* 89 (1), 29–35, 2007.
19. Karsten, U., Sawall, T., and Wiencke, C., A survey of the distribution of UV-absorbing substances in tropical macroalgae, *Phycological Research* 46 (4), 271–279, 1998.
20. de la Coba, F., Aguilera, J., de Galvez, M. V., Alvarez, M., Gallego, E., Figueroa, F. L., and Herrera, E., Prevention of the ultraviolet effects on clinical and histopathological changes, as well as the heat shock protein-70 expression in mouse skin by topical application of algal UV-absorbing compounds, *Journal of Dermatological Science* 55 (3), 161–169, 2009.
21. de la Coba, F., Aguilera, J., Figueroa, F. L., de Galvez, M. V., and Herrera, E., Antioxidant activity of mycosporine-like amino acids isolated from three red macroalgae and one marine lichen, *Journal of Applied Phycology* 21 (2), 161–169, 2009.

22. Shibata, K., Pigments and a UV-absorbing substance in coral and a blue-green alga living on the Great Barrier Reef, *Plant and Cell Physiology* 10 (2), 325–335, 1969.
23. Kedar, L., Kashman, Y., and Oren, A., Mycosporine-2-glycine is the major mycosporine-like amino acid in a unicellular cyanobacterium (*Euhalothece* sp.) isolated from a gypsum crust in a hypersaline saltern pond, *Fems Microbiology Letters* 208 (2), 233–237, 2002.
24. Zhang, L. W., Li, L. Z., and Wu, Q. Y., Protective effects of mycosporine-like amino acids of *Synechocystis* sp. PCC 6803 and their partial characterization, *Journal of Photochemistry and Photobiology B—Biology* 86 (3), 240–245, 2007.
25. Carreto, J. I., Carignan, M. O., Daleo, G., and Demarco, S. G., Occurrence of mycosporine-like amino acids in the red tide dinoflagellate *Alexandrium excavatum*: UV-protective compounds? *Journal of Plankton Research* 12 (5), 909–921, 1990.
26. Marchant, H. J., Davidson, A. T., and Kelly, G. J., UV-B protecting compounds in the marine alga *Phaeocystis pouchetji* from Antarctica, *Marine Biology* 109 (3), 391–395, 1991.
27. Sommaruga, R., Libkind, D., van Broock, M., and Whitehead, K., Mycosporine-glutaminol-glucoside, a UV-absorbing compound of two *Rhodotorula* yeast species, *Yeast* 21 (13), 1077–1081, 2004.
28. Torres, A., Hochberg, M., Pergament, I., Smoum, R., Niddam, V., Dembitsky, V. M., Temina, M., Dor, I., Lev, O., Srebnik, M., and Enk, C. D., A new UV-B absorbing mycosporine with photo protective activity from the lichenized ascomycete *Collema cristatum*, *European Journal of Biochemistry* 271 (4), 780–784, 2004.
29. Newman, S. J., Dunlap, W. C., Nicol, S., and Ritz, D., Antarctic krill (*Euphausia superba*) acquire a UV-absorbing mycosporine-like amino acid from dietary algae, *Journal of Experimental Marine Biology and Ecology* 255 (1), 93–110, 2000.
30. Moeller, R. E., Gilroy, S., Williamson, C. E., Grad, G., and Sommaruga, R., Dietary acquisition of photoprotective compounds (mycosporine-like amino acids, carotenoids) and acclimation to ultraviolet radiation in a freshwater copepod, *Limnology and Oceanography* 50 (2), 427–439, 2005.
31. Carignan, M. O., Cardozo, K. H. M., Oliveira-Silva, D., Colepicolo, P., and Carreto, J. I., Palythine-threonine, a major novel mycosporine-like amino acid (MAA) isolated from the hermatypic coral *Pocillopora capitata*, *Journal of Photochemistry and Photobiology B—Biology* 94 (3), 191–200, 2009.
32. Shick, J. M., Romaine-Lioud, S., Ferrier-Pages, C., and Gattuso, J. P., Ultraviolet-B radiation stimulates shikimate pathway-dependent accumulation of mycosporine-like amino acids in the coral *Stylophora pistillata* despite decreases in its population of symbiotic dinoflagellates, *Limnology and Oceanography* 44 (7), 1667–1682, 1999.
33. Singh, S. P., Klisch, M., Sinha, R. P., and Hader, D. P., Genome mining of mycosporine-like amino acid (MAA) synthesizing and non-synthesizing cyanobacteria: A bioinformatics study, *Genomics* 95 (2), 120–128, 2010.
34. Ferreyra, G. A., Schloss, E. R., and Neale, P. J., Phytoplankton responses to natural ultraviolet irradiance during early spring in the Weddell-Scotia Confluence: An experimental approach, *Antarctic Journal of the United States* 29 (5), 268–270, 1994.
35. Lesser, M. P., Acclimation of phytoplankton to UV-B radiation: Oxidative stress and photoinhibition of photosynthesis are not prevented by UV-absorbing compounds in the dinoflagellate *Prorocentrum micans*, *Marine Ecology-Progress Series* 132 (1–3), 287–297, 1996.
36. Carreto, J. I., Fluence and wavelength dependence of mycosporine-like amino acid synthesis in the dinoflagellate *Alexandrium excavatum*, in *Toxic marine phytoplankton*, ed. E. Graneli, L. Edler, B. Sundström, and D. M. Anderson, 275–279, New York: Elsevier, 1990.
37. Klisch, M., and Hader, D. P., Wavelength dependence of mycosporine-like amino acid synthesis in *Gyrodinium dorsum*, *Journal of Photochemistry and Photobiology B—Biology* 66 (1), 60–66, 2002.
38. Riegger, L., and Robinson, D., Photoinduction of UV-absorbing compounds in *Antarctic diatoms* and *Phaeocystis antarctica*, *Marine Ecology-Progress Series* 160, 13–25, 1997.
39. Singh, S. P., Sinha, R. P., Klisch, M., and Hader, D. P., Mycosporine-like amino acids (MAAs) profile of a rice-field cyanobacterium *Anabaena doliolum* as influenced by PAR and UVR, *Planta* 229 (1), 225–233, 2008.
40. Karsten, U., Franklin, L. A., Luning, K., and Wiencke, C., Natural ultraviolet radiation and photosynthetically active radiation induce formation of mycosporine-like amino acids in the marine macroalga *Chondrus crispus* (Rhodophyta), *Planta* 205 (2), 257–262, 1998.
41. Litchman, E., Neale, P. J., and Banaszak, A. T., Increased sensitivity to ultraviolet radiation in nitrogen-limited dinoflagellates: Photoprotection and repair, *Limnology and Oceanography* 47 (1), 86–94, 2002.
42. Portwich, A., and Garcia-Pichel, F., A novel prokaryotic UVB photoreceptor in the cyanobacterium *Chlorogloeopsis* PCC 6912, *Photochemistry and Photobiology* 71 (4), 493–498, 2000.
43. Klassen, J. L., Phylogenetic and evolutionary patterns in microbial carotenoid biosynthesis are revealed by comparative genomics, *Plos One* 5 (6), e11257, 2010.

44. Britton, G., *Carotenoids handbook*, Basal, Switzerland: Birkhäuser Verlag, 2004.
45. Frank, H. A., and Brudvig, G. W., Redox functions of carotenoids in photosynthesis, *Biochemistry* 43 (27), 8607–8615, 2004.
46. Frank, H. A., and Cogdell, R. J., Carotenoids in photosynthesis, *Photochemistry and Photobiology* 63 (3), 257–264, 1996.
47. Fraser, N. J., Hashimoto, H., and Cogdell, R. J., Carotenoids and bacterial photosynthesis: The story so far, *Photosynthesis Research* 70 (3), 249–256, 2001.
48. Tian, B., Xu, Z. J., Sun, Z. T., Lin, J., and Hua, Y. J., Evaluation of the antioxidant effects of carotenoids from *Deinococcus radiodurans* through targeted mutagenesis, chemiluminescence, and DNA damage analyses, *Biochimica Et Biophysica Acta—General Subjects* 1770 (6), 902–911, 2007.
49. Liu, G. Y., Essex, A., Buchanan, J. T., Datta, V., Hoffman, H. M., Bastian, J. F., Fierer, J., and Nizet, V., *Staphylococcus aureus* golden pigment impairs neutrophil killing and promotes virulence through its antioxidant activity, *Journal of Experimental Medicine* 202 (2), 209–215, 2005.
50. Gruszecki, W. I., and Strzalka, K., Carotenoids as modulators of lipid membrane physical properties, *Biochimica Et Biophysica Acta—Molecular Basis of Disease* 1740 (2), 108–115, 2005.
51. Kupisz, K., Sujak, A., Patyra, M., Trebacz, K., and Gruszecki, W. I., Can membrane-bound carotenoid pigment zeaxanthin carry out a transmembrane proton transfer? *Biochimica Et Biophysica Acta–Biomembranes* 1778 (10), 2334–2340, 2008.
52. Auldridge, M. E., McCarty, D. R., and Klee, H. J., Plant carotenoid cleavage oxygenases and their apocarotenoid products, *Current Opinion in Plant Biology* 9 (3), 315–321, 2006.
53. Fuhrman, J. A., Schwalbach, M. S., and Stingl, U., Opinion—Proteorhodopsins: An array of physiological roles? *Nature Reviews Microbiology* 6 (6), 488–494, 2008.
54. Lanyi, J. K., and Balashov, S. P., Xanthorhodopsin: A bacteriorhodopsin-like proton pump with a carotenoid antenna, *Biochimica Et Biophysica Acta—Bioenergetics* 1777 (7–8), 684–688, 2008.
55. Sharma, A. K., Spudich, J. L., and Doolittle, W. F., Microbial rhodopsins: Functional versatility and genetic mobility, *Trends in Microbiology* 14 (11), 463–469, 2006.
56. Del Campo, J. A., Garcia-Gonzalez, M., and Guerrero, M. G., Outdoor cultivation of microalgae for carotenoid production: Current state and perspectives, *Applied Microbiology and Biotechnology* 74 (6), 1163–1174, 2007.
57. Rajeev, K. J., and Xu, Z. R., Biomedical compounds from marine organisms, *Marine Drugs* 2 (3), 123–146, 2004.
58. Stahl, W., and Sies, H., Carotenoids and flavonoids contribute to nutritional protection against skin damage from sunlight, *Molecular Biotechnology* 37 (1), 26–30, 2007.
59. Aust, O., Stahl, W., Sies, H., Tronnier, H., and Heinrich, U., Supplementation with tomato-based products increases lycopene, phytofluene, and phytoene levels in human serum and protects against UV-light-induced erythema, *International Journal for Vitamin and Nutrition Research* 75 (1), 54–60, 2005.
60. Guerin, M., Huntley, M. E., and Olaizola, M., Haematococcus astaxanthin: Applications for human health and nutrition, *Trends in Biotechnology* 21 (5), 210–216, 2003.
61. Sies, H., and Stahl, W., Nutritional protection against skin damage from sunlight, *Annual Review of Nutrition* 24, 173–200, 2004.
62. Wertz, K., Hunziker, P. B., Seifert, N., Riss, G., Neeb, M., Steiner, G., Hunziker, W., and Goralczyk, R., β-carotene interferes with ultraviolet light A-induced gene expression by multiple pathways, *Journal of Investigative Dermatology* 124 (2), 428–434, 2005.
63. Mayne, S. T., Beta-carotene, carotenoids, and disease prevention in humans, *Faseb Journal* 10 (7), 690–701, 1996.
64. Tapiero, H., Townsend, D. M., and Tew, K. D., The role of carotenoids in the prevention of human pathologies, *Biomedicine and Pharmacotherapy* 58 (2), 100–110, 2004.
65. Cardozo, K. H. M., Guaratini, T., Barros, M. P., Falcao, V. R., Tonon, A. P., Lopes, N. P., Campos, S., Torres, M. A., Souza, A. O., Colepicolo, P., and Pinto, E., Metabolites from algae with economical impact, *Comparative Biochemistry and Physiology C—Toxicology and Pharmacology* 146 (1–2), 60–78, 2007.
66. Coesel, S., Obornik, M., Varela, J., Falciatore, A., and Bowler, C., Evolutionary origins and functions of the carotenoid biosynthetic pathway in marine diatoms, *Plos One* 3 (8), e2896, 2008.
67. Cunningham, F. X., Lee, H., and Gantt, E., Carotenoid biosynthesis in the primitive red alga *Cyanidioschyzon merolae*, *Eukaryotic Cell* 6 (3), 533–545, 2007.
68. Hosokawa, M., Okada, T., Mikami, N., Konishi, I., and Miyashita, K., Bio-functions of marine carotenoids, *Food Science and Biotechnology* 18 (1), 1–11, 2009.
69. Sandmann, G., Molecular evolution of carotenoid biosynthesis from bacteria to plants, *Physiologia Plantarum* 116 (4), 431–440, 2002.

70. Schafer, L., Vioque, A., and Sandmann, G., Functional *in situ* evaluation of photo synthesis-protecting carotenoids in mutants of the cyanobacterium *Synechocystis* PCC6803, *Journal of Photochemistry and Photobiology B–Biology* 78 (3), 195–201, 2005.
71. Holt, N. E., Fleming, G. R., and Niyogi, K. K., Toward an understanding of the mechanism of nonphotochemical quenching in green plants, *Biochemistry* 43 (26), 8281–8289, 2004.
72. Maoka, T., Akimoto, N., Yim, M.-J., Hosokawa, M., and Miyashita, K., New C-37 skeletal carotenoid from the clam, *Paphia amabillis*, *Journal of Agricultural and Food Chemistry* 56 (24), 12069–12072, 2008.
73. Shindo, K., Endo, M., Miyake, Y., Wakasugi, K., Morritt, D., Bramley, P. M., Fraser, P. D., Kasai, H., and Misawa, N., Methyl glucosyl-3,4-dehydro-apo-8′-lycopenoate, a novel antioxidative glyco-C-30-carotenoic acid produced by a marine bacterium *Planococcus maritimus*, *Journal of Antibiotics* 61 (12), 729–735, 2008.
74. Nägeli, C., Gattungen einzelliger Algen, physiologisch und systematisch bearbeitet, Neue Denkschrift, *Allg. Schweiz. Natur Ges.* 10, 1–138, 1894.
75. Nägeli, C. S. Schwenderer, *Das Mikroskop*, 2nd ed., Leipzig: Willhelm Engelmann, 1877.
76. Geitler, L., Cyanophyceae (Blaualgen), in *Kryptogamen-Flora von Deutschland, Österreich und der Schweiz*, ed. L. Rabenhorst, 1–119, Leipzig: Akademische Verlagsgesellschaft, 1932.
77. Desikachary, T. V., *Cyanophyta*, New Delhi, India, Indian Council of Agriculture Research, 1959.
78. Proteau, P. J., Gerwick, W. H., Garciapichel, F., and Castenholz, R., The structure of scytonemin, an ultraviolet sunscreen pigment from the sheaths of cyanobacteria, *Experientia* 49 (9), 825–829, 1993.
79. Hunsucker, S. W., Tissue, B. M., Potts, M., and Helm, R. F., Screening protocol for the ultraviolet-photoprotective pigment scytonemin, *Analytical Biochemistry* 288 (2), 227–230, 2001.
80. Bultel-Ponce, V., Felix-Theodose, F., Sarthou, C., Ponge, J. F., and Bodo, B., New pigments from the terrestrial cyanobacterium *Scytonema* sp. collected on the Mitaraka Inselberg, French Guyana, *Journal of Natural Products* 67 (4), 678–681, 2004.
81. Balskus, E. P., and Walsh, C. T., Investigating the initial steps in the biosynthesis of cyanobacterial sunscreen scytonemin, *Journal of the American Chemical Society* 130 (46), 15260, 2008.
82. Balskus, E. P., and Walsh, C. T., An enzymatic cyclopentyl[b]indole formation involved in scytonemin biosynthesis, *Journal of the American Chemical Society* 131 (41), 14648–14649, 2009.
83. Jones, A. C., Gu, L., Sorrels, C. M., Sherman, D. H., and Gerwick, W. H., New tricks from ancient algae: Natural products biosynthesis in marine cyanobacteria, *Current Opinion in Chemical Biology* 13 (2), 216–223, 2009.
84. Sorrels, C. M., Proteau, P. J., and Gerwick, W. H., Organization, evolution, and expression analysis of the biosynthetic gene cluster for scytonemin, a cyanobacterial UV-absorbing pigment, *Applied and Environmental Microbiology* 75 (14), 4861–4869, 2009.
85. Soule, T., Garcia-Pichel, F., and Stout, V., Gene expression patterns associated with the biosynthesis of the sunscreen scytonemin in *Nostoc punctiforme* ATCC 29133 in response to UVA radiation, *Journal of Bacteriology* 191 (14), 4639–4646, 2009.
86. Soule, T., Palmer, K., Gao, Q., Potrafka, R. M., Stout, V., and Garcia-Pichel, F., A comparative genomics approach to understanding the biosynthesis of the sunscreen scytonemin in cyanobacteria, *BMC Genomics* 10, 336, 2009.
87. Soule, T., Stout, V., Swingley, W. D., Meeks, J. C., and Garcia-Pichel, F., Molecular genetics and genomic analysis of scytonemin biosynthesis in *Nostoc punctiforme* ATCC 29133, *Journal of Bacteriology* 189 (12), 4465–4472, 2007.
88. Garciapichel, F., and Castenholz, R. W., Characterization and biological implications of scytonemin, a cyanobacterial sheath pigment, *Journal of Phycology* 27 (3), 395–409, 1991.
89. Dillon, J. G., Tatsumi, C. M., Tandingan, P. G., and Castenholz, R. W., Effect of environmental factors on the synthesis of scytonemin, a UV-screening pigment, in a cyanobacterium (*Chroococcidiopsis* sp.), *Archives of Microbiology* 177 (4), 322–331, 2002.
90. Lindquist, N., Lobkovsky, E., and Clardy, J., Tridentatols A–C, novel natural products of the marine hydroid *Tridentata marginata*, *Tetrahedron Letters* 37 (51), 9131–9134, 1996.
91. Regalado, E. L., Rodriguez, M., Menendez, R., Concepcion, A. A., Nogueiras, C., Laguna, A., Rodriguez, A. A., Williams, D. E., Lorenzo-Luaces, P., Valdes, O., and Hernandez, Y., Repair of UVB-damaged skin by the antioxidant sulphated flavone glycoside thalassiolin B isolated from the marine plant *Thalassia testudinum Banks ex Konig*, *Marine Biotechnology* 11 (1), 74–80, 2009.
92. Matsunaga, T., Burgess, J. G., Yamada, N., Komatsu, K., Yoshida, S., and Wachi, Y., An ultraviolet (UV-A) absorbing biopterin glucoside from the marine planktonic cyanobacterium *Oscillatoria* sp., *Applied Microbiology and Biotechnology* 39 (2), 250–253, 1993.

93. Salih, A., Photoprotection of symbiotic dinoflagellates by fluorescent pigments in reef corals, in *ACRS proceedings—75th anniversary conference*, Heron Island, October 1997, ed. J. G. Greenwood and N. J. Hall, 217–230, School of Marine Science, University of Queensland, Brisbane, 1998.
94. Ortonne, J. P., Photoprotective properties of skin melanin, *British Journal of Dermatology* 146, 7–10, 2002.
95. Barnett, T. A., and Hageman, J. H., Characterization of a brown pigment from *Bacillus subtilis* cultures, *Canadian Journal of Microbiology* 29 (3), 309–315, 1983.
96. Fuqua, W. C., Coyne, V. E., Stein, D. C., Lin, C. M., and Weiner, R. M., Characterization of mel.4: A gene encoding melanin biosynthesis from the marine bacterium *Shewanella colwelliana, Gene* 109 (1), 131–136, 1991.
97. Fuqua, W. C., and Weiner, R. M., The melA gene is essential for melanin biosynthesis in the marine bacterium *Shewanella colwelliana, Journal of General Microbiology* 139, 1105–1114, 1993.
98. Smit, N., Vicanova, J., and Pavel, S., The hunt for natural skin whitening agents, *International Journal of Molecular Sciences* 10 (12), 5326–5349, 2009.
99. Wilcox, D. E., Porras, A. G., Hwang, Y. T., Lerch, K., Winkler, M. E., and Solomon, E. I., Substrate analogue binding to the coupled binuclear copper active site in tyrosinase, *Journal of the American Chemical Society* 107 (13), 4015–4027, 1985.
100. Mason, H. S., Structures and functions of the phenolase complex, *Nature* 177 (4498), 79–81, 1956.
101. Kang, H. S., Kim, H. R., Byun, D. S., Son, B. W., Nam, T. J., and Choi, J. S., Tyrosinase inhibitors isolated from the edible brown alga *Ecklonia stolonifera, Archives of Pharmacal Research* 27 (12), 1226–1232, 2004.
102. Perez-Bernal, A., Munoz-Perez, M. A., and Camacho, F., Management of facial hyperpigmentation, *American Journal of Clinical Dermatology* 1 (5), 261–268, 2000.
103. Kim, Y. M., Yun, J., Lee, C. K., Lee, H., Min, K. R., and Kim, Y., Oxyresveratrol and hydroxystilbene compounds—Inhibitory effect on tyrosinase and mechanism of action, *Journal of Biological Chemistry* 277 (18), 16340–16344, 2002.
104. Cooksey, C. J., Garratt, P. J., Land, E. J., Pavel, S., Ramsden, C. A., Riley, P. A., and Smit, N. P. M., Evidence of the indirect formation of the catecholic intermediate substrate responsible for the autoactivation kinetics of tyrosinase, *Journal of Biological Chemistry* 272 (42), 26226–26235, 1997.
105. Raper, H. S., The aerobic oxidases, *Physiological Reviews* 8 (2), 245–282, 1928.
106. Mason, H. S., The chemistry of melanin. III. Mechanism of the oxidation of trihydroxyphenylalanine by tyrosinase, *Journal of Biological Chemistry* 172 (1), 83–99, 1948.
107. Schallreuter, K. U., Kothari, S., Chavan, B., and Spencer, J. D., Regulation of melanogenesis—Controversies and new concepts, *Experimental Dermatology* 17 (5), 395–404, 2008.
108. Chang, T. S., An updated review of tyrosinase inhibitors, *International Journal of Molecular Sciences* 10 (6), 2440–2475, 2009.
109. Kim, Y. J., and Uyama, H., Tyrosinase inhibitors from natural and synthetic sources: Structure, inhibition mechanism and perspective for the future, *Cellular and Molecular Life Sciences* 62 (15), 1707–1723, 2005.
110. Parvez, S., Kang, M., Chung, H. S., and Bae, H., Naturally occurring tyrosinase inhibitors: Mechanism and applications in skin health, cosmetics and agriculture industries, *Phytotherapy Research* 21 (9), 805–816, 2007.
111. Imada, C., Sugimoto, Y., Makimura, T., Kobayashi, T., Hamada, N., and Watanabe, E., Isolation and characterization of tyrosinase inhibitor-producing microorganisms from marine environment, *Fisheries Science* 67 (6), 1151–1156, 2001.
112. Tsuchiya, T., Yamada, K., Minoura, K., Miyamoto, K., Usami, Y., Kobayashi, T., Hamada-Sato, N., Imada, C., and Tsujibo, H., Purification and determination of the chemical structure of the tyrosinase inhibitor produced by *Trichoderma viride* strain H1-7 from a marine environment, *Biological and Pharmaceutical Bulletin* 31 (8), 1618–1620, 2008.
113. Faull, J. L., Graemecook, K. A., and Pilkington, B. L., Production of an isonitrille antibiotic by an UV-induced mutant of *Trichoderma harzianum, Phytochemistry* 36 (5), 1273–1276, 1994.
114. Simon, A., Dunlop, R. W., Ghisalberti, E. L., and Sivasithamparam, K., *Trichoderma koningii* produces a pyrone compound with antibiotic properties, *Soil Biology and Biochemistry* 20 (2), 263–264, 1988.
115. Li, X. F., Kim, M. K., Lee, U., Kim, S. K., Kang, J. S., Choi, H. D., and Son, B. W., Myrothenones A and B, cyclopentenone derivatives with tyrosinase inhibitory activity from the marine-derived fungus *Myrothecium* sp., *Chemical and Pharmaceutical Bulletin* 53 (4), 453–455, 2005.
116. Feng, Z., Leutou, A. S., Yang, G., Nenkep, V. N., Siwe, X. N., Choi, H. D., Kang, J. S., and Son, B. W., Bioactive cyclopentenone derivatives from marine isolates of fungi, *Bulletin of the Korean Chemical Society* 30 (10), 2345–2350, 2009.

117. Zhang, D. H., Li, X. G., Kang, J. S., Choi, H. D., and Son, B. W., A new α-pyrone derivative, 6-[(E)-hept-1-enyl]-alpha-pyrone, with tyrosinase inhibitory activity from a marine isolate of the fungus *Botrytis*, *Bulletin of the Korean Chemical Society* 28 (5), 887–888, 2007.
118. Heo, S. J., Ko, S. C., Kang, S. M., Cha, S. H., Lee, S. H., Kang, D. H., Jung, W. K., Affana, B., Oh, C., and Jeon, Y. J., Inhibitory effect of diphlorethohydroxycarmalol on melanogenesis and its protective effect against UV-B radiation-induced cell damage, *Food and Chemical Toxicology* 48 (5), 1355–1361, 2010.
119. Schneider, Y., Vincent, F., Duranton, B., Badolo, L., Gosse, F., Bergmann, C., Seiler, N., and Raul, F., Anti-proliferative effect of resveratrol, a natural component of grapes and wine, on human colonic cancer cells, *Cancer Letters* 158 (1), 85–91, 2000.
120. Nakayama, Y., Takahashi, M., Fukuyama, Y., and Kinzyo, Z., Anti-plasmin inhibitor. 4. An anti-plasmin inhibitor, eckol, isolated from the brown alga *Ecklonia kurome okamura*, *Agricultural and Biological Chemistry* 53 (11), 3025–3030, 1989.
121. Lim, S. N., Cheung, P. C. K., Ooi, V. E. C., and Ang, P. O., Evaluation of antioxidative activity of extracts from a brown seaweed, *Sargassum siliquastrum*, *Journal of Agricultural and Food Chemistry* 50 (13), 3862–3866, 2002.
122. Kakegawa, H., Matsumoto, H., and Satoh, T., Inhibitory effects of some natural products on the activation of hyaluronidase and their anti-allergic action, *Chemical and Pharmaceutical Bulletin* 40 (6), 1439–1442, 1992.
123. Soo-Jin, H., Kim, J. P., Jung, W. K., Lee, N. H., Kang, H. S., Jun, E. M., Park, S. H., Kang, S. M., Lee, Y. J., Park, P. J., and Jeon, Y. J., Identification of chemical structure and free radical scavenging activity of diphlorethohydroxycarmalol isolated from a brown alga, Ishige okamurae, *Journal of Microbiology and Biotechnology* 18 (4), 676–681, 2008.
124. Sano, T., and Kaya, K., Oscillapeptin G, a tyrosinase inhibitor from toxic *Oscillatoria agardhii*, *Journal of Natural Products* 59 (1), 90–92, 1996.
125. Liu, Z. Q., Chemical methods to evaluate antioxidant ability, *Chemical Reviews* 110 (10), 5675–5691, 2010.
126. Kirkwood, T. B. L., and Austad, S. N., Why do we age? *Nature* 408 (6809), 233–238, 2000.
127. Finkel, T., and Holbrook, N. J., Oxidants, oxidative stress and the biology of ageing, *Nature* 408 (6809), 239–247, 2000.
128. Neumann, C. A., Krause, D. S., Carman, C. V., Das, S., Dubey, D. P., Abraham, J. L., Bronson, R. T., Fujiwara, Y., Orkin, S. H., and Van Etten, R. A., Essential role for the peroxiredoxin Prdx1 in erythrocyte antioxidant defence and tumour suppression, *Nature* 424 (6948), 561–565, 2003.
129. Bland, J. S., Oxidants and antioxidants in clinical medicine: Past, present and future potential, *Journal of Nutritional and Environmental Medicine (Abingdon)* 5 (3), 255–280, 1995.
130. Thiele, J. J., Dreher, F., and Packer, L., Antioxidant defense systems in skin, *Cutaneous and Ocular Toxicology* 21, 119–160, 2002.
131. Abdel-Lateff, A., Fisch, K. M., Wright, A. D., and Konig, G. M., A new antioxidant isobenzofuranone derivative from the algicolous marine fungus *Epicoccum* sp., *Planta Medica* 69 (9), 831–834, 2003.
132. Ahn, G. N., Kim, K. N., Cha, S. H., Song, C. B., Lee, J., Heo, M. S., Yeo, I. K., Lee, N. H., Jee, Y. H., Kim, J. S., Heu, M. S., and Jeon, Y. J., Antioxidant activities of phlorotannins purified from *Ecklonia cava* on free radical scavenging using ESR and H2O2-mediated DNA damage, *European Food Research and Technology* 226, 71–79, 2007.
133. Potterat, O., Antioxidants and free radical scavengers of natural origin, *Current Organic Chemistry* 1 (4), 415–440, 1997.
134. Miki, W., Otaki, N., Yokoyama, A., Izumida, H., and Shimidzu, N., Okadaxanthin, a novel C50-carotenoid from a bacterium, *Pseudomonas* sp. KK10206C associated with marine sponge, *Halichondria okadai*, *Experientia* 50 (7), 684–686, 1994.
135. Miyashita, K., Function of marine carotenoids, *Food Factors for Health Promotion* 61, 136–146, 2009.
136. Watanabe, N., Yamamoto, K., Ihshikawa, H., Yagi, A., Sakata, K., Brinen, L. S., and Clardy, J., New chlorophyll-A-related compounds isolated as antioxidants from marine bivalves, *Journal of Natural Products* 56 (3), 305–317, 1993.
137. Yamamoto, K., Sakata, K., Watanabe, N., Yagi, A., Brinen, L. S., and Clardy, J., Chlorophyllonic acid and methyl ester, a new chlorophyll a related compound isolated as an antioxidant from short-necked clam, *Ruditapes philippinarum*, *Tetrahedron Letters* 33 (18), 2587–2588, 1992.
138. Yamamoto, Y., Fujisawa, A., Hara, A., and Dunlap, W. C., An unusual vitamin E constituent (α-tocomonoenol) provides enhanced antioxidant protection in marine organisms adapted to cold-water environments, *Proceedings of the National Academy of Sciences of the United States of America* 98 (23), 13144–13148, 2001.

139. Sakata, K., Yamamoto, K., Ishikawa, H., Yagi, A., Etoh, H., and Ina, K., Chlorophyllone-A. A new pheophorbide-A related compound isolated from *Ruditapes philippinarumas*, an antioxidative compound, *Tetrahedron Letters* 31 (8), 1165–1168, 1990.
140. Ho, C., Food phytochemicals for cancer prevention II, ACS Symposium Series, American Chemical Society, Washington, DC, 1994.
141. Yamamoto, Y., Maita, N., Fujisawa, A., Takashima, J., Ishii, Y., and Dunlap, W. C., A new vitamin E (α-tocomonoenol) from eggs of the Pacific salmon *Oncorhynchus keta*, *Journal of Natural Products* 62 (12), 1685–1687, 1999.
142. Edge, R., McGarvey, D. J., and Truscott, T. G., The carotenoids as anti-oxidants—a review, *Journal of Photochemistry and Photobiology B–Biology* 41 (3), 189–200, 1997.
143. Ramarathnam, N., Osawa, T., Ochi, H., and Kawakishi, S., The contribution of plant food antioxidants to human health, *Trends in Food Science and Technology* 6 (3), 75–82, 1995.
144. Shon, M. Y., Kim, T. H., and Sung, N. J., Antioxidants and free radical scavenging activity of *Phellinus baumii* (Phellinus of Hymenochaetaceae) extracts, *Food Chemistry* 82 (4), 593–597, 2003.
145. Nakamura, T., Nagayama, K., Uchida, K., and Tanaka, R., Antioxidant activity of phlorotannins isolated from the brown alga *Eisenia bicyclis*, *Fisheries Science* 62 (6), 923–926, 1996.
146. Shibata, T., Ishimaru, K., Kawaguchi, S., Yoshikawa, H., and Hama, Y., Antioxidant activities of phlorotannins isolated from Japanese Laminariaceae, *Journal of Applied Phycology* 20 (5), 705–711, 2008.
147. Zhang, G. W., Ma, X. Q., Kurihara, H., Zhang, C. X., Yao, X. S., Su, J. Y., and Zeng, L. M., New hemiketal steroid from the soft coral *Cladiella* sp., *Organic Letters* 7 (6), 991–994, 2005.
148. Abou-Shoer, M. I., Shaala, L. A., Youssef, D. T. A., Badr, J. M., and Habib, A. A. M., Bioactive brominated metabolites from the Red Sea sponge *Suberea mollis*, *Journal of Natural Products* 71 (8), 1464–1467, 2008.
149. Li, K., Li, X. M., Ji, N. Y., and Wang, B. G., Bromophenols from the marine red alga *Polysiphonia urceolata* with DPPH radical scavenging activity, *Journal of Natural Products* 71 (1), 28–30, 2008.
150. Casapullo, A., Cutignano, A., Bruno, I., Bifulco, G., Debitus, C., Gomez-Paloma, L., and Riccio, R., Makaluvamine P, a new cytotoxic pyrroloiminoquinone from *Zyzzya* cf. *fuliginosa*, *Journal of Natural Products* 64 (10), 1354–1356, 2001.
151. Takamatsu, S., Hodges, T. W., Rajbhandari, I., Gerwick, W. H., Hamann, M. T., and Nagle, D. G., Marine natural products as novel antioxidant prototypes, *Journal of Natural Products* 66 (5), 605–608, 2003.
152. Utkina, N. K., Denisenko, V. A., and Krasokhin, V. B., Sesquiterpenoid aminoquinones from the marine sponge *Dysidea* sp., *Journal of Natural Products* 73 (4), 788–791, 2010.
153. Draelos, Z. D., *Cosmetic formulation of skin care products*, New York: Taylor & Francis Group, 2006.
154. Perricone, N., *The wrinkle cure*, New York: Warner Books, 2000.
155. Gautam, R., and Jachak, S. M., Recent developments in anti-inflammatory natural products, *Medicinal Research Reviews* 29 (5), 767–820, 2009.
156. Abad, M. J., Bedoya, L. M., and Bermejo, P., Natural marine anti-inflammatory products, *Mini-Reviews in Medicinal Chemistry* 8 (8), 740–754, 2008.
157. Greaves, M. W., and Camp, R. D. R., Prostaglandins, leukotrienes, phospholipase, platelet activating factor and cytokines: An integrated approach to inflammation of human skin, *Archives of Dermatological Research* 280, S33–S41, 1988.
158. Reilly, D. M., Parslew, R., Sharpe, G. R., Powell, S., and Green, M. R., Inflammatory mediators in normal, sensitive and diseased skin types, *Acta Dermato-Venereologica* 80 (3), 171–174, 2000.
159. Barr, R. M., Wong, E., Mallet, A. I., Olins, L. A., and Greaves, M. W., The analysis of arachidonic acid metabolites in normal, uninvolved and lesional psoriatic skin, *Prostaglandins* 28 (1), 57–65, 1984.
160. Schumacher, M., Cerella, C., Eifes, S., Chateauvieux, S., Morceau, F., Jaspars, M., Dicato, M., and Diederich, M., Heteronemin, a spongean sesterterpene, inhibits TNF α-induced NF-κB activation through proteasome inhibition and induces apoptotic cell death, *Biochemical Pharmacology* 79 (4), 610–622, 2010.
161. Costantino, V., Fattorusso, E., Mangoni, A., Perinu, C., Cirino, G., De Gruttola, L., and Roviezzo, F., Tedanol: A potent anti-inflammatory ent-pimarane diterpene from the Caribbean Sponge *Tedania ignis*, *Bioorganic and Medicinal Chemistry* 17 (21), 7542–7547, 2009.
162. Folmer, F., Jaspars, M., Dicato, M., and Diederich, M., Marine natural products as targeted modulators of the transcription factor NF-κB, *Biochemical Pharmacology* 75 (3), 603–617, 2008.
163. Monti, M. C., Casapullo, A., Cavasotto, C. N., Napolitano, A., and Riccio, R., Scallaradial, a dialdehyde-containing marine metabolite that causes an unexpected noncovalent PLA(2) inactivation, *Chembiochem* 8 (13), 1585–1591, 2007.

164. Xie, Y. L., Liu, L. H., Huang, X. C., Guo, Y. W., and Lou, L. G., Scalaradial inhibition of epidermal growth factor receptor-mediated Akt phosphorylation is independent of secretory phospholipase A(2), *Journal of Pharmacology and Experimental Therapeutics* 314 (3), 1210–1217, 2005.
165. Xie, Y. L., Yu, Z. G., Liu, L. H., Guo, Y. W., and Lou, L. G., Biphasic regulation of extracellular signal-regulated kinases by scalaradial, a secretory phospholipase A(2) inhibitor, *Cancer Biology and Therapy* 5 (8), 988–992, 2006.
166. Mayer, A. M. S., Jacobson, P. B., Fenical, W., Jacobs, R. S., and Glaser, K. B., Pharmacological characterization of the pseudopterosins: Novel anti-inflammatory natural products isolated from the Caribbean soft coral, *Pseudopterogorgia elisabethae*, *Life Sciences* 62 (26), PL401–PL407, 1998.
167. Cobar, O. M., Rodriguez, A. D., Padilla, O. L., and Sanchez, J. A., The calyculaglycosides: Dilophol-type diterpene glycosides exhibiting antiinflammatory activity from the Caribbean gorgonian *Eunicea* sp., *Journal of Organic Chemistry* 62 (21), 7183–7188, 1997.
168. Takaki, H., Koganemaru, R., Iwakawa, Y., Higuchi, R., and Miyamoto, T., Inhibitory effect of norditerpenes on LPS-induced TNF-α production from the Okinawan soft coral, *Sinularia* sp., *Biological and Pharmaceutical Bulletin* 26 (3), 380–382, 2003.
169. Radhika, P., Rao, P. R., Archana, J., and Rao, N. K., Anti-inflammatory activity of a new sphingosine derivative and cembrenoid diterpene (lobohedleolide) isolated from marine soft corals of *Sinularia crassa* TIXIER-DURIVAULT and *Lobophytum* species of the Andaman and Nicobar Islands, *Biological and Pharmaceutical Bulletin* 28 (7), 1311–1313, 2005.
170. Prinsep, M. R., Thomson, R. A., West, M. L., and Wylie, B. L., Tolypodiol, an antiinflammatory diterpenoid from the cyanobacterium *Tolypothrix nodosa*, *Journal of Natural Products* 59 (8), 786–788, 1996.
171. Terracciano, S., Aquino, M., Rodriquez, M., Monti, M. C., Casapullo, A., Riccio, R., and Gomez-Paloma, L., Chemistry and biology of anti-inflammatory marine natural products: Molecules interfering with cyclooxygenase, NF-κB and other unidentified targets, *Current Medicinal Chemistry* 13 (16), 1947–1969, 2006.
172. Posadas, I., Terencio, M. C., De Rosa, S., and Paya, M., Cavernolide: A new inhibitor of human sPLA(2) sharing unusual chemical features, *Life Sciences* 67 (24), 3007–3014, 2000.
173. Nakao, Y., and Fusetani, N., Enzyme inhibitors from marine invertebrates, *Journal of Natural Products* 70, 689–710, 2007.
174. Giannini, C., Debitus, C., Lucas, R., Ubeda, A., Paya, M., Hooper, J. N. A., and D'Auria, M. V., New sesquiterpene derivatives from the sponge *Dysidea* species with a selective inhibitor profile against human phospholipase A(2) and other leukocyte functions, *Journal of Natural Products* 64 (5), 612–615, 2001.
175. Mayer, A. M. S., Rodriguez, A. D., Berlinck, R. G. S., and Hamann, M. T., Marine pharmacology in 2003–4: Marine compounds with anthelmintic antibacterial, anticoagulant, antifungal, anti-inflammatory, antimalarial, antiplatelet, antiprotozoal, antituberculosis, and antiviral activities; affecting the cardiovascular, immune and nervous systems, and other miscellaneous mechanisms of action, *Comparative Biochemistry and Physiology C—Toxicology and Pharmacology* 145 (4), 553–581, 2007.
176. Lucas, R., Giannini, C., D'Auria, M. V., and Paya, M., Modulatory effect of bolinaquinone, a marine sesquiterpenoid, on acute and chronic inflammatory processes, *Journal of Pharmacology and Experimental Therapeutics* 304 (3), 1172–1180, 2003.
177. Posadas, I., Terencio, M. C., Giannini, C., D'Auria, M. V., and Paya, M., Dysidotronic acid, a new sesquiterpenoid, inhibits cytokine production and the expression of nitric oxide synthase, *European Journal of Pharmacology* 415 (2–3), 285–292, 2001.
178. Ferrandiz, M. L., Sanz, M. J., Bustos, G., Paya, M., Alcaraz, M. J., and Derosa, S., Avarol and avarone, 2 new antiinflammatory agents of marine origin, *European Journal of Pharmacology* 253 (1–2), 75–82, 1994.
179. Ortiz, A. R., Pisabarro, M. T., and Gago, F., Molecular model of the interaction of bee venom phospholipase A2 with manoalide, *Journal of Medicinal Chemistry* 36 (13), 1866–1879, 1993.
180. Newman, D. J., and Cragg, G. M., Advanced preclinical and clinical trials of natural products and related compounds from marine sources, *Current Medicinal Chemistry* 11 (13), 1693–1713, 2004.
181. D'Acquisto, F., Lanzotti, V., and Carnuccio, R., Cyclolinteinone, a sesterterpene from sponge *Cacospongia linteiformis*, prevents inducible nitric oxide synthase and inducible cyclo-oxygenase protein expression by blocking nuclear factor-κB activation in J774 macrophages, *Biochemical Journal* 346, 793–798, 2000.
182. De Rosa, M., Giordano, S., Scettri, A., Sodano, G., Soriente, A., Pastor, P. G., Alcaraz, M. J., and Paya, M., Synthesis and comparison of the antiinflammatory activity of manoalide and cacospongionolide B analogues, *Journal of Medicinal Chemistry* 41 (17), 3232–3238, 1998.
183. Cholbi, R., Ferrandiz, M. L., Terencio, M. C., DeRosa, S., Alcaraz, M. J., and Paya, M., Inhibition of phospholipase A(2) activities and some inflammatory responses by the marine product ircinin, *Naunyn-Schmiedebergs Archives of Pharmacology* 354 (5), 677–683, 1996.

184. Shen, Y. C., Cheng, S. Y., Kuo, Y. H., Hwang, T. L., Chiang, M. Y., and Khalil, A. T., Chemical transformation and biological activities of ambrein, a major product of ambergris from *Physeter macrocephalus* (sperm whale), *Journal of Natural Products* 70 (2), 147–153, 2007.
185. De Marino, S., Iorizzi, M., Zollo, F., Debitus, C., Menou, J. L., Ospina, L. F., Alcaraz, M. J., and Paya, M., New pyridinium alkaloids from a marine sponge of the genus *Spongia* with human phospholipase A(2) inhibitor profile, *Journal of Natural Products* 63 (3), 322–326, 2000.
186. Kita, M., Ohishi, N., Washida, K., Kondo, M., Koyama, T., Yamada, K., and Uemura, D., Symbioimine and neosymbioimine, amphoteric iminium metabolites from the symbiotic marine dinoflagellate *Symbiodinium* sp., *Bioorganic and Medicinal Chemistry* 13 (17), 5253–5258, 2005.
187. Pereira, L., Amado, A. M., Critchley, A. T., van de Velde, F., and Ribeiro-Claro, P. J. A., Identification of selected seaweed polysaccharides (phycocolloids) by vibrational spectroscopy (FTIR-ATR and FT-Raman), *Food Hydrocolloids* 23 (7), 1903–1909, 2009.
188. Eswaran, K., Mairh, O. P., and Rao, P. V. S., Inhibition of pigments and phycocolloid in a marine red alga *Gracilaria edulis* by ultraviolet-B radiation, *Biologia Plantarum* 45 (1), 157–159, 2002.
189. Chiovitti, A., Kraft, G. T., Bacic, A., and Liao, M. L., Gelling polysaccharide from Australian seaweeds: Research and potential, *Marine and Freshwater Research* 52 (7), 917–935, 2001.

3 Chemical Properties of Chitosan as a Marine Cosmeceutical

Laura Calderón Muñoz, Elena Lecumberri Lima, Ruth Expósito Harris, María Ángeles López Mengíbar, Niuris Acosta Contreras, and Ángeles María Heras Caballero

CONTENTS

3.1 INTRODUCTION

Chitosan is a natural polymer that is obtained from chitin, which is one of the most abundant biopolymers in nature. Chitin forms structural components in the exoskeleton of arthropods, including crustaceans and mollusks. Chitin is composed of linearly β-(1→4)-linked 2-acetamido-2-deoxy-β-D-glucose (N-acetyl glucosamine) residues. Chitosan is a linear polymer of α-(1→4)-linked 2-amino-2deoxy-β-D-glucopyranose and is easily obtained by partial deacetylation of chitin; therefore it is composed of N-acetyl glucosamine and glucosamine units (Sudha et al., 2005) (Figure 3.1). Primary amino groups of chitosan have a pKa of about 6.2. Below the pKa amino groups are protonated, making chitosan a soluble cationic polyelectrolyte—what could

(a)

(b)

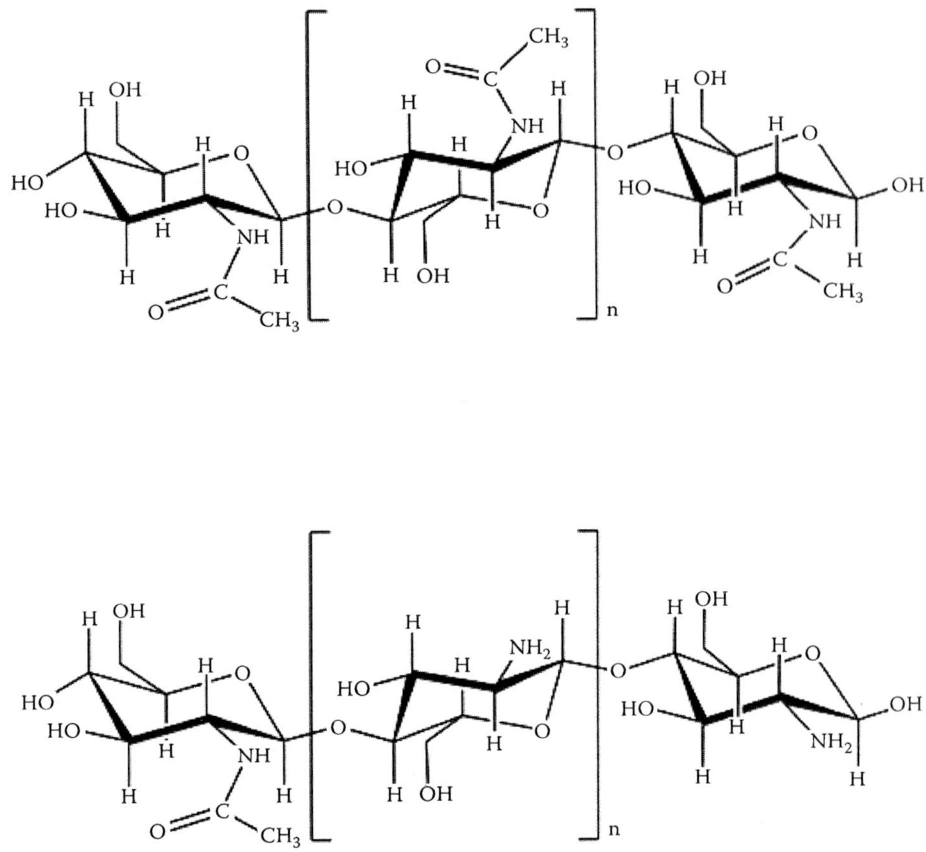

FIGURE 3.1 Chemical structure of 100% acetylated chitin (a) and chitosan (b).

constitute an advantage versus other β-1,4-linked polysaccharides (e.g., cellulose and chitin). Above the pKa, amino groups of chitosan become deprotonated, triggering polymer precipitation. Due to these properties and its ability to interact with a variety of molecules, chitosan has a number of commercial applications in the areas of bioadhesives (Dudhani and Kosaraju, 2010), waste treatment (Crini, 2005), food preservatives (Shahidi et al., 2002), food additives (Shahidi et al., 1999), pharmaceuticals (Paños et al., 2008), nutraceuticals (Chen and Subirade, 2005), and cosmetics (Boucher and Brunet, 2005).

Chitosan is prepared by hydrolysis of acetamide groups of chitin. Due to the resistance of these groups imposed by the *trans* arrangement of the C2–C3 substituents in the sugar ring, chitin hydrolysis is conducted by a severe alkaline treatment (Horton and Lineback, 1965). Usually, sodium or potassium hydroxides are used at a concentration of 30–50% w/v at high temperature (100°C). Several alternative processing methods have also been developed to reduce the long processing times and large amounts of alkali typically needed to deacetylate chitin to an acid-soluble derivative. Examples of these include the use of successive alkali treatments using thiophenol in DMSO (Domard and Rinadudo, 1983), thermomechanical processes using a cascade reactor operated under low alkali concentration (Pelletier et al., 1990), flash treatment under saturated steam (Focher et al., 1990), use of microwave dielectric heating (Goycoolea et al., 1998), and intermittent water washing (Mima et al., 1983).

The main factors that affect the extent of deacetylation are concentration of the alkali, previous treatment, particle size, and density of chitin. Among others, the last two factors highly affect the penetration rate of alkali into the amorphous and crystalline regions of the polymer. In practice, the maximal deacetylation degree (DD) that can be achieved in a single alkaline treatment is about 75–85% (Roberts, 1998). The hydrolysis conditions must be carefully selected to obtain a chitosan with a certain degree of deacetylation. The degree of deacetylation in the chitosan chain has a large impact on the chitosan properties, thus determining its application.

3.2 PHYSICOCHEMICAL CHARACTERIZATION OF CHITOSAN

Numerous species of chitosan can be obtained due to different chitin sources and differences in deacetylation conditions. The presence of free amino groups is responsible for the interaction of chitosan with biological systems, and the distribution of the deacetylated groups along the chitosan molecule may define these interactions. The amino functionality gives rise to chemical reactions such as acetylation, quaternization, reactions with aldehydes and ketones alkylation, grafting, chelation of metals, etc. On the other hand, hydroxyl functional groups also give various reactions such as o-acetylation, H bonding with polar atoms, grafting, etc. The wide variety of products that can be obtained as a result of these chemical reactions would exert valuable properties such as antibacterial, antifungal, antiviral, antiacid, antiulcer, nontoxic, nonallergenic, total biocompatibility and biodegradability, etc. (Pillai et al., 2009).

In contrast to chitin, the presence of free amine groups along the chitosan chain allow this macromolecule to dissolve in diluted aqueous acidic solvents due to the protonation of these groups, rendering the corresponding chitosan salt in solution. For instance, chitosans are soluble in water when interchain hydrogen bond formation is prevented by partial random reacetylation of the amino groups, by insertion of bulky substituents and side chains, or by glycosylation at C6 via oxazoline derivatives (Ravi Kumar, 2000). So, there are important experimental variables that should be taken into account when working with chitosan solutions such as the nature of the salt counterion, degree of acetylation, molecular weight, pH, ionic strength, and the addition of a nonaqueous solvent. This chapter is focused on the physicochemical characteristics that have a direct influence on the cosmeceutical properties of chitosan.

3.2.1 Viscosity

Viscosity is a liquid resistance to flow. A sample viscosity test may yield information on the relative differences in molecular weight (MW) among chitosan samples. The viscosity depends on several parameters, such as temperature (an increase in temperature results in a decrease in the viscosity), ionic strength (the viscosity of a chitosan solution is very sensitive to the ionic environment), chitosan concentration, and MW.

The intrinsic viscosity describes the ability of a polymer to form viscous solutions in water and is directly proportional to the average MW of the polymer. The intrinsic viscosity is determined by measuring the relative viscosity in an Ubbelohde capillary viscometer. In these viscometers, the time that the dissolution of chitosan takes to pass between two fixed marks is measured and related to viscosity by the Poiseuille equation (Agulló et al., 2004). The measurements should be performed at a constant temperature at low chitosan concentration.

3.2.2 Molecular Weight

The MW of a raw chitosan preparation is influenced by variable conditions employed in the deacetylation process. The MW of chitosan can be determined by several methods, such as light scattering spectrophotometry (Muzzarelli et al., 1987), gel permeation chromatography (Yomota et al., 1993), and viscosimetry (Ravi Kumar, 2000). The most commonly used technique for viscosimetry is described below.

Viscosimetry is not an absolute technique and requires the determination of some parameters to obtain MW values applying the Mark-Houwink equation (3.2). Briefly, a chitosan sample is dissolved in solutions of acetic acid and NaCl with a determinate ionic strength. The viscosity is measured in a viscosimeter coupled with a thermostatic bath. First, the time of pure solvent flow is measured and soon after, the time of chitosan solution flow is measured. Measurement of several dilutions is carried out until the flow time of chitosan approaches the flow time of the solvent (Kaplan, 1998).

$$[\eta\text{red}] = \frac{T - T_0}{T_0[Q]} \tag{3.1}$$

Equation 3.1 is used for the reduced viscosity determination, in which [ηred] is the reduced viscosity, T is the chitosan solution time of flow, T_0 is the solvent time of flow, and $[Q]$ is the concentration of chitosan solution in time T_0.

Plotting the calculated reduced viscosity from Equation 3.1 versus concentration, the intrinsic viscosity is determined. Once determined, the intrinsic viscosity and the chitosan molar mass, M, can be determined using Equation 3.2, where [ηint] is the intrinsic viscosity, and K and a are constants that can be obtained from the literature.

$$[\eta\text{int}] = KM^a \tag{3.2}$$

3.2.3 Deacetylation Degree

The deacetylation degree (DD) of chitosan is influenced by the preparation procedure. Two different methods for obtaining chitosan with varying degrees of acetylation are known. These are the heterogeneous deacetylation of solid chitin and the homogeneous deacetylation of preswollen chitin in an aqueous medium. In both heterogeneous and homogeneous conditions, the deacetylation reaction involves the use of concentrated alkali solutions and long processing times that can vary depending on the heterogeneous or homogeneous conditions from 1 to nearly 80 hours. The preparation procedure selected has an impact on the extent of moisture absorption, charge distribution, intrinsic viscosity, and chitosan solubility in aqueous solutions (Singla and Chawla, 2001).

A number of analytical tools have been used to define the DD, such as infrared spectroscopy (IR), first derivative UV spectrophotometry (1DUVS), nuclear magnetic resonance spectroscopy (^{1}H-NMR), ^{13}C solid-state NMR spectroscopy (^{13}CNMR), conductometric and potentiometric titration methods, equilibrium dye adsorption, elemental analysis, acid degradation followed by high performance liquid chromatography (HPLC), as well as thermal analysis (differenetial scanning calorimetry [DSC]) (Muzzarelli et al., 1997). Below, the most commonly used techniques are described.

The DD of chitosan can be determined by ^{1}H-NMR and ^{13}CNMR, considered to be the most sensitive techniques (Lamarque et al., 2004). Chitosan salts should be dissolved in D_2O and partially degraded to a degree of depolymerization of 20 to 30 using sodium nitrite before recording proton or carbon NMR spectra. The DD is determined from the integral of the CH_3 signal compared with the integral of H-1 protons considered as an internal standard.

The use of 1DUVS, as proposed by Muzzarelli and Rocchetti (1985), is the simplest and most convenient technique among all the currently available methods. This quantitative method is used to determine the number of amine functional groups in the polymer. The degree of deacetylation is calculated, recording the first derivative of the UV spectra of N-acetyl glucosamine and of chitosan samples at 202 nm. The 1DUVS method requires only very small amounts of sample and relies on simple reagents and instrumentation. Furthermore, the results obtained from 1DUVS are reasonably independent of protein contamination. Many researchers advocate 1DUVS as the standard method for routine determination of chitosan DD due to its high sensitivity, minimal interference of results from contaminants, and ease of performance.

3.3 FUNCTIONAL CHARACTERIZATION OF CHITOSAN AS A POTENTIAL COSMECEUTICAL

The term *cosmeceutical* is derived from the words *cosmetic* and *pharmaceutical*. Cosmeceuticals contain active ingredients such as vitamins, phytochemicals, enzymes, antioxidants, and essential oils and can be applied to products such as creams, lotions, and ointments. The search for new active agent sources, as marine origin ones, is currently the focus of many researchers. The eligibility of chitosan for cosmeceutical purposes is explained by its good film-forming properties, wound healing benefits, and bioadhesive, antioxidant, and antimicrobial properties (Silva-Júnior et al., 2008). More details can be found in the Chapter 25.

Chitosan films are considered very good biomaterials not only for cell cultivation but also for wound healing. Chatelet et al. (2001) demonstrated that chitosan films placed on a wound adhere to fibroblasts and improve keratinocytes proliferation, and therefore epidermis regeneration (Chatelet et al., 2001).

The wound healing effect of chitin and chitosan oligomers is of great interest due to their ability to stimulate fibroblast production by affecting the fibroblast growth factor (Ahmad et al., 2010). Chitosan is a bioadhesive polymer (Janes et al., 2001) with permeation-enhancing properties (Porporatto et al., 2005) that facilitate opening of the epithelial tight junctions (Van der Lubben et al., 2001; Yamamoto et al., 2005). Bioadhesion properties of chitosan are due to the interaction between its protonated amino groups and the negative residues of epithelia layer.

Another outstanding property of chitosan as a potential cosmeceutical is its antioxidant capacity. Chitosan has shown a significant scavenging capacity against different radical species, comparable with commercial antioxidants. Samples prepared from crab shell chitin with DDs of 90, 75, and 50% were evaluated on the basis of their abilities to scavenge 1,1-diphenyl-2-picrylhydrazyl (DPPH) radical, hydroxyl radical, superoxide radical, and alkyl radical (Park et al., 2004). The results revealed that chitosan with a higher DD can be used as a scavenger to control radical-induced damage in a cosmeceutical formulation.

The antimicrobial activity of chitosan against different groups of microorganisms, such as bacteria, yeast, and fungi, has been recognized and is considered to be one of the most important properties, corresponding directly to their possible cosmeceutical applications. Two main mechanisms have been suggested as the cause of the inhibition of microbial cells by chitosan. The interaction with anionic groups on the cell surface, due to its polycationic nature, causes the formation of an impermeable layer around the cell, which prevents the transport of essential solutes. The second proposed mechanism involves the inhibition of RNA and protein synthesis by permeation into the cell nucleus (Helander et al., 2001). The antimicrobial characteristics of chitosan present a profitable potential for cosmeceutical applications.

Many of these properties are highly dependent on the cationic character and pH-sensitive solubility of this polymer and allow chitosan to be formed into various shapes (beads, films, hydrogels, or particles). Hydrogels are hydrophilic polymer structures with a remarkable capacity to absorb water. Chitosan-based hydrogels may play an outstanding role as moisturizing agents, constituting a valuable alternative to creams for skin hydration (Byrne and Salian, 2008).

3.3.1 Gelification Capacity

Chitosan forms biocompatible and biodegradable hydrogels via covalent reactions with dialdehyde groups. Poor water solubility had greatly limited its application in the pharmaceutical and biomedical fields (Zvonar et al., 2010). Chitosan can form either chemical or physical hydrogel (Muzzarelli, 2009b; Matalanis et al., 2010; Coviello et al., 2007). Chemical hydrogels are obtained by cross-linking chitosan strands with irreversible covalent links. Physical hydrogels are stabilized by various kinds of reversible links, as in ionically cross-linked hydrogels, polyelectrolyte complexes, or entangled gels.

Chitosan is a polysaccharide that requires specific pH conditions to be water soluble. Different chemical modifications of chitosan can improve its water solubility or allow it to bear specific targeting functions (Zhang et al., 2010). While chitin is completely insoluble in aqueous media, chitosan can be dissolved under acidic conditions that provide sufficient protonation of its amino groups. However, chitosan has an apparent pKa of 6.2, which makes it soluble in a few diluted acidic solutions, such as acetic acid, etc. The resulting aqueous solutions are stable as long as the pH is below 6.2 (Pillai et al., 2009). Neutralization of chitosan solutions by common alkali solutions leads to the formation of a hydrated gel-like precipitate when the pH exceeds the pKa. Current processes for forming chitosan gels via pH sensitivity (Chen et al., 2004) include (1) solubilization of chitosan in an acidic aqueous medium followed by precipitation in an alkaline solution (NaOH, KOH, or NH_4OH), (2) treating chitosan solutions with gaseous NH_3, or (3) dialyzing chitosan against alkaline media (Chenite et al., 2006).

3.3.2 Film-Forming Ability

The solubility of chitosan in aqueous, acidic medium is an asset at the stage of preparing a film-forming solution. Chitosan films have excellent mechanical properties that depend on the MW, grade of crystallinity, and ashes content (Sánchez-González et al., 2010).

Chitosan films are formed from concentrated solutions of polymer (usually 3% w/w). These solutions are filtered or centrifuged to remove insoluble substances and then dried by evaporation of the solvent. Finally, the neutralization and washing processes are carried out. Another alternative is the direct coagulation of the polymer with a nonsolvent such as acetone or a basic solution.

The solubility of chitosan films can be limited by combining chitosan with some proteins or by cross-linking reactions. The formation of the covalent bonds between γ-carboxy-amide groups of peptide and primary amino groups of a number of cross-linking agents such as glutaraldehyde or tripolyphosphate are catalyzed by transglutaminase (TG) (Kołodziejska et al., 2006).

Chitosan films are considered excellent interfaces for oxygen permeability in relation to commercial polymers. In addition, chitosan films show optimal adhesion to different types of surfaces, good elasticity, and water resistance. All these valuable properties are important parameters when assessing their applicability in the cosmeceutical field.

3.3.3 Encapsulation Capacity

Chitosan can also be incorporated, in the form of nanoparticles (de la Fuente et al., 2010; Grenha et al., 2007) or microparticles (Varshosaz, 2007; Guliyeva et al., 2006), into cosmeceutical products as a vehicle for the encapsulation of active ingredients. Chitosan encapsulation provides protection of certain ingredients that are sensitive to environmental factors and improve their transport across biological membranes. This polymer is widely used in drug delivery systems (Paños et al., 2008; Illum et al. 2001; Huang and Fu, 2010) that were previously almost exclusive to injectable and oral drug administration and are now implemented in topical formulations.

3.4 RELATION BETWEEN PHYSICOCHEMICAL AND FUNCTIONAL PROPERTIES OF CHITOSAN

The physicochemical properties of chitosan depend on a number of parameters, such as the molecular weight (from approximately 10,000 to 1 million Daltons), deacetylation degree (in the range of 50–95%), sequence of the amino and acetamido groups, polydispersity, crystallinity, and purity of the product (Aranaz et al., 2009; Rinaudo, 2006; Gupta and Jabrail, 2006; Guo et al., 2008). The molecular weight and the degree of deacetylation of a chitosan are two important characteristics

TABLE 3.1
Physicochemical Characteristics Associated with the Potential Use of Chitosan in Several Applications

Application	Physicochemical Characteristics
Emulsifying agent	Low DD for emulsion stability
	High viscosity
Drug delivery systems	High DD
	High MW
Wound healing	High DD chitosan preferred over chitin
	Low MW samples (oligomers)

that greatly affect its chemical and physiological properties. It has been reported that the DD is one of the most important chemical characteristics and could influence the performance of chitosan in many of its applications (Schipper et al., 1996; Ziani et al., 2008). Those chitins with a nitrogen content higher than 7% and a degree of acetylation lower than 0.40 can be considered chitosans (Kumar et al., 2004). These main physicochemical characteristics determine the functional properties of chitosan, and therefore it is important to select them carefully to apply the suitable chitosan for each purpose. Table 3.1 lists certain chitosan physicochemical properties that have been associated with particular applications.

3.4.1 Relationship between the Physicochemical Characteristics of Chitosan and Emulsifying Activity

Chitosan produces w/o/w emulsions without adding any surfactant, because this biopolymer is actually a mixture of molecules within a DD range and MW. Thus, chitosan molecules with low DD may stabilize the water droplets inside the oil drops, while the hydrophilic ones stabilize the oil drops in the multiple emulsions formed (Jumaa et al., 2002). Speiciene et al. (2007) studied the emulsion-stabilizing properties of various chitosans in the presence of whey protein isolate (WPI). A low deacetylated chitosan appeared to be better for the stabilization of the emulsion. The authors reported that higher viscous chitosans, those with higher MW, led to increased stability of the emulsions. Thus, there is a positive correlation between the stability of the emulsions and the viscosity of chitosan that is associated with MW. The undesirable side effect derived from the use of a too low MW chitosan to prepare an emulsified formulation would be the fatal alteration of the emulsion stability explained by a loss of interfacial coadsorption efficiency (Speiciene et al., 2007).

3.4.2 Relationship between the Physicochemical Characteristics of Chitosan and Film Capacity

A remarkable group of chitosan derivatives are its hydrolyzed products known as chitooligosaccharides (COS). COS have improved water solubility due to their shorter chain lengths. The low viscosity and increased solubility of these compounds have attracted the interest of many researchers to study these chitosan derivatives. Chitosan, like other polysaccharides, was susceptible to a variety of degradation mechanisms, including oxidative-reductive depolymerization as well as acid, alkaline, and enzymatic hydrolysis. Both chemical and enzymatic methods are widely used in COS production. Enzymatic processes are preferable over chemical methods due to minimized adverse chemical modifications of final products and improved biological activities (Mengíbar et al., 2010).

However, chemical methods are preferred to carry out the depolymerization of chitosan at industrial scale because of the high cost associated with the use of hydrolytic enzymes.

Chitosan oligomers exert wound healing properties that have been related to their ability to stimulate fibroblast production by affecting the fibroblast growth factor (Ahmad et al., 2010). Recent studies have shown that besides chitin and chitosan, their oligomers effectively enhance wound healing (Muzzarelli, 2009a; Xia et al., 2011). Wound break strength and collagenase activity of the chitosan group (D-glucosamine (GlcN), COS, chitosan) were higher than those of the chitin group (N-acetyl-D-glucosamine (GlcNAc), chitin-oligosaccharide (NACOS), chitin). The break strength was stronger and more activated fibroblasts were observed at a higher DD. Administration of chitin in an animal model showed that chitin induced early macrophage migration into the wound cavity area and produced a large amount of type IV collagen without production of type I collagen (Shigemasa et al., 1998). On the other hand, Minami et al. (2006) studied the possibility of using chitosan and chitin on deep skin trauma reaching to the muscle layer for regeneration of granulating tissue, which is an important substratum for epithelium cell attachment, proliferation, and advanced movement into the wound area. The wound healing effect of chitosan oligomers is of great interest because *in vivo* lysozyme degrades the polymer to these smaller molecules.

3.4.3 Relationship between the Physicochemical Characteristics of Chitosan and Encapsulation Capacity

With regard to active agents' vehiculization, the selection of the suitable type of chitosan with certain characteristics allows the development of customized delivery systems. Chitosan-based release devices may prolong the duration of active agent activity, improving its efficiency and reducing side effects under desired conditions. Hence, an important potential application of chitosans may be the development of delivery systems such as nanoparticles, hydrogels, microspheres, and films for the controlled release of cosmeceutical active agents. As a result of its cationic character, chitosan is able to react with polyanions, giving rise to polyelectrolyte complexes (Paños et al., 2008; Harris et al., 2008; Bernkop-Schnürch et al., 2006). Three main characteristics of chitosan to be considered are MW, DD, and purity. When chitosan chains become shorter (low MW chitosan), it can be dissolved directly in water, which is particularly useful for specific applications in the biomedical and cosmetic fields, when the pH should stay at around 7.0. A higher degree of deacetylation and a higher molecular weight of chitosan are necessary to obtain delivery systems with high efficiency of encapsulation. It has been shown that controlled release systems prepared by ionic gelation with tripolyphosphate and chitosan show higher encapsulation efficiency with higher molecular weight and a higher degree of deacetylation of chitosan (Alasarra and Bitegeri, 2002). The higher the DD is, the higher the number of free amino groups that can interact with negative charges of the tripolyphosphate, increasing the degree of covalent cross-linking and the encapsulation efficiency. Some examples illustrating the influence of MW on chitosan properties and hence on chitosan potential applications are discussed below.

3.4.3.1 Effect of Physicochemical Characteristics of Chitosan on Nanoparticle Formation

Yan and co-workers (2007) studied the effect of MW (20, 38, 300, and 1,100 kDa) on the properties of oleoyl-chitosan and observed that MW affects both the viscosity and morphology of chitosan. At low MW, spherical nanoparticles were observed, but at high MW no nanoparticles were observed, probably due to the high viscosity of the derivative (Yan et al., 2007). The strong relation of chitosan morphology with MW was also probed by Mincheva et al. (2008) with carboxyethyl. At 45 kDa only spherical morphologies were observed, regardless of the pH and polymer concentration studied. On the other hand, chitosan derivatives with a chitosan MW of 670 kDa showed spherical morphologies at pH 5.0 and 5.6. Meanwhile, at pH 5.2 a spherical morphology was only observed at low polymer concentration. This factor rose as another determinant for chitosan morphology as increasing polymer concentrations were associated with rod- and worm-like structures (Mincheva et al., 2008).

3.4.3.2 Effect of Physicochemical Characteristics of Chitosan on Microsphere Formation

Microspheres prepared with a high MW chitosan seem to gel faster than those prepared with a low MW because they have different activation energies of gelation, as shown in Kofuji et al. (2005). On the other hand, chitosans with lower DD values are more soluble than polymers with higher DD. The lower solubility of chitosans with a smaller number of acetyl groups is caused by difficulties in attaching protons to amino groups that are close to already protonated amino groups, due to electrostatic repulsive forces (Kołodziejska et al., 2006). Hence, as probed in the aforementioned works, chitosan MW affects the chemistry (degree of substitution) and morphology of the self-assembly derivatives.

3.5 CONCLUSION

As discussed in this chapter, the search for new active compounds to implement cosmeceutical formulations has gained interest in recent years and is the focus of many researchers. Natural marine origin biopolymer chitosan, extracted from crustacean waste, has been proposed for several applications, including cosmetic, pharmaceutical, and thus cosmeceutical purposes. Chitosan is a versatile polymer with a wide range of functional properties. Due to the presence of reactive groups (NH_2, OH) on its backbone, chitosan can be easily modified to improve these properties. As explained in this chapter, the unique physicochemical characteristics of this polymer provide a large number of potential applications. Chitosans have an intrinsic variability due to their natural origin. Moreover, depending on the manufacturing process, chitosans with different functionalities can be obtained. Hence, chitosan and derivatives appear as a valuable active compound suitable to be included in cosmeceutical formulations.

REFERENCES

Ahmad, R., M.S Bahari, A.S. Halim, K. Hashim, A.H.A. Rashid, N. Yusof, and S. Shamsuddin. 2010. *In vitro* evaluation of novel chitosan derivatives sheet and paste cytocompatibility on human dermal fibroblasts. *Carbohydr Polym* 79(4) (3/17): 1094–100.

Agulló, E., R. Mato, C. Peniche, C. Tapia, A. Heras, J. San Román, W. Argüelles, et al. 2004. *Quitina y quitosano: Obtención, caracterización y aplicaciones*, ed. Ana pastor de Abram. Lima, Perú: Fondo editorial de la Pontificia Universidad Católica del Perú.

Alasarra, I.A, and S. Bitegeri. 2002. Molecular weight and degree of deacetylation effects on lipase-loaded chitosan bead characteristics. *Biomaterials* 23: 3637–3644.

Aranaz, I., M. Mengibar, R. Harris, I. Panos, B. Miralles, N. Acosta, G. Galed, and A. Heras. 2009. Functional characterization of chitin and chitosan. *Curr Chem Biol* 3(2): 203–30.

Bernkop-Schnürch, A., A. Weithaler, K. Albrecht, and A. Greimel. 2006. Thiomers: Preparation and *in vitro* evaluation of a mucoadhesive nanoparticulate drug delivery system. *Int J Pharm* 317(1) (7/6): 76–81.

Boucher, I., and S. Brunet. 2005. Chitosan oligosaccharides and uses thereof. U.S. Patent 2005/0004047A1.

Byrne, M.E., and V. Salian. 2008. Molecular imprinting within hydrogels II: Progress and analysis of the field. *Int J Pharm* 364: 188–212.

Chatelet, C., et al. 2001. Influence of the degree of acetylation on some biological properties of chitosan films. *Biomaterials* 22: 261–68.

Chen, L., and M. Subirade. 2005. Chitosan/β-lactoglobulin core–shell nanoparticles as nutraceutical carriers. *Biomaterials* 26(30) (10): 6041–53.

Chen, S.-C., Y.-C. Wu, F.-L. Mi, Y.-H. Lin, L.-C. Yu, and H.-W. Sung. 2004. A novel pH-sensitive hydrogel composed of N,O-carboxymethyl chitosan and alginate cross-linked by genipin for protein drug delivery. *J Controlled Release* 96(2) (4/28): 285–300.

Chenite, A., S. Gori, M. Shive, E. Desrosiers, and M.D. Buschmann. 2006. Monolithic gelation of chitosan solutions via enzymatic hydrolysis of urea. *Carbohydr Polym* 64: 419–424.

Coviello, T., P. Matricardi, C. Marianecci, and F. Alhaique. 2007. Polysaccharide hydrogels for modified release formulations. *J Controlled Release* 119(1): 5–24.

Crini, G. 2005. Recent developments in polysaccharide-based materials used as adsorbents in waste water treatment. *Prog Polym Sci* 30(1): 38–70.

de la Fuente, M., M. Raviña, P. Paolicelli, A. Sanchez, B. Seijo, and M.J. Alonso. 2010. Chitosan-based nanostructures: A delivery platform for ocular therapeutics. *Adv Drug Deliv Rev* 62(1) (1/31): 100–17.

Domard, A., and M. Rinadudo. 1983. Preparation and characterization of fully deacetylated chitosan. *Int J Biol Macromol* 5: 49–52.

Dudhani, A.R., and S.L. Kosaraju. 2010. Bioadhesive chitosan nanoparticles: Preparation and characterization. *Carbohydr Polym* 81(2) (6/11): 243–51.

Focher, B., P.L. Beltrane, A. Naggi, and G. Torri. 1990. Alkaline N-deacetylation of chitin enhanced by flash treatments. Reaction kinetics and structure modifications. *Carbohydr Polym* 12: 405–18.

Grenha, A., C.I. Grainger, L.A. Dailey, B. Seijo, G.P. Martin, C. Remuñán-López, and B. Forbes. 2007. Chitosan nanoparticles are compatible with respiratory epithelial cells *in vitro*. *Eur J Pharm Sci* 31(2) (6): 73–84.

Goycoolea, F.M., I. Higuera-Ciapara, J.L. Hernandez, and K.D. García. Preparation of chitosan from squid (*Loligo* spp.) pen by a microwave-accelerated thermochemical process. In A. Domard, G.A.F. Roberts, K.M. Värum, eds., *Advances in Chitin Science*, Lyon: Jacques André 1998: pp. 78–83.

Guliyeva, A., F. Öner, S. Ozsoy, and R. Hazirolu. 2006. Chitosan microparticles containing plasmid DNA as potential oral gene delivery system. *Eur J Pharm Biopharm* 62: 17–25.

Guo, Z., R. Xing, S. Liu, Z. Zhong, X. Ji, L. Wang, and P. Li. 2008. The influence of molecular weight of quaternized chitosan on antifungal activity. *Carbohydr Polym* 71(4) (3/7): 694–97.

Gupta, K.C, and F.H Jabrail. 2006. Effects of degree of deacetylation and cross-linking on physical characteristics, swelling and release behavior of chitosan microspheres. *Carbohydr Polym* 66(1): 43–54.

Harris, R., I. Paños, N. Acosta, and A. Heras. 2008. Preparation and characterization of chitosan microspheres for controlled release of tramadol. *J Controlled Release* 132(3) (12/18): e76–77.

Helander, I., E. Nurmiaho-Lassila, R. Ahvenainen, J. Rhoades, and S. Roller. 2001. Chitosan disrupts the barrier properties of the outer membrane of gram-negative bacteria. *Int J Food Microbiol* 71: 235–44.

Horton, D., and D.R. Lineback. 1965. N-deacetylation, chitodan from chitin. In *Methods in carbohydrate chemistry*, ed. R.L. Whistler and M.L. Wolfson, 403. New York: Academic Press.

Huang, S., and X. Fu. 2010. Naturally derived materials-based cell and drug delivery systems in skin regeneration. *J Controlled Release* 142 (2) (3/3): 149–59.

Illum, L., I. Jabbal-Gill, M. Hinchcliffe, A.N Fisher, and S.S Davis. 2001. Chitosan as a novel nasal delivery system for vaccines. *Adv Drug Deliv Rev* 51(1–3) (9/23): 81–96.

Janes, K.A, P. Calvo, and M.J. Alonso. 2001. Polysaccharide colloidal particles as delivery systems for macromolecules. *Adv Drug Deliv Rev* 47(1): 83–97.

Jumaa, M., F.H. Furkert, and B.W. Müller. 2002. A new lipid emulsion formulation with high antimicrobial efficacy using chitosan. *Eur J Pharm Biopharm* 53 (1) (1): 115–23.

Kaplan, R.S 1998. Innovation action research: Creating new management theory and practice. *J Management Accounting Res* (10): 89–118.

Kofuji, K., C.J. Qian, M. Nishimura, I. Sugiyama, Y. Murata, and S. Kawashima. 2005. Relationship between physicochemical characteristics and functional properties of chitosan. *Eur Polym J* 41(11): 2784–91.

Kołodziejska, I., B. Piotrowska, M. Bulge, and R. Tylingo. 2006. Effect of transglutaminase and 1-ethyl-3-(3-dimethylaminopropyl) carbodiimide on the solubility of fish gelatin–chitosan films. *Carbohydr Polym* 65(4) (9/13): 404–9.

Kumar, M.N V.R, R.A A. Muzzarelli, C. Muzzarelli, H. Sashiwa, and A.J Domb. 2004. Chitosan chemistry and pharmaceutical perspectives. *Chem Rev* 104: 6017–84.

Lamarque, G., C. Viton, and A. Domard. 2004. Comparative study of the first heterogeneous deacetylation of α- and β-chitins in a multistep process. *Biomacromolecules* 5(3), 992–1001.

Matalanis, A., U. Lesmes, E.A. Decker, and D.J. McClements. 2010. Fabrication and characterization of filled hydrogel particles based on sequential segregative and aggregative biopolymer phase separation. *Food Hydrocolloids* 24(8)(12): 689–701.

Mengíbar, M., M. Ganan, B. Miralles, A.V Carrascosa, A.J. Martínez-Rodriguez, M.G. Peter, and A. Heras. 2010. Antimicrobial activity of chitosan oligosaccharides (COS) against *Campylobacter Jejuni*. *Carbohydr Polym*, 84(2): 844–848.

Mima, S., M. Miya, R. Iwamoto, and S. Yoshikawa. 1983. Highly deacetylated chitosan and its properties. *J Appl Polym Sci* 28: 1909–17.

Minami, S., Y. Okamura, and Y. Okamoto. 2006. *Abstract book of 10th International Conference of the European Chitin Society*, Montpellier, 36.

Mincheva, R., F. Bougard, D. Paneva, M. Vachaudez, N. Manolova, and P. Dubois. 2008. Self-assembly of N-carboxyethylchitosan near the isoelectric point. *J Polym Sci* 46: 6712–21.

Muzzarelli, R.A.A. 2009a. Chitins and chitosans for the repair of wounded skin, nerve, cartilage and bone. *Carbohydr Polym* 76(2): 167–82.

Muzzarelli, R.A.A. 2009b. Genipin-crosslinked chitosan hydrogels as biomedical and pharmaceutical aids. *Carbohydr Polym* 77(1) (5/22): 1–9.

Muzzarelli, R.A.A., C. Lough, and M. Emanuelli. 1987. The molecular weight of chitosans studied by laser light-scattering. *Carbohydr Res* 164(7/1): 433–42.

Muzzarelli, R.A.A., and R. Rocchetti. 1985. Determination of the degree of acetylation of chitosans by first derivative ultraviolet spectrophotometry. *Carbohyd Polym* 5(6): 461–72.

Muzzarelli, R.A.A., R. Rocchetti, V. Stanic, and M. Weckx. 1997. Methods for the determination of the degree of acetylation of chitin and chitosan. In R.A.A. Muzzarelli and M.G. Peter, eds., *Chitin Handbook*, pp. 109–15. Italy: E.C Society.

Paños, I., N. Acosta, and A. Heras. 2008. New drug delivery systems based on chitosan. *Curr Drug Discov Technol* 5: 333–41.

Park, P.J., J.Y. Je, H.G. Byun, S.H. Moon, and S.K. Kim. 2004. Antimicrobial activity of hetero-chitosans and their oligosaccharides with different molecular weights. *J Microbiol Biotechnol* 14(2): 317–23.

Pelletier, A., I. Lemire, J. Sygusch, E. Chornet, and R.P. Overend. 1990. Chitin/chitosan transformation by thermo-mechano-chemical treatment including characterization by enzymatic depolymerization. *Biotechnol Bioeng* 36: 310–15.

Pillai, C.K.S., W. Paul, and C.P Sharma. 2009. Chitin and chitosan polymers: Chemistry, solubility and fiber formation. *Progr Polym Sci* 34(7) (7): 641–78.

Porporatto, C., I.D Bianco, and S.G Correa. 2005. Local and systemic activity of the polysaccharide chitosan at lymphoid tissues after oral administration. *J Leukocyte Biol* 78(1): 62–69.

Ravi Kumar, M.N.V. 2000. A review of chitin and chitosan applications. *Reactive Funct Polym* 46 (1) (11): 1–27.

Rinaudo, M. 2006. Chitin and chitosan: Properties and applications. *Progr Polym Sci* 31 (7) (7): 603–32.

Roberts, G. 1998. *Chitin chemistry*. London: Macmillan.

Sánchez-González, L., M. Cháfer, A. Chiralt, and C. González-Martínez. 2010. Physical properties of edible chitosan films containing bergamot essential oil and their inhibitory action on penicillium italicum. *Carbohydr Polym* 82 (2) (9/5): 277–83.

Schipper, N.G M., K.M Varum, and P. Artursson. 1996. Chitosans of absorption enhancers of poorly absorbable drugs: Influence of molecular weight and degree of acetylation. *Eur J Pharm Sci* 4(Suppl 1) (9): S153.

Shahidi, F., J.K.V. Arachchi, and Y.J. Jeon. 1999. Food applications of chitin and chitosans. *Trends Food Sci Technol* 10: 37–51.

Shahidi, F., J. Kamil, Y.J. Jeon, and S.K. Kim. 2002. Antioxidant role of chitosan in a cooked cod (*Gadus morhua*) model system. *J Food Lipids* 9: 57–64.

Shigemasa, Y., M. Morimoto, H. Saimoto, Y. Okamoto, and S. Minami. 1998. *Proceedings of the 3rd Asia Pacific Chitin and Chitosan Symposium*, Taiwan, ed. R.H. Chen and H. Chen, 47.

Silva-Júnior, A.A., M.V. Scarpa, K.C. Pestana, L.P. Mercuri, J.R. de Matos, and A.G. de Oliveira. 2008. Thermal analysis of biodegradable microparticles containing ciprofloxacin hydrochloride obtained by spray drying technique. *Thermochimica Acta* 467(1–2) (1/30): 91–98.

Singla, A.K, and M. Chawla. 2001. *Chitosan: Some pharmaceutical and biological aspects—An update*. Vol. 53. Oxford: Blackwell.

Speiciene, V., F. Guilmineau, U. Kulozik, and D. Leskauskaite. 2007. The effect of chitosan on the properties of emulsions stabilized by whey proteins. *Food Chem* 102(4): 1048–54.

Sudha, S., V.R. Giri Dev, R. Neelakandan, and M. Senthil Kumar. 2005. Chitosan—A versatile polymer for textile applications. *J Textile Assoc* 66: 161–168.

Van der Lubben, I.M., J.C. Verhoef, G. Borchard, and H.E. Junginger. 2001. Chitosan for mucosal vaccination. *Adv Drug Deliv Rev* 52(2): 139–44.

Varshosaz, J. 2007. The promise of chitosan microspheres in drug delivery systems. *Expert Opin Drug Deliver* 4: 263–73.

Xia, W., P. Liu, J. Zhang, and J. Chen. 2011. Biological activities of chitosan and chitooligosaccharides. *Food Hydrocolloids* 25(2) (3): 170–79.

Yamamoto, H., Y. Kuno, S. Sugimoto, H. Takeuchi, and Y. Kawashima. 2005. Surface modified PLGA nanosphere with chitosan improved pulmonary delivery of calcitonin by mucoadhesion and opening of the intercellular tight junctions. *J Controlled Release* 102(2): 373–81.

Yan, Y.L., G.C. Xi, S.L. Cheng, S.C. Dong, J.P. Hyun, and C.M. Lee. 2007. Effect of the molecular mass and degree of substitution of oleoylchitosan on the structure, rheological properties, and formation of nanoparticles. *J Agric Food Chem* 55: 4842–47.

Yomota, C., T. Miyazaki, and S. Okada. 1993. Determination of the viscometric constants for chitosan and the application of universal calibration procedure in its gel permeation chromatography. *Colloid Polym Sci* 271:76–82.

Zhang, X., J. Liu, H. Qiao, H. Liu, J. Ni, W. Zhang, and Y. Shi. 2010. Formulation optimization of dihydroartemisinin nanostructured lipid carrier using response surface methodology. *Powder Technol* 197(1–2) (1/10): 120–28.

Ziani, K., J. Oses, V. Coma, and J.I Maté. 2008. Effect of the presence of glycerol and Tween 20 on the chemical and physical properties of films based on chitosan with different degree of deacetylation. *LWT—Food Sci Technol* 41(10) (12): 2159–65.

Zvonar, A., K. Berginc, A. Kristl, and M. Gašperlin. 2010. Microencapsulation of self-microemulsifying system: Improving solubility and permeability of furosemide. *Int J Pharm* 388(1–2) (3/30): 151–58.

4 Marine Phytochemical Compounds and Their Cosmeceutical Applications

Abdul Bakrudeen Ali Ahmed and Rosna Mat Taha

CONTENTS

ABSTRACT

Cosmeceuticals consist of a large number of phytochemical components combined together that have been isolated from various marine resources. The structural diversity of these marine compounds comes from different modification reactions of the main derivatives. The recent progress opens an avenue toward understanding how marine natural products and their potential for development identify novel potent phytochemical compounds for cosmeceutical purposes. These cosmetic substitutes are being developed to enhance the restoration of ionic balance and prevent the skin from drying. This chapter assembles major research findings of marine-based phytochemical compounds, which is of direct relevance to pharmaceuticals, functional foods, and cosmetic skin care products.

4.1 INTRODUCTION

Marine (or blue) biotechnology encompasses the applications of biotechnology tools on marine resources. It encompasses those efforts that involve the marine resources of the world, as either the source or target of biotechnology applications. Biotechnology is the application of science and technology to living organisms, as well as parts, products, and models, and alters living or nonliving materials for the production of knowledge, goods, and human services. In the case of marine biotechnology, the living organisms derive from marine sources. Biotechnology is defined as the industrial use of living organisms or biological techniques developed through basic research; marine biotechnology is an emerging discipline based on the use of marine natural resources. The ocean covers about 70% of the earth's surface, providing a diverse living environment for invertebrates (Lalli and Parsons 1993). The marine environment is a rich source of both biological and chemical diversity. This diversity has been the source of unique chemical compounds with the potential for industrial development as pharmaceuticals, cosmetics, nutritional supplements, molecular probes, fine chemicals, and agrochemicals. The marine environment is an exceptional reservoir of bioactive natural products, many of which exhibit structural/chemical features not found in terrestrial natural products (Carte 1996). Marine organisms have evolved biochemical and physiological mechanisms that include the production of bioactive compounds for such purposes as reproduction, communication, and protection against predation, infection, and competition (Halvorson 1998). Therefore, marine natural products will play a major role in drug discovery in the future.

To date, researchers have isolated approximately 7,000 marine natural products, 25% of which are from algae, 33% from sponges, 18% from coelenterates (sea whips, sea fans, and soft corals), and 24% from representatives of other invertebrate *phyla*, such as ascidians (also called tunicates), opisthobranch mollusks (nudibranchs, sea hares, etc.), echinoderms (starfish, sea cucumbers, etc.), and bryozoans (moss animals) (Kijjoa and Sawangwong 2004). Initial attempts were determining the chemical nature of marine organisms and were simply extensions of tried and true phytochemical techniques. Thus, easily accessible organisms (generally sponges and encrusting organisms such as ascidians) were collected by hand using snorkel or simple scuba systems, and then their chemical components were extracted and identified. Any biological activity was found as an afterthought in these initial experiments (though as shown above, active compounds could be found by these techniques that would ultimately be useful as treatments for human diseases). A corollary to the more systematic searching for marine-derived products was that very sensitive analytical tools had to be used, as, in general, the amount of bioactive materials that could be recovered was exceedingly small.

4.1.1 Cosmeceuticals

Seaweeds are one of the potential contributions for cosmetic compounds production. In this chapter, the majority of the cosmetic compounds are seaweeds. The total world seaweed production is estimated to be 8.5 million metric tons. Of this, 88.65% (i.e., 7.5 million metric tons) is produced by cultivation in an area of 200×10^3 ha, while the remaining 0.96 million metric tons are exploited from the natural seaweed beds the world over (FAO 2003). The seaweed industry uses 7.5–8 million metric tons of wet seaweeds annually either from the wild or from cultivated crop. The estimated value of a wide variety of products derived from seaweeds is US$5–6 billion (FAO 2004). In earlier days, seaweeds were used for food preparation, and later they were used to yield industrial, medicinal, pharmaceutical, and cosmetic products (Dhargalkar and Verlecar 2009). Cosmeceuticals, derived from the words *cosmetic* and *pharmaceutical*, have drug-like benefits and contain active ingredients such as vitamins, phytochemical components, enzymes, antioxidants, and essential oils. Cosmeceuticals have attracted and increased the attention of the world, because of their beneficial effects on human health. Hence, bioactive substances derived from marine organisms have diverse functional roles as secondary metabolites, and these properties can be applied to the development of novel pharmaceuticals and cosmeceuticals. Recently, extensive studies have been conducted on the general aspects of the chemical structures, physical and biochemical properties, and biotechnological applications of bioactive substances derived from marine organisms. Global industry analysis of the advancement in scientific disciplines such as molecular biology, bioinformatics, and genomics has also contributed to world market growth. Marine biomaterials represent the largest segment of the market, including the thickener carrageen, often used in cosmetic products. In addition, the marine bioactive substances used for their functional properties make up the second largest market segment and would include all algae and seaweed-based actives used in cosmetics, nutraceutical, and pharmaceutical products. Increasing numbers of highly targeted products are hitting the market, from ingredients specifically designed to fight double chins to products designed for postmenopausal skin.

4.1.2 Marine Natural Product Discovery

The phytochemical drug discovery process can be divided into four steps: drug target identification, target validation, lead compounds identification, and optimization. Marine natural products have often opened up completely new therapeutic approaches. They have contributed to identifying and understanding novel biochemical pathways and proved to make not only valuable marine drugs available but also essential tools in biochemistry and molecular cell biology (Grabley and Sattler 2003). Marine organisms contain various chemical components, such as saponins, alkaloids, volatile oils, flavonoids, and anthraquinones. In the development of rapid and reproducible analytical techniques, the combination of HPLC with different detectors, such as diode array detector (DAD), refractive index detector (RID), evaporative light scattering detector (ELSD), and mass spectrometric detector (MAD), have been widely developed. It is important to use phytochemical methods to screen and analyze bioactive components, not only for the quality control of crude drugs, but also for the elucidation of their therapeutic mechanisms.

Some of the compounds from marine invertebrates initially discovered were either too toxic or not effective in treating diseases for pharmaceutical purposes, but were found to be useful as biological tools, cosmetic ingredients, or agrochemicals (Fenical 1997). To date, examples of marine-derived drugs include an antibiotic from fungi, two closely related compounds from a sponge that treat cancer and the herpes virus, and a neurotoxin from a snail that has painkiller properties, making it 10,000 times more potent than morphine without any side effects. However, there are several more marine-derived compounds currently in clinical trials, and it is likely that many more will advance to the clinic as more scientists look to the sea for these biotechnological uses. In addition to new medicines, other uses for marine-derived compounds include cosmetics (algae, crustacean, and sea fan compounds), nutritional supplements (algae and fish compounds), artificial bone (corals), and industrial applications (fluorescent compounds from jellyfish, novel glues from mussels, and heat-resistant enzymes from deep-sea bacteria). These products have received increasing attention from chemists and pharmacologists during the last two decades. They have been exploited for a variety of purposes, including use as food, fragrances, pigments, insecticides, and medicines. Through an improved biological screening method the role of these products in drug discovery has been greatly enhanced in the last few years. These products show an interesting array of diverse and novel chemical structures with potent biological activities. It is apparent that they have significant potential as pharmaceuticals, cosmetics, nutraceuticals, research tools, and others. Over the past 50 years, marine organisms have provided key structures and compounds that proved their potential for industrial development as cosmetics, nutritional supplements, fine chemicals, agrochemicals, and therapeutic agents for a variety of diseases (Tziveleka et al. 2003).

Chemistry is a central, useful, and creative science, assuming in many aspects the characteristics of an interdisciplinary science, which supplies the essential means and the fundamental language for the correct understanding of other scientific disciplines (Breslow 1997). Life is in fact based on chemistry. Birth, growth, reproduction, aging, diseases, and death depend on chemical transformations executed by the primary and secondary metabolisms of living organisms. In contrast, secondary metabolism, predominantly found in plants, microorganisms, fungi, marine organisms, and to a lower extent, animals, was proved to include processes nonessential for the continuity of the life cycle or for growth and development. Thousands of novel compounds have been isolated from various marine organisms and tested for cosmeceuticals properties, many of which are commercially available. The screening of natural products derived from marine species for cosmetics has yielded a considerable number of active crude aqueous and organic solvent extracts. The present review therefore aims toward the assessment of marine resources and cosmeceutical production. Attempts have also been made to quantify the potential marine organisms for commercial exploitation for their unique use as bioactive compounds with cosmetic properties. The present communication discussing the methodologies is proposed to develop and manage this resource for sustainable utilization for marine cosmeceutical production.

4.2 PHYTOCHEMICAL COMPOUNDS WITH COSMECEUTICAL APPLICATIONS

4.2.1 Mycosporine

Mycosporines and mycosporine-like amino acids (MAAs) are low molecular weight water-soluble molecules absorbing UV radiation in the wavelength range of 310 ~ 365 nm secondary metabolites involved in the protection of marine and fresh water organisms against solar radiation (Figure 4.1). They are accumulated in a wide range of prokaryotic microorganisms (cyanobacteria), eukaryotic microorganisms (microalgae, yeast, and fungi), marine microalgae, corals, and other marine organisms. MAA plays a role in skin protection against UV radiation. These natural products are characterized by a cyclohexenone or cyclohexenimine chromophore conjugated with one or two amino acids, which present an absorption maximum ranging from 310 to 360 nm (Nakamura and Kobayashi 1982). Shallow-water corals produce large quantities of MAAs, and their apparent ability to withstand long-term environmental UV exposure suggests utilization of their UV-absorbing chromophore for sunscreen application, 1-alkyl-3-alkanoyl-1,4,5,6 tetrahydropyridines, based on their natural UV protection. Several sunscreen candidates have been examined for sunscreen suitability, and a limited selection is being tested for potential use in human skin care and cosmetic products (Dunlap et al. 1998).

4.2.2 Pseudopterosins

These compounds are part of a complex mixture reported (Look et al. 1986) from Caribbean gorgonian *Pseudopterogorgia elisabethae* and have the distinction of being the first commercialized human use marine natural product. As a particularly purified defined mixture, they are constituents of the cosmetic "antiwrinkle" sold by Estee Lauder under the brand name Resilience. However, the pseudopterosins mechanism was further delineated (Mayer et al. 1998). The pseudopterosins (Figure 4.2) are tricyclic diterpene glycosides isolated from the Caribbean Sea whip (gorgonian) *Pseudopterogorgia elisabethae* (Gorgoniidae). They are potent anti-inflammatory and analgesic agents and appear to inhibit eicosanoid biosynthesis by inhibition of both PLA_2 and 5-lopoxygenase. Interestingly, the pseudopterosins that are found to inhibit only pseudopterosins may well be a function of the glycoside moiety and a novel example of drug targeting (Potts and Faulkner 1992).

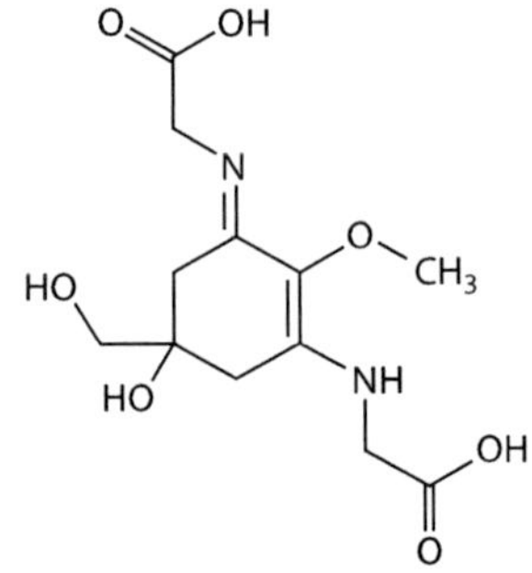

FIGURE 4.1 Structure of mycosporine derivatives.

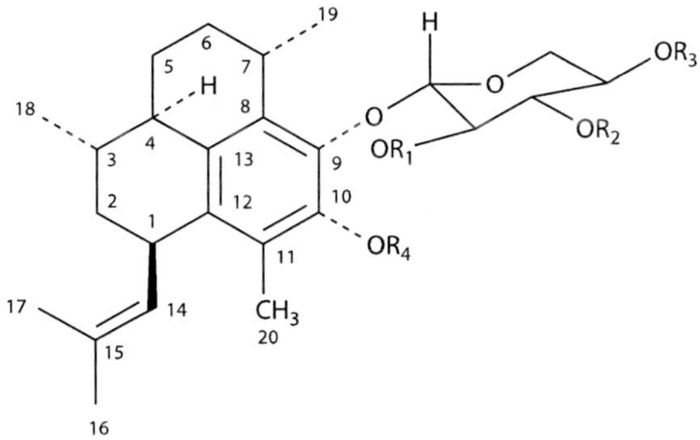

Pseudopterosin A: $R_1 = R_2 = R_3 = R_4 = H$

Pseudopterosin B: $R_1 = Ac, R_2 = R_3 = R_4 = H$

Pseudopterosin C: $R_2 = Ac, R_1 = R_3 = R_4 = H$

Pseudopterosin C: $R_3 = Ac, R_1 = R_2 = R_4 = H$

FIGURE 4.2 Structure of pseudopterosins.

A simpler modification of the psedopterosins may have entered phase I clinical trials as an anti-inflammatory agent, though the structure and company were not listed in the reference to the trial (Proksch et al. 2002).

Since then, the pseudopterosins have been licensed to a small pharmaceutical firm, Osteo Arthritis Science, Inc., for medical use as potential anti-inflammatory drugs. The company has completed preclinical tests of one pseudopterosin, a potent topical anti-inflammatory compound, and filed an Investigational New Drug (IND) application with the U.S. Food and Drug Administration. Clinical trials on human subjects for irritant contact dermatitis are anticipated.

4.2.3 Neo Fatty (Carboxylic) Acid Derivatives

Neo acids and derivatives have a low order of toxicity and are used by different chemical companies for synthesis of novel biologically active compounds in the fields of agronomy, medicine, and cosmetics. The development of surfactants based on natural renewable resources is a concept that is gaining recognition in detergents, cosmetics, and green chemistry. Acid- or base-catalyzed hydrolysis of complex lipids yields the free neo fatty acid (Figure 4.3). These long- or short-chain acids are generally referred to by their common names, which in most cases reflect their sources. Natural neo fatty acid can be saturated, unsaturated, or branched. The saturated (branched) acids have higher melting points than unsaturated acids of corresponding size (Wickson and Moore 1964). Neo fatty (carboxylic) acid derivatives are used in the cosmetic, agronomic, and pharmaceutical industries. Different pivalic acid derivatives or compounds with a tertiary butyl group are produced and used in the cosmetic, agronomic, and perfume industries. Skin creams and lotions contain a pivaloyl ascorbic acid and pivaloyl derivates (De Silva and Scheuer 1980) as well as derivatives of vitamin E (vitamin E acetate and vitamin E nicotinate). These cosmetics have little irritating effect on skin, and they condition the skin for a long period of time (Deems et al. 1987). Two seco-theapiranes (Glaser and Jacobs 1986) with pronounced black currant notes were synthesized as odor and odorous substances and used in cosmetics (Kernan et al. 1987). Derivatives of neo fatty acids (Deems et al.

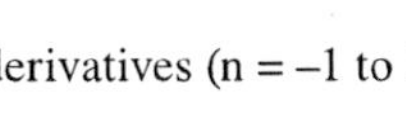

FIGURE 4.3 Structure of neo fatty acid derivatives (n = –1 to 13).

1987; Kernan et al. 1987) are used in the cosmetic industry. An amide of a trialkylacetic acid amide of neo fatty acid (Yang and Andersen 2002) and a polyamine were used with a particulate carrier (e.g., detergent builder or filler) or liquid medium (e.g., aqueous nonionic surfactant solution) in a composition that was added to fabrics in a laundry bath to impart antistatic properties to fabrics after laundering and automatic machine drying (Yang and Andersen 2002). Neo fatty acid constituents are the most abundant class of natural compounds and are components of complex lipids. Neo fatty acids differ in their number of olefinic bonds, the extent of branching, the length of the hydrocarbon chain, and the number of functional groups (Minnikin et al. 2002; Dembitsky 2005).

Neo fatty (carboxylic) acids as well as other metabolites containing a tertiary butyl group belong to a rare group of natural products. In the last 100 years fewer than 300 representatives from this rare and unique group of secondary metabolites have been identified. However, many different natural and synthesized compounds with tert-butyl groups show high biological activities. 2,4-D di-tert-butylcyclohexanones have been synthesized and their components used, as in perfume. Neo fatty acid derivative has a mild, soft, floral-woody, and violet-like scent. These neo-cis isomer derivatives have a highly diffusible cyclamen-like floral scent; the neo trans-isomers derivatives have an orris-like woody scent (Takei et al. 1994). The most biologically active forms of the insect attractant trimedlure, a synthetic lure used to attract male Mediterranean fruit flies, are the tert-Bu esters of cis-4- and trans-5-chloro-trans-2 methyl-cyclohexanecarboxylic acids (Coulson and O'Donnell 2000; Shen and Burgoyne 2002).

This new class of biodegradable and biocompatible products is a response to the increasing consumer demand for products that are greener, milder, and more efficient. *Juniperus occidentalis* and *Chamaecuparis lawsoniana* are commercial products for pest repellents and cosmetics (Duringer et al. 2010).

4.2.4 Phlorotannins

As shown in this review, there are significant numbers of very interesting molecules that have come from marine sources, among them phlorotannins, an important compound. Phlorotannin derivatives (Figure 4.4) from *Ecklonia cava* and has been present in due to its limited origin (only Korea and Japan) *Ecklonia cava* algae is becoming the focus for targeting effective antioxidants for oxidative stress in the human body due to the presence of diverse natural products with unique structures possibly caused by an extreme marine environment. Based on the literature, it could be suggested that phlorotannins derived from *Ecklonia cava* compounds have worthy potential for application as antioxidants, in the functional food, cosmetics, and pharmaceutical industries. Meanwhile, additional studies on the mechanisms and *in vivo* are highly warranted to achieve a better understanding of important antioxidant properties of the isolated phlorotannins from *Ecklonia cava*. It could be

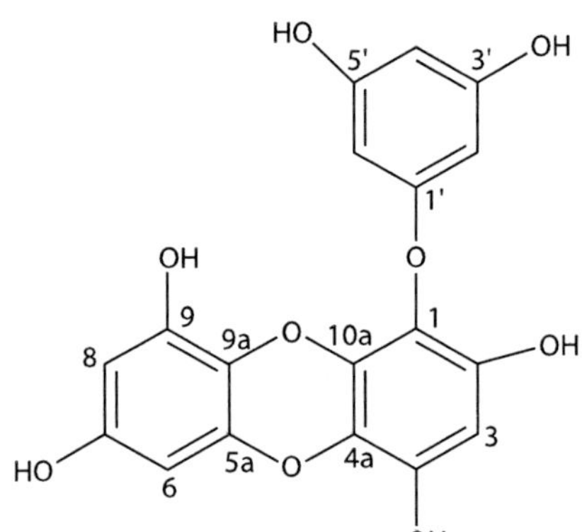

FIGURE 4.4 Structure of phlorotannin derivatives.

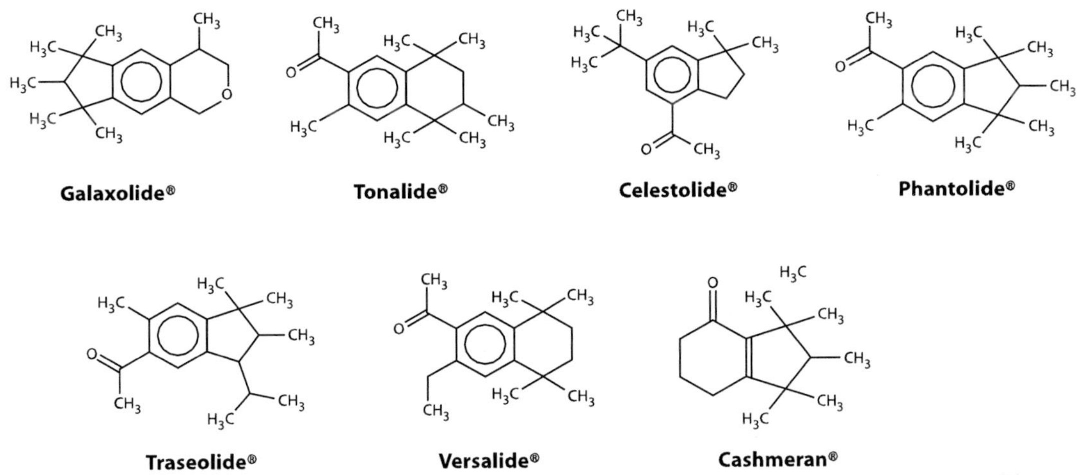

FIGURE 4.5 Structure of polycyclic musk compounds.

suggested that phlorotaninns would be potential candidates for the development of unique natural antioxidants for future industrial applications as functional foods, cosmetics, and pharmaceuticals (Li et al. 2009).

4.2.5 Polycyclic Musk Compounds

Polycyclic musks from the marine environment are used in almost all scented consumer products, such as perfumes, cosmetics, and laundry detergents. Their production has been increased continuously during the last few years with a worldwide production volume today of about 6,000 tons per year. Today, almost all of the polycyclic musks used in cosmetics and fragrance products are of synthetic origin (Rimkus 1999). Polycyclic musks are used in almost all consumer products, such as perfumes, cosmetics, soaps, shampoos, laundry detergents, fabric softeners, household cleaners, air fresheners, and other household products (Rimkus 1999). The synthetic musks are still widely used as artificial fragrances added to perfume, household, hygienic and cosmetic products (Kallenborn et al. 2001). Recently isolated polycyclic musk compound structures and descriptions are given in Figure 4.5.

In contrast to most of the conventional environmental contaminants, synthetic musk is a relatively heterogeneous group of chemicals with just one common general property. The odorous character of all compounds makes them favorable to produce synthetic musks for application in perfumes and other cosmetic products (Kallenborn et al. 2001). The sum of nitro musks usually contributes about 6–20% of the total synthetic musk burden. This value is surprisingly high and indicates a larger proportion of nitro musks used as fragrances in cosmetic and hygienic products offered on the Norwegian market (Kallenborn et al. 2001).

4.2.6 β-Benzylmercaptoethylamine (β-BMEA) Derivatives

The β-Benzylmercaptoethylamine (β-BMEA) derivatives (Figure 4.6) are the most widely used synthons for the incorporation of the cysteamine moiety into many natural and nonnatural products. The utility of this functionality arises from distinct advantages, including ease of preparation, stability, and compatibility with diverse reaction conditions, commercial availability, and scope of functional group manipulation through the reactive amine terminal, thioether linkage, or substituents in the aryl ring (Tochrop et al. 2009).

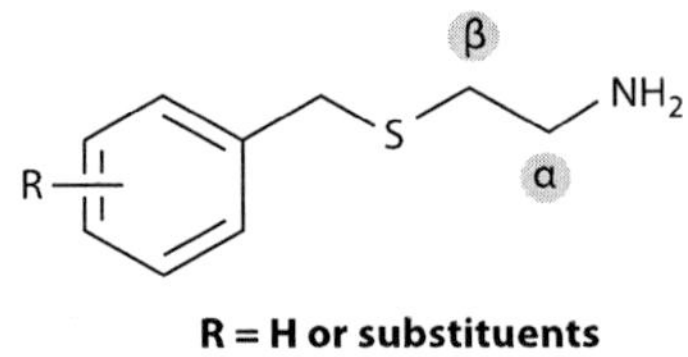

FIGURE 4.6 General structure of β-BMEA derivatives.

Apart from their wide-ranging applications in organic synthesis, this functionality has been further utilized in a range of applications that span materials to cosmetics. To incorporate the cysteamine moiety into many natural and nonnatural products, β-BMEA derivatives have been used as efficient and important synthons in various fields of chemistry. These derivatives have also provided access to many heterocycles and biologically active compounds. Their use in the synthesis of a high number of 99^{m} TC binding ligands clearly shows their value in the nuclear medicine field. Further, the presence of a thioether donor as well as a reactive amine terminal has made this functionality broadly important as polydentate ligands in inorganic chemistry. Truly, this functionality is broadly applicable and synthetically versatile. This report is meant to highlight this utility and versatility and serve as a resource for the further use of the β-BMEA (Tochrop et al. 2009).

4.2.7 Petroleum Derivatives

Petroleum is a viscous liquid mixture that contains thousands of compounds mainly consisting of carbon and hydrogen. All petroleum products are derived from crude oil whose major constituents are hydrocarbons. Petroleum components can be separated into four fractions, the saturated, aromatic, resin, and asphaltene fractions, by absorption chromatography. Each of these fractions contains a large number of compounds, and these products are the major source of energy for industry and daily life. Petroleum is also the raw material for many chemical products such as plastics, paints, and cosmetics (Karlsen and Larter 1991; Harayama et al. 1999).

4.2.8 Other Cosmetic Compounds

A major cosmetic ingredient from the Antarctic seaweeds seems to be the latest craze of cosmetic formulation. In addition, cosmetic markers are looking for pure and uncontaminated natural ingredients. The resistance of Antarctic seaweeds to the cold environment is boosted by their specific composition of polyunsaturated fatty acids and sugars, which allow them to resist differences in osmotic pressure as well as low temperature. The Caribbean gorgonian, *Pseudopterogorgonian elisabethae*, is an example of a source of marine natural product used in the cosmetic industry. The extract from this gorgonian shows anti-inflammatory activity, which nowadays is used as an ingredient in cosmetic skin care products (Proksch et al. 2002). According to Green tech company (French supplier of cosmetics), the extracts obtained from such seaweeds are highly concentrated in active substances, giving them protective, moisturizing, and soothing properties (Dhargalkar and Verlecar 2009).

The cold water conditions and suitable habitat that have persisted for at least 14 million years in the southern hemisphere have resulted in the strong adaptation to low temperature, especially in endemic Antarctic seaweeds (Wiencke et al. 1994). Adaptation to survive the cycle of freezing and thawing at low temperatures could be an essential quality of these seaweeds as a potential source for cosmetic and pharmaceutical industries (Wiencke and Dieck 1989). Paris-based laboratories Kurbiel launched their polar skin ethics range that features marine algae and Antarctic seaweed, which grow in the freezing waters of Antarctica. These seaweeds develop active ingredients that block the effects of metalloproteinase, an enzyme that accelerates the skin aging process. These products are dedicated to men between the ages of 25 and 50 (Gallon 2006). Sargafuran is one of the active compounds isolated from *Sargassum macrocarpum* (brown alga), and this might be useful as a lead compound to develop new types of anti-propionibacterium acne substances and new skin care cosmetics to prevent or improve acne (Kamei et al. 2009).

Red algae species such as *Ceramium rubrum* and *Porphyra leucosticta* show that due to a high percentage of vitamin C and natural carotenoids, *Porphyra leucosticta* could become a valuable source of raw material to obtain such compounds, which have multiple uses in the pharmaceutical field and cosmetics and food industries (Sava et al. 2009). *Ceramium rubrum*, the red alga, is well known as an agar source and can be used to obtain chlorophyll a, this green pigment being a useful therapeutic agent that can also be used in the cosmetics industry (as a deodorant) (Sava et al. 2009).

Marine phytochemical compounds application to drugs has been explored since the early 1970s, but only a small number of marine natural products (MNPs) have been applied to pharmaceuticals (Newman and Cragg 2004). Recently, increasing numbers of MNPs have been developed for cosmetics and nutraceuticals (Venugopal 2008). The number of MNPs that have been applied to the biotechnological industry is very limited, although nearly 20,000 new compounds were discovered from marine organisms (Fusetani 2010). Marine sponges produce secondary metabolites that can be used as a natural source for the design of new drugs and cosmetics (Caralt et al. 2010).

4.3 CONCLUSION

As demonstrated in this review, there is a potential for marine natural products as sources and/or leads to drugs that cover a wide range of cosmeceutical applications. While the marine world offers an extremely rich bioresource for novel compounds, it also represents a great challenge that requires input from various scientific areas to bring the marine chemical diversity to its therapeutic potential. The cosmetic preparations were used for the purpose of worship and sensual enjoyment. Moreover, marine extracts, in whole or part, have been used for various ailments of the skin, hair, and for overall appearance. The market research shows an upward trend in the marine trade, with the marine cosmetic industry playing a major role in fueling this worldwide demand for marine sources. The recent interest of consumers in marine cosmetics has been stimulated by the decline of faith in modern cosmetics and the reference to successful historical use by different cultures. These reasons have contributed to the increased acceptance as well as manufacture of marine cosmetics. Moreover, a few high-value-added compounds recovered from such types of byproducts are economically more attractive than the target cosmetic products themselves. Development of novel and clean technologies aimed at more efficient recovery of bioactive nutraceutical compounds from marine byproducts will lead to the development of more profitable processes, thus giving rise to many great opportunities for the marine industry.

REFERENCES

Breslow, R. 1997. *Chemistry today and tomorrow: The central, useful and creative science*. Washington, DC: American Chemical Society.

Caralt, S.D., Sanchez-Fontenla, J., Uriz, M.J., and Wijffels, R.H. 2010. *In situ* aquaculture methods for *Dysidea avara* (Demospongiae Porifera) in the Northwestern Mediterranean. *Mar. Drugs*, 8: 1731–1742.

Carte, B.K. 1996. Biomedical potential of marine natural products. *Bio. Sci.*, 46: 271–286.

Coulson, F.R., and O'Donnell, S.R. 2000. The effects of contignasterol (IZP-94,005) on allergen induced plasma protein exudation in the tracheobronchial airways of sensitized guinea-pigs *in vivo*. *Inflamm. Res.*, 49: 123–127.

Deems, R.A., Lombardo, D., Morgan, D.P., Mihelich, E.D., and Dennis, E.A. 1987. The inhibition of phospholipase A2 by manoalide and manoalide analogues. *Biochim. Biophys. Acta*, 917: 258–268.

Dembitsky, V.M. 2005. Astonishing diversity of natural surfactants. 3. Carotenoid glyxosides and isoprenoid glycolipids. *Lipids*, 40: 535–557.

De Silva, E.D., and Scheuer, P. 1980. Manoalide, an anitibiotic seterterpenoid from the marine sponge *Luffariella variabilis* (Polejaffe), *Tetrahedron Lett.*, 21: 1611–1614.

Dhargalkar, V.K., and Verlecar, X.N. 2009. Southern ocean seaweeds: A resource for exploration in food and drugs. *Aquaculture*, 287: 229–242.

Dunlap, W.C., Chalker, B.E., Bandaranayake, W.M., and Wu Won, J.J. 1998. Nature's sunscreen from the Great Barrier Reef, Australia. *Int. J. Cosmet. Sci.*, 20: 41–51.

Duringer, J.M., Swan, L.R., Walker, D.B., and Craig, A.M. 2010. Acute aquatic toxicity of western juniper (*Juniperus occidentalis*) foliage and Port Orford cedar (*Chamaecyparis lawsoniana*) heartwood oils. *Environ. Monit. Assess.*, 170: 585–598.

FAO. 2003. *Year book fishery statistics*. Vol. 96: 112–114, 393–397.

FAO. 2004. *The state of world fisheries and aquaculture*. Part 3. Highlights of special FAO studies. Publ. FAO, UN, 1–36.

Fenical, W. 1997. New pharmaceuticals from marine organisms. *Mar. Biotechnol.*, 15: 339–341.

Fusetani, N. 2010. Biotechnological potential of marine natural products. *Pure Appl. Chem.*, 82: 17–26.

Gallon, V. 2006. *E-News*. Newsletter 144. www.Beauty-on-line.com.

Glaser, K.B., and Jacobs, R.S. 1986. Molecular pharmacology of manoalide. inactivation of bee venom phospholipase A2. *Biochem. Pharmacol.*, 35: 449–453.

Grabley, S., and Sattler, I. 2003. *Natural products for lead identification: Nature is a valuable resource for proving tools, in modern method of drug discovery*, ed. A. Hillisch and R. Hilgenfeld, Birthauser Verlag, Switzerland, 87–107.

Halvorson, H.O. 1998. Aquaculture, marine sciences and oceanography: A confluence connection. *New Engl. J. Higher Ed. Econ. Dev.*, 13: 28–42.

Harayama, S., Kishira, H., Kasai, Y., and Shutsubo, K. 1999. Petroleum biodegradation in marine environments. *J. Mol. Microbiol. Biotechnol.*, 1: 63–70.

Kallenborn, R., Gatermann, R., Nygard, T., Knutzen, J., and Schlabach, M. 2001. Synthetic musks in Norwegian marine fish samples collected in the vicinity of densely populated areas. *Fresenius Environ. Bull.*, 10: 832–842.

Kamei, Y., Sueyoshi, M., Hayashi, K.I., Terada, R., and Nozaki, H. 2009. The novel anti-*Propionibacterium acnes* compound, Sargafuran, found in the marine brown alga *Sarfassum macrocarpum*. *J. Antibiot.*, 62: 259–263.

Karlsen, D.A., and Larter, S.R. 1991. Analysis of petroleum fractions by TLC-FID: Applications to petroleum reservoir description. *Org. Geochem.*, 17: 603–617.

Kernan, M.R., Faulkner, D.J., and Jacobs, R.S. 1987. The luffariellins, novel anti-inflammatory sesterterpenes of chemotaxonomic importance from the marine sponge *Luffariella variabilis*. *J. Org. Chem.*, 52: 3081–3083.

Kijjoa, A., and Sawangwong, P. 2004. Drugs and cosmetics from the sea. *Mar. Drugs*, 2: 73–82.

Lalli, C.M., and Parson, T.R. 1993. *Biological oceanography*, 1–10. Pergmon Press, Oxford, New York.

Li, Y., Qian, Z.J., Ryu, B., Lee, S.H., Kim, M.M., and Kim, S.K. 2009. Chemical components and its antioxidant properties *in vitro*: An edible marine brown alga, *Ecklonia cava*. *Bioorg. Med. Chem.*, 17: 1963–1973.

Look, S.A., Fenical, W., Jacobs, R.S., and Clardy, J. 1986. The pseudopterosins: A new class of anti-inflammatory and analgesic natural products from the sea whip *Pseudopterogorgia elisabethae*. *Proc. Natl. Acad. Sci. U.S.A.*, 83: 6238–6240.

Mayer, A.M.S., Jacobson, P.B., Fenical, W., Jacobs, R.S., and Glaser, K.B. 1998. Pharmacological characterization of the pseudopterosins: Novel anti-inflammatory natural products isolated from the Caribbean soft coral, *Pseudopterogorgia elisabethae*. *Life Sci.*, 62: PL401–PL407.

Minnikin, D.E., Kremer, L., Dover, L.G., and Besra, G.S. 2002. The methyl-branched fortifications of *Mycobacterium tuberculosis*. *Chem. Biol.*, 9: 545–553.

Nakamura, H., and Kobayashi, J. 1982. Separation of mycosporine-like amino acids in marine organisms using reverse phase high-performance liquid chromatography. *J. Chromat.*, 250: 113–118.

Newman, D.J., and Cragg, G.M. 2004. Marine natural products and related compounds in clinical and advanced preclinical trials. *J. Nat. Prod.*, 67: 1216–1238.

Potts, B.C.M., and Faulkner, D.J. 1992. Phospholipase A2 inhibitors from marine organisms. *J. Nat. Prod.*, 55: 1701–1717.

Proksch, P., Edrada, R.A., and Ebel, R. 2002. Drugs from the seas—Current status and microbiological implications. *Appl. Microbiol. Biotechnol.*, 59: 125–134.

Rimkus, G.G. 1999. Polycyclic musk fragrances in the aquatic environment. *Toxicol. Lett.*, 111: 37–56.

Sava, D., Rotaru-Stancic, M., Doroftei, E., and Arcus, M. 2009. Pharmaceutical importance of some multicellular red algae species from the Romanian Black Sea shore. *Annals of RSCB*, XIV(2): 297–300.

Shen, Y., and Burgoyne, D.L. 2002. Efficient synthesis of IPL 576, 092: A novel anti-asthma agent. *J. Org. Chem.*, 67: 3908–3910.

Takei, M., Burgoyne, D.L., and Andersen, R.J. 1994. Effect of contignasterol on histamine release induced by anti-immunoglobulin E from rat peritoneal mast cells. *J. Pharm. Sci.*, 83: 1234–1235.

Tochrop, G.P., Sadhukhan, S., Koner, R.R., and Ghosh, S. 2009. The syntheses and applications of β-benzylmercaptoethylamine derivative. *Tetrahedron*, 65: 10515–10534.

Tziveleka, L.A., Vagias, C., and Roussis, V. 2003. Natural products with anti-HIV activity from marine organisms. *Curr. Top. Med. Chem.*, 3: 1512–1535.

Venugopal, V. 2008. *Marine products for healthcare: Functional and bioactive nutraceutical compounds from the ocean*. Boca Raton, FL: CRC Press.

Wickson, E.J., and Moore, R.R. 1964. Now: Neo-acids go commercial, hydrocarbon process. *Pet. Refine.*, 43: 185–190.

Wiencke, C., Bartsch, I., Peters, A.F., and Breeman, A.M. 1994. Temperature requirement and biogeography of Antarctic, Arctic and amphiequatorial seaweeds. *Bot. Mar.*, 37: 247–259.

Wiencke, C., and Dieck, I.T. 1989. Temperature requirements for growth and temperature tolerance of macroalgae endemic to the Antarctic region. *Mar. Ecol. Prog. Ser.*, 59: 157–170.

Yang, L., and Andersen, R.J. 2002. Absolute configuration of the anti-inflamatory [*sic*] sponge natural product contignasterol. *J. Nat. Prod.*, 65: 1924–1926.

5 Effect of Marine Cosmeceuticals on the Pigmentation of Skin

Sumathy Babitha and Eun-Ki Kim

CONTENTS

5.1 INTRODUCTION

The term *cosmeceutical* was coined in 1980 by the dermatologist Albert Kligman, by blending the words *cosmetics* and *pharmaceutical* (Kligman, 2005). Cosmeceuticals are cosmetic products with biologically active ingredients having medical or drug-like benefits by treating and preventing imperfections of the skin. Some of the most popular cosmeceutical ingredients include grape seed extracts, *Aloe vera*, mushrooms, olive oil, green tea, licorice, coffee, berry extracts, vitamin E, etc. (Baumann, 2007). In recent years, marine derivatives have been revolutionizing skincare, and there has been an increasing interest in the untapped potential of marine-based ingredients as cosmeceuticals. Marine microorganisms have proven to be a rich source of biologically active secondary metabolites (Kim et al., 2008).

In Asian countries, the demand for skin glowing products is rooted in the intention to brighten and lighten the general skin tone as well to eliminate local hyperpigmentation. Depigmenting agents can have potent effects by acting on one or more steps in the melanogenic pathway, melanosome transfer or posttransfer pigment processing, and degradation. Several depigmenting agents modulate skin pigmentation by influencing the transcription and activity of tyrosinase-related melanogenic enzymes, tyrosinase-related protein-1 (TYRP-1), tyrosinase-related protein-2 (TYRP-2), or peroxidase. The transcriptional level is the first stage by which the expression of tyrosinase and related melanogenic enzymes may be modulated. Tyrosinase inhibition is the most common approach to achieve skin hypopigmentation, as this enzyme catalyzes the rate-limiting step of pigmentation. Despite the large number of tyrosinase inhibitors *in vitro*, only a few are able to show induced effects in clinical trials (Kim et al., 2002). In this chapter, we review some potential marine organisms with their effects on pigmentation of skin, focusing mainly on tyrosinase inhibitors.

5.2 MACROALGAE

The term *macroalgae* refers to a diverse family of marine plants. The macroalagal family can be broadly divided into three main groups: phaeophyta (brown algae), rhodophyta (red algae), and chlorophyta (green algae). Used as emollients, skin conditioning agents, and viscosity controlling ingredients, functional products from macroalgae have been used for decades in cosmetics mainly due to their physicochemical properties. Their extracts, such as agar and carrageenan, have been used for nutritional and nutraceutical benefits. In an attempt to find new antibrowning and whitening agents, Cha et al. (2011) investigated 43 indigenous marine algae for tyrosinase-inhibiting activity. It was reported from their results that extracts from *Ecklonia cava* and *Sargassum silquastrum* exhibited excellent inhibitory effects on the pigmentation of zebra fish, most likely due to their potential tyrosinase inhibitory activity. It has been shown that the inhibitory effect was due to dieckol isolated from *Ecklonia cava*. Similarly, potential whitening effects of diphlorethohydroxycarmalol isolated from *Ishige okamurae* have been reported by Heo et al. (2009, 2010). Dieckol and diphlorethohydroxycarmalol also demonstrated strong protective properties against UV-B radiation-induced DNA damage via damaged tail intensity and morphological changes in fibroblasts. Hence, these results indicated that dieckol isolated from *E. cava* and diphlorethohydroxycarmalol isolated from *I. okamurae* have potential whitening effects and prominent protective effects on UV-B radiation-induced cell damage, which might be used in the pharmaceutical and cosmeceutical industries (Yoon et al., 2009). *Hizikia fusiformis* is one of the most common edible brown macroalgae belonging to the Sargassaceae family. In addition, there were reports on the isolation of tyrosinase inhibitory flavonoid glycoside from this marine macroalgae (Ranathunga et al., 2006).

It has been reported that phloroglucinol derivatives, a common secondary metabolite constituent of brown algae, show tyrosinase inhibitory activity due to their ability to chelate copper in this enzyme (Kang et al., 2004). β-Carotene from several species of marine algae is also an active anti-melanogenic compound where β-carotene directly inhibited tyrosinase (Yuto, 2001). Fucoxanthin, a carotenoid derived from the edible sea algae *Laminaria japonica*, inhibited tyrosinase activity, melanogenesis in melanoma, and UV-B-induced skin pigmentation. It was speculated that this effect of fucoxanthin may be due to the suppression of prostaglandin (PG) E2 synthesis and melanogenic stimulant receptors (neurotrophin, PGE2, and melanocyte-stimulating hormone expression) (Shimoda et al., 2010).

5.3 MICROALGAE

Microalgae are small, unicellular, and colorful organisms that occur in water and most likely are photoautotrophic. This diversity makes microalgae a potentially rich source for various chemical products with applications in nutritional, cosmetic, pharmaceutical, and even medicinal industries. Microalgae constitute a reservoir of natural substances of bioactive functions. It was found that tyrosinase activities were inhibited by the extract of several microalgae (Matsukawa et al., 1997). Microalgae are also an abundant resource for zeaxanthin. Antityrosinase zeaxanthin from *Nannochloropsis oculata* has recently been reported (Shen et al., 2011). Many algae-based whitening creams with purified extract of *Nannochloropsis culata* have been on the market.

5.4 SEA CUCUMBER

Sea cucumber is a seabed-dwelling marine echinoderm that feeds on microscopic algae, absorbing nutrients from the organic matter. Sea cucumber has been used for thousands of years by Asians as a culinary delicacy. In spite of using them as food, there is also an emerging market for the use of sea cucumbers in the pharmaceutical and cosmetic industries. Studies conducted on Red Sea cucumber (*Stichopus japonicus*) showed remarkable inhibition of melanogenesis (Yoon et al., 2010; Lee et al., 2010). It has been reported that *S. Japonicus* extract inhibited the expression of tyrosinase and tyrosinase-related proteins (TRP-1 and TRP-2).

5.5 MARINE FUNGUS

Marine fungi have recently drawn much attention as an important source of unique biologically active secondary metabolites such as myrothenones, gliotoxin derivatives, and asperflavin ribofuranoside, suggesting that these organisms would be valuable producers of potential therapeutic agents (Blunt et al., 2006). Almost one-third of all known higher marine fungi are associated with algae. Recently, the fungus *Botrytis* sp. (family Sclerotiniaceae), isolated from the surface of the marine red alga *Hyalosiphonia caespitosa*, has shown potent tyrosinase-inhibiting activity (Zhang et al., 2007). Another marine algicolous fungus of genus *Myrothecium* was found to contain 6-n-pentyl-α-pyrone, which was a potent tyrosinase inhibitor (Li et al., 2005). Among other marine fungi, *Trichoderma* inhibits melanin synthesis due to its ability to act as a tyrosinase inhibitor (Imada et al., 2001).

5.6 MARINE SPONGE

Marine sponges that produce secondary metabolites have long been used as a natural source for the design of new drugs and cosmetics. Recently, a steroidal detergent isolated from a marine sponge has inhibited maturation of tyrosinase to a form that is associated with melanin synthesis (Townsend et al., 1992).

5.7 CONCLUSION

The increasing faith of consumers in natural cosmetics has given birth to the urge to explore natural wealth so that potent therapeutic natural compounds may be utilized for cosmetic applications. Identification of the cosmeceutical potential of marine-based ingredients is a growing field, and use of marine microorganisms as depigmenting agents is a new approach for developing more commercial applications. So far, a large number of tyrosinase inhibitors have been isolated from marine organisms, but very limited numbers have been used as cosmeceuticals. With so many new marine species still to be discovered, much research effort will be needed to appreciate the vast potentialities marine life has to offer us.

REFERENCES

Baumann, L. S. 2007. Less-known botanical cosmeceuticals. *Dermatol Ther.* 20: 330–342.

Blunt, J. W., Copp, B. R., Munro, M. H. G., Northcote, P. T., and Prinsep, M. 2006. Marine natural products. *Nat. Prod. Rep.* 23: 26–78.

Cha, S. H., Ko, S. C., Kim, D., and Jeon, Y. J. 2011. Screening of marine algae for potential tyrosinase inhibitor. *J. Dermatol.*, in press.

Heo, S. J., Ko, S. C., Cha, S. H., Kang, D. H., Park, H. S., Choi, Y. U., Kim, D., Jung, W. K., and Jeon, Y. J. 2009. Effect of phlorotannins isolated from *Ecklonia cava* on melanogenesis and their protective effect against photo-oxidative stress induced by UV-B radiation. *Toxicol in Vitro* 23: 1123–1130.

Heo, S. J., Ko, S. C., Kang, S. M., Cha, S. H., Lee, S. H., Kang, D. H., Jung, W. K., Affan, A., Oh, C., and Jeon, Y. J. 2010. Inhibitory effect of diphlorethohydroxycarmalol on melanogenesis and its protective effect against UV-B radiation-induced cell damage. *Food Chem. Toxicol.* 485: 1355–1361.

Imada, C., Sugimoto, Y., Makimura, T., Kobayashi, T., Hamada, N., and Watanabe, E. 2001. Isolation and characterization of tyrosinase inhibitor-producing microorganisms from marine environment. *Fish Sci.* 67: 1151–1156.

Kang, H. S., Kim, H. R., Byun, D. S., Son, B. W., Nam, T. J., and Choi, J. S. 2004. Tyrosinase inhibitors isolated from the edible brown alga *Ecklonia stolonifera*. *Arch. Pharm.* 27: 1226–1232.

Kligman, A. M. 2005. Cosmeceuticals: A broad-spectrum category between cosmetics and drugs. In Elsner, P., and Maibach, H., eds., *Cosmeceuticals and active cosmetics. Drug versus cosmetics*, 1–9. Boca Raton, FL: Taylor and Francis.

Kim, S. K., Ravichandran, Y. D., Khan, S. B., and Kim, Y. T. 2008. Prospective of the cosmeceuticals derived from marine organisms. *Biotechnol. Bioproc. Eng.* 13: 511–523.

Kim, Y. M., Yun, J., Lee, C. K., Lee, H. H., Min, K. R., and Kim, Y. 2002. Oxyresveratrol and hydroxystilbene compounds. Inhibitory effect on tyrosinase and mechanism of action. *J. Biol. Chem.* 277: 16340–16344.

Lee, M. O., Oh, H. G., Park, S. H., Lee, H. A., Sul, J. D., Song, J., and Kim, O. 2010. Skin whitening effects of *Sanguisorba officinalis* and *Stichopus japonicas*. *Lab Anim Res.* 26: 127–132.

Li, X., Kim, M. K., Lee, U., Kim, S. K., Kang, J. S., Choi, H. D., and Son, B. W. 2005. Myrothenones A and B, cyclopentenone derivatives with tyrosinase inhibitory activity from the marine-derived fungus *Myrothecium* sp. *Chem. Pharm. Bull.* 53: 453–455.

Matsukawa, R., Dubinsky, Z., Masaki, K., Takeuchi, T., and Karube, I. 1997. Enzymatic screening of microalgae as a potential source of natural antioxidants. *Appl. Biochem. Biotechnol.* 66: 239–247.

Ranathunga, S., Rajapakse, N., and Kim, S. K. 2006. Tyrosinase inhibitory flavonoid glycoside derived from an edible brown alga, *Hizikia fusiformis*, and its structure-activity relationships. *Marin Baio Tekunoroji Gakkai Taikai Koen Youshishu* 9: 133.

Shen, C. T., Chen, P. Y., Wu, J. J., Lee, T. M., Hsu, S. L., Chang, C. M. J., Young C. C., and Shieh, C. J. 2011. Purification of algal anti-tyrosinase zeaxanthin from *Nannochloropsis oculata* using supercritical anti-solvent precipitation. *J. Supercrit. Fluids* 55: 955–962.

Shimoda, H., Tanaka, J., Shan, S. J., and Maoka, T. 2010. Anti-pigmentary activity of fucoxanthin and its influence on skin mRNA expression of melanogenic molecules. *J. Pharm. Pharmacol.* 62: 1137–1145.

Townsend, E., Moni, R., Quinn, R., and Parsons, P. G. 1992. Reversible depigmentation of human melanoma cells by halistanol trisulphate, a novel marine sterol. *Melanoma Res.* 1: 349–357.

Yoon, N. Y., Eom, T. K., Kim, M. M., and Kim, S. K. 2009. Inhibitory effect of phlorotannins isolated from *Ecklonia cava* on mushroom tyrosinase activity and melanin formation in mouse B16F10 melanoma cells. *J. Agric. Food Chem.* 57: 4124–4129.

Yoon, W. J., Kim, M. J., Koh, H. B., Lee, W. J., Lee, N. H., and Hyun, C. G. 2010. Effect of Korean red sea cucumber (*Stichopus japonicus*) on melanogenic protein expression in murine B16 melanoma. *Int. J. Pharmacol.* 6: 37–42.

Yuto, K. 2001. Applications of fat soluble vitamins. Melanin synthesis inhibiting function of β-carotene. *Fragr. J.* 29: 28–32.

Zhang, D., Li, X., Kang, J. S., Choi, H. D., and Son, B. W. 2007. A new α-pyrone derivative, 6-[(E)-Hept-1-enyl]-α-pyrone, with tyrosinase inhibitory activity from a marine isolate of the fungus *Botrytis*. *Bull. Korean Chem. Soc.* 28: 887–888.

6 Photoprotective and Cosmeceutical Prospects from Marine Origin

Ramjee Pallela and Se-Kwon Kim

CONTENTS

6.1 INTRODUCTION

Human skin is continuously exposed to ultraviolet (UV) radiation, the major cause of skin disorders such as sunburn, photodamage, and nonmelanoma skin cancer (Sies and Stahl, 2004). UV irradiation is the main etiological agent for most types of skin cancer and a key factor responsible for photoaging and photodamage (González et al., 2008). The solar UV spectrum that reaches the earth's surface has been divided into UV-B (290–320 nm) and UV-A (320–400 nm). UV-B is absorbed in the epidermis and can cause acute sunburn, DNA mutation, or even cancer, whereas the longer wavelengths of the UV-A region can penetrate much deeper into the skin. Hence, photoprotection is critical to avoid these undesired effects. The increase in solar ultraviolet radiation on the earth's surface due to the deterioration in the stratospheric ozone layer has increased the interest in searching for natural photoprotective compounds from various sources (Rastogi et al., 2010). Micronutrients such as carotenoids, tocopherols, ascorbates, flavonoids, and n-3 fatty acids can act as UV absorbers, as antioxidants, or can modulate signaling pathways elicited upon UV exposure. Skin protection depends solely on endogenous defense when there is no topical application of such compounds. Hence, as additional supplements, cosmetics or cosmeceuticals can be given to the human body in order to bring resilience from the external environment.

In 1980, Albert Kligman recognized that even water could alter the structure and function of skin and coined the term *cosmeceutical* to indicate a topical substance that is sold as a cosmetic product but has performance characteristics that suggest pharmaceutical action (Elsner and Maibach, 2005). Cosmeceuticals are cosmetic products that have drug-like properties and enhance or protect the functional appearance or order of the human body, which include antiphotoaging and photoprotective creams, moisturizers, lotions, shampoos, etc. These represent a category of products placed between nonprescription and prescription products that treat various topical diseases (Amer and Maged, 2009). However, the results of these treatments are sometimes disappointing, creating a need for more effective and safer agents that are less irritating. Presently, there is a universal trend not only favoring natural treatments of various skin diseases but also enhancing human beauty by means of photoprotection.

Photoprotection is a group of mechanisms that is developed to minimize damage, when an organism is exposed to UV irradiation. These mechanisms can be controlled or organized by certain organic and inorganic compounds or substances (e.g., melanin) produced by different terrestrial and aquatic sources (Pallela et al., 2010). A number of photoprotective compounds from different organisms have been developed to counteract photodamage, e.g., scytonemins (cyanobacteria), mycosporines (fungi), mycosporine-like amino acids (MAAs; in cyanobacteria, algae, and animals), phenyl propanoids and flavonoids (higher plants), melanins (humans and other animals and even some bacteria), and several other UV-absorbing substances of unknown chemical structures (Gröniger et al., 2000; Rastogi et al., 2010; Sinha et al., 1998, 2007). Besides terrestrial organisms, aquatic organisms have a variety of physiological and biochemical mechanisms for reducing the damage incurred due to exposure to UV radiation, including screening (physical and chemical), quenching (antioxidation and pigment cycles), and repair (protein and nucleic acid repair) (Banaszak, 2003). Among natural photoprotecting substances, marine-originated compounds have gained much advantage in terms of safety and abundancy (Congfen and Yinmao, 2009; Kim et al., 2008). However, the production of photoprotective compounds is affected by several environmental factors, such as different wavelengths of UVR, desiccation, nutrients, salt concentration, light as well as dark periods, and still there is controversy about the biosynthesis of various photoprotective compounds. For example, the synthesis of scytonemin, a predominant UV-A-photoprotective pigment in cyanobacteria, has been found to be affected by factors such as osmotic stress, salt stress, and temperature (Rastogi et al., 2010). It was previously well demonstrated that seaweed extracts can help to protect skin from UV radiation (Bulteau et al., 2006; Ryu et al., 2009). In addition to seaweeds, several other marine organisms are involved in the production of photoprotective cosmeceuticals (Pallela et al., 2010).

The present chapter elucidates various marine photoprotecting substances that are used as cosmeceuticals, and their potential roles in treating various diseases. The preliminary portion of the chapter narrates the important photoprotecting compounds, and the latter describes the cosmeceutical applications of those marine-originated substances. The structures of many such compounds have been covered in various chapters of this book.

6.2 MARINE COMPOUNDS OF PHOTOPROTECTION

6.2.1 Secondary Metabolites

6.2.1.1 Mycosporine-Like Amino Acids (MAAs)

Mycosporine-like amino acids (MAAs) are a family of secondary metabolites, which involve minimizing the damage from exposure to solar UV radiation by direct screening. MAAs are intracellular water-soluble compounds with high molar extinction coefficients at UV-B and UV-A regions with an absorption maxima between 309 and 360 nm, which is a characteristic of UV protection and the subsequent photoprotectivity.

There are plenty of MAAs; more than 20 have been reported from various terrestrial and aquatic sources. The occurrence of UV-absorbing compounds was first reported in zooxanthellate scleractinian corals at the Great Barrier Reef (Shibata, 1969). However, modern analytical approaches prove these compounds to be MAAs, which are mainly reported from cyanobacteria, algae, fungi, and bacteria (Pallela et al., 2010; Shick and Dunlap, 2002).

6.2.1.2 Terpenes and Carotenoids

Isoprenoid compounds are commonly referred to as terpenes and carotenoids. Marine organisms must be considered one of the prime sources of terpenes and terpenoids, which are actually of primary and secondary metabolites, derived from the five-carbon isoprene entity (Gross and König 2006). The combinatorial modifications of these terpenes form diversified terpenoid derivatives, which show different chemical and biological properties. These have been ethnopharmacologically applied for centuries, where steroidal terpenoids were the first marine isoprenes that were discovered.

Marine terpenoids are also listed as C_{10}, C_{15}, C_{20}, C_{25}, and C_{30} ringed isoprenoids and sterols. More importantly, isoprenoids cover most of the marine monoterpenoids, diterpenoids, sesquiterpenoids, sesterterpenes, triterpenoids, oligoglycosides, marine sterols, etc. These terpenoids are very isolated from sponges, corals, algae, fungi, bacteria, etc. Among these marine resources, sponges are becoming the more important research targets for various terpenoids.

Carotenoids like astaxanthin, β-carotene, zeaxanthin, and lutein were shown to modulate UV-A-induced gene expression in human keratinocytes and are also believed to play a key role in the amelioration/prevention of several human pathological processes such as UV-mediated photooxidation of skin, inflammation, prostate and mammary carcinogenesis, age-related diseases, etc. (Cardozo et al., 2007).

6.2.2 Vitamins

Systemic photoprotection by means of dietary supplements is gaining momentum, especially in the form of vitamins and micronutrients available from marine resources (Venugopal, 2008). Ounce for ounce, along with herbs, sea vegetables are higher in vitamins and minerals than any other food on the planet. However, the quantity of these vitamins from marine sources is very minute when compared to other organisms. There are numerous vitamins, and vitamin-composed supplements have been reported from marine sources even as early as the 1990s; few of the photoprotective vitamins are listed as below.

6.2.2.1 Flavonoids

Flavonoids are the most common group of polyphenolic compounds that can be observed in the human diet and are found universally in various plants. Marine microbes, along with marine flora and fauna, have been well studied for various flavonoid compounds of photoprotectivity. For example, actinoflavoside, a novel glycoside composed of a rare 2,3,6-tridcoxy-3-amino-ribopyranoside (ristosamino) aminosugar and an aglycone reminiscent of the plant flavonoids, was isolated from marine *Streptomyces* sp. (Jiang et al., 1997).

6.2.2.2 Retinoids

Retinoids or the vitamin-based retinoic acid, e.g., 9-cir-retinoic acid, are available in marine bacteria, which show potent pharmacological effects that pass clinical trials (Schwartsmann, 2000).

6.2.2.3 Tocopherol

Tocopherols (vitamin E) are lipid-soluble skin-protecting antioxidants, which are synthesized only by photosynthetic plants. Animals are unable to synthesize them, and hence they must obtain tocopherols from plant sources (Durmaz, 2007). Vitamin E is a generic term used to refer to eight homologues: α-tocopherol, β-tocopherol, γ-tocopherol, δ-tocopherol, α-tocotrienol, β-tocotrienol, γ-tocotrienol, and δ-tocotrienol. Among them, α-tocopherol is the most abundant form that has the highest antioxidant activity *in vivo*, and also has an important role in cell membrane electron transport reactions and stabilization, which is related to membrane permeability and fluidity (Carballo-Cárdenas et al., 2003).

Marine algae are a good source of these photoprotecting molecules, e.g., marine-originated novel vitamin E is a different kind of tocomonoenol, named marine-derived tocopherol (MDT) (Gotoh et al., 2009). The marine species *Dunaliella tertiolecta* and *Tetraselmis suecica*, which are widely used in aquaculture as feed for fish and mollusk larvae, produce relatively high concentrations of α-T (α-tocopherol) and other vitamins (Kim et al., 2008). It was reported that this MDT is broadly distributed with α-T in the tissue of marine fish, as this unusual vitamin E constituent (α-tocomonoenol) enhances antioxidant protection in marine organisms of cold water environments (Yamamoto et al., 2001). This is indicative of exploring cosmeceutical principles from marine organisms.

6.2.2.4 Other Vitamins

Vitamin D is very important in the physiological basis of human life and is generally synthesized by epidermal photoreaction supported by the levels of calcium and phosphorous in the body (Dalgård et al., 2010). The risk of an insufficient vitamin D synthesis can be counteracted by a high intake of marine food. Reports indicate that ergosterols increased in patients who are supplemented with a marine diet, which is an indicating principle of applying marine foods for various cosmeceutical purposes.

6.2.3 Carbohydrates

The enormous production of various polysaccharides from marine plants and animal organisms or by marine bacteria is constantly evolving (Laurienzo, 2010). Marine polysaccharides are abundantly available in most of the algal communities, and the high solubility and affinity of these macromolecules to water is a characteristic feature to explore them as cosmeceuticals.

6.2.3.1 Chitin, Chitosan, and Their Derivatives

Chitin is extracted from the exoskeleton of marine organisms, mainly crabs and shrimp, whereas chitosan is a copolymer of β-(1→4)-linked 2-acetamido-2-deoxy-D-glucopyranose and 2-amino-2-deoxy-D-glucopyranose, obtained by the deacethylation of chitin. Scientific advances have led to the development of chitin, chitosan, and their derivatives, chitooligosaccharides (COSs), in medical, pharmaceutical, and cosmeceutical applications (Kim, 2010). In fact, it is reported by Global Industry Analysts, Inc. that the global market for chitin derivatives is expected to reach $63 billion, while the global chitosan market could exceed $21 billion by the year 2015, which is indicative of an increase in chitin and chitosan applications, including nutraceutical and cosmeceutical uses. Several innovative products of chitosan have entered the cosmeceutical market as topical as well as oral supplements, which have potential antiphotoaging and anti-inflammatory effects.

6.2.3.2 Carrageenan

Carrageenans are extracted from red seaweeds and are available in three distinct chemical forms: kappa (κ) carrageenans form strong, brittle gels; iota (ι) carrageenans form soft elastic gels; and lambda (λ) carrageenans are nongelling thickeners. These are well processed toward various cosmetic preparations; e.g., carrageenan's gelling ability can be applied in numerous cosmetic products to impart texture and consistency (CP Kelco, 2004–2007). Diverse products can be generated based on the type and preparation condition, such as lotions for skin, hair care ingredients, eye makeup constituents, tooth paste binders, shaving foams, air freshener gels, etc. (Laurent and Vieira, 2009).

6.2.3.3 Chondroitin Sulfate

In addition to medical and pharmacological applications, chondroitin sulfate is gaining more importance nowadays for its cosmetic preparations, e.g., low molecular weight chondroitin sulfate compound, extracted from cartilages of marine animals by a process conducted exclusively in the aqueous phase (Landrein et al., 2001).

6.2.3.4 Biopterin Glucose

The chemical formulations of a special compound called biopterin glucose have been isolated from marine phytoplankton, which possibly participates as one of the important cosmetic ingredients (Matsunaga et al., 1993). There could be many prospects from this UV-A-absorbing chromophore, as it can be utilized as pharmaceuticals, functional food, supplements, food, cosmetics, veterinary drugs, veterinary supplements, or animal feed (Hiroyuki et al., 2010). Although Van Baalen and O'Donnell (1972) have suggested that tetrahydropteridines that are similar to biopterins are possible chromophores for UV-mediated cell death, the physiological functions and the exact molecular mechanism of this molecule have yet to be understood.

6.2.3.5 Fucoidan

Macroalgae and macroalgal blends have been included in cosmetic formulations such as emollients, skin conditioning agents, and viscosity controlling ingredients in the past few decades (Fitton et al., 2007). *Fucoidan* or *fucan* is a generic term that is used for a class of carbohydrates isolated from marine algae and even from certain echinoderms. Fucoidan, which is similar to heparin, involves the cascade of coagulation and potentiation of growth factor activity (Berteau and Mulloy, 2003). These principles of physiological augmentation will indirectly contribute to the development of useful pharmacological and cosmetic preparations (Mizutani et al., 2010).

6.2.4 Proteins, Peptides, and Enzymes

Collagen, one of the important structural and functional proteins in the human body, has been explored from various terrestrial, avian, and marine resources. Based on the safety principle of marine collagens, oceans are being explored for the isolation and application of collagens in skin and bone tissue engineering (Xu et al., 2010). Marine collagen exhibits antioxidant properties that have been used in skin care products to prevent or even repair the damage caused by environmental factors, such as UV rays and low humidity, as well as damage associated with the aging process (Xinrong et al., 2009; Xu et al., 2010). Various marine organisms such as fish and sponges produce useful collagens that can be used in pharmaceutical and cosmeceutical preparations (Kim and Mendis, 2006; Swatschek et al., 2002).

6.2.5 Lipids

6.2.5.1 Unsaturated Fatty Acids

Comprehensive reports on polyunsaturated fatty acids (PUFAs) from marine sources revealed that PUFAs like eicosapentaenoic (Arepalli et al., 2009) and docosahexaenoic acids (DHAs) upregulate mitochondrial biogenesis and induce beta-oxidation in white fat, thereby reducing the adipose tissue mass, preferentially in the abdomen (Flachs et al., 2005). This application can be implemented in the facial region so that the level of fat can be controlled or regulated for better appearance. Despite this cosmetic effect of augmenting external appearance, certain oils from marine sources involve tremendous health benefits.

Marine-derived oils like fish oils can be used in soap preparation, with a significant purpose of lubrication and skin protection from external chemicals and dangerous rays. Even oleic acid, a potential modulator of oxidative stress, could be added to PUFA-rich marine oils to prevent oxidative stress-linked corneal pathology (Dutot et al., 2009). Macroalgae have many prospects for PUFA production, especially green algae, a good source of 16:4 n-3, which needs further investigation to appreciate its possible bioconversion in animals or its putative cosmetic or pharmacological properties (Fleurence et al., 1994).

In a similar way, marine microalgae such as *Tetraselmis suecica*, *Porphyridium cruentum*, and *Chaetoceros calcitrans* produce large amounts of arachidonic and linolenic acid, and *Nannochloropsis* sp. produce eicosapentenoic acid, whereas *Isochrysis galbana* produces a large amount of docosahexaenoic acid (Kim et al., 2008).

6.2.6 Phenolic Compounds

Nowadays, phenolic compounds from marine origin have a significant contribution in cosmetics development. The photoprotective effect of phlorotannins from *Eckloina cava* against photooxidative stress induced by UV-B radiation was reported recently (Heo et al., 2009). These results concluded that among the isolated phlorotannins, dieckol showed prominent inhibitory activity against melanogenesis and effectively reduced UV-B radiation-induced cellular damage, which further indicated that dieckol from *E. cava* can be used as an effective natural source in cosmeceutical or pharmaceutical formulations.

6.3 COSMECEUTICAL APPLICATIONS OF MARINE PHOTOPROTECTIVE AGENTS

A cream containing MAAs from the red alga *Porphyra umbilicalis* efficiently protects the skin against UV-A exposure (Schmid et al., 2003).

Several biologically active terpenoids possessing cosmeceutical significance are already in preclinical or clinical development (Newman and Cragg, 2004). One of the interesting marine cosmeceutical products is a family of diterpene glycosides, pseudopterosins (A~D). Previously, it has been proved that an extract from *Crithmum maritimum*-containing minerals, essential oils, polyphenols, flavonoids, and vitamin C activates the protein synthesis of connective tissues, such as collagen and elastin, resulting in the improvement of tone and elasticity in the skin (Kim et al., 2008).

Marine phospholipids comprising flavonoids and polyunsaturated fatty acids have been reported to have tremendous applications in food and cosmetic preparations (Sampalis, 2004).

Recently it was found that daily topical application of the aqueous ethanolic extract of the marine sea grass *Thalassia testudinum* on mice skin exposed to UV-B radiation resulted in the recovery of skin macroscopic alterations (Regalado et al., 2009).

It was proved that α-tocopherol has a major function in the prevention of light-induced pathologies of the skin and eyes (Chiu and Kimball, 2003).

Pyrenoine is a cosmetic agent from the brown alga *Fucus spiralis* with antioxidative properties suitable for antiaging and skin protection products. *F. spiralis* is regularly exposed to intertidal sun and hence develops a high content of phloroglucinol derivatives to protect itself (Adina, 2010). The marine zooplankton *Artemia salina* is the source of GP4G IS from ISP-Vincience, which increases cellular defenses against UV stress and shows multiple photoprotective effects. In addition, it involves the increase in cellular protection against UV aggressions and DNA damage, stimulates HSP70, implicated in the fight against cellular stress, and offers an anti-inflammatory action by decreasing the level of inflammatory cytokines such as IL-1a, IL-8, and TNFα.

It has been reported that astaxanthin (AST)-containing algal extracts show tremendous modulatory effects on UV-A-irradiated human skin fibroblasts, human melanocytes, and human intestinal Caco-2 cells (Lyons and O'Brien, 2002). These studies indicated that these extracts depicted tremendous alterations in the antioxidants like SOD and GSH, and also showed a protective effect against UV-A-induced DNA damage when used in significant concentrations (equivalent to around 10 μM AST). It was thus concluded that algal extracts have a suggestive role as potential antioxidants. Another species of the same alga, *Sargassum siliquastrum,* can also produce xanthin compounds, and more recently it became known that fucoxanthin exerts a potential photoprotective effect against UV-B-induced cell damage (Heo and Jeon, 2009).

The clinical applications of many photoprotective ingredients are still debated to determine their positive role of photoprotection in end users. One such example is the study on the spawn of the sea hare *Aplysia dactylomela*, where the role of MAAs as protective sunscreens was investigated (Carefoot et al., 1998). Although the isolation and characterization of photoprotective and antiphotoaging molecules from marine origin are being explored, the clinical significance and development of these molecules as cosmetic and cosmeceutical formulations is still in the budding stage. Hence, the present chapter directs readers to the implication of marine diversity to cosmeceutical marketing.

6.4 CONCLUSION

Advances in molecular biology and culturing technologies are bridging the gap between the challenges pertaining to the exploitation of the marine environment as a potential source of natural protective compounds. The photoprotective and antiphotoaging mechanisms of marine compounds as strong cosmeceuticals have already been studied from different marine sources like algae, fungi, lichens, bacteria, phytoplanktons, animals, and plants. Several marine-based pharmaceuticals are under active commercial development for which aquaculture is providing an ever-greater proportion of the seafood in developed countries. In order to reveal the enormity of marine sources in terms of photoprotective compounds that promote DNA protection by reducing oxidative stress after UV exposure, tremendous research has been pending from many decades. There is a need to properly understand *in situ* and *ex situ* approaches to marine floral and faunal culture for the development of novel photoprotective molecules, by considering global warming and other catastrophic causes of ozone destruction.

Considering the present aspects on marine cosmetology, it is worthwhile to state that research on antiphotoaging and photoprotective natural products from such deceptively simple marine organisms will pave the way for novel products of cosmeceutical value.

REFERENCES

Adina. 2010. http://www.cosmeticingredients.co.uk/products.asp?prod=558.

Amer, M., Maged, M. 2009. Cosmeceuticals versus pharmaceuticals. *Clinics in Dermatology* 27: 428–430.

Arepalli, S., Sridhar, V., Venkateswara Rao, J., Kavin Kennady, P., Venkateswarlu, Y. 2009. Furano-sesquiterpene from soft coral, *Sinularia kavarittiensis*: induces apoptosis via the mitochondrial-mediated caspase-dependent pathway in THP-1, leukemia cell line. *Apoptosis* 14: 729–740.

Banaszak, A.T. 2003. Photoprotective physiological and biochemical responses of aquatic organisms. In *UV effects in aquatic organisms and ecosystems*, Helbling, E.W., Zagarese, H.E., Zagarese, H. (Eds.). Cambridge, UK: Royal Society of Chemistry, pp. 329–356.

Berteau, O., Mulloy, B. 2003. Sulfated fucans, fresh perspectives: structures, functions, and biological properties of sulfated fucans and an overview of enzymes active toward this class of polysaccharide. *Glycobiology* 13: 29R–40R.

Bulteau, A., Moreau, M., Saunois, A., Nizard, C., Friguet, B. 2006. Algae extract-mediated stimulation and protection of proteasome activity within human keratinocytes exposed to UVA and UVB irradiation. *Antioxidants and Redox Signaling* 8: 136–143.

Carballo-Cárdenas, E.C., Tuan, P.M., Janssen, M., Wijffels, R.H. 2003. Vitamin E (α-tocopherol) production by the marine microalgae *Dunaliella tertiolecta* and *Tetraselmis suecica* in batch cultivation. *Biomolecular Engineering* 20: 139–147.

Cardozo, K.H.M., Guaratini, T., Barros, M.P., Falcão, V.R., Tonon, A.P., Lopes, N.P., Campos, S., Torres, M.A., Souza, A.O., Colepicolo, P., Pinto, E. 2007. Metabolites from algae with economical impact. *Comparative Biochemistry and Physiology Part C: Toxicology and Pharmacology* 146: 60–78.

Carefoot, T., Harris, M., Taylor, B., Donovan, D., Karentz, D. 1998. Mycosporine-like amino acids: possible UV protection in eggs of the sea hare *Aplysia dactylomela. Marine Biology* 130: 389–396.

Chiu, A., Kimball, A. 2003. Topical vitamins, minerals and botanical ingredients as modulators of environmental and chronological skin damage. *British Journal of Dermatology* 149: 681–691.

CP Kelco. 2004–2007. http://www.in-cosmetics.com/page.cfm/Action=Exhib/ExhibID=39.

Dalgård, C., Petersen, M.S., Schmedes, A.V., Brandslund, I., Weihe, P., Grandjean, P. 2010. High latitude and marine diet: vitamin D status in elderly Faroese. *British Journal of Nutrition* 104: 914–918.

Durmaz, Y. 2007. Vitamin E (α-tocopherol) production by the marine microalgae *Nannochloropsis oculata* (Eustigmatophyceae) in nitrogen limitation. *Aquaculture* 272: 717–722.

Dutot, M., Liang, H., Martin, C., Rousseau, D., Grynberg, A., Warnet, J.M., Rat, P. 2009. Per os administered refined olive oil and marine PUFA-rich oils reach the cornea: possible role on oxidative stress through caveolin-1 modulation. *Nutrition and Metabolism* 6: 48–57.

Elsner, P., Maibach, H. (Eds.). 2005. *Cosmeceuticals and active cosmetics*. Boca Raton, FL: Taylor & Francis.

Fitton, J.H., Irhimeh, M., Falk, N. 2007. Macroalgal fucoidan extracts: a new opportunity for marine cosmetics. Cosmetics and Toiletries 122: 55–64.

Flachs, P., Horakova, O., Brauner, P., Rossmeisl, M., Pecina, P., Franssen-van Hal, N., Ruzickova, J., Sponarova, J., Drahota, Z., Vlcek, C., Keijer, J., Houstek, J., Kopecky, J. 2005. Polyunsaturated fatty acids of marine origin upregulate mitochondrial biogenesis and induce β-oxidation in white fat. *Diabetologia* 48: 2365–2375.

Fleurence, J., Gutbier, G., Mabeau, S., Leray, C. 1994. Fatty acids from 11 marine macroalgae of the French Brittany coast. *Journal of Applied Phycology* 6: 527–532.

González, S., Fernández-Lorente, M., Gilaberte-Calzada, Y. 2008. The latest on skin photoprotection. *Clinics in Dermatology* 26: 614–626.

Gotoh, N., Watanabe, H., Oka, T., Mashimo, D., Noguchi, N., Hata, K., Wada, S. 2009. Dietary marine-derived tocopherol has a higher biological availability in mice relative to alpha-tocopherol. *Lipids* 44: 133–143.

Gröniger, A., Sinha, R., Klisch, M., Häder, D. 2000. Photoprotective compounds in cyanobacteria, phytoplankton and macroalgae—A database. *Journal of Photochemistry and Photobiology B: Biology* 58: 115–122.

Gross, H., König, G. 2006. Terpenoids from marine organisms: unique structures and their pharmacological potential. *Phytochemistry Reviews* 5: 115–141.

Hiroyuki, H., Keiko, S., Ichiro, S., Yasuhiro, I. 2010. Compsition containing biopterin and method for using the same. U.S. Patent Application 20100234385.

Heo, S.-J., Jeon, Y.-J. 2009. Protective effect of fucoxanthin isolated from *Sargassum siliquastrum* on UV-B induced cell damage. *Journal of Photochemistry and Photobiology B: Biology* 95: 101–107.

Heo, S.-J., Ko, S.-C., Cha, S.-H., Kang, D.-H., Park, H.-S., Choi, Y.-U., Kim, D., Jung, W.-K., Jeon, Y.-J. 2009. Effect of phlorotannins isolated from *Ecklonia cava* on melanogenesis and their protective effect against photo-oxidative stress induced by UV-B radiation. *Toxicology in Vitro* 23: 1123–1130.

Jiang, Z.-D., Jensen, P.R., Fenical, W. 1997. Actinoflavoside, a novel flavonoid-like glycoside produced by a marine bacterium of the genus *Streptomyces. Tetrahedron Letters* 38: 5065–5068.

Kim, S.-K. 2010. *Chitin, chitosan, oligosaccharides and their derivatives: biological activities and applications*. Boca Raton, FL: CRC-Taylor & Francis.

Kim, S.-K., Mendis, E. 2006. Bioactive compounds from marine processing byproducts—A review. *Food Research International* 39: 383–393.

Kim, S., Ravichandran, Y., Khan, S., Kim, Y. 2008. Prospective of the cosmeceuticals derived from marine organisms. *Biotechnology and Bioprocess Engineering* 13: 511–523.

Landrein, A., Roy, P., Durand, P., Delannoy, C. 2001. Low molecular weight chondroitin sulphate compound having cosmetic activity. U.S. Patent Application 20030032620.

Laurent, L., Vieira, V. 2009. Cosmetic composition comprising at least one Lambda-carrageenan polysaccharide and inorganic particles. U.S. Patent Application 20100186763.

Laurienzo, P. 2010. Marine polysaccharides in pharmaceutical applications: an overview. *Marine Drugs* 8: 2435–2465.

Lyons, N.M., O'Brien, N.M. 2002. Modulatory effects of an algal extract containing astaxanthin on UVA-irradiated cells in culture. *Journal of Dermatological Science* 30: 73–84.

Matsunaga, T., Burgess, J.G., Yamada, N., Komatsu, K., Yoshida, S., Wachi, Y. 1993. An ultraviolet (UV-A) absorbing biopterin glucoside from the marine planktonic cyanobacterium *Oscillatoria* sp. *Applied Microbiology and Biotechnology* 39: 250–253.

Mizutani, S., Deguchi, S., Kobayashi, E., Nishiyama, E., Sagawa, H., Kato, I. 2010. Fucoidan-containing cosmetics. U.S. Patent 7678368.

Newman, D., Cragg, G. 2004. Marine natural products and related compounds in clinical and advanced preclinical trials. *Journal of Natural Products* 67: 1216–1238.

Pallela, R., Yoon N.-Y., Kim, S. 2010. Anti-photoaging and photoprotective compounds derived from marine organisms. *Marine Drugs* 8: 1189–1202.

Rastogi, R., Sinha, R., Singh, S., Häder, D. 2010. Photoprotective compounds from marine organisms. *Journal of Industrial Microbiology and Biotechnology* 37: 537–558.

Regalado, E.L, Rodríguez, M., Menéndez, R., Concepción, A.A., Nogueiras, C., Laguna, A., Rodríguez, A.A., Williams, D.E., Lorenzo-Luaces, P., Valdés, O., Hernandez, Y. 2009. Repair of UVb-damaged skin by the antioxidant sulphated flavone glycoside thalassiolin b isolated from the marine plant *Thalassia testudinum* König. *Marine Biotechnology* 11: 74–80.

Ryu, B.-M., Qian, Z.-J., Kim, M.-M., Nam, K.-W., Kim, S.-K. 2009. Anti-photoaging activity and inhibition of matrix metalloproteinase (MMP) by marine red alga, *Corallina pilulifera* methanol extract. *Radiation Physics and Chemistry* 78: 98–105.

Sampalis, F. 2004. Natural marine source phospholipids comprising flavonoids, polyunsaturated fatty acids and their applications. U.S. Patent Application 20040234587.

Schmid, D., Schürch, C., Zülli, F., Nissen, H., Prieur, H. 2003. Mycosporine-like amino acids: natural UV-screening compounds from red algae to protect the skin against photoaging. *SÖFW-Journal* 129: 38–42.

Schwartsmann, G. 2000. Marine organisms and other novel natural sources of new cancer drugs. *Annals of Oncology* 11: 235–243.

Shibata, K. 1969. Pigments and a UV-absorbing substance in corals and a blue-green alga living in the Great Barrier Reef. *Plant and Cell Physiology* 10: 325–335.

Shick, J., Dunlap, W. 2002. Mycosporine-like amino acids and related glusols: biosynthesis, accumulation, and UV-protective functions in aquatic organisms. *Annual Review of Physiology* 64: 223–262.

Sies, H., Stahl, W. 2004. Nutritional protection against skin damage from sunlight. *Annual Review of Nutrition* 24: 173–200.

Sinha, R., Klisch, M., Gröniger, A., Häder, D. 1998. Ultraviolet-absorbing/screening substances in cyanobacteria, phytoplankton and macroalgae. *Journal of Photochemistry and Photobiology B: Biology* 47: 83–94.

Sinha, R., Singh, S., Häder, D. 2007. Database on mycosporines and mycosporine-like amino acids (MAAs) in fungi, cyanobacteria, macroalgae, phytoplankton and animals. *Journal of Photochemistry and Photobiology B: Biology* 89: 29–35.

Swatschek, D., Schatton, W., Kellermann, J., Müller, W.E.G., Kreuter J. 2002. Marine sponge collagen: isolation, characterization and effects on the skin parameters surface-pH, moisture and sebum. *European Journal of Pharmaceutics and Biopharmaceutics* 53: 107–113.

Van Baalen, C., O'Donnell, R. 1972. Action spectra for ultraviolet killing and photoreactivation in the blue-green alga agmenellum quadruplicatum. *Photochemistry and Photobiology* 15: 269–274.

Venugopal, V. 2008. *Marine sources of vitamins and minerals. Marine products for healthcare*. Boca Raton, FL: CRC-Taylor and Francis, pp. 239–259.

Xinrong, P., Ruiyue, Y., Haifeng, Z., Qiong, L., Zhigang, L., Junbo, W., Yong, L. 2009. Preventive effect of marine collagen peptide on learning and memory impairment in SAMP8 Mice. *Food and Fermentation Industries* 7:1–5.

Xu, Y.J., Han, X.L., Li, Y. 2010. Effect of marine collagen peptides on long bone development in growing rats. *Journal of the Science of Food and Agriculture* 90: 1485–1491.

Yamamoto, Y., Fujisawa, A., Hara, A., Dunlap, W.C. 2001. An unusual vitamin E constituent (α-tocomonoenol) provides enhanced antioxidant protection in marine organisms adapted to cold-water environments. *Proceedings of the National Academy of Sciences of the United States of America* 98: 13144–13148.

7 Prospects of Marine Sponge Collagen and Its Applications in Cosmetology

Janapala Venkateswara Rao, Ramjee Pallela, and G.V.S. Bhanu Prakash

CONTENTS

7.1 INTRODUCTION

Age is a biological phenomenon that promotes skin shrinkage, wrinkles, and sagging. The skin's elasticity and firmness are controlled by thick layers of keratinized cells and the elastic fibers composed of collagen and elastin. Collagen serves as the structural backbone or scaffolding of animals and plays a crucial role in maintaining structural integrity (Kim et al., 2009) in preventing organs/tissues from tear or losing their functional shape, when they are exposed to sudden and wild movements.

As the age progresses, the collagen production in the human body slows down (Diggs, 2008). From the age of 25–30 years onward, the body loses collagen at a rate of approximately 1.5% per annum. Externally, depletion of collagen in the skin is the main reason for the formation of lines, wrinkles, and the appearance of aging, and internally collagen loss causes the weakening of skeletal structure due to the breakdown of connective tissue, including muscle tissue. Based on the importance of collagen functions and gradual decrease of collagen in the adult body, proper collagen balance in the body is highly desirable and crucial. Certain genetic defects like osteogenesis imperfecta (commonly known as brittle bone disease) lead to the deficiency of collagens (Paterson et al., 1993). Other genetic diseases also lead to the deficiency of collagen, and these complications are linked to certain diseases like Bethlam myopathy and Ulrich myopathy, caused by mutations, congenital muscular dystrophies, etc. (Jobsis et al., 1999; Lampe and Bushby, 2005). Hence, it is necessary to supply collagens to combat the imbalances and deficiencies in the human body. Regular intake of collagen supplement makes human skin nourished and reduces joint pain. It is important to know and understand the function of collagen in order to prevent the early signs of aging, through regular intake of collagen supplement.

In the present scenario people with different lifestyles having altered food habits, climate variations, a smoking habit, alcohol consumption, etc., may also adversely affect the normal architecture of skin resulting in altered physical appearance. To regain their normal appearance, most people are dependent on specific or multiple raw materials for cosmetic use. These raw materials have various origins, especially vegetable (algae, microalgae, halophytes) sources, and also animal (collagen, elastin, chitosan, etc.) origin. Primarily they can serve as excipients (gellifying agents), but more precisely as hydrating agents (Coiffard, 2004). There are many industries like BASF, Croda, Dow Chemical, Mascot, and KLK Oleo that are focusing on the supply of ingredients and fine chemicals for various cosmetic and personal care companies.

These plant-based cosmeceuticals have drug-like benefits and contain active ingredients (a.i.) such as vitamins, phytochemicals, enzymes, antioxidants, and essential oils. Nonetheless, the percentages of these active ingredients are comparatively more from marine organisms. Cosmeceuticals from marine organisms have attracted increased attention because of their beneficial effects on human health. Recently, widespread research has been undertaken worldwide on the general aspects of the chemical structures, physical and biochemical properties of isolated bioactive secondary metabolites from marine organisms, and their biotechnological application in the development of novel pharmaceuticals and cosmeceuticals (Kim et al., 2008).

The global cosmetics and pharmaceuticals market had more than $270 billion in sales in 2006, of which about $140 billion is represented by the cosmetics and personal care sectors. These industries are outpacing most others. The largest global segment is skin care and is estimated to have more than $31 billion during this same time period (Briony Davies, 2006). The year-over-year demand for such cosmetics is enormously increased. Euromonitor International forecasts an annual global growth of 3.7% to reach international sales of more than $300 billion by 2010. The largest domestic market is in the United States, where sales in the cosmetics and personal care sectors are nearly $5 billion. The lack of a rejuvenation process certainly helped manufacturers and distributors boost their cosmeceuticals sales.

7.2 A BRIEF HISTORY OF COSMETOLOGY

The oldest actual makeup dates back to 4000 B.C. from ancient Egypt, with today's equivalent of makeup artists and wig stylists producing the look of darkly lined eyes and hairstyles. Ancient civilizations used cosmetics to indicate nobility and wealth status. Cosmetologists in these ancient civilizations were either servants or slaves who applied cosmetics to their masters.

Cosmetology had advanced considerably with oils and lotions by Roman times. They were used in makeup, perfumes, and cleansers as well as in hair to keep it in shape. Greasy lotions made from animal fat were often applied to the face and other parts of the body for hiding defects. Roman preparations were known for being highly toxic in those times. Female slaves were known as cosmetae, who used to dissolve various substances in their own saliva. Roman cosmetic manufacturers were among the most productive of the ancient world. During that time, advancements in hair dyes and wigs allowed anyone to alter their appearance almost immediately and very satisfactorily. Roman wigs were in special demand from generals and aging senators. Other parts of the world such as the Middle East and India had been using henna as a decorative covering in complex designs.

Arab chemists had developed a distillation process by medieval times; this helped in making perfume oils, among other worthy products. However, human vanity used this process for the first time in scale for the pleasure of wealthy merchants who wanted sweet-smelling bodies and clothes. The reformation period in Europe was one of the prominent periods in the modern history of cosmetology. The invention of eau de toilette allowed subtle fragrances for the first time to be sprayed on to the skin rather than thickly brushed.

It is easy to think in the early twenty-first century that cosmetics are only being used by women. In reality they have been used by both genders throughout history. In today's society, cosmetics are available to everyone, regardless of wealth or social status. The history of cosmetology probably began with the first human being to become self-conscious. From the moment humans realized that they wanted to look good, other humans have helped them to improve their appearance. Cosmetology is basically the art of creating physical beauty, which includes grooming the hair, skin, and nails. This artistic field has no exact creation date or country of origin. From ancient Egypt to ancient China, it was found all around the world.

7.2.1 Modern Cosmetology

Historically, beauty has changed as cultures and technologies have evolved. Though it still deals mainly with the skin, hair, and nails, cosmetology now covers an array of job positions and people must take cosmetology courses to practice the trade. Cosmetology can lead to many careers, such as nail technician, makeup artist, hair stylist, wig technician, aesthetician, hair removal specialist, or skin care specialist. The over-the-counter (OTC) world of beauty and well-being is expanding due to new raw materials, advanced technology, a growing consumer market, and a willingness to purchase promises. This need for cosmetics that do something other than adorn the skin has led to a new category of functional cosmetics. Functional cosmetics encompass the realms of colored cosmetics, cosmeceuticals, neurocosmeceuticals, and nutraceuticals. All of these categories have the same goal, namely, improving senescence of the skin, body, or mind. Before proceeding with an analysis of the world of functional cosmetics, it is necessary to distinguish the subtle differences between each of these categories.

7.2.2 Cosmeceuticals

Cosmeceuticals extend beyond cosmetics to enhance functioning, usually returning the structure to a more youthful state. For example, wrinkle-reducing moisturizers, nail growth serums, and skin lightening creams all fall into this category. Cosmeceuticals are somewhat confusing, however, as both prescription and OTC products have been labeled by this term. Drug cosmeceuticals include topical retinoids for improving dermal collagen production, topical minoxidil for enhanced scalp hair growth, and eflornithine for facial hair growth reduction. The second category of cosmeceuticals includes OTC drugs, such as sunscreens and antiperspirants. Sunscreens are some of the most important functional cosmetics currently available and will be discussed at length.

The world of cosmeceuticals is rapidly changing, since it is driven to a large extent by new product introductions designed to capture consumer attention and fads. This accounts for the dizzying array of ingredients, with an even more amazing number of appearance claims. However, it is possible to make some generalizations regarding cosmeceutical development that might be useful to the reader.

A subgroup of functional cosmetics has been recently termed neurocosmeceuticals. These are topical agents applied to the skin to induce a feeling of well-being, playing into the philosophical concept of the mind-body connection. Furthermore, the health of the mind can be improved through the skin. Neurocosmeceuticals claim to improve well-being by altering neurotransmitters through topical applications. Indeed, ingredients have been identified that inhibit the release of substance P (a neuropeptide in sensory nerves) and enhance the release of beta-endorphins. Nutrition is also considered to be a part of the mind-body connection in the form of nutraceuticals. Nutraceuticals have been billed as the oral equivalent of cosmeceuticals, since most dermatologists agree that the best way of improving skin appearance is through improved overall general health. Both neurocosmeceuticals and nutraceuticals are beyond the scope of this discussion, which will mainly focus on topical agents designed to improve the functioning and appearance of the skin, known as cosmeceuticals.

Once the active ingredient is identified and synthesized, it is typically tested *in vitro* to determine an effect on cultured fibroblasts. The active ingredient is then placed in a vehicle suitable for human application and clinical studies are undertaken. Successful human clinical studies pave the way for successful introduction into the marketplace via ingredient licensing arrangements. While the cosmeceutical market may be vast and confusing, it is possible to simplify the discussion by organizing cosmeceutical ingredients in terms of their functionality. There are only so many identified scientifically valid methods of improving the skin. The remainder of the cosmeceutical functional categories actually alter the structure and function of the skin, making them drugs and cosmetics. Basically cosmetology is divided into various branches. Cosmetology includes the study of skin care, cosmetics, hairstyles and hair care, beauty therapy, etc.

7.3 COLLAGEN: AN OVERVIEW

Collagen is a family of highly characteristic fibrous proteins found in all multicellular animals, which are the most abundant proteins found in mammals (Di Lullo Dagger et al., 2002). It is the main protein found in animal connective tissues and is present in dermis, bones, tendons, cartilage (hyaline), ligaments, blood vessels, teeth, cornea, intervertebral discs, vitreous bodies, placenta (fetal), etc. (Burgeson, 1987; Prockop, 1998). Seventy-five percent of human skin is made up of collagen, providing texture, resiliency, and shape; in total about 30% of the human body is collagen (Beckman et al., 2004). It is an important protein that wraps the organs, and part of it holds specialized cells together in discrete units (Kadler, 1994). It serves as the structural backbone or scaffolding of animals, plays a crucial role in maintaining structural integrity (Kim et al., 2009), and prevents organs/tissues from tearing or losing their functional shape when they are exposed to sudden and wild movements. Besides the structural role in mature tissues, collagen plays a regulating role in developing tissues as well (Kosher and Church, 1975). Regular intake of collagen supplements human skin, nourishes, and reduces joint pain.

It is well known that collagen available in food is digested and passes through the body with no benefit. Thus, scientists discovered the hydrolyzed collagen protein (HCP) or collagen hydrolyzate or protein hydrate, which is simply a modified form of collagen, sometimes called gelatin, that is used for many years in aesthetics and medicine; e.g., in some studies, patients were able to reduce or discontinue their use of analgesic medication while consuming HCP (Moskowitz, 2000). HCP helps regulate the body's metabolic state by providing enzymatically digested collagen of a low molecular weight that is rapidly absorbed in the digestive tract (Talbott, 2003). Hence, supplementing with hydrolyzed collagen provides the body with the amino acids necessary to manufacture and support collagen. It is very important to know and understand the function of collagen in order to prevent early signs of aging, through regular intake of collagen supplements.

Collagens have been classified under several subfamilies according to their primary structures or forms of supramolecular organization (Exposito et al., 2002). Presently, there are 29 types of collagens recognized (8 collagens have recently been discovered but their function is still unknown), which are denoted by assigning Roman numerals (Miller and Gay, 1987; Kielty et al., 1993).

The most abundant and essential types of collagens in the human body are grouped according to their physical structure (van der Rest and Garrone, 1991) (Table 7.1). The main types of collagen in connective tissues are types I, II, III, IV, V, VIII, and IX (Kuhn et al., 1987) (Table 7.2). Type I is the principal collagen of skin and bone; by far, it is the most abundant type of collagen in the body, representing around 90% of the total body collagen. Type II is found in the cartilage, whereas type

III is found in skin, blood vessels, and internal organs. Type V is found in bone, skin, tendons, ligaments, and cornea. Type IX collagen often associates with type II and gives structural integrity to the tissues. Other important collagens are types IV and VIII, which are network-forming collagens that polymerize to form sheet-like basal lamina and anchor fibril beneath stratified squamous epithelia (Prockop and Kivirikko, 1995).

7.3.1 Structure of Collagen Molecule

All collagen molecules consist of three polypeptide chains (called alpha chains) folded individually into a left-handed helical conformation and wrapped around each other to form a long, stiff, rope-like, right-handed triple helix. Bundles of these collagen molecules are arranged in fibrils with a molecular weight close to 97.1 Daltons. These fibrils are twisted into a right-handed coil (fiber) with a total weight between 95,000 and 102,000 Daltons (Waterlow et al., 1978; Woodhead-Galloway, 1980; Garlick, 1969; Smith et al., 1983). The triple helical peptides are extremely rich in proline and glycine, both of which are important for the formation of the collagen-specific helical structure (Rossler et al., 1995; Alberts et al., 1994; Zubay, 1998; Lehninger, 1987). Collagen polypeptides contain one or more blocks of Gly-X-Y repeats; the X residue is often proline (Pro) and Y is often hydroxyproline (proline to which an –OH group is added after synthesis of the polypeptide) (Kuhn, 1987). Hydroxyproline is a specific marker amino acid for all the collagen phenotypes, and its estimate in urinary secretion is directly related to the bone resorption rate (Schonau and Rauch, 1997). Collagen contains a considerable amount of glycine (330 residues per 1,000 amino acid residues), which is regularly distributed in the entire molecule, appearing at every third position in the sequence of individual chains, and this distribution permits close connection of polypeptide chains to form a triple helix. Hydroxyproline is a characteristic amino acid present in collagen (135 residues per 1,000 amino acid residues), formed by the hydroxylation of proline as a posttranslational modification. The content of proline and hydroxyproline varies according to the origin of the collagen synthesis. Yet another typical amino acid of collagen is hydroxylysine (56 residues per 1,000 residues of amino acids), which is also present in other types of proteins. Its main role is to bind with the saccharide components. Low content of leucine and isoleucine is typical for collagen peptides, and generally the collagen fibers are established by means of hydrogen bridges between the amino acids and also by the formation of pyridinoline bridges between lysines and hydroxylysines (Eyre et al., 1984).

7.3.2 Collagen-Related Diseases

Most of the commercial brands made of collagens are derived from cattle (bovine) or chicken (avian) products, and these collagens carry the risk of transmission of infectious agents (prions in the case of cattle) and precipitating immunological reactions (Taylor, 1989; Swanson, 2002). The risks of side effects vary, depending on what kind of collagen treatment an individual has undergone. People with autoimmune disorders can expect a flare-up, as the body reacts to foreign substances like collagen injections and causes allergic reactions to the recipient. Previous and present experiences have shown that the influence of collagens derived from bovine sources became problematic, especially with the appearance of fatal collagen diseases (polymyositis and dermatomyositis) that seem to develop in patients using bovine collagen for a prolonged period. In addition, the use of cattle as the main source for collagen has to be reconsidered because of the risks of transmissible spongiform encephalopathies (TSEs), such as bovine spongiform encephalopathy (BSE), the scientific name for mad cow disease. TSEs are a family of diseases of humans and animals characterized by spongy degeneration of the brain with severe and fatal neurological signs and symptoms (WHO, 2002a). In 1996, the first human cases of a new type of TSE, known as variant Creutzfeldt-Jakob disease (vCJD), were observed in young individuals in the UK. The causative agent of vCJD has been

TABLE 7.1
Types of Collagens

Type	Available Tissue	Notes	Gene(s) NCBI Number, AA Present	Disorders	Reference
I	Most of the connective tissue including bones, scar tissue, tendons, skin, artery walls, the endomysium of myofibrils, fibrocartilage, cornea, and the organic part of bones and teeth	This is the most abundant collagen of the human body; the end product when tissue heals by repair	*COL1A1*: NP_000079/1464 (includes 22 aa SP) *COL1A2*: NP_000080/1464 (includes 22 aa SP)	Osteogenesis imperfect, Ehlers-Danlos, infantile cortical hyperostosis aka Caffey's disease, osteoporosis	Chu et al., 1982 Myers et al., 1981
II	Cartilage, nucleus pulposus, vitreous humor (clear gel that fills the space between the lens and the retina of the eyeball of humans and other vertebrates)	Hyaline cartilage makes up 50% of all cartilage protein	*COL2A1*: NP_001835/1487 aa (includes 25 aa SP) *COL2A2*: NP_149162/ Same as a1(II)A but lacks vWC domain	Collagenopathy type II and XI, severe chondrodysplasia osteoarthitis	Miller and Matukas, 1969
III	Extendable connective tissues such as skin, lungs, liver, spleen and vascular system, intestines, and the uterus	This is the collagen of granulation tissue, and is produced quickly by young fibroblasts before the tougher type I collagen is synthesized	*COL3A1*: NP_000081/1466 aa (includes 23 aa SP)	Ehlers-Danlos syndrome type IV	Cameron et al., 2002
IV	Basement membrane, basal lamina, eye lens	Serves as part of the filtration system in capillaries and the glomeruli of nephron in the kidney	*COL4A1*: NP_001836/1669 aa (includes 27 aa SP) *COL4A2*: NP_001837/1712 aa (includes 25 aa SP) *COL4A3*: NP_000082/1670 aa (includes 28 aa SP) *COL4A4*: NP_000083/1690 aa (includes 38 aa SP) *COL4A5*: NP_000486/1685 aa (includes 26 aa SP) *COL4A6*: NP_001838/1691 aa (includes 21 aa SP) (B isoform P_378667)	Alport syndrome, Goodpasture's syndrome	Timpl and Brown, 1996 Timpl et al., 1981
V	Minor component in tissue-connecting collagen I, lung, cornea, bone, fetal membranes	Associated with placenta	*COL5A1*: NP_000084/1838 aa (includes SP) *COL5A2*: NP_000384/1499 aa (includes SP) *COL5A3*: NP_056534/1745 aa (includes 29 aa SP)	Ehlers-Danlos syndrome	Birk, 2001

VI	Most of the connective tissues, interstitial tissue	Associated with type I	*COL6A1*: NP_001839/1028 aa (includes 19 aa SP) *COL6A2*: NP_001840/1019 aa (includes 20 aa SP) 2C2a and 2C2a' isoforms, NP_478054 and NP_478055 *COL6A3* : NP_004360—multiple splicing/3177 (with 25 aa SP)	Ulrich myopathy and Bethlem myopathy	Kielty et al., 1992
VII	Anchoring fibrils, skin, dermal-epidermal junctions, oral mucosa, cervix, epidermis (extend from the basal lamina of epithelial cells and attach to the lamina reticularis)	Forms anchoring fibrils in dermal epidermal junctions	*COL7A1*: P_000085/2944 aa (includes 16 aa SP)	Epidermolysis bullosa dystrophica	Keene et al., 1987
VIII	Endothelium, Descemet's membrane, and other tissues	Endothelial cells	*COL8A1*: NP_065084; NP_001841/744 aa (includes 28 aa SP) *COL8A2*: NP_005193/703 aa (includes SP)	Posterior polymorphous corneal dystrophy 2	Kapoor et al., 1986, 1988 Stephan et al., 2004
IX	Tissue-connecting collagen II, FACIT (fibril-associated collagens with interrupted triple helices) collagen, cartilage, vitreous humor, cornea	Associated with type II and XI fibrils	*COL9A1*: NP_001842/921 aa (includes 23 aa SP) Short form NP_511040/678 aa (includes 23 aa SP) *COL9A2*: NP_001843/689 aa (includes SP *COL9A3*: NP_001844/684 aa (includes SP)	EDM2 (epiphyseal dysplasia, multiple) and EDM3[a]	Olsen, 1997 Shimokomaki et al., 1990
X	Hypertrophic cartilage	Hypertrophic and mineralizing cartilage	*COL10A1*: P_000484/680 aa (includes 18 aa SP)	Schmid metaphyseal dysplasia	Kwan et al., 1991
XI	Tissue-connecting collagen II, vitreous body	Cartilage	*COL11A1*: NP_001845/1806 aa NP_542196/1818 aa (includes 36 aa SP) NP_542197/767 aa (includes 36 aa SP) *COL11A2*:NP_542411/1736 aa (includes 22 aa SP) Isoforms 2 and 3 NP_542412 and NP_542410 *COL11A3*:NP_001835/ Same as α1(II)A	Collagenopathy types II and XI	Mendler et al., 1989
XII	Tissue-connecting collagen I, FACIT collagen, decorin, perichondrium, ligaments, tendon, and glycosaminoglycans	Interacts with type I-containing fibrils	*COL12A1*: NP_004361/3063 aa (includes 23 aa SP) (has NC1 variants) Short form NP_542376/1899 aa (includes same SP as long form) (has NC1 splice variants)	—	Keene et al., 1991 Nishiyama et al., 1994 Zhang et al., 2003

(continued)

TABLE 7.1 (continued)
Types of Collagens

Type	Available Tissue	Notes	Gene(s) NCBI Number, AA Present	Disorders	Reference
XIII	Many tissues (transmembrane collagen), epidermis, hair follicle, endomysium, intestine, chondrocytes, lungs, liver	Interacts with integrin a1b1, fibronectin, and components of basement membranes like nidogen and perlecan	*COL13A1*: NP_005194/717 aa (transmembranous) (20+ splice variants)	—	Latvanlehto et al., 2003
XIV	Tissue-connecting collagen I, dermis, tendon, vessel wall, placenta, lungs, liver	FACIT collagen	*COL14A1*: NP_066933 (has NC1 variants)/1796 aa (includes SP) Short form without N-terminal FNIII domain; also has NC1 variants	—	Young et al., 2000, 2002
XV	Many tissues, fibroblasts, smooth muscle cells, kidney, pancreas	—	*COL15A1*: NP_001846/1388 aa (includes 25 aa SP)	—	Myers et al., 1996 Ramchandran et al., 1999
XVI	Many tissues, fibroblasts, amnion, keratinocytes	—	*COL16A1*: NP_001847/1604 aa (includes SP)	—	Kassner et al., 2003
XVII	Skin hemidesmosomes, dermal-epidermal junctions	Transmembrane collagen, also known as BP180, a 180 kDa protein	*COL17A1*: NP_000485/1497 aa (transmembranous)	Bullous pemphigoid and certain forms of junctional epidermolysis bullosa	Hopkinson et al., 1998
XVIII	Many tissues (liver, lungs, kidney)	Source of endostatin	*COL18A1*: NP_085059/1516 aa (includes 23 aa SP) Short form NP_569712/1336 aa (includes 33 aa SP)	—	Sasaki et al., 1998
XIX	Rhabdomyosarcoma cells	FACIT collagen	*COL19A1*: NP_001849/1142 aa (includes 23 aa SP)	—	Myers et al., 2003
XX	Corneal epithelium, embryonic skin, sternal cartilage, tendon; may be associated with type I collagen fibrils	—	*COL20A1*: NP_001004392/ Not in human; ch = 1472 aa (without SP)	—	Koch et al., 2001
XXI	May be fibril associated, widespread expression pattern, blood vessel wall	FACIT collagen	*COL21A1*: NP_110447/957 aa (includes 22 aa SP)	—	Fitzgerald and Bateman, 2001
XXII	Located in specific tissue junctions and may be associated with microfibrils	—	*COL22A1*: NP_690848/1626 aa (includes SP)	—	Koch et al., 2004

XXIII	Transmembrane collagen identified in cell culture	MACIT (membrane-associated collagens with interrupted triple helices) collagen	*COL23A1*: NP_775736/540 aa (transmembranous)	—	Banyard et al., 2003
XXIV	Expressed in tissues containing type I collagen	—	*COL24A1*: NP_690850/1714 aa (includes SP)	—	Koch et al., 2003
XXV	Transmembrane collagen, cleaved form present in Alzheimer's amyloid plaques in neurons	—	*COL25A1*: NP_942014/654 aa (transmembranous) Isoform 2 is NP_115907/642 aa	—	Hashimoto et al., 2002
XXVI	Expressed in testis and ovary of adult tissues	—	*COL26A1*: NP_597714/439 aa (includes SP)	—	Sato et al., 2002
XXVII	Widespread expression particularly in cartilage	—	*COL27A1*: NP_116277/1860 aa (includes 41 aa SP)	—	Boot-Handford et al., 2003 Pace et al., 2003
XXVIII	Peripheral nervous system nodes	—	*COL28A1*: NP_001032852/1125 aa (includes SP)	—	Veit et al., 2006
XXIX	Epidermal collagen, skin, lung, small intestine, colon, testis	—	*COL29A1*: ABW81241/2615 aa	Atopic dermatitis	Söderhäll et al., 2007

Source: Sato et al., 2002; Gelse et al., 2003; Koch et al., 2003, 2004; Pace et al., 2003; Söderhäll et al., 2007; Tong et al., 2010; Gordon and Hahn, 2010.

[a] Multiple epiphyseal dysplasia is a clinically and genetically heterogeneous skeletal disorder characterized by joint pain and stiffness, mild short stature, and degenerative joint disease. Onset of the disorder is usually in childhood.

TABLE 7.2
Most Documented Collagens Useful for Humans

Type	Tissue Distribution	Ultrastructure	Synthesis	Function
I	Skin, bone, tendon, fascia	Skin, bone, tendon, fascia	Fibroblasts, osteoblasts, chondroblasts	Mechanical stability and resistance
II	Cartilage	Very thin fibers	Chondroblasts	Tensile strength
III	Skin, vessels, internal organs, smooth muscle	Loosely packed thin fibrils	Smooth muscle, fibroblasts, hepatoblasts	Flexibility
IV and VIII	Network-forming collagens	Thin amorphous sheets	Endothelial, epithelial cells	Support and filtration
V	Bone, skin, tendons, ligaments, cornea	Resembles type IV, but fibrillar in nature	Endothelial origin	Similar to type III collagen
IX	Hyaline cartilage, intervertebral discs, vitreous humor	Long triple helices connected by a flexible kink	Chondrocytes	Stability and resistance to cartilage framework

Source: Pallela, R., Isolation and Characterization of Native Collagens from Marine Demosponges, PhD thesis, Zoology Department, Osmania University, Hyderabad, India, 2009.

shown to be indistinguishable from the BSE agent, and infection was presumed to have been caused by eating BSE-contaminated food (WHO, 2002b).

7.4 MARINE SPONGES AS COLLAGEN SUPPLEMENTS

Consequently, in search of safer, nonimmunogenic collagens, marine invertebrates are competing worldwide with other animals and certain collagens from jellyfish (Song et al., 2006), and sponges have already been identified for the presence of nonimmunogenic collagens for the benefit of humankind. In sponges, *Chondrosia reniformis* is a rich source of collagen, useful in pharmaceutical technology, cosmetics, and nutrition. The isolation procedure for collagen of *C. reniformis* yields 30% of freeze-dried collagen from the total freeze-dried sponge (Swatschek et al., 2002). According to Connes et al. (1972), sponge species of the family Suberitidae are found to be rich in collagen and display a high potency in tissue regeneration, especially after injuries (Diaz, 1977; Eckert et al., 2006).

Members of phylum Porifera are one of the better-known, diverse, multicellular invertebrates, characterized by a pronounced plasticity in the determination of cell lineages, and are conspicuous and abundant members of marine benthic communities ranging from the euryhaline, to the estuarine, to the intertidal, to the deep sea (Koltun, 1970; Hechtel, 1976; Bergquist, 1978; Hooper and van Soest, 2002). With respect to their body organization, cell differentiation, and embryogenesis, sponges have been considered separate from other metazoans and grouped in Parazoa or Archaemetazoa (Willmer, 1994). Porifera is subdivided into three major classes. One is the aptly named *Calcarea* of hard-bodied animals having stiff carbonate structures (colonies). Other classes include the *Hexactinellida* (glass sponges), and the third, what most hobbyists consider the true sponges in the class, Demospongiae. Although phylogenetically crucial, Demospongiae have been overlooked for developmental studies because their adult body plan is simple and appears to have little in common with the rest of the Metazoa (Brusca and Brusca, 2003).

Sponges serve as sources of bioactive compounds, are capable of synthesizing many secondary metabolites and have been a prolific source of new molecules with pharmacological activity (Sipkema et al., 2005; Gross and Konig, 2006). Recently, many therapeutic agents have been isolated through interdisciplinary approaches adopted by biologists and chemists (Frincke and Faulkner,

1982; Bakus et al., 1985; Sakemi et al., 1988; Venkateswara Rao et al., 1998). The production of marine sponge biomass for the important secondary metabolites and proteins through a variety of molecular approaches is one of the main goals of marine biotechnology. Due to the increased number of sponge secondary metabolites and the importance of the major structural protein (collagen), interest in sponge cultivation has increased over the last few years.

7.4.1 Methodological Approaches to Select Potential Sponges for Collagen

Based on the current status and future needs, one has to explore the possibility of identifying suitable marine sponges as a direct collagen source, which is ultimately assumed to be a substitute for human collagen. The structural composition of demosponges (including amount of spicules, proteinaceous extracellular matrix, total tissue, protein (spongin and collagen) content, etc.) is the most important factor to identifying the potential sponge candidate for the isolation of collagen (Gross et al., 1956). According to our study, the experimented demosponges (collected from the Gulf of Mannar (n = 18)) composed of 15–70% of freeze-dried sponge tissue matter and the remaining portion are of siliceous spicules.

The amount of total protein among the sponges differed greatly, ranging from 2.74 to 33.31% in relation to the tissue content. The maximum amount of protein was observed in *Callyspongia fibrosa* (33.31%, which is equivalent to 209.55 mg/g per 629.03 mg/g tissue). However, the sponge *Pseudoceratina purpurea* maintains a maximum amount of tissue content but possesses a comparatively lesser amount of relative protein (2.74%, equivalent to 24.16 mg/g of protein per 882.69 mg/g tissue). It could be due to the presence of nonproteinaceous or amorphous material in the tissue mass of the sponge body (Pallela, 2009). Hence, the first step is to make a short list of sponges based on the protein content. The percent amount of collagen present in the total content of protein in sponges was the second important factor to short-list the most potential sponge candidates. Collagen contains a high concentration of the amino acid 4-hydroxyproline (Hyp); e.g., up to 14% of the dry weight of collagen is Hyp, compared to 1% found in elastin (Etherington and Sims, 1981). It is well known that Hyp has been historically used as an estimate of collagen content from the tissues of plant and animal sources for many years (Flores, 1980; Vazquez-Ortiz and Gonzalez-Mendez, 1996; Bilgen et al., 1999). Hence, the amount of collagen present in each sponge was quantified by Hyp estimation using either spectrophotometric or HPLC methods. Among the 18 sponges, the content of hydroxyproline (Hyp) significantly varied, ranging from 6 to 16%, which coincides with the experimental data from the previous study on collagen content (approximately 10% of the total organic matter) of Demospongiae (Wiens et al., 1999).

The orientation of spongin along with the fine fibrillar material embedded in the extracellular matrix of sponges is a major feature that forms the basis of differentiation of marine sponges (Diehl-Seifert et al., 1985). In our earlier studies, spatial coexistence of typically arranged/distributed spicules along with spongin in the extracellular matrix of sponge species was analyzed by scanning electron microscopy (SEM) (Figure 7.1). It is evident from our studies that the structural architectures of sponges may differ from each other. Hence before short-listing the sponges, we should study the in-depth structural architecture as one of the parameters.

For further confirmation to identify the potential sponge, the histological occurrence of collagen in marine sponge tissues (fine cryosections) should be studied with trichrome staining, which is a standard protocol for identifying collagen in sponge tissues.

Generally, characteristic insolubility and mineralization cause methodological problems when isolating and studying collagens from marine sponges (Ehrlich et al., 2005; Imhoff and Garrone, 1983), and often these sponge collagen fibers are covered with amorphous cuticle, which acts as a shield to protect the native collagen fibrils and is the unique characteristic feature of marine collagens (Figure 7.2). Hence the outer envelope (amorphous cuticle) can be dissolved by using other acids and alkali to expose the collagen fibrils prepacked to form bundles (Garrone, 1978; Junqua et al., 1974).

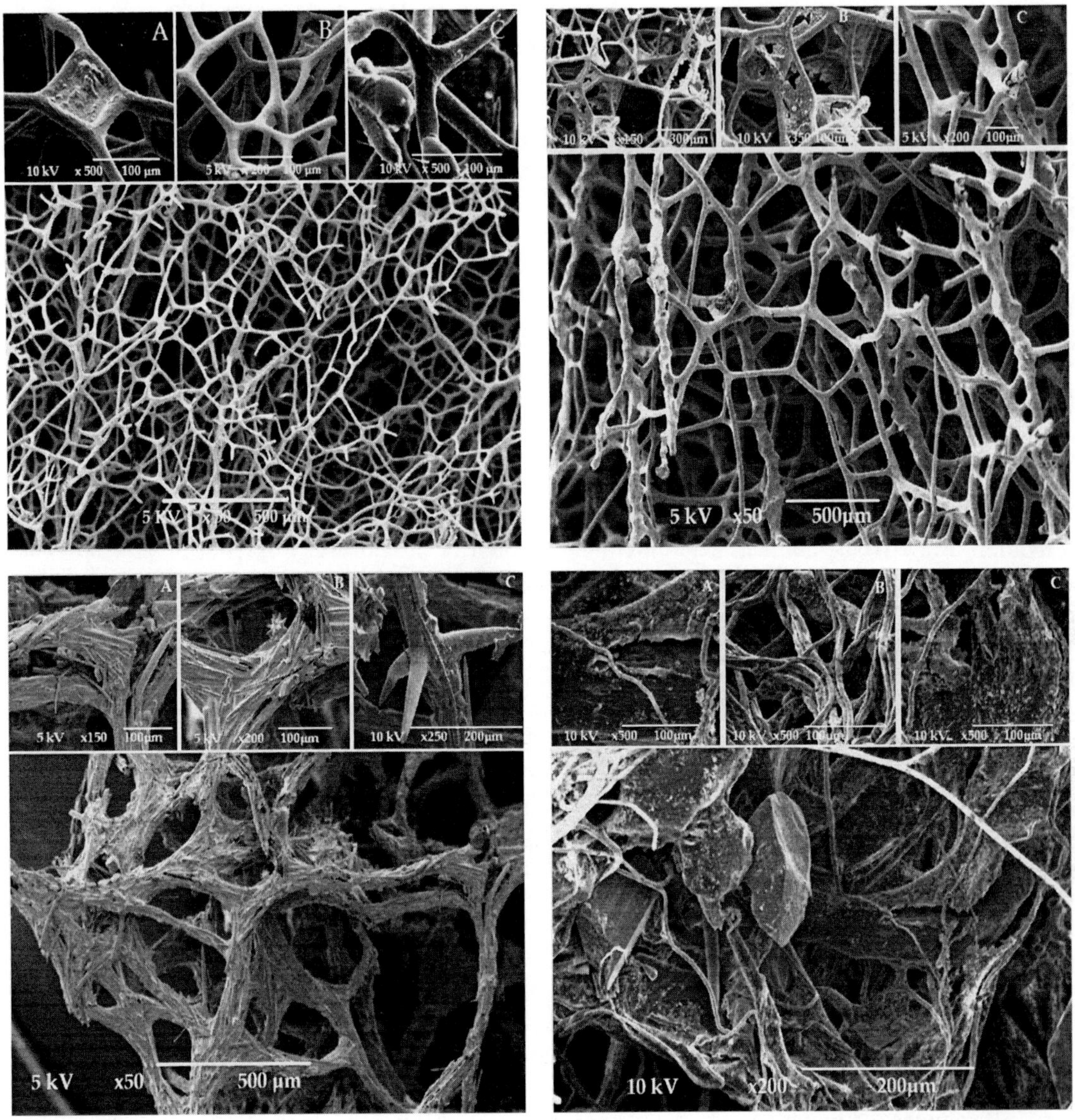

FIGURE 7.1 Scanning electron micrographs of some sponges, i.e., *Spongia officinalis*, *Stelospongia cribriformis*, *Dysidea fragilis*, and *Ircinia fusca*. (From Pallela, R., Isolation and Characterization of Native Collagens from Marine Demosponges, PhD thesis, Zoology Department, Osmania University, Hyderabad, India, 2009.)

The last step to identify the potential sponge is to study the repeated band periodicity under transmission electron microscopy (TEM), which is a characteristic feature of collagen in the isolated filaments (Figure 7.3).

Our atomic force microscope (AFM) studies indicated that the fibrils of sponge *Ircinia fusca* possess repeated band periodicity of 67 nm with a width of 300 nm and height of 20 nm, which resembles human native type I collagen fibrils (Figure 7.4). A similar ultrastructure and organization of collagen fibrils has been elucidated recently in another marine sponge, *Chondrosia reniformis* Nardo (Heinemann et al., 2007).

On the other hand, these fibril 1 fractions were subjected to Western blot studies, which revealed that these fibrils had immunocompatibility with anticollagen type IV antibodies, whereas the fibril

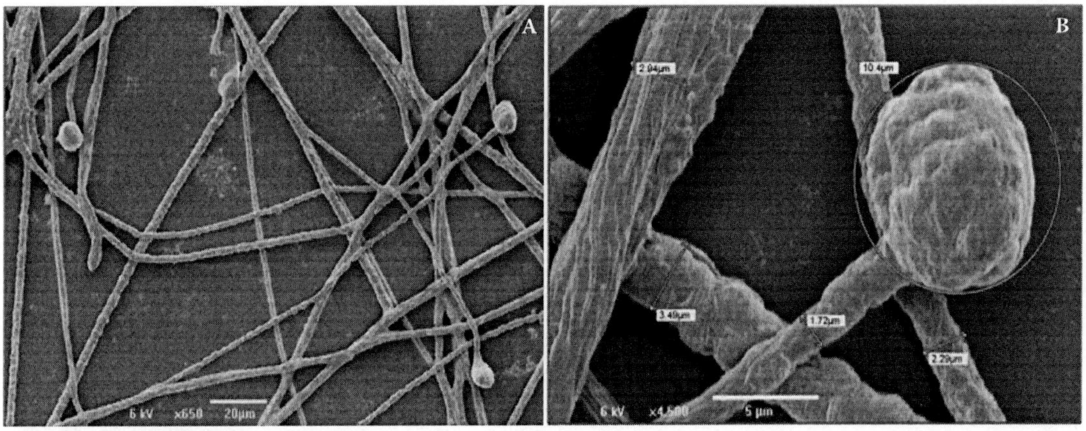

FIGURE 7.2 Scanning electron micrographs of Irciniid filaments of *Ircinia fusca*. (A) Terminally knobbed filaments of differential thickness. (B) Magnified images of the filaments, showing annular grooves formed by the tightly packed fibrils and the nodular, wavy surface of the terminal knob. (From Pallela, R., Bojja, S., and Janapala, V. R. (2011). Biochemical and biophysical characterization of collagens of marine sponge, *Ircinia fusca* (Porifera: Demospongiae:Irciniidae). *International Journal of Biological Macromolecules*, 49: 85–92.)

2 fraction with anticollagen type 1 antibodies. Boute et al. (1996) earlier reported about the type IV collagen in sponges, narrating it to be a missing link of basement ubiquity. This indicates that the purified native collagens of marine sponges (*I. fusca*) are identical to human type I and IV collagens and may be safe and competitive alternatives to harmful bovine originated collagens (Figure 7.5).

7.5 APPLICATIONS OF COLLAGENS

Collagen is well recognized as a useful and important biopolymer in the field of biomaterial/tissue engineering (Lazovic et al., 2005). It is a highly versatile material, extensively used in the medical, dental, and pharmacological fields. The main applications of collagen are as drug delivery systems, i.e., collagen shields in ophthalmology, cross-linked collagen matrix for burns/wounds, mini-pellets and tablets for protein delivery, gel formulation for sustained drug delivery,

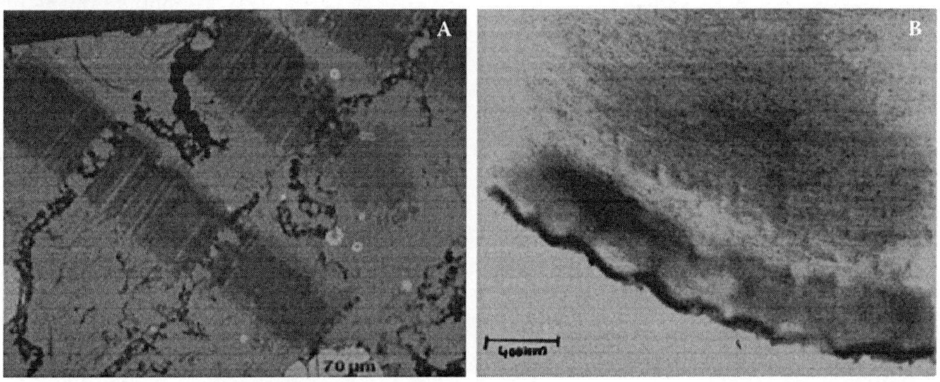

FIGURE 7.3 Transmission electron micrographs of isolated Irciniid filaments of *Ircinia fusca*. (A) Longitudinal section of the filament showing the isometric fibrillar repetition of band periodicity, as observed in typical collagen fibril. (B) Cross section of the filament showing granular layered appearance of microfibrillar structures covered by outer layer of cuticle. (From Pallela, R., Bojja, S., and Janapala, V. R. (2011). Biochemical and biophysical characterization of collagens of marine sponge, *Ircinia fusca* (Porifera: Demospongiae:Irciniidae). *International Journal of Biological Macromolecules*, 49: 85–92.)

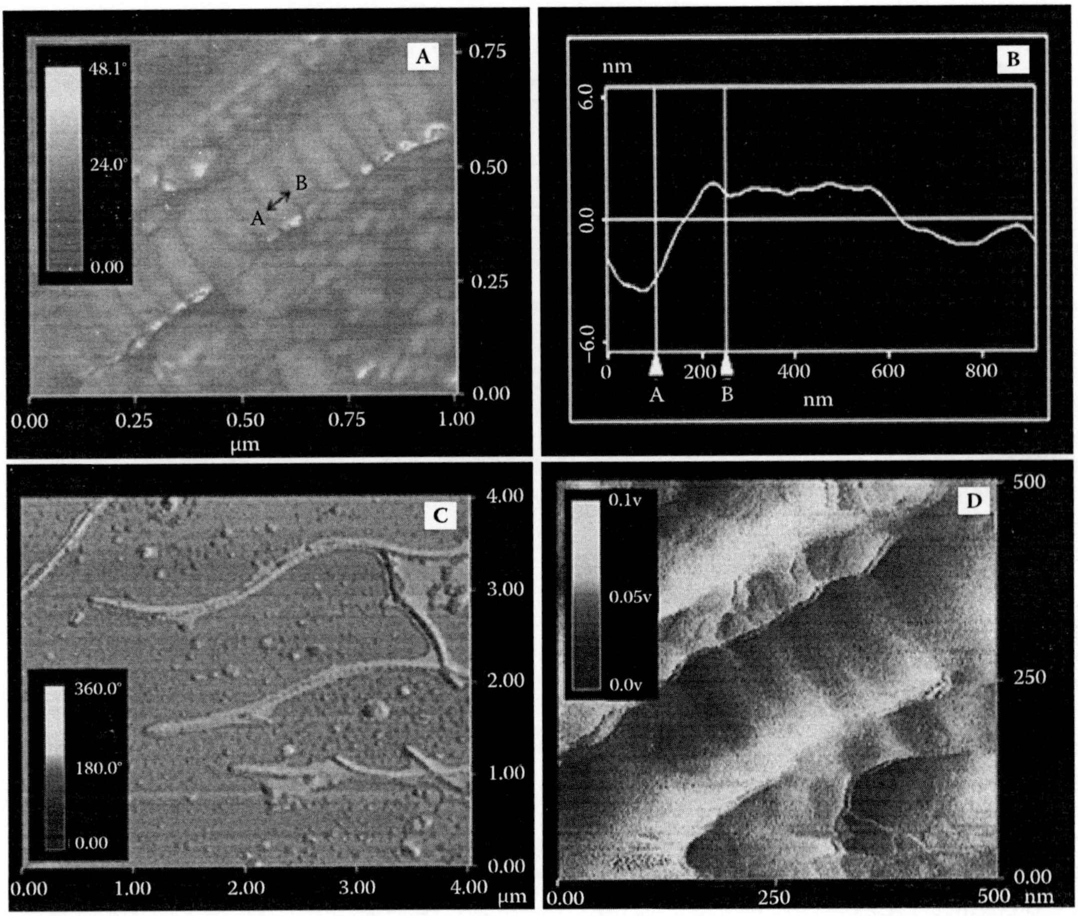

FIGURE 7.4 Atomic force micrographs of the purified fibrillar collagens from *Ircinia fusca*. (A) Deflection mode height image of a collagen fibril with a typical band spacing of 67 nm (scan size 1 × 1 µm). (B) Amplitude signal showing the distance between the protrusions of two bands. (C) Low scan phase image of the fibrils showing the repeated band periodicity (scan size 4 × 4 µm). (D) Amplitude image of collagen fibrils with repeated bands (scan size 0.5 × 0.5 µm). (From Pallela, R., Bojja, S., and Janapala, V. R. (2011). Biochemical and biophysical characterization of collagens of marine sponge, *Ircinia fusca* (Porifera: Demospongiae:Irciniidae). *International Journal of Biological Macromolecules*, 49: 85–92.)

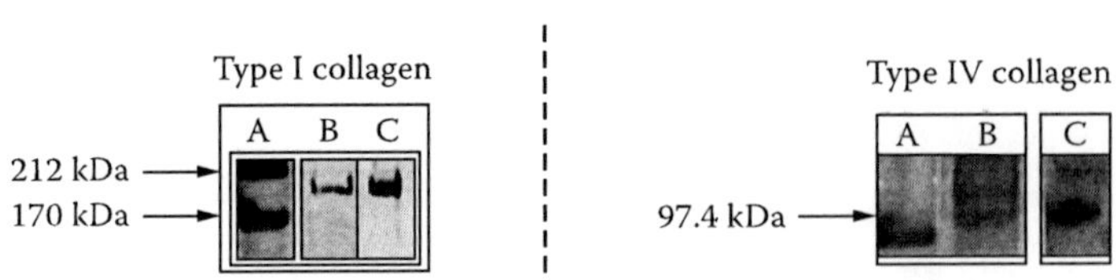

FIGURE 7.5 Western blot analyses of the purified collagens from *Ircinia fusca*, showing specificity with monoclonal type I and type IV collagen antibodies. (A) Protein marker, (B) collagen band on polyacrylamide gel electrophoresis (PAGE) gel, and (C) electroblotted collagen on polyvinylidene fluoride (PVDF) membrane. (From Pallela, R., Bojja, S., and Janapala, V. R. (2011). Biochemical and biophysical characterization of collagens of marine sponge, *Ircinia fusca* (Porifera: Demospongiae:Irciniidae). *International Journal of Biological Macromolecules*, 49: 85–92.)

etc. (Friess, 1998). Collagen is used for the construction of artificial skin substitutes for patients who have undergone severe skin burns and other accidents. Collagens can be produced transgenically from plants (by modifying polysaccharides through fermentation of yeast) and naturally from animals. Attempts have been made to produce human collagens that rely on the hydroxylation machinery naturally present in plants, but resulted in collagen that is poor in proline hydroxylation (Merle et al., 2002). Tobacco plants provide a useful expression system for the large-scale production of recombinant human collagen (Ruggiero et al., 2000). However, these plant collagens used in cosmetics may be biologically inactive, and hence extensive research has been done to isolate animal-originated collagen for human therapy. Collagens are used for tissue engineering, including skin replacement, bone substitutes, and artificial blood vessels and valves. The industrial use of collagen has been expanded and now includes foods, cosmetics, and biomedical materials (Chvapil, 1979; Nimni et al., 1988). Traditionally, collagen has been animal sourced, principally from cows that prove to be a major source for a number of applications in the medical industry. There are plenty of supplements found in various health food stores, in which two types of collagens, type I and type II, are available. These types of collagen are the ones that are present in our hair, nails, skin, muscles, bones, eyes, teeth, and tendons. They are also responsible for strong cartilage support, and by consuming these types of collagen as oral supplements, we can replenish the body collagen that is naturally lost with age (Hall, 2009). Collagen implants developed by many industries are natural products derived from highly purified bovine collagen that is injected just under the surface of the skin and is well known for cosmetic surgery. Injectable collagen formulations derived from human dermis, for example, Cosmoderm®, will enhance the fullness of the cheeks and other facial contours. Injectable fillers are one of the most popular facial rejuvenation techniques, and implantable collagen hydrogels have been examined as agents for delivery of chemotherapeutic agents (Rao, 1995; Tracey, 2004; Eckart, 2006). The actual effects of collagen supplements are debatable. The collagens may be either of bovine or porcine origin, used along with a mixture of fibroblasts, growth factors, silicone, and glycosaminoglycans for various pharmaceutical applications. Recent development of liquid collagen supplement (Hydroderm®) has been isolated from marine origin, and it proved to deliver collagen through the upper layer of the skin deep into the lower layers of the epidermis. In addition, collagen is used extensively in guided tissue regeneration (GTR), a procedure that attempts to reconstitute lost tissues by collagen substitutes. The first report of a human tooth treated by guided tissue regeneration was by Nyman et al. (1982).

7.5.1 Scope and Present Scenario of Collagen and Its Uses in Cosmetology

One of the most exciting and rewarding areas of advanced materials science research is that which involves the applications of materials to health care, and especially to reconstructive surgery. At a time when technological research should ideally contribute both to wealth creation and improvements to the quality of life, it is obvious that developments that lead to the implantation of advanced materials and devices within the body in order to treat disease and injury should have a very high priority.

Collagen varies insignificantly among different species; however, structural changes have a significant effect on the functional properties of this protein. Collagen of higher vertebrates is usually more cross-linked and has a higher denaturation temperature, which makes vertebrate collagen less elastic and well toned. The collagen isolated from fish was found to be significantly less cross-linked than the bovine or porcine collagens, and thus its solubility is much higher than others (Senaratne et al., 2006). Chicken collagen II is similar to human collagen II and is extracted from the breastbone of chickens for the use in nutritional supplements (Rister, 2003). However, clinical trials have found that only the very lowest doses of chicken collagen have a beneficial effect. Obtaining collagen from porcine origin requires a very offensive chemical or enzymatic processing, resulting in significant degradation of this protein. Although there are many studies about collagen in marine vertebrates and invertebrates, the main sources of industrial collagen are limited to those from bovine and pig skins (Nagai et al., 1999, 2000, 2001, 2002; Nagai and Suzuki, 2000a, 2000b, 2000c, 2002a, 2002b).

Moreover, the bovine protein may be a carrier of prions (potentially infectious proteins), commonly observed in the majority of the higher animals (mammals, reptiles, and amphibians), and harmful only when their natural conformation is transformed into infectious particles (Hartwig, 2003). The infectious prions cause lethal nervous system diseases in both animals and humans, e.g. BSE and TSE as described above. In addition, collagens from these sources show certain immunogenic reactions from allergies to myopathies and encephalopathies (Lynn et al., 2004). Collagen derived from sharks can also be another source to fulfill collagen needs of human beings (Nomura, 2004). Marine-derived collagen has been found to be ~60% purer than bovine collagen, and marine organisms like sponges offer a very attractive alternative to existing vertebrate collagens. Marine collagens are naturally more expensive than the cheaper alternative, but probably worth the extra expense. Because of the positive aspects of the marine collagens over the other sources, few institutes (e.g., KliniPharm GmbH, Frankfurt, Germany) have been dedicated to marine sponge collagen research and development and marketing of natural products for therapy in medicine, cosmetics, and nutrition (Swatschek et al., 2002; Nicklas et al., 2009).

Major advances in surgery often come from a cooperation of fundamental and clinical research. To solve the problem of a significant shortage of tissues and organs suitable for clinical transplantation (Koshland, 1990), tissue engineering was developed for providing alternative therapies that can maintain, restore, or improve lost tissue functions by developing biological artificial tissue substitutes (Langer and Vacanti, 1993). Tissue engineering is a relatively new and emerging interdisciplinary field that applies the knowledge of bioengineering, life science, and the clinical sciences for creating new functional substitutes for damaged host tissues (Nerem, 1992; Patrik et al., 1999). Thus, it has been proposed as a therapeutic approach to treat patients suffering from the loss or failure of organs and tissues.

With the approach of tissue engineering, many human tissues were developed by using various kinds of biomaterials (Mooney and Mikos, 1999), which includes the replacement of skin, soft tissue (Hollander et al., 2001), hard tissues (Hutmacher et al., 2001), blood vessel, and heart valve (Sodian et al., 2000; Watanabe et al., 2001). As for the ideal biomaterials that serve as a temporal scaffold in the body, they should be biocompatible, biodegradable, highly porous with a large surface-to-volume ratio, mechanically strong, and capable of being formed into desired shapes, which play a positive role in manipulating host cell functions by secreting their own extracellular matrix proteins to gradually form normal, completely natural tissue. Collagen as a natural extracellular matrix component meets all the demands above. Therefore, it is not surprising that collagen is the most commonly used biomaterial in skin, connective tissue, peripheral nerve, and blood vessel tissue engineering (Doillon et al., 1994b; Friess, 1998; Sheridan and Tompkins, 1999).

After synthetic biomaterials are implanted, a critical question is how to supply the cells that grow into the biomaterials with sufficient oxygen and nutrients to sustain their survival and proliferation, and allow for the integration of the developing tissue with the surrounding tissue. A rapid and high level of vascularization in transplanted matrices is essential in tissue engineering approaches to meet this challenge.

7.5.2 Plastic Surgery

The aim of plastic surgery is the restoration of form and function with resultant improvement in the patient's quality of life and aesthetic outcome. Accordingly, two main tasks of plastic surgery are:

1. To improve the quality of wound healing by repairing the area of a tissue deficit
2. To reduce as much as possible the destruction in the donor area

Wound healing is fundamental to surgery. As to plastic surgery, one of the main objectives is to reduce the scar formation of healing, which will influence both functional restoration and the aesthetic outcome. At the same time, restoration with less or no destruction still challenges plastic surgery. Projection indicates that the gap between tissue supply and demand will continue to widen in the future. Naturally, the question arises whether a tissue substitute may be developed that can not only replace the deficit tissues, but also improve the wound healing simultaneously, to meet the above demands. Fortunately, the rapid development of modern science and the new findings in biological materials and factors involved with cell development gave birth to the interdisciplinary field of research termed tissue engineering.

7.5.3 Tissue Engineering

Wound care has become one of the first fields to see the benefit of new technology tissue engineering (TE). Tissue engineering involves the *in vitro* or *in vivo* generation of organoids such as cartilage, skin, or nerves. More ambitious projects seek to ameliorate the quality of life of diseased or injured patients and reduce the economic burden of treatment. Bioartificial organs involve an *in vitro*-prepared tissue-material interface fabricated into a durable device.

Tissue engineering is an interdisciplinary field that involves cell biology, materials science, reactor engineering, and clinical research with the goal of creating new tissues and organs. The term *tissue engineering* was initially defined by the attendees of the first National Scientific Foundation (NSF)-sponsored meeting in 1988 as "application of the principles and methods of engineering and life sciences toward fundamental understanding of structure function relationship in normal and pathological mammalian tissues and the development of biological substitutes for the repair or regeneration of tissue or organ function" (Skalak and Fox, 1988). According to Langer and Vacanti (1993), who summarized the early developments in this field, *tissue engineering* is defined as an interdisciplinary field that applies the principles of engineering and life sciences toward the development of biological substitutes that restore, maintain or improve tissue or organ function.

Tissue engineering is a relatively young field that aims at repairing, regenerating, or replacing damaged tissues with cellularized constructs grown in the laboratory (Langer and Vacanti, 1993; Vacanti and Langer, 1999). The arsenal of TE comprises cell biology, needed for finding a suitable cell source and for ensuring conditions for cell growth; biomaterial chemistry, needed for preparing biocompatible support for anchorage-dependent cells; and physiology, needed for maintaining a biomimetic, organ-specific environment *in vitro* (Vunjak-Novakovic, 2003). The promise of solving the problem of transplantable organ shortage is only one of the engines that drive TE research. Several laboratories and biotechnology companies have already developed organ modules that are used for tissue repair and regeneration. Moreover, functional subunits of human organs are viewed as a viable alternative for animal testing of new drugs (Griffith and Naughton, 2002).

Tissue engineering is an emerging interdisciplinary field that applies the principles of biology and engineering to the development of viable substitutes that restore, maintain, or improve the function of human tissues or organs. This form of therapy differs from standard therapies in that the engineered tissue becomes integrated with the patient's tissue, affording a potentially permanent and specific cure of the disease state. Although cells have been cultured, or grown, outside the body for many years, the possibility of growing complex, three-dimensional tissues, literally replicating the design and function of human tissue, is a recent development. The intricacies of this process require input from many fields, including engineering, hence the term *tissue engineering.* Tissue expansion is a new surgical technique that allows stretching of the skin by approximately a factor of 2, but some complications can arise, including rupture or infection of the expander and necrosis of the expanded skin before transplantation. Skin substitutes that contain cultured cells can provide large quantities of grafts for wound treatment, but can restore only some anatomic structures and physiologic functions of skin. The full potential of skin substitute engineering will not be realized until the complex structure of native skin is ideally reproduced.

Tissue engineering requires a porous, biodegradable scaffold to replicate the natural extracellular matrix (ECM), which serves to organize cells spatially, provide them with environmental signals, and direct site-specific cellular regulation (Wang et al., 2006). Pore size, pore number, and surface area are widely recognized as important parameters for scaffolds used in tissue engineering. Other architectural features such as pore shape and pore wall morphology of the scaffold material have also been suggested as important for cell seeding, migration, growth, and new tissue formation in three dimensions (Chen et al., 2002; Yang et al., 2001). Collagen-based materials are ideal for soft tissue engineering because of their superior biocompatibility (Soiderer et al., 2004).

Using collagen as a biomaterial for providing scaffold used in tissue engineering introduces some disadvantages compared to other materials. These include its low biomechanical stiffness and rapid biodegradation. Further, the high rate of enzymatic degradation of natural collagen *in vivo* makes stabilization of collagen-based biomaterials necessary (Lee et al., 2001a; Ma et al., 2004). Moreover, for tissue engineering, previous studies have demonstrated that high mechanical strength is important for biodegradable polymer scaffolds (Hoerstrup et al., 2000; Shinoka et al., 1995, 1998; Stock et al., 2000) or allogenic acellular matrix scaffolds (Steinhoff et al., 2000). Major disadvantages of biodegradable polymer scaffolds include the requirement of reconstruction of the natural three-dimensional structures and the absence of important ECM proteins in the synthetic polymers (Cheresh et al., 1989; Joshi et al., 1993). Thus acellularized xenogenic tissues that maintain natural three-dimensional structures are a promising alternative to biodegradable polymer scaffolds. Acellularized biological matrices have been successfully used for tissue engineering of blood vessels, heart valves, and urinary bladder (Atala, 2000; Nerem and Seliktar, 2001; Steinhoff et al., 2000).

7.5.4 Tissue Engineering of Human Skin

Improved understanding of biological processes has facilitated the development of new classes of biomaterials, polymers, and diagnostic and analytical reagents. This emerging technology can be used to introduce better-functioning materials that can be processed from naturally occurring or synthetic resources, or a combination of these. Cellular and other biologic components may also be added. These material components are usually called matrices or scaffolds. Their function is to direct the growth of cells migrating from surrounding tissue or of cells seeded within these porous structures. They provide a suitable substrate for cell attachment, proliferation, and differentiated function and cell migration. These matrices/scaffolds can be permanent or biodegradable (temporary). Although there are many biocompatible materials that could potentially be used to construct matrices/scaffolds, a biodegradable material is desirable, since the role of these matrices/scaffolds is usually temporary. Many natural or synthetic biodegradable polymers have been developed (e.g., collagen, hyaluronic acid derivatives, poly alpha-hydroxyesters, and polyanhydrides), and many more are currently under development.

7.5.5 Tissue Engineering and Angiogenesis

Tissues with a high cell density require a vascular network of arteries, veins, and capillaries for the delivery of nutrients to each cell. Thus, the development of efficient methods for enhancing angiogenic effects of biomaterials is critical for a successful outcome. Generally, three approaches may lead to this goal: (1) incorporation of angiogenic factors in biomaterials, (2) seeding of endothelial cells (ECs) along with other cell types in biomaterials (Park et al., 2002), and (3) prevascularization of matrices prior to cell seeding. It should be kept in mind that some fundamental guidelines must be followed (Soker et al., 2000):

1. The support matrix for biomaterials must be compatible with EC growth and capillary formation. It should have a high degree of porosity to allow the penetration of blood vessels into the implant.

2. Angiogenic growth factors should be applied for enhancing positive effects.
3. The combination of EC seeding with insertion of a constitutive source for angiogenic factor secretion could be advantageous.
4. The angiogenic process in biomaterials should be controlled before a functional vascular network is obtained, which is important since it was reported that sustained overproduction of vascular endothelial growth factor (VEGF) may result in deformed, nonfunctional blood vessels (Springer et al., 1998).

7.5.6 Collagen Matrices as Tissue Substitutes

Among biological materials, collagen, particularly type 1 collagen, is a major component of various connective tissues, such as duramater, fascia, intestine, tail tendons, etc. The replacement of human tissues with biomaterials results in the improvement of the wound healing process because of the presence of collagen. Therefore, the application of collagen-derived products as biomaterials has tremendous advantages in biomedicine because of these products' natural structure as a biological support for cells and scaffold for tissue repair or regeneration, their biodegradability that obviates removal of implants, and their biocompatibility. It is not surprising that collagen has been used to design numerous biomaterials in plastic surgery, such as wound dressings, artificial dermis, tissue-engineered devices, and injectable materials (Hutmacher et al., 2001; Sodian et al., 2000; Watanabe et al., 2001; Yannas et al., 1982).

The research on collagen matrices as a scaffold for tissue engineering may be classified into four aspects:

1. Cell seeding experiments: Various types of cells, such as HUVEC cells (Wissink et al., 2000b), human menses cells (Nakatake et al., 2001), hepatocytes (Glicklis et al., 2000), chondrocytes (Glicklis et al., 2000), and fibroblasts (Doillon et al., 1994a), have been seeded into collagen matrices for testing the ability of functioning as a scaffold. Generally these cells grow and proliferate successfully within collagen scaffolds. Furthermore, stem cells, such as mesenchymal stem cells (Hori et al., 2002), have also been applied in order to differentiate into the desired mature cells. Some gene transfected cells, such as bone morphogenetic protein-2-expressing muscle-derived cells (Lee et al., 2001b), have been used in order to express desired proteins or growth factors favorable to tissue regeneration.
2. Incorporating growth factor experiments: The application of growth factors to biomaterials has become an important aspect of tissue engineering because it allows modulation of cellular function and tissue formation at the affiliated site (Lu et al., 2000). As for the biomaterials based on collagen sponge, they have been applied as a sustained released vehicle for various kinds of growth factors, including epidermal growth factor (EGF), basic fibroblast growth factor (bFGF) (Fujisato et al., 1996), and VEGF (Dellian et al., 1996).
3. Experiments focusing on incorporating bioactive molecules: Some bioactive molecules (e.g., heparin, hyaluronic acids, and glycosaminoglycans) may have substantial effects on cell growth and may also stabilize the activity of growth factors. Therefore, they were also incorporated into the collagen sponges and their effects were evaluated *in vitro* and *in vivo* (Park et al., 2002; Pieper et al., 2000; Wissink et al., 2001, 2000a).
4. Experiments directed toward the modification of structural characteristics of the collagen matrices. The internal structure of the matrices may have effects on tissue in growth and response. Some investigations are focusing on modifying the porous structure, for instance, by treating them with alkaline (Rocha et al., 2002).

As a result of the above-mentioned investigations, a number of devices have been developed for commercial products for wound healing or as an artificial skin (Gao et al., 2002).

Based on the information provided in the above, marine sponges are the most suitable substitutes for isolation of collagens and their application in cosmetology.

REFERENCES

Alberts, B., Bray, D., Lewis, J., Raff, M., Roberts, K., and Watson, J. D. (1994). In *The molecular biology of the cell*. Garland Publishers, New York, 971–990.

Atala, A. (2000). Tissue engineering for bladder substitution. *World Journal of Urology*, 18: 364–370.

Bakus, G. J., Shulte, B., Jhu, S., Wright, M., Green, G., and Gomez, P. (1985). Antibiosis and antifouling in marine sponges: laboratory vs. field studies. In *New perspectives in sponge biology*, ed. K. Rutzler. 3rd International Sponge Conference. Smithsonian Institution Press, Washington, DC, 102–108.

Banyard, J., Bao, L., and Zetter, B. R. (2003). Type XXIII collagen, a new transmembrane collagen identified in metastatic tumor cells. *Journal of Biological Chemistry*, 278: 20989–20994.

Beckman, M. J., Shields, K. J., and Diegelmann, R. F. (2004). Collagen. In *Encyclopedia of biomaterials and biomedical engineering*, ed. G. E. Wnek and G. L. Bowlin. Marcel Dekker, New York, 324–335.

Bergquist, P. R. (1978). *Sponges*. University of California Press, Los Angeles, 268.

Bilgen, G., Oktay, G., Tokgoz, Z., Gune, G., and Yalcin, S. (1999). Collagen content and electrophoretic analysis of type I collagen in breast skin of heterozygous naked neck and normally feathered commercial broilers. *Turkish Journal of Veterinary and Animal Sciences*, 23: 483–487.

Birk, D. E. (2001). Type V collagen: heterotypic type I/V collagen interactions in the regulation of fibril assembly. *Micron*, 32: 223–237.

Boot-Handford, R. P., Tuckwell, D. S., Plumb, D. A., Rock, C. F., and Poulsom, R. (2003). A novel and highly conserved collagen [pro(alpha)1(XXVII)] with a unique expression pattern and unusual molecular characteristics establishes a new clade within the vertebrate fibrillar collagen family. *Journal of Biological Chemistry*, 278: 31067–31077.

Boute, N., Exposito, J. Y., Boury-Esnault, N., Vacelet, J., Noro, N., Miyazaki, K., Yoshizato, K., and Garrone, R. (1996) Type IV collagen in sponges, the missing link in basement membrane ubiquity. *Biology of Cell*, 88(1-2): 37–44.

Brusca, R. C., and Brusca, G. J. (2003). *The invertebrates*, 2nd ed. Sinauer Associates, Sunderland, MA, 936.

Burgeson, R. E. (1987). The collagens of skin. *Current Problems in Dermatology*, 17: 61–75.

Cameron, G. J., Alberts, I. L., Laing, J. H., and Wess, T. J. (2002). Structure of type I and type III heterotypic collagen fibrils: an x-ray diffraction study. *Journal of Structural Biology*, 137: 15–22.

Chen, G., Ushida, T., and Tateishi, T. (2002). Scaffold design for tissue engineering. *Macromolecular Bioscience*, 2: 67–77.

Cheresh, D. A., Berliner, S. A., Vicente, V., and Ruggeri, Z. M. (1989). Recognition of distinct adhesive sites on fibrinogen by related integrins on platelets and endothelial cells. *Cell*, 58: 945–953.

Chu, M. L., Myers, J. C., Bernard, M. P., Ding, J. F., and Ramirez, F. (1982). Cloning and characterization of five overlapping cDNAs specific for the human pro alpha 1(I) collagen chain. *Nucleic Acids Research*, 10: 5925–5934.

Chvapil, M. (1979). Industrial uses of collagen. In *Fibrous proteins, scientific, industrial and medical aspects*, Vol. 1, ed. D. A. D. Parry and L. K. Creamer. Academic Press, London, 247–269.

Coiffard, L. J. M. (2004). Contribution of marine products for cosmetology. *Nouvelles Dermatologiques*, 2: 479–481.

Connes, R., Diaz, J. P., and Paris, J. (1972). Seasonal variations and cellular populations of sponge, *Suberites massa* Nardo. Cytological and histological study, 3rd ser. *Bulletin of the Natural History Museum (Paris)*, 84: 1013–1038.

Davies, B. (2006). State of the industry: will megabrands rule? *Euromonitor International*, August 15. Available at http://www.gcimagazine.com/marketstrends/regions/world/3574517.html.

Dellian, M., Witwer, B. P., Salehi, H. A., Yuan, F., and Jain, R. K. (1996). Quantitation and physiological characterization of angiogenic vessels in mice: effect of basic fibroblast growth factor, vascular endothelial growth factor/vascular permeability factor, and host microenvironment. *American Journal of Pathology*, 149: 59–71.

Di Lullo Dagger, G. A., Sweeney, S. M., Korkko, J., Ala-Kokko, L., and Antonio, J. D. S. (2002). Mapping the ligand-binding sites and disease-associated mutations on the most abundant protein in the human, type I collagen. *Journal of Biological Chemistry*, 277(6): 4223–4231.

Diaz, J. P. (1977). Histological and cytological processing post-traumatic stress among demosponge *Suberites massa* Nardo. *Zoology*, 308: 375–396.

Diehl-Seifert, B., Kurelec, B., Zahn, R. K., Dorn, A., Jerecevic, B., Uhlenbruck, G., and Muller, W. E. G. (1985). Attachment of sponge cells to collagen substrata: effect of a collagen assembly factor. *Journal of Cell Science*, 79: 271–285.

Diggs, J. (2008). The cross-linkage theory of ageing. In *Encyclopedia of aging and public health*, ed. S. Loue, S. Martha, and K. Siran. Springer, New York, 250.

Doillon, C. J., Cote, M. F., Pietrucha, K., Laroche, G., and Gaudreault, R. C. (1994a). Porosity and biological properties of polyethylene glycol-conjugated collagen materials. *Journal of Biomaterials Science, Polymer Edition*, 6: 715–728.

Doillon, J., Deblois, C., Cote, M. F., and Fournier, N. (1994b). Bioactive collagen sponge as connective tissue substitute. *Materials Science and Engineering*, C2: 43–49.

Eckart, H. (2006). Skin rejuvenation without a scalpel. I. Fillers. *Journal of Cosmetic Dermatology*, 5: 157–167.

Eckert, C., Schroder, H. C., Brandt, D., Perovic-Ottstadt, S., and Muller, W. E. G. (2006). Histochemical and electron microscopic analysis of spiculogenesis in the demosponge *Suberites domuncula. Journal of Histochemistry and Cytochemistry*, 54: 1031–1040.

Ehrlich, H., Hanke, T., Simon, P., Goebel, C., Heinemann, S., Born, R., and Worch, H. (2005). Demineralization of natural silica based biomaterials: New strategy for the isolation of organic scaffold structures. *BIOmaterialien*, 6: 297–302.

Etherington, D. J., and Sims, T. J. (1981). Detection and estimation of collagen. *Journal of the Science of Food and Agriculture*, 32: 539–546.

Exposito, J. Y., Cluzel, C., Garrone, R., and Lethias, C. (2002). Evolution of collagens. *Anatomical Records*, 268: 302–316.

Eyre, D. R., Paz, M. A., and Gallop, P. M. (1984). Cross-linking in collagen and elastin. *Annual Review of Biochemistry*, 53: 717–748.

Fitzgerald, J., and Bateman, J. F. (2001). A new FACIT of the collagen family: COL21A1. *FEBS Letter*, 505: 275–280.

Flores, J. (1980). Quality control of meat products. *Magazine of Agrochemical and Food Technology*, 20: 180–188.

Friess, W. (1998). Collagen—biomaterial for drug delivery. *European Journal of Pharmaceutics and Biopharmaceutics*, 45: 113–136.

Frincke, J. M., and Faulkner, D. J. (1982). Antimicrobial metabolites of the sponge *Reniera* sp. *Journal of American Chemical Society*, 104: 265–269.

Fujisato, T., Sajiki, T., Liu, Q., and Ikada, Y. (1996). Effect of basic fibroblast growth factor on cartilage regeneration in chondrocyte-seeded collagen sponge scaffold. *Biomaterials*, 17: 155–162.

Gao, C., Wang, D., Yuan, J., and Shen, J. (2002). [Materials and structure design of artificial dermis equivalent based on collagen]. *Sheng Wu Yi Xue Gong Cheng Xue Za Zhi*, 19: 127–131.

Garlick, J. D. (1969). Buried bone. In *Science in archaeology: a survey of progress and research*, ed. D. Brothwell and E. Higgs. Thames and Hudson, Great Britain, 503–512.

Garrone, R. (1978). Phylogenesis of connective tissue In: *Morphological aspects and biosynthesis of sponge intercellular matrix*, ed. Robert, L., *Frontiers of matrix biology*, S. Karger, Basel, Vol. 5, pp 176–179.

Gelse, K., Poschl, E., and Aigner, T. (2003). Collagens—structure, function, and biosynthesis. *Advanced Drug Delivery Reviews*, 55: 1531–1546.

Glicklis, R., Shapiro, L., Agbaria, R., Merchuk, J. C., and Cohen, S. (2000). Hepatocyte behavior within three-dimensional porous alginate scaffolds. *Biotechnology and Bioengineering*, 67: 344–353.

Gordon, M. K., and Hahn, R. A. (2010). Collagens. *Cell and Tissue Research*, 339: 247–257.

Griffith, L. G., and Naughton, G. (2002). Tissue engineering—current challenges and expanding opportunities. *Science*, 295: 1009–1014.

Gross, H., and Konig, G. M. (2006). Terpenoids from marine organisms: unique structures and their pharmacological potential. *Phytochemistry Reviews*, 5: 115–141.

Gross, J., Sokal, Z., and Rougvie, M. (1956). Structural and chemical studies on the connective tissue of marine sponges. *Journal of Histochemistry and Cytochemistry*, 4: 227–246.

Hall, G. (2009). Do collagen supplements really help prevent and treat the effects of aging? http://ezinearticles.com/?Do-Collagen-Supplements-Really-Help-Prevent-And-Treat-The-Effects-Of-Aging?&id=339175.

Hartwig, N. R. (2003). What you should know about BSE. *BEEF Questions and Answers—2004*, 9: 4–5.

Hashimoto, T., Wakabayashi, T., Watanabe, A., Kowa, H., Hosoda, R., Nakamura, A., Kanazawa, I., Arai, T., Takio, K., Mann, D. M., and Iwatsubo, T. (2002). CLAC: a novel Alzheimer amyloid plaque component derived from a transmembrane precursor, CLAC-P/collagen type XXV. *EMBO Journal*, 21: 1524–1534.

Hechtel, G. J. (1976). Zoogeography of Brazilian marine Demospongiae. In *Aspects of sponge biology*, ed. F. W. Harrison and R. R. Cowden. Academic Press, New York, 237–260.

Heinemann, S., Ehrlich, H., Douglas, T., Heinemann, C., Worch, H., Schatton, W., and Hanke, T. (2007). Ultrastructural studies on the collagen of the marine sponge *Chondrosia reniformis* Nardo. *Biomacromolecules*, 8(11): 3452–3457.

Hoerstrup, S. P., Sodian, R., Daebritz, S., Wang, J., Bacha, E. A., Martin, D. P., Moran, A. M., Guleserian, K. J., Sperling, J. S., Kaushal, S., Vacanti, J. P., Schoen, F. J., and Mayer, J. E., Jr. (2000). Functional living trileaflet heart valves grown *in vitro*. *Circulation*, 102: III-44–III-49.

Hollander, D. A., Soranzo, C., Falk, S., and Windolf, J. (2001). Extensive traumatic soft tissue loss: Reconstruction in severely injured patients using cultured hyaluronan based three-dimensional dermal and epidermal autografts. *The Journal of Trauma*, 50: 1125–1136.

Hooper, J. N. A., and van Soest, R. W. M. (2002). *Systema Porifera. A guide to the classification of sponges*, ed. J. N. A. Hooper and R. W. M. van Soest. Kluwer Academic, New York, 1810.

Hopkinson, S. B., Findlay, K., deHart, G. W., and Jones, J. C. (1998). Interaction of BP180 (type XVII collagen) and alpha6 integrin is necessary for stabilization of hemidesmosome structure. *Journal of Investigative Dermatology*, 111: 1015–1022.

Hori, Y., Nakamura, T., Kimura, D., Kaino, K., Kurokawa, Y., Satomi, S., and Shimizu, Y. (2002). Experimental study on tissue engineering of the small intestine by mesenchymal stem cell seeding. *Journal of Surgical Research*, 102: 156–160.

Hutmacher, D. W., Kirsch, A., Ackermann, K. L., and Hurzeler, M. B. (2001). A tissue engineered cell-occlusive device for hard tissue regeneration—a preliminary report. *International Journal of Periodontics and Restorative Dentistry*, 21: 49–59.

Imhoff, J. M., and Garrone, R. (1983). Solubilization and characterization of Chondrosia reniformis sponge collagen. *Connective Tissue Research*, 11(2-3): 193–197.

Jobsis, G. J., Boers, J. M., Barth, P. G., and de Visser, M. (1999). Bethlam myopathy: a slowly progressive congenital muscular dystrophy with contractures. *Brain*, 122: 649–655.

Joshi, P., Chung, C., Aukhil, I., and Erickson, H. (1993). Endothelial cells adhere to the RGD domain and the fibrinogen-like terminal knob of tenascin. *Journal of Cell Science*, 106: 389–400.

Junqua, S., Robert, L., Garrone, R., de Ceccatty, M. P., and Vacelet, J. (1974). Biochemical and morphological studies on collagens of horny sponges. *Ircinia* filaments compared to spongines. *Connective Tissue Research*, 2(3): 193–203.

Kadler, K. (1994). Extracellular matrix 1: fibril forming collagens. *Protein Profile*, 1: 519–638.

Kapoor, R., Bornstein, P., and Sage, E. H. (1986). Type VIII collagen from bovine Descemet's membrane: structural characterization of a triple-helical domain. *Biochemistry*, 25: 3930–3937.

Kapoor, R., Sakai, L. Y., Funk, S., Roux, E., Bornstein, P., and Sage, E. H. (1988). Type VIII collagen has a restricted distribution in specialized extracellular matrices. *Journal of Cell Biology*, 107: 721–730.

Kassner, A., Hansen, U., Miosge, N., Reinhardt, D. P., Aigner, T., Bruckner-Tuderman, L., Bruckner, P., and Grassel, S. (2003). Discrete integration of collagen XVI into tissue-specific collagen fibrils or beaded microfibrils. *Matrix Biology*, 22: 131–143.

Keene, D. R., Lunstrum, G. P., Morris, N. P., Stoddard, D. W., and Burgeson, R. E. (1991). Two type XII-like collagens localize to the surface of banded collagen fibrils. *Journal of Cell Biology*, 113: 971–978.

Keene, D. R., Sakai, L. Y., Lunstrum, G. P., Morris, N. P., and Burgeson, R. E. (1987). Type VII collagen forms an extended network of anchoring fibrils. *Journal of Cell Biolology*, 104: 611–621.

Kielty, C. M., Hopkinson, I., and Grant, M. E. (1993). Collagen: the collagen family: structure, assembly and organization in the extracellular matrix. In *Connective tissue and its heritable disorders. Molecular, genetics, and medical aspects*, ed. P. M. Royce and B. Steinmann. Wiley-Liss, New York, 103–147.

Kielty, C. M., Whittaker, S. P., Grant, M. E., and Shuttleworth, C. A. (1992). Type VI collagen microfibrils: evidence for a structural association with hyaluronan. *Journal of Cell Biology*, 118: 979–990.

Kim, S. E., Sung, Y. H., Shin, M. S., Kim, C. J., Park, J. H., Lee, B. J., Yi, J. W., and Han, S. Y. (2009). Preventive effect of pentoxifylline on cyclosporine A-induced collagen synthesis in calf pulmonary artery endothelial cells. *Journal of Korean Surgical Society*, 76: 135–143.

Kim, S. K., Ravichandran, Y. D., Khan, S. B., and Kim, Y. T. (2008). Prospective of the cosmeceuticals derived from marine organisms. *Biotechnology and Bioprocess Engineering*, 13: 511–523.

Koch, M., Foley, J. E., Hahn, R., Zhou, P., Burgeson, R. E., Gerecke, D. R., and Gordon, M. K. (2001). Alpha 1(Xx) collagen, a new member of the collagen subfamily, fibril-associated collagens with interrupted triple helices. *Journal of Biological Chemistry*, 276: 23120–23126.

Koch, M., Laub, F., Zhou, P., Hahn, R. A., Tanaka, S., Burgeson, R. E., Gerecke, D. R, Ramirez, F., and Gordon, M. K. (2003). Collagen XXIV, a vertebrate fibrillar collagen with structural features of invertebrate collagens: selective expression in developing cornea and bone. *Journal of Biological Chemistry*, 278: 43236–43244.

Koch, M., Schulze, J., Hansen, U., Ashwodt, T., Keene, D. R., Brunken, W. J., Burgeson, R. E., Bruckner, P., and Bruckner-Tuderman, L. (2004). A novel marker of tissue junctions, collagen XXII. *Journal of Biological Chemistry*, 279: 22514–22521.

Koltun, V. M. (1970). *Sponges of the Arctic and Antarctic: a faunistic review*. Symposia of the Zoological Society of London 25. Academic Press, London, 285–297.

Kosher, R. A., and Church, R. L. (1975). Stimulation of *in vitro* somite chondrogenesis by procollagen and collagen. *Nature*, 258: 327–330.

Koshland, D. E., Jr. (1990). Society, cure thyself. *Science*, 248: 9–16.

Kuhn, K. (1987). The classical collagens: types I, II and III. In *Structure and function of collagen types*, ed. R. Mayne and R. E. Burgeson. Academic Press, Orlando, FL, 1–42.

Kuhn, K., Glanville, R. W., Fessler, J. H., Fessler, L. I., Timpl, R., Engel, J., Burgeson, R. E., Sage, H., Bornstein, P., van der Rest, M., Mayne, R., Schmid, T. M., Linsenmayer, T. F., Eyre, D., and Wu, J. J. (1987). Structure and function of collagen types. Academic Press, San Diego, CA.

Kwan, A. P., Cummings, C. E., Chapman, J. A., and Grant, M. E. (1991). Macromolecular organization of chicken type X collagen *in vitro*. *Journal of Cell Biology*, 114: 597–604.

Lampe, A. K., and Bushby, K. M. (2005). Collagen VI related muscle disorders. *Journal of Medicinal Genetics*, 42: 673–685.

Langer, R., and Vacanti, J. P. (1993). Tissue engineering science. *Science*, 260: 920–926.

Latvanlehto, A., Snellman, A., Tu, H., and Pihlajaniemi, T. (2003). Type XIII collagen and some other transmembrane collagens contain two separate coiled-coil motifs, which may function as independent oligomerization domains. *Journal of Biological Chemistry*, 278: 37590–37599.

Lazovic, G., Colic, M., Grubor, M., and Jovanovic, M. (2005). The application of collagen sheet in open wound healing. *Annals of Burns and Fire Disasters*, XVIII: 151–156.

Lee, C. H., Singla, A., and Lee, Y. (2001a). Biomedical applications of collagen. *International Journal of Pharmaceutics*, 221: 1–22.

Lee, J. Y., Musgrave, D., Pelinkovic, D., Fukushima, K., Cummins, J., Usas, A., Robbins, P., Fu, F. H., and Huard, J. (2001b). Effect of bone morphogenetic protein-2-expressing muscle-derived cells on healing of critical-sized bone defects in mice. *Journal of Bone and Joint Surgery, American*, 83: 1032–1039.

Lehninger, A. L. (1987). *Biochemie*. VCH Verlagsgesellschaft mbH, Weinheim, Germany, 107–108.

Lu, L., Stamatas, G. N., and Mikos, A. G. (2000). Controlled release of transforming growth factor beta1 from biodegradable polymer microparticles. *Journal of Biomedical Materials Research*, 50: 440–451.

Lynn, A. K., Yannas, I. V., and Bonfield, W. J. (2004). Antigenicity and immunogenicity of collagen. *Journal of Biomedical Materials Research. Part B: Applied Biomaterials*, 71: 343–354.

Ma, L., Gao, C., Mao, Z., Zhou, J., and Shen, J. (2004). Enhanced biological stability of collagen porous scaffolds by using amino acids as novel cross-linking bridges. *Biomaterials*, 25: 2997–3004.

Mendler, M., Eich-Bender, S. G., Vaughan, L., Winterhalter, K. H., and Bruckner, P. (1989). Cartilage contains mixed fibrils of collagen types II, IX, and XI. *Journal of Cell Biology*, 108: 191–197.

Merle, C., Perret, S., Lacour, T., Jonval, V., Hudaverdian, S., Garrone, R., Ruggiero, F., and Theisen, M. (2002). Hydroxylated human homotrimeric collagen I in *Agrobacterium tumefaciens*-mediated transient expression and in transgenic tobacco plant. *FEBS Letters*, 515: 114–118.

Miller, E. J., and Gay, S. (1987). The collagens: an overview and update. In *Methods in enzymology*, Vol. 144, ed. L. W. Cunningham. Academic Press, Orlando, FL, 3–41.

Miller, E. J., and Matukas, V. J. (1969). Chick cartilage collagen: a new type of alpha 1 chain not present in bone or skin of the species. *Proceedings of National Academy of Sciences USA*, 64: 1264–1268.

Mooney, D. J., and Mikos, A. G. (1999). Growing new organs. *Scientific American*, 280, 60–65.

Moskowitz, R. W. (2000). Role of collagen hydrolyzate in bone and joint disease. *Seminars in Arthritis and Rheumatism*, 30: 87–99.

Myers, J. C., Chu, M. L., Faro, S. H., Clark, W. J., Prockop, D. J., and Ramirez, F. (1981). Cloning a cDNA for the pro-alpha 2 chain of human type I collagen. *Proceedings of National Academy of Sciences USA*, 78: 3516–3520.

Myers, J. C., Dion, A. S., Abraham, V., and Amenta, P. S. (1996). Type XV collagen exhibits a widespread distribution in human tissues but a distinct localization in basement membrane zones. *Cell and Tissue Research*, 286: 493–505.

Myers, J. C., Li, D., Amenta, P. S., Clark, C. C., Nagaswami, C., and Weisel, J. W. (2003). Type XIX collagen purified from human umbilical cord is characterized by multiple sharp kinks delineating collagenous subdomains and by intermolecular aggregates via globular, disulfide-linked, and heparin-binding amino termini. *Journal of Biological Chemistry*, 278: 32047–32057.

Nagai, N., Hosokawa, M., Itohara, S., Adachi, E., Matsushita, T., Hosokawa, N., and Nagata, K. (2000). Embryonic lethality of molecular chaperone hsp_{47} knockout mice is associated with defects in collagen biosynthesis. *Journal of Cell Biology*, 150: 1499–1506.

Nagai, T., Nagamori, K., Yamashita, E., and Suzuki, N. (2002). Collagen of octopus *Callistoctopus arakawai* arm. *International Journal of Food Science and Technology*, 37: 285–289.

Nagai, T., Ogawa, T., Nakamura, T., Ito, T., Nakagawa, H., Fujiki, K., Nakao, M., and Yano, T. (1999). Collagen of edible jellyfish exumbrella. *Journal of the Science of Food and Agriculture*, 79: 855–858.

Nagai, T., and Suzuki, N. (2000a). Isolation of collagen from fish waste material-skin, bone and fins. *Food Chemistry*, 68: 277–281.

Nagai, T., and Suzuki, N. (2000b). Partial characterization of collagen from purple sea urchin (*Anthocidaris crassispina*) test. *International Journal of Food Science and Technology*, 35: 497–501.

Nagai, T., and Suzuki, N. (2000c). Preparation and characterization of several fish bone collagens. *Journal of Food Biochemistry*, 24: 427–436.

Nagai, T., and Suzuki, N. (2002a) Collagen of the skin of ocellate puffer fish (*Takifugu rubripes*). *Food Chemistry*, 78: 173–177.

Nagai, T., and Suzuki, N. (2002b). Preparation and partial characterization of collagen from paper nautilus (*Argonauta argo*, Linnaeus) outer skin. *Food Chemistry*, 76: 149–153.

Nagai, T., Yamashita, E., Taniguchi, K., Kanamori, N., and Suzuki, N. (2001). Isolation and characterization of collagen from the outer skin waste material of cuttlefish (*Sepia lycidas*). *Food Chemistry*, 72: 425–429.

Nakatake, Y., Hoshikawa, M., Asaki, T., Kassai, Y., and Itoh, N. (2001). Identification of a novel fibroblast growth factor, FGF-22, preferentially expressed in the inner root sheath of the hair follicle. *Biochimica et Biophysica Acta*, 1517: 460–463.

Nerem, R. M. (1992). Tissue engineering in the USA. *Medical and Biological Engineering and Computing*, 30: 8–12.

Nerem, R. M., and Seliktar, D. (2001). Vascular tissue engineering. *Annual Review of Biomedical Engineering*, 3: 225–243.

Nicklas, M., Schatton, W., Heinemann, S., Hanke, T., and Kreuter, J. (2009). Enteric coating derived from marine sponge collagen. *Drug Development and Industrial Pharmacy*, 35: 1384–1388.

Nimni, M. E., Cheung, D. T., Strates, B., Kodama, M., and Sheikh, K. (1988). Bioprosthesis derived from crosslinked and chemically modified collagenous tissues. In *Collagen*, Vol. 3, ed. M. E. Nimni. CRC Press, Boca Raton, FL, 1–39.

Nishiyama, T., McDonough, A. M., Bruns, R. R., and Burgeson, R. E. (1994). Type XII and XIV collagens mediate interactions between banded collagen fibers *in vitro* and may modulate extracellular matrix deformability. *Journal of Biological Chemistry*, 26: 28193–28199.

Nomura, Y. (2004). Properties and utilization of shark collagen. In *More efficient utilization of fish and fisheries products, developments in food science*, ed. M. Sakaguchi. Elsevier, Amsterdam, 147.

Nyman, S., Gottlow, J., Karring, T., and Lindhe, J. (1982). The regenerative potential of the periodontal ligament. An experimental study in the monkey. *Journal of Clinical Periodontology*, 9: 257–265.

Olsen, B. R. (1997). Collagen IX. *International Journal of Biochemistry and Cell Biology*, 29, 555–558.

Pace, J. M., Corrado, M., Missero, C., and Byers, P. H. (2003). Identification, characterization and expression analysis of a new fibrillar collagen gene, *COL27A1*. *Matrix Biology*, 22: 3–14.

Pallela, R., Bojja, S., and Janapala, V. R. (2011). Biochemical and biophysical characterization of collagens of marine sponge, *Ircinia fusca* (Porifera: Demospongiae:Irciniidae). *International Journal of Biological Macromolecules*, 49: 85–92.

Park, S. N., Park, J. C., Kim, H. O., Song, M. J., and Suh, H. (2002). Characterization of porous collagen/hyaluronic acid scaffold modified by 1-ethyl-3-(3-dimethylaminopropyl) carbodiimide cross-linking. *Biomaterials*, 23: 1205–1212.

Paterson, C. R., Burns, J., and McAllion, S. J. (1993). Osteogenesis imperfecta: the distinction from child abuse and the recognition of a variant form. *American Journal of Medicinal Genetics*, 45: 187–192.

Patrik, C. W. J., Mikos, A. G., and McIntire, L. W. (1999). Prospects of tissue engineering. *Frontiers in Tissue Engineering*, 3–11.

Pieper, J. S., Hafmans, T., Veerkamp, J. H., and van Kuppevelt, T. H. (2000). Development of tailor-made collagen-glycosaminoglycan matrices: EDC/NHS crosslinking, and ultrastructural aspects. *Biomaterials*, 21: 581–593.

Prockop, D. J. (1998). What holds us together? Why do some of us fall apart? What can we do about it? *Matrix Biology*, 16: 519–528.

Prockop, D. J., and Kivirikko, K. I. (1995). Collagens: molecular biology, diseases, and potentials for therapy. *Annual Review of Biochemistry*, 64: 403–434.

Ramchandran, R., Dhanabal, M., Volk, R., Waterman, M. J. F., Segal, M., Lu, H., Knebelmann, B., and Sukhatme, V. P. (1999). Antiangiogenic activity of restin, NC10 domain of human collagen XV: comparison to endostatin. *Biochemistry Biophysical Research Communication*, 255: 735–739.

Rao, K. P. (1995). Recent developments of collagen-based materials for medical applications and drug delivery systems. *Journal of Biomaterial Science Polymers Edition*, 7: 623–645.

Rister, R. S. (2003). Chicken collagen. In *Healing without medication, a comprehensive guide to the complementary techniques anyone can use to achieve real healing*. Part II. *Healing tools*. Basic Health Publications, New York, 561.

Rocha, L. B., Goissis, G., and Rossi, M. A. (2002). Biocompatibility of anionic collagen matrix as scaffold for bone healing. *Biomaterials*, 23: 449–456.

Rossler, B., Kreuter, J., and Scherer, D. (1995). Collagen microparticles: preparation and properties. *Journal of Microencapsulation*, 12: 49–57.

Ruggiero, F., Exposito, J. Y., Bournat, P., Gruber, V., Perret, S., Comte, J., Olagnier, B., Garrone, R., and Theisen, M. (2000). Triple helix assembly and processing of human collagen produced in transgenic tobacco plants. *FEBS Letters*, 469: 132–136.

Sakemi, S., Ichiba, T., Kohmoto, S., Saucy, G., and Higa, T. (1988). Isolation and structure elucidation of onnamide A, a new bioactive metabolite of a marine sponge, *Theonella* sp. *Journal of the American Chemical Society*, 110: 4851–4853.

Sasaki, T., Fukai, N., Mann, K., Gohring, W., Olsen, B. R., and Timpl, R. (1998). Structure, function and tissue forms of the C-terminal globular domain of collagen XVIII containing the angiogenesis inhibitor endostatin. *The EMBO Journal*, 17: 4249–4256.

Sato, K., Yomogida, K., Wada, T., Yorihuzi, T., Nishimune, Y., Hosokawa, N., and Nagata, K. (2002). Type XXVI collagen, a new member of the collagen family, is specifically expressed in the testis and ovary. *Journal of Biological Chemistry*, 277: 37678–37684.

Schonau, E., and Rauch, F. (1997). Markers of bone and collagen metabolism. Problems and perspectives in pediatrics. *Hormone Research*, 48: 50–59.

Senaratne, L. S., Park, P. J., and Kim, S. K. (2006). Isolation and characterization of collagen from brown backed toadfish (*Lagocephalus gloveri*) skin. *Bioresource Technology*, 97: 191–197.

Sheridan, R. L., and Tompkins, R. G. (1999). Skin substitutes in burns. *Burns*, 25: 97–103.

Shimokomaki, M., Wright, D. W., Irwin, M. H., van der Rest, M., and Mayne, R. (1990). The structure and macromolecular organization of type IX collagen in cartilage. *Annals of the New York Academy of Science*, 580: 1–7.

Shinoka, T., Breuer, C. K., Tanel, R. E., Zund, G., Miura, T., Ma, P. X., Langer, R., Vacanti, J. P., and Mayer, J. J. E. (1995). Tissue engineering heart valves: valve leaflet replacement study in a lamb model. *Annals of Thoracic Surgery*, 60: 513–516.

Shinoka, T., Shum-Tim, D., Ma, P. X., Tanel, R. E., Isogai, N., Langer, R., Vacanti, J. P., and Mayer, J. E., Jr. (1998). Creation of viable pulmonary artery autografts through tissue engineering. *Thoracic and Cardiovascular Surgery*, 115: 536–546.

Sipkema, D., Osigna, R., Schatton, W., Mendola, D., Tramper, J., and Wijffels, R. (2005). Large-scale production of pharmaceuticals by marine sponges: sea, cell, or synthesis? *Biotechnology and Bioengineering*, 90: 201–222.

Skalak, R., and Fox, C. F. (1988). Tissue engineering: proceedings of a workshop held at Granlibakken, Lake Tahoe, California. Liss, 1988 UCLA symposia on molecular and cellular biology, new ser., New York, 107.

Smith, E. L., Hill, R. L., Lehman, I. R., Lefkowitz, R. J., Handler, P., and White, A. (1983). Mammalian biochemistry. In *Principles of biochemistry*, 7th ed. McGraw-Hill, New York, 498–513.

Söderhäll, C., Marenholz, I., Kerscher, T., Rüschendorf, T., Esparza-Gordillo, J., Worm, M., Gruber, C., Mayr, G., Albrecht, M., Rohde, K., Schulz, H., Wahn, U., Hubner, N., and Lee, Y. (2007). Variants in a novel epidermal collagen gene (*COL29A1*) are associated with atopic dermatitis. *PLoS Biology*, 5: 1952–1961.

Sodian, R., Sperling, J. S., Martin, D. P., Egozy, A., Stock, U., Mayer, J. E., Jr., and Vacanti, J. P. (2000). Fabrication of a trileaflet heart valve scaffold from a polyhydroxyalkanoate biopolyester for use in tissue engineering. *Tissue Engineering*, 6: 183–188.

Soiderer, E. E., Lantz, G. C., Kazacos, E. A., Hodde, J. P., and Wiegand, R. E. (2004). Morphologic study of three collagen materials for body wall repair. *Journal of Surgical Research*, 118: 161–175.

Soker, S., Machado, M., and Atala, A. (2000). Systems for therapeutic angiogenesis in tissue engineering. *World Journal of Urology*, 18: 10–18.

Song, E., Kim, S. Y., Chun, T., Byun, H. J., and Lee, Y. M. (2006). Collagen scaffolds derived from a marine source and their biocompatibility. *Biomaterials*, 27: 2951–2961.

Springer, M. L., Chen, A. S., Kraft, P. E., Bednarski, M., and Blau, H. M. (1998). VEGF gene delivery to muscle: potential role for vasculogenesis in adults. *Molecular Cell*, 2: 549–558.

Steinhoff, G., Stock, U., Karim, N., Mertsching, H., Timke, A., Meliss, R. R., Pethig, K., Haverich, A., and Bader, A. (2000). Tissue engineering of pulmonary heart valves on allogenic acellular matrix conduits: *in vivo* restoration of valve tissue. *Circulation*, 102: 50–55.

Stephan, S., Sherratt, M. J., Hodson, N., Shuttleworth, C. A., and Kielty, C. M. (2004). Expression and supramolecular assembly of recombinant alpha1(viii) and alpha2(viii) collagen homotrimers. *Journal of Biological Chemistry*, 279: 21469–21477.

Stock, U. A., Nagashima, M., Khalil, P. N., Nollert, G. D., Herden, T., Sperling, J. S., Moran, A., Lien, J., Martin, D. P., Schoen, F. J., Vacanti, J. P., and Mayer, J. E., Jr. (2000). Tissue-engineered valved conduits in the pulmonary circulation. *Journal of Thorac Cardiovascular Surgery*, 119: 732–740.

Swanson, J. C. (2002). *Bovine spongiform encephalopathy*, 2nd ed. Animal Welfare Information Center, National Agricultural Library, Beltsville, MD.

Swatschek, D., Schatton, W., Kellerman, J., Muller, W. E. G., and Kreuter, J. (2002). Marine sponge collagen: isolation, characterization and effects on the skin parameters surface pH, moisture and sebum. *European Journal of Pharmaceutics and Biopharmaceutics*, 53: 107–113.

Talbott, S. M. (2003). Supplements for joint health. In *A guide to understanding dietary supplements*. Haworth Press, New York, 259.

Taylor, D. M. (1989). Scrapie agent decontamination: implications for bovine spongiform encephalopathy. *Veterinary Record*, 124: 291–292.

Timpl, R., and Brown, J. C. (1996). Supramolecular assembly of basement membranes. *BioEssays* 18: 123–132.

Timpl, R., Wiedemann, H., van Delden, V., Furthmayr, H., and Kuhn, K. (1981). A network model for the organization of type IV collagen molecules in basement membranes. *European Journal of Biochemistry*, 120: 203–211.

Tong, Y., Xu, Y., Scearce-Levie, K., Ptacek, L. J., and Fu, Y. H. (2010). *COL25A1* triggers and promotes Alzheimer's disease-like pathology *in vivo*. *Neurogenetics* 11: 41–52.

Tracey, H. (2004). Dermal fillers: the next generation. *Plastic Surgical Nursing*, 24: 14–19.

Vacanti, J. P., and Langer, R. (1999). Tissue engineering: the design and fabrication of living replacement devices for surgical reconstruction and transplantation. *Lancet*, 354: SI32–SI34.

van der Rest, M., and Garrone, R. (1991). Collagen family of proteins. *The Federation of American Societies for Experimental Biology (FASEB) Journal*, 5: 2814–2823.

Vazquez-Ortiz, F. A., and Gonzalez-Mendez, N. F. (1996). Determination of collagen as a quality index in bologna from northwestern Mexico. *Journal of Food Composition and Analysis* 9: 269–276.

Veit, G., Kobbe, B., Keene, D. R., Paulsson, M., Koch, M., and Wagener, R. (2006). Collagen XXVIII, a novel von Willebrand factor A domain-containing protein with many imperfections in the collagenous domain. *Journal of Biological Chemistry*, 281: 3494–3504.

Venkateswara Rao, J., Desaiah, D., Vig, P. J. S., and Venkateswarlu, Y. (1998). Marine biomolecules inhibit rat brain nitric oxide synthase. *Toxicology*, 129: 103–110.

Vunjak-Novakovic, G. (2003). The fundamentals of tissue engineering: scaffolds and bioreactors. In *Tissue engineering of cartilage and bone. Novartis Foundation Symposia*. Novartis Foundation. John Wiley & Sons, New York, 34–51.

Wang J. H., Hung, C. H., and Young, T. H. (2006). Proliferation and differentiation of neural stem cells on lysine-alanine sequential polymer substrates. *Biomaterials*, 27: 3441–3450.

Watanabe, M., Shin'oka, T., Tohyama, S., Hibino, N., Konuma, T., Matsumura, G., Kosaka, Y., Ishida, T., Imai, Y., Yamakawa, M., Ikada, Y., and Morita, S. (2001). Tissue-engineered vascular autograft: inferior vena cava replacement in a dog model. *Tissue Engineering*, 7: 429–439.

Waterlow, J. C., Garlick, P. J., and Millward, D. J. (1978). *Protein turnover in mammalian tissues and in the whole body*. North-Holland Publishing Company, Amsterdam, 117–176.

WHO. (2002a). Fact sheet on bovine spongiform encephalopathy, no. 113. http://www.who.int/mediacentre/factsheets/fs113/en/.

WHO. (2002b). Fact sheet on variant Creutzfeldt-Jakob disease, no. 180. http://www.who.int/mediacentre/factsheets/fs180/en/.

Wiens, M., Koziol, C., Batel, R., and Muller, W. E. G. (1999). Prolidase in the marine sponge *Suberites domuncula*: enzyme activity, molecular cloning, and phylogenetic relationship. *Marine Biotechnology* 1: 191–199.

Willmer, P. G. (1994). *Invertebrate relationships*. Cambridge University Press, Cambridge, UK.

Wissink, M. J., Beernink, R., Pieper, J. S., Poot, A. A., Engbers, G. H., Beugeling, T., van Aken, W. G., and Feijen, J. (2001). Binding and release of basic fibroblast growth factor from heparinized collagen matrices. *Biomaterials*, 22: 2291–2299.

Wissink, M. J., Beernink, R., Scharenborg, N. M., Poot, A. A., Engbers, G. H., Beugeling, T., van Aken, W. G., and Feijen, J. (2000a). Endothelial cell seeding of (heparinized) collagen matrices: effects of bFGF pre-loading on proliferation (after low density seeding) and pro-coagulant factors. *Journal of Controlled Release*, 67: 141–155.

Wissink, M. J., van Luyn, M. J., Beernink, R., Dijk, F., Poot, A. A., Engbers, G. H., Beugeling, T., van Aken, W. G., and Feijen, J. (2000b). Endothelial cell seeding on cross-linked collagen: effects of cross-linking on endothelial cell proliferation and functional parameters. *Journal of Thrombosis and Haemostasis*, 84: 325–331.

Woodhead-Galloway, J. (1980). *Collagen: the anatomy of a protein*. Edward Arnold, London, 20–21.

Yang, S., Leong, K. F., Du, Z., and Chua C. K. (2001). The design of scaffolds for use in tissue engineering. Part I. Traditional factors. *Tissue Engineering*, 7: 679–689.

Yannas, I. V., Burke, J. F., Orgill, D. P., and Skrabut, E. M. (1982). Wound tissue can utilize a polymeric template to synthesize a functional extension of skin. *Science*, 215: 174–176.

Young, B. B., Gordon, M. K., and Birk, D. E. (2000). Expression of type XIV collagen in developing chicken tendons: association with assembly and growth of collagen fibrils. *Developmental Dynamics* 217: 430–439.

Young, B. B., Zhang, G., Koch, M., and Birk, D. E. (2002). The roles of types XII and XIV collagen in fibrillogenesis and matrix assembly in the developing cornea. *Journal of Cellular Biochemistry*, 87: 208–220.

Zhang, G., Young, B. B., and Birk, D. E. (2003). Differential expression of type XII collagen in developing chicken metatarsal tendons. *Journal of Anatomy*, 202: 411–420.

Zubay, G. L. (1998). In *Biochemistry*, 4th ed., ed. I. O. Dubuque. Wm. C. Brown, 90–92.

8 Skin Photoprotection by Marine Carotenoids

Kazuo Miyashita, Nana Mikami, M. Airanthi K. Widjaja-Adhi, Masayuki Abe, and Masashi Hosokawa

CONTENTS

8.1 INTRODUCTION

Skin is the largest organ of the body and comprises a surface area that protects the internal parts of the body by acting as an effective barrier against the detrimental effects of environmental and xenobiotic agents. Among them, solar ultra violet (UV) radiation is recognized as a major environmental contaminant and a global health concern, presenting acute and chronic risks to the skin (González et al., 2008; Nichols and Katiyar, 2010; Shapira, 2010; Svobodobá and Vostálová, 2010; Verschooten et al., 2006; Wang et al., 2010). The solar UV spectrum can be divided into three segments based on the wavelengths of the radiation: short wave (UVC; 200–290 nm), midwave (UVB; 290–320 nm), and long wave (UVA; 320–400 nm). Nearly 100% of the UVC is blocked by the ozone layer; therefore UVB and UVA are major the sources inducing skin damage.

UV exposure leads to a series of biochemical events. As a primary reaction, light of an appropriate wavelength interacts with suitable chromophores (Krutmann, 2000; Kulms and Schwarz, 2000). The chromophores act as photosensitizers to trigger photosensitized oxidation in skin tissue. The photooxidative processes are initiated in the presence of reactive oxygen species (ROS), mainly singlet oxygen ($^1O_2$) or secondary lipid peroxyl radicals, damaging biological molecules and impairing their function. The cellular damage includes modified gene expression, activation or inactivation of kinase-dependent regulatory pathways, immune and inflammatory events, as well as induction of apoptosis.

Sunburn is the most common type of skin damage. This reddening of the skin is due to the stimulation of inflammatory pathways and is accompanied by tenderness, sometimes painful blistering, and second-degree burns. The modification of immune factors contributes to the pathogenesis of solar UV-induced skin cancer in mice and probably in humans as well (Urbach, 1991; Yoshikawa et al., 1990). UV radiation also plays a major role in the acceleration of photoaging. Clinically, these changes consist of wrinkling, dryness, telangiectasia, sagging, and pigmentation. Histologically, aging skin is reflected in the disorganization of collagen bundles, loss of elastin fibers, flattening of the dermal-epidermal junction, and dilation of blood vessels.

A nutritional approach to protection against UV irradiation damage has been focused on antioxidants such as carotenoids, tocopherols, and polyphenols. Carotenoids are among the most efficient natural scavengers of $^1O_2$ (Cantrell et al., 2003; Stahl and Sies, 1997). Thus, much attention has been paid to antioxidant activities of carotenoids. Carotenoids are accessory light harvesting pigments and play an essential role in the protection of plants against excess light and photooxidative stress (Demmig-Adams and Adams, 2002). Carotenoids have a long central chain of conjugated double bonds. The extended system of conjugated double bonds in the carotenoids' backbone is largely responsible for their color and antioxidant properties (Krinsky, 1998; Stahl and Sies, 1997; Young and Lowe, 2001). In this chapter, we describe skin photoprotection by carotenoids, especially focusing on those of marine origin.

8.2 SKIN DAMAGE BY UV IRRADIATION

UVA comprises the largest spectrum of solar UV radiation (90–95%) and can penetrate deeper into the epidermis and dermis of the skin. Thus, UVA is recognized as the main UV irradiation range to induce significant and permanent skin damage (Bachelor and Bowden, 2004; Bossi et al., 2008; Bruls et al., 1984). Exposure to UVA induces the generation of $^1O_2$ as the main ROS together with other kinds of reactive oxygen species, such as hydroxy radical (HO•), superoxide radical ($O_2^{\bullet-}$), and hydrogen peroxide (H_2O_2). UVA also produces reactive nitrogen species (RNS), such as nitrogen oxide (NO), nitrogen dioxide (NO_2), and peoxinitrite ($OONO^-$) (Wang et al., 2010). ROS and RNS initiate lipid oxidation to produce lipid hydroperoxides, which are easily decomposed to form free radicals. These active molecules cause damage to skin cellular macromolecules, such as proteins, lipids, and DNA (DiGiovanni, 1992).

UVB radiation can also induce both direct and indirect adverse biological effects, including oxidative stress, DNA damage, premature aging of the skin (De Gruijl and van der Leun, 1994; Ichihashi et al., 2003; Mukhtar and Elmets, 1996), and multiple effects on the immune system (Kripke, 1990; Meunier et al., 1998; Parrish, 1983), which together play important roles in the generation and maintenance of UV-induced neoplasms (Hruza and Pentland, 1993; Katiyar and Mukhtar, 2001). UVB is generally accepted to be the most carcinogenic portion of sunlight (Donawho and Kripke, 1991; Katiyar et al., 1997; Kligman et al., 1980; Marrot and Meunier, 2008; Ziegler et al., 1994).

UV exposure enhances blood flow and infiltration of inflammatory blood leukocytes (macrophages and neutrophils) to the skin tissue. UV light also stimulates activity of phospholipase and cyclooxygenase-2 (COX-2) to produce prostaglandin (PG) from arachidonic acid. Increased generation of PG and NO by UV exposure further enhances leukocyte infiltration and lipid peroxidation. Furthermore, ROS produced by UV exposure can affect various cell pathways and gene expression, such as PG synthesis, expression of inflammatory interleukin (IL), tumor necrosis factor-α (TNF-α), nuclear factor-kappaB (NF_B), matrix metalloproteinases (MMPs), mitogen-activated protein kinases (MAPKs), cyclindependent kinases (CDKs), tumor suppressor p53, and pro- and antiapoptotic pathway genes (Svobodobá and Vostálová, 2010). These signaling molecules may result in skin inflammation, immunosuppression, premature skin aging (photoaging), or carcinogenesis.

Production of reactive oxygen species and free radicals is neutralized by the body's innate defense mechanism through a series of enzymatic (e.g., superoxide dismutase, catalase, glutathione reductase, and peroxidase) and nonenzymatic antioxidants. The body also has a repair system for oxidatively damaged DNA. Moreover, endogenous photoprotective ability can also be enhanced by dietary antioxidants. Thus, dietary antioxidants have been paid much attention from this viewpoint.

8.3 CAROTENOIDS AS EFFECTIVE ANTIOXIDANTS IN PHOTOPROTECTION

Carotenoids represent a large group of isoprenoid structures with many different structural characteristics and biological activities. They are the most important pigments that are responsible for various colors of plants and animals. The number of naturally occurring carotenoids reported continues to rise and has now reached more than 700. Animals including humans are incapable of synthesizing carotenoids, and many are colored by carotenoids derived through their diets. More than 40 carotenoids may be available from the diet and absorbed, metabolized, or utilized by the human body.

The best-known biological function of carotenoids is their established role as pro-vitamin A. Carotenoids such as α- and β-carotene and β-cryptoxanthin can be converted to retinoic acid. Retinoic acid in its all-trans or 9-cis configuration is a highly potent activity of the retinoic acid receptors (RARs) and the retinoid-X receptors (RXRs). By activation of these nuclear receptors retinoic acids can influence the transcription of various retinoid response genes (De Luca, 1991). In addition, dietary carotenoids, including non-pro-vitamin A carotenoids, are considered to play a role in the prevention of common chronic diseases such as cardiovascular disease, age-related macular degeneration, and cancers (Cooper et al., 1999a, 1999b).

Oxidative damage to biomolecules has been widely postulated to be involved in the cause and progression of several diseases, including cancer and cardiovascular diseases. Skin damage is also mainly induced by ROS produced after UV irradiation. Carotenoids have been implicated as important dietary nutrients having antioxidant potential, being the most efficient natural scavengers of $^1O_2$ (Cantrell et al., 2003; Stahl and Sies, 1997). $^1O_2$ is the main ROS molecule responsible for UV light-induced skin damage. Thus, the strong ability of carotenoids to quench $^1O_2$ has been suggested as being the main mechanism by which they afford their beneficial effects on skin protection.

The quenching of $^1O_2$ by carotenoids has been attributed mainly to a physical mechanism, where the excess energy of singlet oxygen is transferred to the carotenoid. The carotenoid with added energy is excited to a triplet state and, upon losing the energy as heat relaxes, to a singlet state without change in the structure as follows:

$$^1O_2 + {}^1Car \rightarrow {}^3O_2 + {}^3Car$$

$$^3Car \rightarrow {}^1Car + heat$$

The $^1O_2$ quenching rates of carotenoids are characterized by the rate constant k_q, with the larger the k_q values, the faster the quenching reaction. The efficacy of carotenoids for physical quenching is related to the number of conjugated double bonds present in the molecule, which determines their lowest triplet energy level.

As the number of conjugated double bonds increases, the energies of the excited states decrease, and this is reflected in the dependence of the $^1O_2$ quenching rate constant on the carotenoid chain length. Thus, the ability to quench $^1O_2$ increases with an increasing number of conjugated double bonds (Edge et al., 1997). Conn et al. (1991) showed that the quenching rate constant ($k_q/M^{-1}s^{-1}$) increased with an increasing number of conjugated double bonds of β-carotene and its related compounds (Table 8.1). The same effect of the number of conjugated double bonds has been also found in carotene and its related compound (Edge et al., 1997) (Table 8.2).

Although the efficacy of deactivation of $^1O_2$ by carotenoids basically increased with the number of conjugated double bonds, the quenching ability varied with chain structure, functional groups, and solvent viscosity (Conn et al., 1991; Hirayama et al., 1994; Mascio et al., 1991). Mascio et al. (1991) compared the $^1O_2$ quenching ability of lycopene, astaxanthin, and β-carotene (Figure 8.1). The number of conjugated double bonds of lycopene and β-carotene is 11, while that of astaxanthin is 13 due to the presence of two conjugated keto groups in its molecule. However, the quenching rate

TABLE 8.1
Relationship between Carotene Structure and Singlet Oxygen Quenching Rate Constant

Carotenoid Structure	Number of Conjugated Double Bounds	10^9 kq ($M^{-1}s^{-1}$) In Benzene	10^9 kq ($M^{-1}s^{-1}$) In Toluene
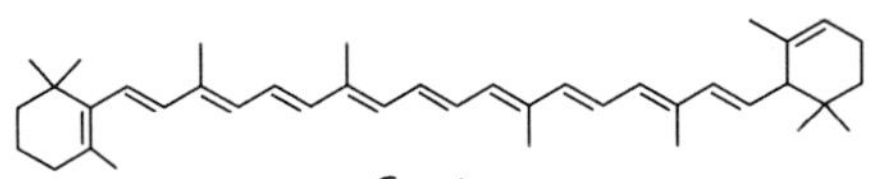 α-Carotene	10	–	8.2
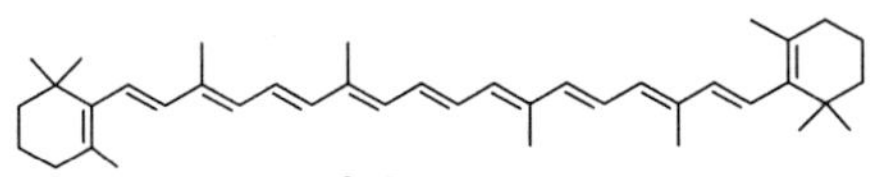 β-Carotene	11	13	14
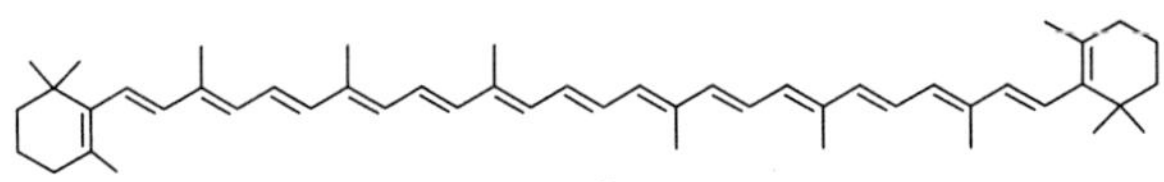 Docapreno-β-Carotene	15	20	21
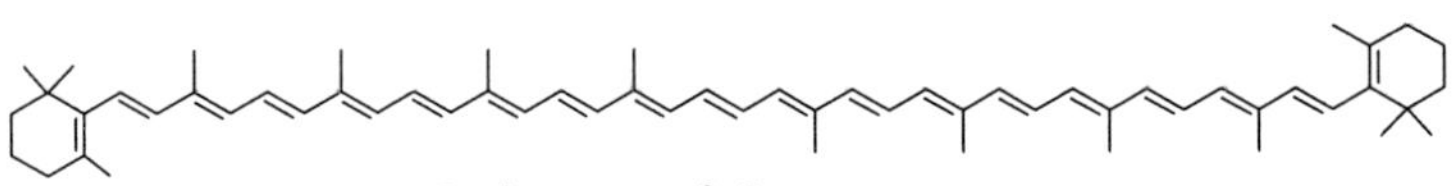 Dodecapreno-β-Carotene	19	23	29

Source: From Conn et al. 1991.

constants ($k_q/M^{-1}s^{-1}$) of these carotenoids in ethanol/chloroform/water were 3.1×10^{10} for lycopene, 2.4×10^{10} for astaxanthin, and 1.4×10^{10} for β-carotene.

Fukuzawa et al. (1998) observed that $^1O_2$ quenching of carotenoids is 40–80 times higher than that of α-tocopherol in ethanol, but only six times higher in liposomes. On the other hand, Cantrell et al. (2003) reported that lycopene and β-carotene showed the fastest $^1O_2$ quenching rate constants ($2.3–2.5 \times 10^{10}/M^{-1}s^{-1}$), with lutein the least efficient ($1.1 \times 10^{10}/M^{-1}s^{-1}$), when the ability of six kinds of dietary carotenoids to quench $^1O_2$ was compared in a model membrane system (unilamellar liposomes). Thus, the $^1O_2$ quenching rates of carotenoids in a biological system will depend on factors such as concentration of carotenoids in membranes, membrane localization of active groups, solubility of generation site of $^1O_2$ in membranes, and the mobility of carotenoids in membranes.

Another role of carotenoids as antioxidants is attributed to the scavenging of free radicals. Free radicals obtain the electron from other molecules or form an adduct with the other molecules. The electron-rich status of carotenoids makes them more suitable for reaction with free radicals, thus avoiding the use of cellular components by free radicals for reactions. Krinsky and Yeum (2003) reviewed the carotenoid-radical interactions and suggested following three possible interactions:

1. Adduct formation:

$$\text{Car} + \text{R}\bullet \rightarrow \text{R-Car}\bullet$$

2. Electron transfer:

$$\text{Car} + \text{R}\bullet \rightarrow \text{Car}^{\bullet +} + \text{R}^-$$

TABLE 8.2
Singlet Oxygen Quenching Rate Constant of α-Carotene and Its Related Compounds

Carotenoid Structure	Number of Conjugated Double Bonds	10^9 kq ($M^{-1}s^{-1}$) In Benzene
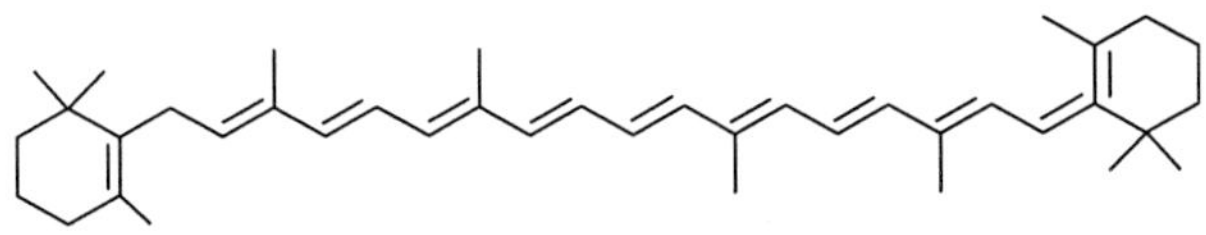 7, 7'-dihydo-β-Carotene	8	0.3
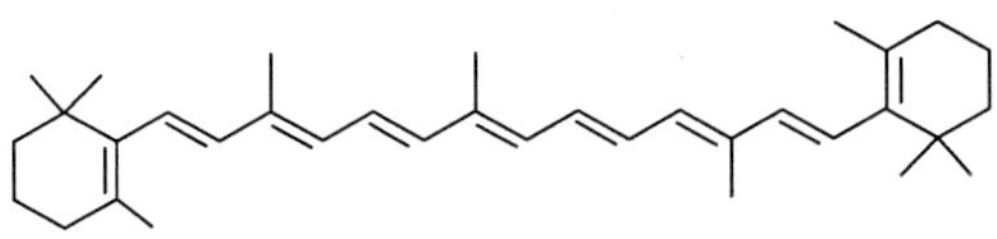 Septapreno-β-Carotene	9	1.38
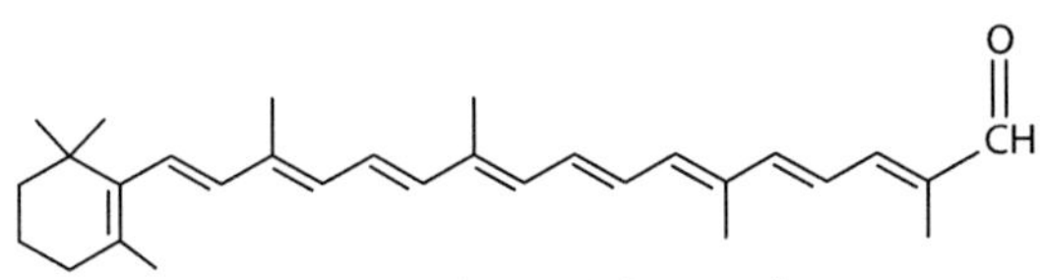 β-apo-8'-Carotenal	10	5.27
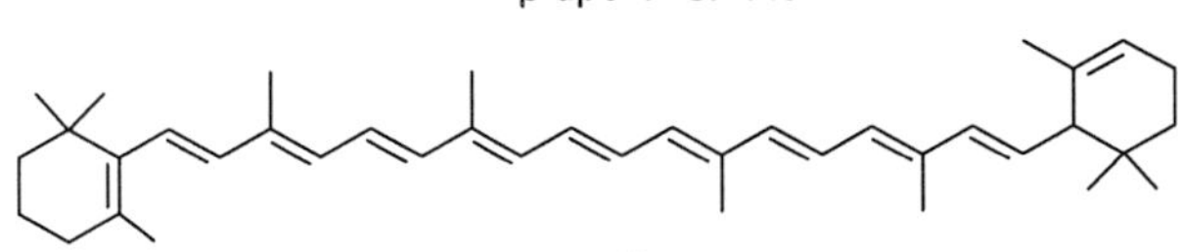 α-Carotene	11	12.0

Source: From Edge et al. 1997.

3. Allylic hydrogen abstraction:

$$\text{Car} + \text{R}\bullet \rightarrow \text{Car}\bullet + \text{RH}$$

In addition, dietary carotenoids react with a wide range of free radicals, such as $CCl_3O_2\bullet$, $RSO_2\bullet$, $NO_2\bullet$, and various aryl-peroxy radicals, via electron transfer, producing the radical cation of the carotenoid.

8.4 ANTIOXIDANT ACTIVITY OF MARINE CAROTENOIDS

The unique and phenomenal biodiversity of the marine environment makes a large pool of novel and bioactive molecules, including carotenoids from both algae and marine animals. Palermo et al. (1991) reported the presence of β-carotene, zeaxanthin, fucoxanthin, and fucoxanthinol in red algae. In brown algae fucoxanthin is the dominant carotenoid (Dembitsky and Maoka 2007) (Figure 8.1). Fucoxanthin is the most abundant carotenoid, contributing more than 10% of the estimated total production of carotenoids in nature (Matsuno, 2001). Marine photosynthetic organisms such as micro/macroalgae synthesize β-carotene *de novo* from iroprenyl diphosphate via phytoene and lycopene, and then alter it to produce other derivatives. Marine animals do not synthesize carotenoids *de novo*, and those found in animal bodies either are the result of the direct accumulation of carotenoids from food or are partly modified through metabolic reactions.

Carotenoids are responsible for the color of many important marine products, such as fish, shellfish, and seaweed. Commonly found carotenoids from fish are tunaxanthin in yellow fish, astaxanthin in red fish, zeaxanthin in anchovies, flatfish, and shark, tunaxanthin, lutein, and zeaxanthin in

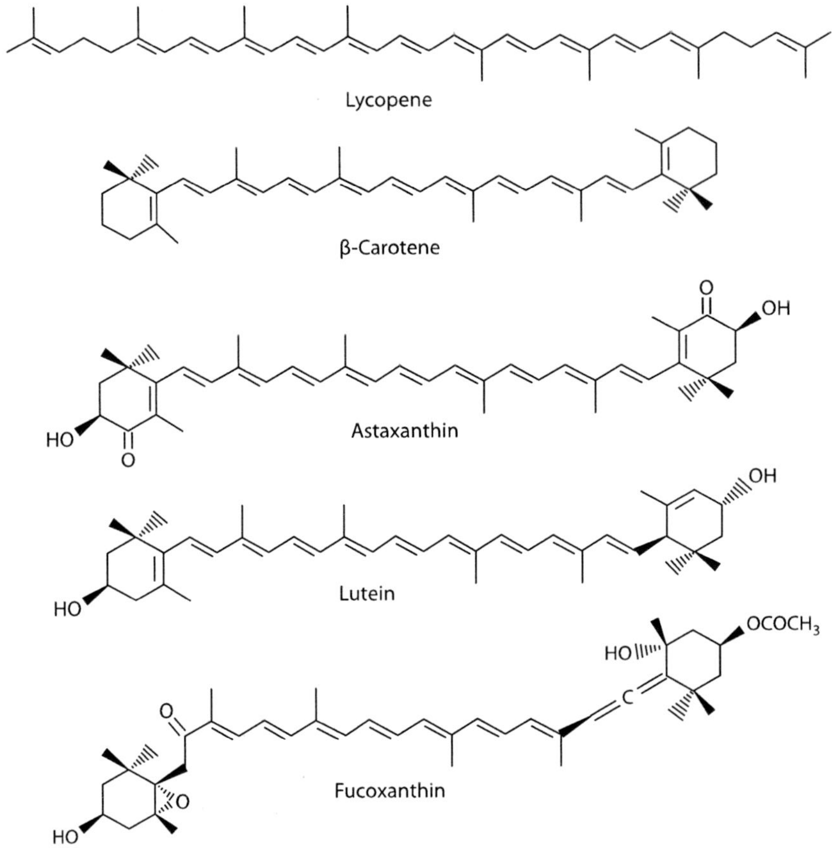

FIGURE 8.1 Chemical structure of major carotenoids.

brackish water fish, and lutein and zeaxanthin in fresh water fish (Goodwin, 1986; Matsuno, 2001). Crustaceans such as shrimp, prawn, lobster, krill, and crab contain astaxanthin as their main pigment present in free forms, esterified, or as a form bound to macromolecules such as protein or chitin.

Crustaceans absorb pigments from the diet and deposit them as such or transfer them metabolically to keto or hydroxy derivatives (Davies, 1985). Astaxanthin and its esters are the main pigment in crustaceans irrespective of the species and the environment from which they are harvested (Sachindra et al., 2005). Overall astaxanthin is the main carotenoid pigment found in aquatic animals, while fucoxanthin is specifically abundant in micro/macroalgae.

Astaxanthin has been found to be more effective than β-carotene in preventing fatty acid preoxidation in chemical solutions (Terao, 1989) and delaying lipid peroxidation in a membrane model (Lim et al., 1992). Astaxanthin is a better agent to destroy free radicals than other carotenoids (Lawlor and O'Brien, 1995). In a study using fibroblast cells harvested from chicken embryos, astaxanthin effectively decreased the oxidative stress induced by the addition of paraquat. Goto et al. (2001) examined the effects of β-carotene and astaxanthin on ADP/Fe^{2+}-induced lipid peroxidation in liposomes under various conditions and found a higher antioxidant activity of astaxanthin than of β-carotene. They showed that astaxanthin trapped radicals not only at the conjugated polyene chain but also in the terminal ring moiety, in which the hydrogen atom at the C3 methine was suggested to be the radical trapping site. This trapping activity of the terminal ring is due to the higher antioxidant activity of astaxanthin at the surface and inside the phospholipid membrane.

Naguib (2000) reported that astaxanthin exhibited the highest antioxidant activity toward peroxyl radicals compared to a range of other carotenoids assessed using novel fluorometric assays. The relative reactivities of astaxanthin, α-carotene, lutein, β-carotene, and lycopene were determined to be 1.3, 0.5, 0.4, 0.2, and 0.4, respectively. Matsushita et al. (2000) compared the antioxidant activity of polar carotenoids including astaxanthin and astaxanthin-β-glucoside on phosphatidylcholine (PC) in liposomes and found that astaxanthin and astaxnathin-β-glucosides are highly active antioxidants. This result indicates that the antioxidant activity of carotenoids on PC liposomes depends not only on their ability to scavenge free radicals but also on their location and orientation in a PC liposome system and the extent of their incorporation into PC bilayers.

Although it is the most abundant carotenoid found in natural resources (Matsuno, 2001), investigations on the antioxidative activity of fucoxanthin are limited. Nomura et al. (1997) demonstrated the 1,1-diphenyl-2-picrylhydrazyl (DPPH) quenching activity of fucoxanthin isolated from the diatom *Phaeodactylum tricornutum*. Electron spin resonance (ERS) analysis showed the quenching ability of fucoxanthin against both organic radicals DPPH and 12-doxyl-steraic acid (12DS) (Nishino, 1998). Yan et al. (1999) demonstrated the strong DPPH radical scavenging activity of organic extracts from different edible seaweeds, and fucoxanthin was identified as the active compound.

Fucoxanthin has a unique structure, including an unusual allenic bond and 5,6-monoepoxide in its molecule (Figure 8.1). This structure will affect its antioxidant activity. Carotenoids can be divided into two classes:, hydrocarbons (carotenes) and their oxygenated derivatives (xanthophylls). Xanthophyll is structurally similar to carotenes but possesses an additional hydroxyl and ketone group on each β-ionone ring. The ability of carotenoid to quench $^1O_2$ increases with an increasing number of conjugated double bands (Conn et al., 1991), whereas the free radical trapping activity of carotenoids increases with the presence of a functional group in the terminal rings, as seen in astaxanthin and fucoxanthin (Miller et al., 1996).

Antioxidant activity of fucoxanthin will be derived not only from the long central chain of conjugated double bonds, but from its characteristic functional groups of the terminal rings. Murakami et al. (2000) screened 19 natural carotenoids for their structure-function relationship with respect to radical scavenging activity using human promyelocytic HL-60 cells. They found that the presence of the allenic bond as seen in fucocoxanthin and halocynthiaxanthin is an increasing factor for inhibition of superoxide and NO generation, while the presence of a 4-oxo-β-end group in their structure in astaxanthin and canthaxanthin enhances NO generation.

8.5 MICROALGAL CAROTENOIDS

The chemical structures of the six carotenoids found in the highest concentrations in human blood have been reported to be β-carotene, lycopene, α-carotene, lutein, zeaxanthin, and β-cryptoxanthin (Basu et al., 2001). Carotenoid levels in human organisms depend on their contributions and contents of foods. Foods in Western societies contain higher α- and β-carotene and lycopene, while lutein and zeaxanthin are much lower (Rühl, 2007). Carotenoid levels in human skin are reported to be in the range of 0.2–0.6 nmol/g wet tissue (Peng et al., 1993); however, there are significant differences in the level and distribution of carotenoids of skin areas (Stahl et al., 1998). Stahl et al. (1998) carried out repeat supplementation of β-carotene in 12 females for 12 weeks, and showed that the carotenoid level increased in all areas of the skin, with the highest levels being in the forehead and palm of the hand. They also reported an increase in skin carotenoid content after intake of a carotenoid supplement (25 mg of carotenoids/day) for 12 weeks by healthy volunteers (Stahl et al., 2000).

The photooxidative skin damage level decreases with increasing the carotenoid content after oral intake (Roberts et al., 2009; Stahl and Sies, 2007; Terao et al., 2011). Thus, carotenoid supplements are widely used for photoprotection of the skin against UV light-induced disorders. Lycopene, β-carotene, and lutein have been reported to be absorbed by humans and protect the lipid oxidation of the skin (Darvin et al., 2011; Palombo et al., 2007), photooxidative damage of human fibroblasts (Eichler et al., 2002; Offord et al., 2002), and erythema (Stahl et al., 2000, 2001). Microalgae are a major natural source of many kinds of bioactive compounds, including carotenoids. Among the above, three kinds of carotenoids, β-carotene, and lutein are produced by microalgae.

The biotechnology of microalgae has gained considerable progress and relevance in recent decades, with carotenoid production representing one of its most successful domains. The halophilic green biflagellate microalga, *Dunaliella salina*, has long been recognized as an efficient biological source of β-carotene (Figure 8.1) (Ben-Amotz and Avron, 1990). β-Carotene accumulation in oil globules is found in the interthylakoid spaces of the chloroplast of this microalgae, and it depends on high salinity, stress temperature, high light intensity, and nitrogen limitation. Under these conditions, up to 12% of the algal dry weight is β-carotene (Del Campo et al., 2007).

Lutein (Figure 8.1) is present in dark, leafy green vegetables, such as spinach and kale, as well as in corn, egg yolk, and some other foods with yellow color. Lutein has been studied widely and proven to show diverse beneficial effects on human health, particularly on optimizing eye health together with skin care. A major source of lutein for the market is represented by the petals from marigold flowers (Piccaglia et al., 1998). More than 95% of the lutein in these plant sources is in esterified form. Thus, the conventional plant pigment workup process comprises chemical saponification. Microalgae such as *Muriellopsis* sp. can produce lutein, where lutein usually appears in the free nonesterified form. Some other microalgae (*Scenedesmus almeriensis*, *Chlorella protothecoides*, and *Chlorella zofingiensis*) also accumulate lutein (Del Campo et al., 2000, 2004).

Astaxanthin (Figure 8.1) is another carotenoid that is produced on an industrial scale using microalgae (Ernst, 2002). Astaxanthin has been demonstrated to have up to several-fold stronger antioxidant activity than vitamin E and β-carotene (Goto et al., 2001; Naguib, 2000; Terao, 1989). Astaxanthin can be significantly more effective than β-carotene, canthaxanthin, and lutein at preventing UV light photooxidation of lipids from human dermal fibroblasts (Camera et al., 2008; O'Connor and O'Brien, 1998). Astaxanthin has been reported to be mainly accumulated in the skin of the tail of rats (Petri and Lundebye, 2007). When human dermal fibroblasts were exposed to moderate doses of UVA after incubation with astaxanthin, canthaxanthin, and β-carotene, astaxanthin exhibited the strongest photoprotective effect among them and uptake of astaxanthin by fibroblasts was higher than that of the other two carotenoids (Camera et al., 2008). These findings suggest that astaxanthin has excellent potential as an oral sun protectant. However, there have been few human clinical trials on the photoprotection of astaxanthin, while the effect of astaxanthin has been well documented in the *in vitro* system (Daubrawa et al., 2005; Lyons and O'Brien, 2002; Merwald et al., 2005; Suganuma et al., 2010).

Only two sources of microalgal origin compete with synthetic astaxanthin, which presently dominates the market: the heterobasidious yeast *Phaffia rhodozyma* and the microalgae *Hematococcus pluvialis* have been widely studied for more than two decades under the viewpoint of astaxanthin production. Unicellular microalgae *H. pluvialis* produces configurational isomers of astaxanthin (Kobayashi et al., 1997). The main astaxanthin isomer (3S,3S′) produced by the this microalgae is identical to that present in wild salmon.

8.6 FUCOXANTHIN

Fucoxanthin is metabolized to other carotenoids in marine organisms (Matsuno et al., 1985a, 1985b, 1986; Ookubo and Matsuno, 1985). In the sea squirt, *Halocynthia roretzi*, fucoxanthin is metabolized into fucoxanthinol and halocynthiaxanthin (Ookubo and Matsuno, 1985). Sugawara et al. (2002) also demonstrated that fucoxanthin is hydrolyzed to fucoxanthinol during absorption by Caco-2 human intestine cells and mouse. Fucoxanthin and their metabolites such as fucoxanthinol and halocynthiaxanthin have an ability to quench $^1O_2$ and scavenge different free radicals (Sachindra et al., 2007).

DPPH radical and ABTS radical scavenging assay is one of the popular indirect methods of determining the antioxidative capacity of compounds (Roginsky and Lissi, 2005; Shon et al., 2003). The DPPH radical scavenging activity of fucoxanthin and fucoxanthinol was higher than that for halocynthiaxanthin, with concentrations (μM) required for 50% scavenging (EC_{50}) being 164.60, 153.78, and 826.39, respectively (Sachindra et al., 2007). ABTS radical scavenging activity of fucoxanthinol (EC_{50} 2.49 μM) was stronger than fucoxanthin (EC_{50} 8.94 μM). Further, hydroxyl radical scavenging activity as measured by a chemiluminescnce technique showed that the scavenging activity by fucoxanthin was 7.9 times higher than that by fucoxanthinol, 16.3 times higher than that by halocynthiaxanthin, and 13.5 times higher than that by α-tocopherol (Sachindra et al. 2007).

A similar trend was observed when the hydroxyl radical scavenging was assessed by the ERS technique. The signal intensity of hydroxyl radical adducts was lower in the presence of fucoxanthin (56.6% of control) than fucoxanthinol (81.7%) and α-tocopherol (92.5%) at a concentration of 2.5 mg/ml, while halocynthiaxanthin did not show any reduction in signal intensity, indicating no scavenging activity. The ERS analysis of superoxide radical scavenging activity also showed the superiority of fucoxanthin over two other carotenoids tested.

Although fucoxanthin is a major carotenoid found in nature (Matsuno, 2001) and its antioxidant activity has been reported (Miyashita, 2009), there has been little research on the application of fucoxanthin to skin protection. Shimoda et al. (2010) examined inhibitory effects of fucoxanthin on tyrosinase activity, melanin formation in B16 melanoma, and skin pigmentation in UVB-irradiated guinea pigs. Fucoxanthin inhibited tyrosinase activity, melanogenesis in melanoma, and UVB-induced skin pigmentation. Fucoxanthin intake also significantly suppressed mRNA expression of COX-2, endothelin receptor A, p75 neurotrophin receptor, prostaglandin E receptor 1, melanocortin 1 receptor, and tyrosinase-related protein 1.

An anti-inflammatory effect of fucoxanthin has also been reported in a mouse macrophage cell, RAW 264.7 (Shiratori et al., 2005). When the cells were pretreated with fucoxanthin and subsequently incubated with lipopolysaccharide as an inflammation inducer, fucoxanthin addition resulted in a significant decrease in expression of COX and nitric oxide synthase (iNOS) proteins, and thus induced the reduction of PGE2, NO, and TNF-α concentrations. The same results have been obtained in another study (Heo et al., 2010).

UVB irradiation induces COX-2 overexpression and a subsequent increase in the production of PG metabolites in the skin, which is a characteristic response of keratinocytes to acute or chronic exposure to UVB radiation. An overexpression of COX-2 found in chronically UVB-irradiated skin is strongly related to the development of cutaneous edema, erythema, and bifold skin thickness (a biomarker of inflammation). It also plays crucial roles in skin tumor promotion (Mukhtar and Elmets, 1996; Nichols and Katiyar, 2010). Thus the anti-inflammatory effect of fucoxanthin and its metabolites may show its possible application to cosmeceutical material.

8.7 CONCLUSION

Solar radiation offers several health benefits, such as the synthesis of vitamin D. On the other hand, it is very clear that excessive solar exposition is harmful for the eyes and the skin, inducing photoaging and skin cancer. The incidence of skin cancer and other UV-related pathologies has reached high levels at present and is expected to rise in the future. The use of photoprotective clothing and the correct and regular application of sunscreens can offer pretty good protection, certainly against sunburn. Furthermore, recent understanding of the potential protective contribution of nutritional chemoprevention expands the scope of sun protection strategies beyond the currently recommended practices of sun avoidance, body covering, and sunscreen application.

It is well known that carotenoids provide photoprotection against excess light and photoxidative damage in photosynthetic systems. Thus, much attention has been paid to carotenoids as good photoprotective nutrients. The quenching effect on $^1O_2$ and scavenging activity on free radicals of carotenoids have been suggested as being the main mechanisms by which they afford their beneficial health effects. However, it would be difficult to explain all the physiological effects of carotenoids only by their antioxidant activity. Modulation effects of carotenoids on gene and protein expression that are related to photoprotection are likely to be more important. Carotenoid functions as antioxidants must be reconciled with carotenoid functions in signal transduction and redox modulation of key signaling proteins relating to photoprotection.

The marine carotenoids discussed in this chapter show significant antioxidant and anti-inflammatory effects. These effects may contribute to their skin photoprotection and inhibit adverse processes induced or mediated by solar UV radiation. It is suggested that routine consumption or topical treatment of carotenoids such as lycopene, β-carotene, and lutein may provide efficient protection against the harmful effects of solar UV radiation in humans. However, relatively little information is available for major marine carotenoids, astaxanthin and fucoxanthin, though both marine carotenoids are expected to have higher photoprotective potential than of lycopene, β-carotene, and lutein. More research is required on astaxanthin and fucoxanthin based on human trials and laboratory studies conducted using *in vitro* and *in vivo* systems.

REFERENCES

Bachelor, M.A., and Bowden, G.T. 2004. UVA-mediated activation of signaling pathways involved in skin tumor promotion and progression. *Semin. Cancer Biol.* 14:131–138.

Basu, H.N., Del Vecchio, A.J., Flider, F., and Orthoefer, F.T. 2001. Nutritional and potential disease prevention properties of carotenoids. *J. Am. Oil Chem. Soc.* 78:665–675.

Ben-Amotz, A., and Avron, M. 1990. The biotechnology of cultivating the halotolerant alga *Dunaliella*. *Trends Biotechnol.* 8:121–125.

Bossi, O., Gartsbein, M., Leitges, M., Kuroki, T., Grossman, S., and Tennenbaum, T. 2008. UV irradiation increases ROS production via PKCδ signaling in primary murine fibroblasts. *J. Cell Biochem.* 105:194–207.

Bruls, W.A., Slaper, H., van der Leun, J.C., and Berrens, L. 1984. Transmission of human epidermis and stratum corneum as a function of thickness in the ultraviolet and visible wavelengths. *Photochem. Photobiol.* 40:485–494.

Camera, E., Mastrofrancesco, Fabbri, C., Daubrawa, Picardo, M., Sies, H., and Stahl, W. 2008. Astaxanthin, canthaxanthin and β-carotene differently affect UVA-induced oxidative damage and expression of oxidative stress-responsive enzymes. *Exp. Dermatol.* 18:222–231.

Cantrell, A., McGarvey, D.J., Truscott, T.G., Rancan, F., and Bohm, F. 2003. Singlet oxygen quenching by dietary carotenoids in a model membrane environment. *Arch. Biochem. Biophys.* 412:47–54.

Conn, P.F., Schalch, W., and Truscott, G.T. 1991. The singlet oxygen and carotenoid interaction. *J. Photochem. Photobiol. B Biol.* 11:41–47.

Cooper, D.A., Eldridge, A.L., and Peters, J.C. 1999a. Dietary carotenoids and lung cancer: a review of recent research. *Nutr. Rev.* 57:133–145.

Cooper, D.A., Eldridge, A.L., and Peters, J.C. 1999b. Dietary carotenoids and certain cancers, heart disease and age related macular degeneration: a review of recent research. *Nutr. Rev.* 57:201–214.

Darvin, M.E., Fluhr, J.W., Meinke, M.C., Zastrow, L., Sterry, W., and Lademann, J. 2011. Topical beta-carotene protects against infra-red-light–induced free radicals. *Exp. Dermatol.* 20:125–129.

Daubrawa, F., Sies, H., and Stahl, W. 2005. Astaxanthin diminishes gap junctional intercellular communication in primary human fibroblasts. *J. Nutr.* 135:2507–2511.

Davies, B.H. 1985. Carotenoid metabolism in animals: a biochemist's view. *Pure Appl. Chem.* 57:679–684.

De Gruijl, F.R., and van der Leun, J.C. 1994. Estimate of the wavelength dependency of ultraviolet carcinogenesis in humans and its relevance to the risk assessment of stratospheric ozone depletion. *Health Phys.* 67:319–325.

Del Campo, J.A., García-González, M., and Guerrero, M.G. 2007. Outdoor cultivation of microalgae for carotenoid production: current state and perspectives. *Appl. Microbiol. Biotechnol.* 74:1163–1174.

Del Campo, J.A., Moreno, J., Rodríguez, H., Vargas, M.A., Rivas, J., and Guerrero, M.G. 2000. Carotenoid content of chlorophycean microalgae. Factors determining lutein accumulation in *Muriellopsis* sp. (Chlorophyta). *J. Biotechnol.* 76:51–59.

Del Campo, J.A., Rodríguez, H., Moreno, J., Vargas, M.A., Rivas, J., and Guerrero, M.G. 2004. Accumulation of astaxanthin and lutein in *Chlorella zofingiensis* (Chlorophyta). *Appl. Microbiol. Biotechnol.* 64:848–854.

De Luca, L.M. 1991. Retinoids and their receptors in differentiation, embryogenesis, and neoplasia. *FASEB J.* 5:2924–2933.

Dembitsky, V.M., and Maoka, T. 2007. Allenic and cumulenic lipids. *Prog. Lipid Res.* 46:328–375.

Demmig-Adams, B., and Adams III, W.W. 2002. Antioxidants in photosynthesis and human nutrition. *Science* 298:2149–2153.

DiGiovanni, J. 1992. Multistage carcinogenesis in mouse skin. *Pharmacol. Ther.* 54:63–128.

Donawho, C.K., and Kripke, M.L. 1991. Evidence that the local effect of ultraviolet radiation on the growth of murine melanomas is immunologically mediated. *Cancer Res.* 51:4176–4181.

Edge, R., McGarvey, D.J., and Truscott, T.G. 1997. The carotenoids as anti-oxidants—a review. *J. Photochem. Photobiol. B Biol.* 41:189–200.

Eichler, O., Sies, H., and Stahl, W. 2002. Divergent optimum levels of lycopene, β-carotene and lutein protecting against UVB irradiation in human fibroblasts. *Photochem. Photobiol.* 75:503–506.

Ernst, H. 2002. Recent advances in industrial carotenoid synthesis. *Pure Appl. Chem.* 74:2213–2226.

Fukuzawa, K., Inokami, Y., Tokumura, A., Terao, J., and Suzuki, A. 1998. Rate constants for quenching singlet oxygen and activities for inhibiting lipid peroxidation of carotenoids and α-tocopherol in liposomes. *Lipids* 33:751–756.

González, S., Fernández-Lorente, M., and Gilaberte-Calzada, Y. 2008. The latest on skin photoprotection. *Clin. Dermatol.* 26:614–626.

Goodwin, T.W. 1986. Metabolism, nutrition, and function of carotenoids. *Ann. Rev. Nutr.* 6:273–297.

Goto, S., Kogure, K., Abe, K., Kimata, Y., Kitahama, K., Yamashita, E., and Terada, H. 2001. Efficient radical trapping at the surface and inside the phospholipid membrane is responsible for highly potent antiperoxidative activity of the carotenoid astaxanthin. *Biochim. Biophys. Acta* 1512:251–258.

Heo, S.-J., Yoon, W.-J., Kim, K.-N., Ahn, G.-N., Kang, S.-M., Kang, D.-H., Affan, A., Oh, C., Jung, W.-K., and Jeon, Y.-J. 2010. Evaluation of anti-inflammatory effect of fucoxanthin isolated from brown algae in lipopolysaccharide-stimulated RAW 264.7 macrophages. *Food Chem. Toxicol.* 48:2045–2051.

Hirayama, O., Nakamura, K., Hamada, S., and Kobayasi, Y. 1994. Singlet oxygen quenching ability of naturally occurring carotenoids. *Lipids* 29:149–150.

Hruza, L.L., and Pentland, A.P. 1993. Mechanisms of UV-induced inflammation. *J. Invest. Dermatol.* 100:35S–41S.

Ichihashi, M., Ueda, M., and Budiyanto, A. 2003. UV-induced skin damage. *Toxicology* 189:21–39.

Katiyar, S.K., Korman, N.J., Mukhtar H., and Agarwal, R. 1997. Protective effects of silymarin against photocarcinogenesis in a mouse skin model. *J. Natl. Cancer Inst.* 89:556–566.

Katiyar, S.K., and Mukhtar, H. 2001. Green tea polyphenol (–)-epigallocatechin-3-gallate treatment to mouse skin prevents UVB induced infiltration of leukocytes, depletion of antigen presenting cells and oxidative stress. *J. Leukoc. Biol.* 69:719–726.

Kligman, L.H., Akin, F.J., and Kligman, A.M. 1980. Sunscreens prevent ultraviolet photocarcinogenesis. *J. Am. Acad. Dermatol.* 3:30–35.

Kobayashi, M., Kakizono, T., Nishio, N., Nagai, S., Kurimura, Y., and Tsuji, Y. 1997. Antioxidant role of astaxanthin in the green alga *Haematococcus pluvialis*. *Appl. Microbiol. Biotechnol.* 48:351–356.

Krinsky, N.I. 1998. The antioxidant and biological properties of thecarotenoids. *Ann. NY Acad. Sci.* 854:443–447.

Krinsky, N.I., and Yeum, K.J. 2003. Carotenoid-radical interactions. *Biochem. Biophys. Res. Commun.* 305:754–760.

Kripke, M.L. 1990. Photoimmunology. *Photochem. Photobiol.* 52:919–924.

Krutmann, J. 2000. Ultraviolet A radiation-induced biological effects in human skin: relevance for photoaging and photodermatosis. *J. Dermatol. Sci.* 23:S22–S26.

Kulms, D., and T. Schwarz, T. 2000. Molecular mechanisms of UV-induced apoptosis. *Photodermatol. Photoimmunol. Photomed.* 16:195–201.

Lawlor, S.M., and O'Brien, N.M. 1995. Astaxanthin: antioxidant effects in chicken embryo fibroblasts. *Nutr. Res.* 15:1695–1704.

Lim, B.P., Nagao, A., Terao, J., Tanaka, K., Suzuki, T., and Takama, K. 1992. Antioxidant activity of xanthophylls on peroxyl radical-mediated phospholipid peroxidation. *Biochim. Biophys. Acta* 1126:178–184.

Lyons, N.M., and O'Brien, N.M. 2002. Modulatory effects of an algal extract containing astaxanthin on UVA-irradiated cells in culture. *J. Dermatol. Sci.* 30:73–84.

Marrot, L., and Meunier, J.R. 2008. Skin DNA photodamage and its biological consequences. *J. Am. Acad. Dermatol.* 58(Suppl):S139–S148.

Mascio, P.D., Murhy, M.E., and Sies, H. 1991. Antioxidant defense systems: the role of carotenoids, tocopherols, and thiols. *Am. J. Clin. Nutr.* 53:194S–200S.

Matsuno T. 2001. Aquatic animal carotenoids. *Fisheries Sci.* 67:771–783.

Matsuno, T., Maoka, T., Shiba, K., and Ookubo, M. 1986. Isolation of fucoxanthinol from short-necked clam *Tapes philippinarum* [Asari in Japanese]. *Bull. Jap. Soc. Sci. Fish.* 52:167.

Matsuno, T., Ookubo, M., and Komori, T. 1985a. Carotenoids of tunicates. III. The structural elucidation of two new marine carotenoids, amarouciaxanthin A and B. *J. Nat. Prod.* 48:606–613.

Matsuno, T., Sakaguchi, S., Ookubo, M., and Maoka, T. 1985b. Isolation and identification of amaroucisxanthin A from the bivalve *Paphia euglypta* (Sudaregai in Japanese). *Bull. Jap. Soc. Sci. Fish.* 51:1909.

Matsushita, Y., Suzuki, R., Nara, E., Yokoyama, A., and Miyashita, K. 2000. Antioxidant activity of polar carotenoids including astaxanthin-β-glucoside from marine bacterium on PC liposomes. *Fish. Sci.* 66:980–985.

Merwald, H., Klosner, G., Kokesch, C., Der-Petrossian, M., Herbert Hönigsmann, H., and Trautinger, F. 2005. UVA-induced oxidative damage and cytotoxicity depend on the mode of exposure. *J. Photochem. Photobiol. B* 79:197–207.

Meunier, L., Raison-Peyron, N., and Meynadier, J. 1998. UV-induced immunosuppression and skin cancers. *Rev. Med. Interne.* 19:247–254.

Miller, N.J., Sampson, J., Candeias, L.P., Bramley, P.M., and Rice-Evans, C.A. 1996. Antioxidant activities of carotenes and xanthophylls. *FEBS Lett.* 384 240–242.

Miyashita, K. 2009. Function of marine carotenoids. *Forum Nutr.* 61:136–146.

Mukhtar, H., and Elmets, C.A. 1996. Photocarcinogenesis: mechanisms, models and human health implications. *Photochem. Photobiol.* 63:355–447.

Murakami, A., Nakashima, M., Koshiba, T., Maoka, T., Nishino, H., Yano, M., Sumida, T., Kim, O.K., Koshimizu, K., and Ohigashi, H. 2000. Modifying effects of carotenoids on superoxide and nitric oxide generation from stimulated leukocytes. *Cancer Lett.* 149:115–123.

Naguib, Y.M.A. 2000. Antioxidant activities of astaxanthin and related carotenoids. *J. Agric. Food Chem.* 48:1150–1154.

Nichols, J.A., and Katiyar, S.K. 2010. Skin photoprotection by natural polyphenols: anti-inflammatory, antioxidant and DNA repair mechanisms. *Arch. Dermatol. Res.* 302:71–83.

Nishino, H. 1998. Cancer prevention by carotenoids. *Mut. Res.* 402:159–163.

Nomura, T., Kikuchi, M., Kubodera, A., and Kawakami Y. 1997. Proton-donative antioxidant activity of fucoxanthin with 1,1-diphenyl-2-picrylhydrazyl (DPPH). *Biochem. Mol. Biol. Int.* 42:361–370.

O'Connor, I., and O'Brien, N. 1998. Modulation of UVA light-induced oxidative stress by beta-carotene, lutein and astaxanthin in cultured fibroblasts. *J. Dermatol. Sci.* 16:226–230.

Offord, E.A., Gautier, J.-C., Avantia, O., Scaletta, C., Runger, F., Krämer, K., and Applegate, L.A. 2002. Photoprotective potential of lycopene, β-carotene, vitamin E, vitamin C and carnosic acid in UVA-irradiated human skin fibroblasts. *Free Radic. Biol. Med.* 32:1293–1303.

Ookubo, M., and Matsuno, T. 1985. Carotenoids of sea squirts. II. Comparative biochemical studies of carotenoids in sea squirts. *Comp. Biochem. Physiol.* 81B:137–141.

Palermo, J.A., Gros, E.G., and Seldes, A.M. 1991. Carotenoids from tree red algae of the corallinaceae. *Phytochemistry* 30:2983–2986.

Palombo, P., Fabrizi, G., Ruocco, V., Ruocco, E., Fluhr, J., Roberts, R., and Morganti, P. 2007. Beneficial long-term effects of combined oral/topical antioxidant treatment with the carotenoids lutein and zeaxanthin on human skin: a double-blind, placebo-controlled study. *Skin Pharmacol. Physiol.* 20:199–210.

Parrish, J.A. 1983. Photoimmunology. *Adv. Exp. Med. Biol.* 160:91–108.

Peng, Y.-M., Peng, Y.-S., and Lin, Y. 1993. A nonsaponfication method for the determination of carotenoids, retinoids, and tocopherols in solid human tissues. *Can. Epidemiol. Biol. Prev.* 2:139–144.

Petri, D., and Lundebye, A.-K. 2007. Tissue distribution of astaxanthin in rats following exposure to graded levels in the feed. *Comp. Biochem. Physiol. C Toxicol. Pharmacol.* 145:202–209.

Piccaglia, R., Marotti, M., and Grandi, S. 1998. Lutein and lutein ester content in different types of *Tagetes patula* and *T. erecta*. *Ind. Crops Prod.* 8:45–51.

Roberts, R.L., Green, J., and Lewis, B. 2009. Lutein and zeaxanthin in eye and skin health. *Cli. Dermatol.* 27:195–201.

Roginsky, V., and Lissi, E.A. 2005. Review of methods to determine chain-breaking antioxidant activity in food. *Food Chem.* 92:235–254.

Rühl, R. 2007. Effects of dietary retinoids and carotenoids on immune development. *Proc. Nutr. Soc.* 66:458–469.

Sachindra, N.M., Bhaskar, N., and Mahendrakar, N.S. 2005. Carotenoids in different body components of Indian shrimps. *J. Sci. Food Agric.* 85:167–172.

Sachindra, N.M., Sato, E., Maeda, H., Hosokawa, M., Niwano, Y., Kohno, M., and Miyashita, K. 2007. Radical scavenging and singlet oxygen quenching activity of marine carotenoid fucoxanthin and its metabolites. *J. Agric. Food Chem.* 55:8516–8522.

Shapira, N. 2010. Nutritional approach to sun protection: a suggested complement to external strategies. *Nutr. Rev.* 68:75–86.

Shimoda, H., Tanaka, J., Shan, S.J., and Maoka, T. 2010. Anti-pigmentary activity of fucoxanthin and its influence on skin mRNA expression of melanogenic molecules. *J. Pharm. Pharmacol.* 62:1137–1145.

Shiratori, K., Ohgami, K., Ilieva, I., Jin. X.-H., Koyama, Y., Miyashita, K., Yoshida, K., Kase, S., and Ohno, S. 2005. Effects of fucoxanthin on lipopolysaccharide-induced inflammation *in vitro* and *in vivo*. *Exp. Eye Res.* 81:422–428.

Shon, M.Y., Kim, T.H., and Sung, N.J. 2003. Antioxidants and free radical scavenging activity of *Phellinus baumii* (Phellinus of Hymenochaetaceae) extracts. *Food Chem.* 82:593–597.

Stahl, W., Heinrich, U., Jungmann, H., Sies, H., and Tronnier, H. 2000. Carotenoids and carotenoids plus vitamin E protect against ultraviolet light-induced erythema in humans. *Am. J. Clin. Nutr.* 71:795–798.

Stahl, W., Heinrich, U., Jungmann, H., von Laar, J., Schietzel, M., Sies, H., and Tronnier, H. 1998. Increased dermal carotenoid levels assessed by noninvasive reflection spectrophotometry correlate with serum levels in women ingesting betatene. *J. Nutr.* 128:903–907.

Stahl, W., Heinrich, U., Wiseman, S., Eichler, O., Sies, H., and Tronnier, H. 2001. Dietary tomato paste protects against ultraviolet light-induced erythema in humans. *J. Nutr.* 131:1449–1451.

Stahl, W., and Sies, H. 1997. Antioxidant defense: vitamins E and C and carotenoids. *Diabetes* 46:S14–S18.

Stahl, W., and Sies, H. 2007. Carotenoids and flavonoids contribute to nutritional protection against skin damage from sunlight. *Mol. Biotechnol.* 37:26–30.

Suganuma, K., Nakajima, H., Ohtsuki, M., and Imokawa, G. 2010. Astaxanthin attenuates the UVA-induced up-regulation of matrixmetalloproteinase-1 and skin fibroblast elastase in human dermal fibroblasts. *J. Dermatol. Sci.* 58:136–142.

Sugawara, T., Baskaran, V., Tsuzuki, W., and Nagao, A. 2002. Brown algae fucoxanthin is hydrolyzed to fucoxanthinol during absorption by Caco-2 human intestinal cells and mice. *J. Nutr.* 132:946–951.

Svobodobá, A., and Vostálová, J. 2010. Solar radiation induced skin damage: review of protective and preventive options. *Int. J. Radiat. Biol.* 86:999–1030.

Terao, J. 1989. Antioxidant activity of β-carotene-related carotenoids in solution. *Lipids* 24:659–661.

Terao, J., Minami, Y., and Bando, N. 2011. Singlet molecular oxygen-quenching activity of carotenoids: relevance to protection of the skin from photoaging. *J. Clin. Biochem. Nutr.* 48:57–62.

Urbach, F. 1991. Incidences of nonmelanoma skin cancer. *Dermatol. Clin.* 9:751–755.

Verschooten, L., Claerhout, S., Van Laethem, A., Agostinis, P., and Garmyn, M. 2006. New strategies of photoprotection. *Photochem. Photobiol.* 82:1016–1023.

Wang, S.Q., Balagula, Y., and Osterwalder, U. 2010. Photoprotection: a review of the current and future technologies. *Dermatol. Ther.* 23:31–47.

Yan, X., Chuda, Y., Suzuki, M., and Nagata, T. 1999. Fucoxanthin as the major antioxidant in *Hijikia fusiformis*, a common edible seaweed. *Biosci. Biotechnol. Biochem.* 63:605–607.

Yoshikawa, T., Rae, V., and Bruins-Slot, W. 1990. Susceptibility to effects of UVB radiation on induction of contact hypersensitivity as a risk factor for skin cancer in humans. *J. Invest. Dermatol.* 95:530–536.

Young, A.J., and Lowe, G.M. 2001. Antioxidant and prooxidant properties of carotenoids. *Arch. Biochem. Biophys.* 385:20–27.

Ziegler, A., Jonason, A.S., Leffell, D.J., Simon, J.A., Sharma, H.W., Kimmelman, J., Remington, L., Jacks, T., and Brash, D.E. 1994. Sunburn and p53 in the onset of skin cancer. *Nature* 372:773–776.

9 Role of Astaxanthin in Cosmeceutical Applications

Pamela Chávez-Crooker, Johanna Obreque, Jeanette Vera, and Karla Moya

CONTENTS

9.1 WHAT IS ASTAXANTHIN?

Astaxanthin (AX) is a ketocarotenoid, a red pigment that belongs to the family of the xanthophylls, an oxygenated derivative of carotenoids. AX is closely related to other carotenoids, such as β-caroteno, zeaxanthin, and lutein. Astaxanthin cannot be synthesized by animals and must be acquired from the diet, but it is synthesized by plants, bacteria, a few fungi, algae, and microalgae (Yuan et al., 2011). Therefore, through the food chain, AX can be found in birds such as flamingoes and quail, but is mainly found in aquatic animals such as salmonids, shrimp, lobsters, and crayfish, contributing to the pinkish red color of the flesh.

In most aquatic animals in which it is found, astaxanthin has several functions, such as protection against UV light effects; protection against oxidation of essential polyunsaturated fatty acids, immune response, and pigmentation; communication; and reproductive behavior and improved reproduction (Lorenz and Cysewski, 2000). All these functions are based on the high antioxidant activity of the AX molecule. It has been reported to be 550 times higher than vitamin E and 6,000 times higher than vitamin C (Nishida et al., 2007).

An antioxidant molecule can remove free radicals from an organism by reacting with them to produce innocuous compounds or disrupting oxidation reactions. This activity prevents other molecules or tissues from being damaged. The AX molecule (Figure 9.1) is highly antioxidant because it has unique chemical properties based on its molecular structure. Each ionone ring contains hydroxyl and keto moieties, which are capable of resonance-stabilizing carbon-centered radicals (Martin et al., 1999). These terminal rings can scavenge radicals both at the surface and in the interior of the phospholipid membrane, preventing membrane proxidation (Goto et al., 2001).

In humans, AX can have several beneficial effects, such as anti-inflammatory, antibacterial, inmunomodulatory, photoprotective, neuroprotective, anticarginogenic, antiatherogenic, protective effects against ethanol and drug toxicities, hepatoprotective, antidiabetic, dental pulp protective, cardiovascular protective, ocular protective, skin protective, and fertility improvement activities (Yuan et al., 2011). All these extraordinary properties open up a new potential application for nutraceuticals and as constitutional ingredients for cosmetics (Schmidt et al., 2011).

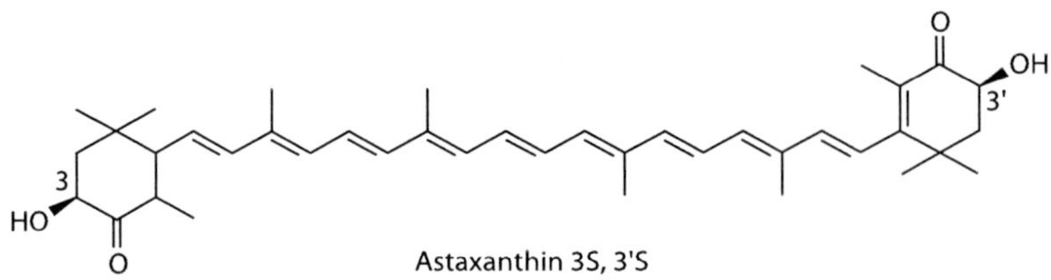

FIGURE 9.1 Astaxanthin chemical structure. The presence of hydroxyl and keto endings on each ionone ring explains unique properties compared to other carotenoids. (Adapted from Guerin, M., Huntley, M.E., and Olaizola, M., *Trends Biotechnol.*, 2003;21:210–16.)

9.2 BIOLOGICAL SOURCES OF ASTAXANTHIN

Astaxanthin is produced mainly by crustaceans (crabs, shrimp, and lobster), the yeast *Phaffia rhodozyma* now known as *Xanthophyllomyces dendrorhous*, and the microalgae *Haematococcus pluvialis*. In this chapter, only microbial sources will be described.

Only a few microorganisms are able to synthetize AX (Neil and Leenheer, 1991; Yokoyama and Miki, 1995); among them the marine bacteria *Paracoccus marcusii*, *Paracoccus carotinifaciens*, *Paracoccus* spp., and *Agrobacterium aurantiacum* have been reported as AX producers (Lee et al., 2004).

Natural or artificial AX was used as pigment for animal feeding for several years. However, during the last decade many benefits have been reported for animal and human health, adding more value to natural AX. Recently, nutraceutical and medical applications have led to the formulation of food supplements. The main advantage of natural AX esters in nutraceutical applications is the higher stability of the esterified form, providing a longer storage time without being oxidized. Natural AX offers a great opportunity to explore new applications.

In order to make a more efficient process for AX production in natural sources, the biosynthesis of AX has been well studied and described (Fan et al., 1995; Lee and Kim, 2006). In brief, AX is synthesized through the isoprenoid pathway. Figure 9.2 shows the AX pathway from *Haematococcus*. The pathway is indicated by acetyl-CoA and proceeds through phytoene, lycopene, β-carotene, and cathaxanthin before the last oxidative step to astaxanthin (Goodwin, 1984; Davis, 1985; Grung et al., 1992; Lorenz and Cysewski, 2000). To induce this pathway cells must be cultivated under nutrient starvation (mainly nitrogen and phosphate), sunlight, high temperature, or salt concentration (Schmidt et al., 2011).

Recently, genetics tools have been suggested for AX yield improvements. Genetic engineering can help to increase astaxanthin titers by redirecting the flow of precursors into a specific pathway and improving enzymatic activities for efficient conversions of intermediates for an end product. In the case of *P. rhodozyma*, up- and downregulation of key enzymes in the terpenoid pathway to increase astaxanthin biosynthesis has been reported (Schmidt et al., 2011). Another example of genetic modification has been reported for *H. pluvialis*, where the enzyme phytoene desaturase was modified to accelerate AX biosynthesis, improving final yields compared to a wild type (Steinbrenner and Sandmann, 2006). These techniques are very promising to improve AX yield in cells and a great contribution for a more cost competitive process.

9.3 BIOLOGICAL FUNCTIONS

Originally, AX was used as a feed additive in aquaculture for appropriate healthy pigmentation, growth, and reproduction of salmonids (Higuera-Ciapara et al., 2006). AX is also used in the diet of chickens to produce a stronger pigmentation of yolk and flesh.

Recently, it has been shown that AX has beneficial health effects, enhancing immune response and antioxidant activity (Lorenz and Cysewski, 2000). Astaxanthin can prevent the pro-inflammatory

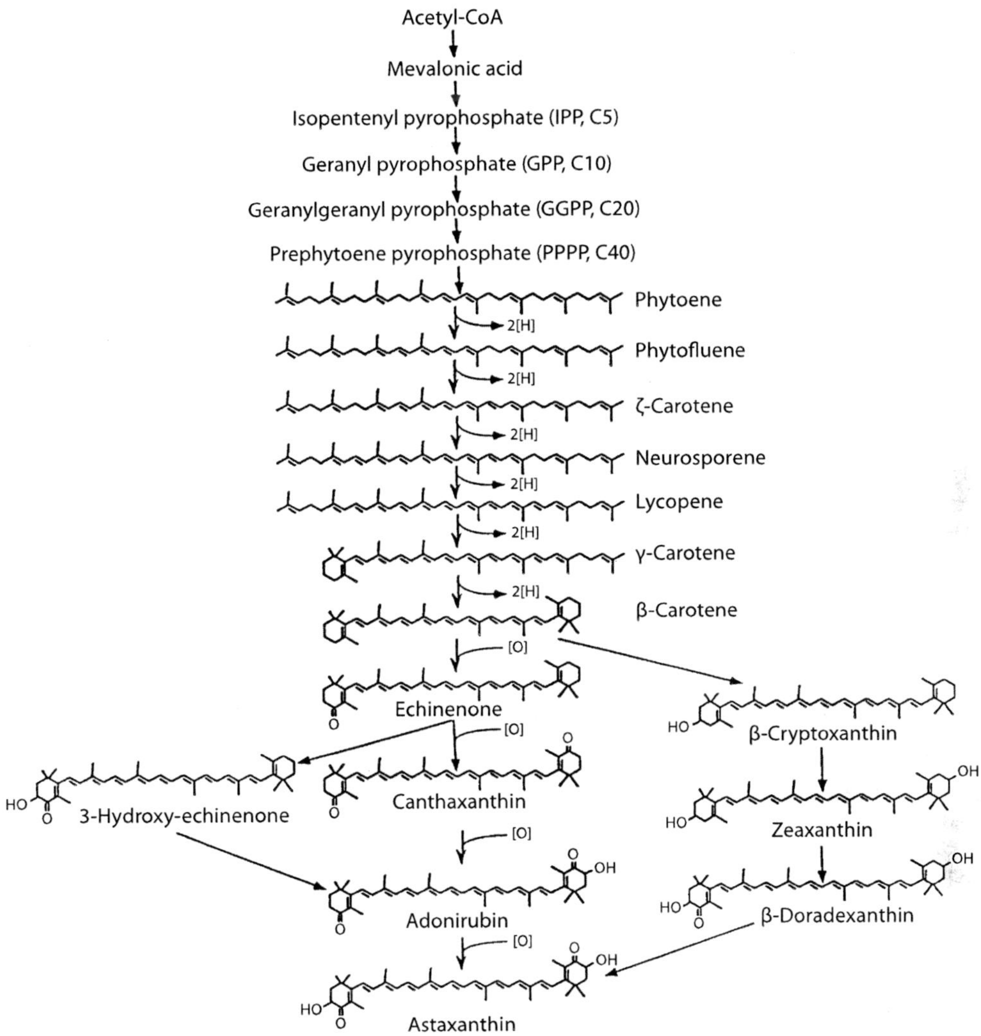

FIGURE 9.2 Astaxanthin pathway of *Haematococcus pluvialis*. (NatuRose™ Technical Bulletin #060, revised March 30, 1999; contact: Dr. R. Todd Lorenz Cyanotech Corporation; extracted from Lorenz, R.T., and Cysewski, G.R., *Trends Biotechnol.*, 2000;18:160–67. With permission.)

process from inhibiting the expression of nitric oxide synthase and cyclooxygenase-2, resulting in a decrease of nitric oxide and prostaglandin E_2 (PGE_2) in cell-mediated immune responses (Lee et al., 2003). The same effect is observed for pro-inflammatory cytokine tumor necrosis factor-alpha (TNF-α) and interleukin 6 (IL-6) (Ohgami et al., 2003; Macedo et al., 2010). In addition, AX also can modulate the immunological system by increasing T-helper cell antibodies and immunoglobulins (IgA, IgG, and IgM). The immunomodulatory and anti-inflammatory activity of AX would likely influence the etiology of cancer and inflammatory diseases such as asthma (Yuan et al., 2011). Therefore, AX can be used as medication for the treatment of autoimmune diseases such as multiple sclerosis, rheumatoid arthritis, and Crohn's disease (Lignell and Botiger, 2001).

On the other hand, AX has been reported to affect *H. pylori*, an important human pathogenic strain affecting the gastric system, by two mechanisms. The first one is controlling gastric inflammation, and the second one is by affecting colonization levels (Guerin et al., 2003; Yuan et al., 2011). Clinical studies have reported (Kupcinskas et al., 2008) that although AX did not have any curative effect in functional dyspepsia patients, it helped to decrease reflux symptoms, and this effect was greater in *H. pylori*-infected patients.

Astaxanthin also has been reported to have neuroprotective effects. AX can prevent neuronal apoptosis, mitochondria abnormalities, and intracellular reactive oxygen species generation (Liu et al., 2009). In addition, it can protect against neurodegenerative diseases such as Parkinson's, Alzheimer's, and cerebral infarction (Ikeda et al., 2008; Liu et al., 2009, Chan et al., 2009; Chang et al., 2010; Shen et al., 2009). AX can also help to prevent memory deficits in vascular dementia (Hussein et al., 2006).

Astaxanthin also can help visual improvement. Only 4 to 12 mg per day for 28 days can significantly improve uncorrected far visual acuity (Nakamura et al., 2004). AX might increase retinal capillary blood flow in normal patients without affecting intraocular pressure (Nagaki et al., 2005). Also, AX suppressed the development of choroidal neovascularization, responsible for vision loss and blindness (Izumi-Nagai et al., 2008). Moreover, AX has been reported to have greater protective activity than α-tocopherol against UVB insult on human lens epithelial cells (Chitchumroonchokchai et al., 2004).

In the liver, AX inhibits lipid peroxidation, preventing liver damage. In particular, it could protect liver ischemia-reperfusion injury by reducing oxidant-induced protein carbonyl formation and conversion of xanthin dehydrogenase to xanthine oxidase (Curek et al., 2010; Yuan et al., 2011).

Sugar levels in blood can also be lowered by AX as well as nonesterified fatty acid and triglycerids. Also, AX can increase levels of high-density lipoprotein cholesterol and adiponectin. All these factors can help to improve insulin resistance in rats (Hussein et al., 2007).

Antioxidant activity also limits formation of atheroma in the vascular system, preventing atherosclerosis (Iwamoto et al., 2000). Studies have shown that AX can reduce arterial blood pressure in hypertensive rats in 14 days (Hussein et al., 2005).

Extracts from *H. pluvialis* can induce apoptosis in colon cancer cell growth, suggesting anticancer activity (Palozza et al., 2009). AX can suppress fibrosarcoma cell growth and stimulate immunity against tumor antigen (Jyonouchi et al., 2000). Dietary astaxanthin is also effective against mammary cancer by reducing the growth of mammary tumors up to 50% in mice (Chew et al., 1999). Other anticancer activities have been reported for gastric, lung, colon, breast, and prostate cancers, leukemia, oral fibrosarcoma, and embryonic fibroblast (Yuan et al., 2011). These findings open a great area for further studies in cancer therapy.

9.4 COSMECEUTICAL APPLICATIONS

Astaxanthin can protect skin against UV rays, composed of two types of waves: UVA and UVB. The UVB rays are the main cause of wrinkle formation and melanin production. However, UVA rays are responsible for much damage through photoaging. UVA rays affect deep into the dermis and damage collagen fibers (Lee et al., 2003). The AX mechanism of protection is unique since it is the only carotenoid that can protect cell membranes inside and outside.

In general, oral administration of AX improves the skin's moisture, reduces fine lines and wrinkles, and restores skin elasticity, and hence is of great interest for cosmeceutical applications.

Astaxanthin should be highly recommended as an active ingredient for sunscreen and after sun gels in low concentrations (between 2.5 and 10 μM) (Arakane, 2002). Also, it can be used in toothpaste for plaque control, hand cream for antimicrobial activity, cosmetic creams for antiaging, antiwrinkle, and antiacne effects, and lotions for antibacterial effects. In hair care, shampoo or conditioner could help sunlight exposure and chemical damage. Another promising application is antipigmentation properties. AX reduces melanin production by 40% in skin cells (Arakane, 2002), protecting from flakes, age spots, and blemishes. Further, it can be used in products for skin conditioning to avoid dryness and decrease swelling under the eyes.

Although AX seems to have multiple applications in cosmeceuticals and several other kinds of products can be formulated, only a few products have come on the market, and this could be attributed to the lack of enough natural sources of AX. Future biotechnology tools will allow us to further improve the efficiency of this important molecule to be used in the cosmeceutical industry.

REFERENCES

Arakane K. Superior skin protection via astaxanthin. *Carotenoid Research*, Vol. 5, April 2002.

Chan KC, Mong MC, Yin MC. Antioxidative and anti-inflammatory neuroprotective effects of astaxanthin and canthaxanthin in nerve growth factor differentiated PC12 cells. *J. Food Sci.* 2009;74(7):H225–31.

Chang CH, Chen CY, Chiou JY, Peng RY, Peng CH. Astaxanthine secured apoptotic death of PC12 cells induced by beta-amyloid peptide 25–35: its molecular action targets. *J. Med. Food.* 2010;13:548–56.

Chew BP, Park JS, Wong MW, Wong TS. A comparison of the anticancer activities of dietary beta-carotene, canthaxanthin and astaxanthin in mice *in vivo*. *Anticancer Res.* 1999;19:1849–53.

Chitchumroonchokchai C, Bomser JA, Glamm JE, Failla ML. Xanthophylls and alpha-tocopherol decrease UVB-induced lipid peroxidation and stress signaling in human lens epithelial cells. *J. Nutr.* 2004;134:3225–32.

Curek GD, Cort A, Yucel G, Demir N, Ozturk S, Elpek GO, Savas B, Aslan M. Effect of astaxanthin on hepatocellular injury following ischemia/reperfusion. *Toxicology* 2010;267:147–53.

Davis BH. Carotenoid metabolism in animals: a biochemist's view. *Pure Appl. Chem.* 1985;57:679–84.

Fan L, Vonshak A, Gabbay R, Hirshberg J, Cohen Z, Boussiba S. The biosynthetic pathway of astaxanthin in a green alga *Haematococcus pluvialis* as indicated by inhibition with diphenylamine. *Plant Cell Physiol.* 1995;36:1519–24.

Goodwin TW. *The biochemistry of the carotenoids*. Vols. 1 and 2. Chapman & Hall, London, 1984.

Goto S, Kogure K, Abe K, Kimata Y. Kitahama K, Yamashita E, Terada H. Efficient radical trapping at the surface and inside the phospholipid membrane is responsible for highly potent antiperoxidative activity of the carotenoid astaxanthin. *Biochim. Biophys. Acta* 2001;1512:251–58.

Grung M, D'Souza FML, Borowitzka M, Liaaen-Jensen S. Algal carotenoids 51. Secondary carotenoids 2. *Haematococcus pluviales* aplanospores as a source of (3S,3´S)-astaxanthin esters. *J. Appl. Phycol.* 1992;4:165–71.

Guerin M, Huntley ME, Olaizola M. *Haematococcus* astaxanthin: applications for human health and nutrition. *Trends Biotechnol.* 2003;21:210–16. Review.

Higuera-Ciapara I, Félix-Valenzuela L, Goycoolea FM. Astaxanthin: a review of its chemistry and applications. *Crit. Rev. Food Sci. Nutr.* 2006;46:185–96. Review.

Hussein G, Nakagawa T, Goto H, Shimada Y, Matsumoto K, Sankawa U, Watanabe H. Astaxanthin ameliorates features of metabolic syndrome in SHR/NDmcr-cp. *Life Sci.* 2007;80:522–29.

Hussein G, Sankawa U, Goto H, Matsumoto K, Watanabe H. Astaxanthin, a carotenoid with potential in human health and nutrition. *J. Nat. Prod.* 2006;69:443–49. Review.

Hussein G, Nakamura M, Zhao Q, Iguchi T, Goto H, Sankawa U, Watanabe H. Antihypertensive and neuroprotective effects of astaxanthin in experimental animals. *Biol. Pharm. Bull.* 2005;28:47–52.

Ikeda Y, Tsuji S, Satoh A, Ishikura M, Shirasawa T, Shimizu T. Protective effects of astaxanthin on 6-hydroxydopamine-induced apoptosis in human neuroblastoma SH-SY5Y cells. *J. Neurochem.* 2008;107:1730–40.

Iwamoto T, Hosoda K, Hirano R, Kurata H, Matsumoto A, Miki W, Kamiyama M, Itakura H, Yamamoto S, Kondo K. Inhibition of low-density lipoprotein oxidation by astaxanthin. *J. Atheroscler. Thromb.* 2000;7:216–22.

Izumi-Nagai K, Nagai N, Ohgami K, Satofuka S, Ozawa Y, Tsubota K, Ohno S, Oike Y, Ishida S. Inhibition of choroidal neovascularization with an anti-inflammatory carotenoid astaxanthin. *Invest. Ophthalmol. Vis. Sci.* 2008;49:1679–85.

Jyonouchi H, Sun S, Iijima K, Gross MD. Antitumor activity of astaxanthin and its mode of action. *Nutr. Cancer* 2000;36:59–65.

Kupcinskas L, Lafolie P, Lignell A, Kiudelis G, Jonaitis L, Andersen LP, Watstrong T. Efficacy of the natural antioxidant astaxanthin in the treatment of functional dyspepsia in patients with or without *Helicobacter pylori* infection: a prospective, randomized, double blind, and placebo-controlled study. *Phytomedicine* 2008;15:391–99.

Lee JH, Kim YT. Cloning and characterization of the astaxanthin biosynthesis gene cluster from the marine bacterium *Paracoccus haeundaensis*. *Gene* 2006;370:86–95.

Lee JH, Kim YS, Choi TJ, Lee WJ, Kim YT. *Paracoccus haeundaensis* sp. nov., a gram-negative, halophilic, astaxanthin-producing bacterium. *Int. J. Syst. Evol. Microbiol.* 2004;54:1699–702.

Lee SJ, Bai SK, Lee KS, Namkoong S, Na HJ, Ha KS, Han JA, Yim SV, Chang K, Kwon YG, Lee SK, Kim YM. Astaxanthin inhibits nitric oxide production and inflammatory gene expression by suppressing I (kappa) B kinase-dependent NF-kappaB activation. *Mol. Cells* 2003;16:97–105.

Lignell A, Botiger P. 2001. Use of xanthropylls, astaxanthin e.g. for treatment of autoimmune diseases, chronic viral and intracellular bacterial infections. Patent WO01/24787 A1.

Liu X, Shibata T, Hisaka S, Osawa T. Astaxanthin inhibits reactive oxygen species-mediated cellular toxicity in dopaminergic SH-SY5Y cells via mitochondria-targeted protective mechanism original research. *Brain Res.* 2009;1254:18–27.

Lorenz RT, Cysewski GR. Commercial potential for *Haematococcus* microalgae as a natural source of astaxanthin. *Trends Biotechnol.* 2000;18:160–67.

Macedo RC, Bolin AP, Marin DP, Otton R. Astaxanthin addition improves human neutrophils function: *in vitro* study. *Eur. J. Nutr.* 2010;49:447–57.

Martin HD, Jager C, Ruck C, Schmidt M. Anti- and prooxidant properties of carotenoids. *J. Prakt. Chem.* 1999;341:302–8.

Nagaki Y, Mihara M, Takahashi J, Kitamura A, Horita Y, Sugiura Y, Tsukahara H. The effects of astaxanthin on retinal capillary blood flow in normal volunteers. *J. Clin. Ther. Med.* 2005;21:537–42.

Nakamura A, Isobe R, Otaka Y, Abematsu Y, Nakata D, Honma C, Sakurai S, Shimada Y, Horiguchi M. Changes in visual function following peroral astaxanthin. *Jpn. J. Clin. Ophthalmol.* 2004;58(6):1051–54.

Neil HJ, Leenheer AP. Microbial sources of carotenoid pigments uses in foods and feeds. *J. Appl. Bacteriol.* 1991;70:181–91.

Nishida Y, Yamashita E, Miki W. Comparison of astaxanthin's singlet oxygen quenching activity with common fat and water soluble antioxidants. Results presented at the 21st Annual Meeting on Carotenoid Research, Osaka, Japan, September 6–7, 2007.

Ohgami K, Shiratori K, Kotake S, Nishida T, Mizuki N, Yazawa K, Ohno S. Effects of astaxanthin on lipopolysaccharide-induced inflammation *in vitro* and *in vivo*. *Invest. Ophthalmol. Vis. Sci.* 2003;44(6):2694–701.

Palozza P, Torelli C, Boninsegna A, Simone R, Catalano A, Mele MC, Picci N. Growth-inhibitory effects of the astaxanthin-rich alga *Haematococcus pluvialis* in human colon cancer cells. *Cancer Lett.* 2009;283:108–17.

Schmidt I, Schewe H, Gassel S, Jin C, Buckingham J, Hümbelin M, Sandmann G, Schrader J. Biotechnological production of astaxanthin with *Phaffia rhodozyma/Xanthophyllomyces dendrorhous*. *Appl. Microbiol. Biotechnol.* 2011;89:555–71.

Shen H, Kuo CC, Chou J, Delvolve A, Jackson SN, Post J, Woods AS, Hoffer BJ, Wang Y, Harvey BK. Astaxanthin reduces ischemic brain injury in adult rats. *FASEB J.* 2009;23:1958–68.

Steinbrenner J, Sandmann G. Transformation of the green alga *Haematococcus pluvialis* with a phytoene desaturase for accelerated astaxanthin biosynthesis. *Appl. Environ. Microbiol.* 2006;72:7477–84.

Yokoyama A, Miki W. Composition and presumed biosynthetic pathway of carotenoids in the astaxanthin-producing bacterium *Agrobacterium aurantiacum*. *FEMS Microbiol. Lett.* 1995;128:139–44.

Yuan JP, Peng J, Yin K, Wang JH. Potential health-promoting effects of astaxanthin: a high-value carotenoid mostly from microalgae. *Mol. Nutr. Food Res.* 2011;55:150–65.

10 Effect of Marine Collagen Peptide on Skin Condition

Kenji Sato and Yasutaka Shigemura

CONTENTS

10.1 INTRODUCTION

Collagen is defined as proteinous components with a triple helical structure in the extracellular matrix. Collagen forms a family consisting of closely related but different gene products and divided into two groups: fibrillar and nonfibrillar collagens (van der Rest and Garrone, 1991; Heino, 2007). In mammalian and avian tissues, the fibrillar collagen includes type I, II, III, V, and XI collagens (Sato et al., 1989; Zylberberg et al., 1992). Type I and II collagens are major constituents in skin, bone, tendon (type I), and cartilage (type II) in vertebrates, including fish, and type V collagen is codistributed with type I collagen as minor components in fish (Sato et al., 1989). In cartilage, type XI collagen is codistributed with type II collagen (Kelly et al., 1988). Thus, type I and II collagens are mainly used for collagen-related food ingredients. The collagens specifically consist of hydroxyproline (Hyp) and hydroxylysine (Hyl), posttranslational modified amino acids of proline (Pro) and lysine (Lys), respectively. The triple helical domain of the fibrillar collagen has a Gly-X-Y repeating motif, where X and Y are frequently occupied by Pro and Hyp, respectively (van der Rest and Garrone, 1991). The triple helical structure will be lost by heating the solution and converted into a globular structure. The denatured form of collagen is referred to as gelatin, which can be dissolved in hot water, yielding a viscous solution, and gel formation occurs by cooling the gelatin solution. Gelatin has long been used as a food ingredient as well as in folk medicine in Asia to improve blood circulation and arrest bleeding (Yao et al., 1989). In Western countries, gelatin consumption has been believed to improve joint condition by reducing pain (Moskowitz, 2000). Recently, partially hydrolyzed gelatin products have been prepared by enzymatic hydrolysis in order to increase their solubility in cold water. These are referred to as collagen peptides, and in some cases, simply as collagen in commercial products. Large amounts of collagen peptide preparations have been obtained from bone and skin of cattle, pig, chicken, etc. However, after the prevalence of bovine spongiform encephalopathy, much interest has been focused on production of collagen peptides from fish scales, skin, bone, etc. Episodes suggesting that ingestion of the collagen peptide would improve subjective skin conditions are widely prevalent in Japan and other countries. In Japan, annual sales of collagen peptide as a food supplement reached approximately US$300 million in 2007. Recent human trials using placebo control also have revealed that daily ingestion of collagen peptides increases skin surface water content and improves skin condition (Ohara et al., 2009; Koyama, 2009). Some animal experiments also have revealed the beneficial effects of collagen peptides on skin (Matsuda et al., 2006; Tanaka et al., 2009; Zhuang et al., 2009). In contrast, it has been believed that collagen peptides might be degraded into amino acids during digestion and absorption processes. Collagen predominantly consists of nonessential amino acids: Gly, Ala, Pro, Gln, Asn, etc. (van der Rest and Garrone, 1991). Hence the beneficial effect of collagen peptide could not be explained by supplementation of amino acids for collagen synthesis. In 2005, our group demonstrated the occurrence of 25–60 μM levels of food-derived collagen peptides in human peripheral blood after ingestion of chicken collagen peptides (Iwai et al., 2005). This finding was a breakthrough for elucidation of the mechanism of the beneficial effect of collagen peptide by ingestion.

In this chapter, recent knowledge on beneficial effects on skin and the metabolic fate of food collagen peptides is reviewed and a possible mechanism for the effect of collagen peptide on skin is discussed.

10.2 EFFECT OF COLLAGEN PEPTIDE INGESTION ON SKIN CONDITION

As mentioned above, there are cases suggesting improvement of skin condition by ingestion of collagen peptide, including fish collagen peptide. As shown in Table 10.1, Matsumoto et al. (2006) reported subjective improvement of the skin condition of women volunteers after ingestion of fish collagen peptide for 6 weeks. The subjects positively responded to ingestion of collagen peptide in unexpectedly high rates. However, placebo control was not used in this study. Therefore, this study was followed by a double-blind placebo-controlled study by the same group (Ohara et al., 2009), in which healthy women volunteers aged 25 to 45 were recruited. However, the authors could not find significant improvement in objective and subjective conditions of skin in the groups supplemented with fish collagen peptide compared to the placebo group, when all subjects were analyzed (Figure 10.1, left side). On the other hand, significantly higher water content on the skin surface was observed in the group that consumed 5 and 10 g of collagen peptide/day than that of the placebo group, when only subjects more than 30 years old were analyzed (Figure 10.1, right side). Recently, Koyama (2009) demonstrated that women who ingested 5 or 10 g of pig skin collagen peptide for 2 weeks and longer perceived improvement of their skin condition at a higher ratio than that of the placebo group (Figure 10.2). In this study, women aged between 40 and 54 years old were

TABLE 10.1
Subjective Improvement of Symptoms on Skin after Ingestion of a Fish Collagen Peptide for 6 Weeks (Number of Subjects)

Symptom	Subjects with Symptom	Degree of Improvement		
		Improved	Unchanged	Exacerbated
Face dryness	25	23	2	0
Difficulty of applying makeup	22	21	1	0
Rash caused by makeup	5	5	0	0
Flush	13	10	3	0
Pimple	15	13	2	0
Wrinkle	25	15	9	1
Speckle	25	18	7	0
Under-eye dark circle	20	19	1	0
Dropsy	15	10	4	1

Note: This study did not use placebo control (Matsumoto et al., 2006).

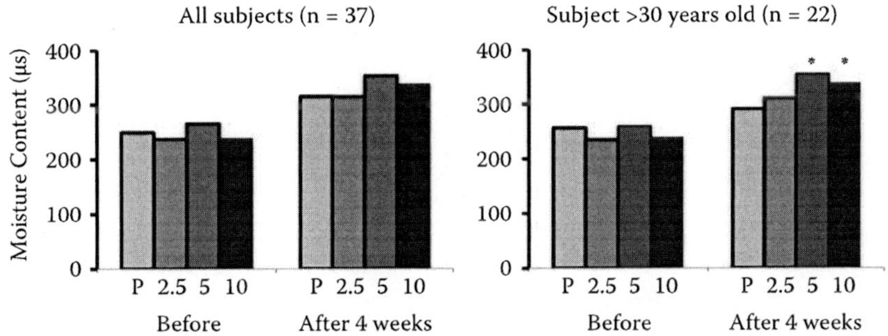

FIGURE 10.1 Moisture content of stratum corneum of cheek of women before and after ingestion of fish collagen peptide for 4 weeks. Moisture content was estimated by determining conductance. P: placebo group; 2.5, 5, and 10 g collagen peptides were daily supplemented. All subjects and the subjects older than 30 years were analyzed. * represents significant difference from the placebo group ($p < 0.05$). Refer to the paper by Matsumoto et al. (2006) for details.

recruited. These facts suggest that ingestion of collagen peptide might have a positive impact on the skin condition of women with a specific background. Further studies are necessary to determine the background of subjects who positively responded to ingestion of collagen peptide. In addition to improvement of skin condition, it has been demonstrated that the supplementation of collagen peptide can moderate joint pain (Moskowitz, 2000; Trč and Bohmová, 2010).

In an animal experiment, Tanaka et al. (2009) demonstrated that supplementation of fish collagen peptide can suppress UV-B-induced hyperplasia of epidermis and decrease skin hydration. Zhuang et al. (2009) also demonstrated that supplementation of jellyfish collagen peptide suppressed UV-B-induced reduction of antioxidative enzymes and glutathionine content in mice skin. They also found that jellyfish collagen peptide suppressed the decrease of lipid, ceramide, and collagen contents in skin. On the basis of these facts, Zhuang et al. (2009) suggested that a protective effect of jellyfish collagen peptide on skin against UV irradiation might involve antioxidative properties and synthesis of endogenous collagen.

In addition to improvement in skin condition, some animal studies have also demonstrated that ingestion of gelatin or collagen peptide increases bone mineral density of animals with malnutrition and enhances the healing process of bone fractures in comparison to milk proteins (Koyama et al., 2001; Wu et al., 2004; Nomura et al., 2005; Tsuruoka et al., 2007; Hata et al., 2008; Watanabe-Kamiyama et al., 2010).

Together with the results of human trials, it could be assumed that ingestion of gelatin and collagen peptide might enhance the healing process of the damaged extracellular matrix by enhancing

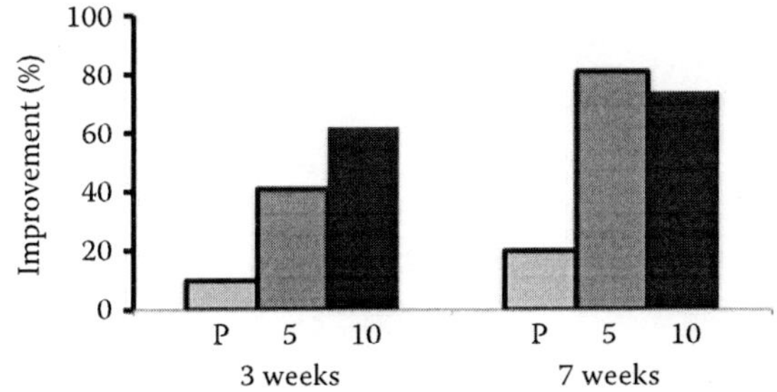

FIGURE 10.2 Improvement of subjective conditions of skin by ingestion of porcine skin collagen peptides after 3 and 7 weeks. Women aged from 40 to 54 years old ingested 5 or 10 g of porcine skin collagen peptide daily. P: placebo group. Percentages who perceived subjective improvement are shown. Refer to the paper by Koyama (2009) for details.

synthesis of extracellular matrix components or growth of the cells, which are responsible for the synthesis of extracellular matrix components.

10.3 METABOLIC FATE OF COLLAGEN PEPTIDE IN HUMANS

It has been demonstrated that the half-life of di- or tripeptide in human blood is a few minutes (Hubl et al., 1989; Druml et al., 1991). In addition, plasma levels of some food-derived peptides with inhibitory activity against angiotensin-converting enzyme have been demonstrated to be less than 1 nM (Foltz et al., 2007). Such low concentrations and short half-lives of food-derived peptide in human blood have cast serious suspicion on the proposed beneficial functions and their mechanisms of food-derived peptides. However, only apparently active peptides, which had been identified by *in vitro* activity-guided fractionation without considering bioavailability, were examined for content in blood. Hence, there is a possibility that occurrence of other food-derived peptides in the human circulation system has been overlooked. Therefore, comprehensive analyses of food-derived peptides in blood and tissues should be carried out. As mentioned in the introduction section, collagen and its derivatives specifically consist of hydroxyproline (Hyp). Therefore, collagen peptide content in blood can be estimated by determining the plasma level of peptide from Hyp, which can be obtained by subtracting free Hyp from total Hyp in HCl hydrolysate. As shown in Table 10.2, approximately 25–140 μM levels of collagen peptide are present 1 h after the ingestion of collagen peptides at 10–25 g/60 kg of body weight (Iwai et al., 2005; Ohara et al., 2007). Significant amounts of collagen peptide can be detected in human peripheral blood (Iwai et al., 2005; Ohara et al., 2007; Ichikawa et al., 2010) even 4 h after ingestion. These studies clearly demonstrate that unexpectedly high levels of food-derived collagen peptide are absorbed into the human blood circulation system and survive a few hours.

Until now, Pro-Hyp, Pro-Hyp-Gly, Gly-Pro-Hyp, Ala-Hyp, Ala-Hyp-Gly, Ser-Hyp, Ser-Hyp-Gly, Leu-Hyp, Ile-Hyp, and Phe-Hyp have been identified in human peripheral blood (Iwai et al., 2005; Ohara et al., 2007; Ichikawa et al., 2010). Pro-Hyp is a major food-derived collagen peptide among these peptides. Only negligible amounts of these peptides are present in human blood before the ingestion of collagen peptides, except for the blood of patients suffering from bone metastasis of cancer (Mazzi et al., 1996). Hence these peptides can be considered food-derived collagen peptides.

As shown in Table 10.2, Hyp-containing peptide content is significantly higher in the blood after ingestion of fish scale-derived collagen peptides than of porcine skin-derived collagen peptides at the same dose, and human subjects, which could be explained by the specific occurrences of Ala-Hyp, Leu-Hyp, Ile-Hyp, Phe-Hyp, Pro-Hyp-Gly, Ala-Hyp-Gly, and Ser-Hyp-Gly after ingestion of fish scale-derived collagen peptides rather than the increased amount of Pro-Hyp (Ohara et al., 2007).

TABLE 10.2
Maximum Level of Hydroxyproline (Hyp)-Containing Peptide in Human Peripheral Blood after Ingestion of Collagen Peptides

Source of Collagen Peptide	Dose (g/60 kg body weight)	Hyp-Containing Peptide (μM)
Chicken feet (type I)	23	60
Chicken cartilage (type II)	11	25
Porcine skin (type I)	25	80
Fish scale (type I)	25	140

Source: Data for chicken collagens: Iwai, K., et al., *J. Agric. Food Chem.*, 53, 6531–6536, 2005. Data for porcine and fish collagens: Ohara, H., et al., *J. Agric. Food Chem.*, 55, 1532–1535, 2007.

Note: These values are the means of seven (chicken) and five (porcine and fish) subjects.

To elucidate the mechanism for incorporation of these collagen peptides, an *in vitro* study used the porcine brush-border membrane vesicles (BBMVs) system, on which peptidases and H^+-coupled peptide transporter 1 (PEPT1) are located. Aito-Inoue et al. (2007) demonstrated that Gly-Pro-Hyp is cleaved into Gly and Pro-Hyp and incorporated into the vesicles. The incorporation of Pro-Hyp is enhanced by pH gradient and suppressed by competitive substrate (Gly-Pro) for PEPT1. Therefore, it could be concluded that the dipeptides, such as Pro-Hyp, are incorporated into enterocytes mainly via PEPT1. However, a transport system such as the paracellular diffusion pathway might also contribute to incorporation of collagen peptides, especially tripeptide (Shimizu et al., 2010).

Pro-Hyp and other food-derived collagen peptides resist peptidases in human blood (Iwai et al., 2005). It has been demonstrated that Pro-Hyp is present in human urine after ingestion of gelatin (Prockop et al., 1962). Hence these facts indicate that collagen peptides in food (average molecular weight from 300 to 5,000 Da) are degraded into di- and tripeptides by gastrointestinal digestion and absorbed into the blood system, at least, via the PEPT1 system. Further, the collagen peptides in blood might target cells and remaining collagen peptides are finally excreted into urine.

10.4 POSSIBLE MECHANISM FOR BENEFICIAL EFFECT OF COLLAGEN PEPTIDE INGESTION

Some collagen-derived oligopeptides such as Pro-Hyp-Gly and Pro-Hyp have demonstrated that they exert a chemotactic action on fibroblasts, peripheral blood neutrophils (Poltlethwaite et al., 1978; Laskin et al., 1986), and monocytes (Poltlethwaite and Kang, 1976) in a Boyden chemotaxis chamber in mM levels. These cells play a significant role in wound healing and inflammation (Martin and Leibovich, 2005). Therefore, it could be speculated that food-derived collagen peptides such as Pro-Hyp might enhance wound healing by stimulating the migration of fibroblasts and other cells to the injured tissues (Iwai et al., 2005). However, it has been demonstrated that fibroblasts stop their growth in the extracellular matrix and on collagen gel, while they grow rapidly in plastic wells and reach confluence within a few days (Kono et al., 1990). Therefore there exists a possibility that the fibroblasts in the Boyden chemotaxis chamber might not reflect the *in vivo* response of fibroblasts. On collagen gel, fibroblasts stop growing, which can mimic the fibroblast in extracellular matrix (Kono et al., 1990). We demonstrated that Pro-Hyp enhances the growth of mouse primary cultured fibroblasts in a dose-dependent manner on the collagen gel in the presence of growth factors, as shown in Figure 10.3 (Shigemura et al., 2009). On the other hand, Pro-Hyp does not enhance growth of fibroblast on the collagen gel in the absence of growth factors. Our group also demonstrated that Pro-Hyp increases the number of fibroblasts migrating from the skin disk, as shown in Figure 10.4. These facts clearly indicate that

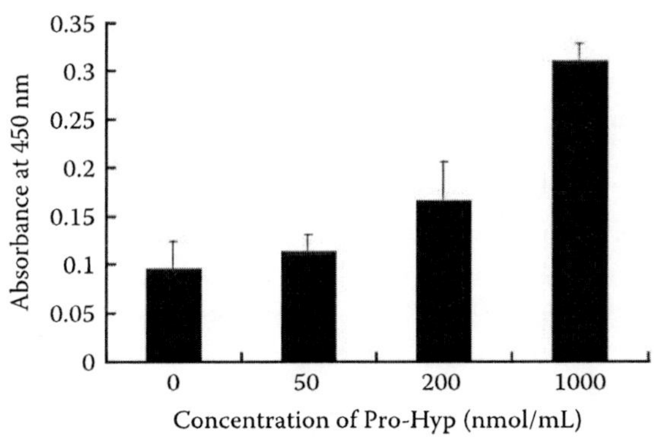

FIGURE 10.3 Enhancement of growth of mouse skin primary cultured fibroblast on collagen gel by addition of Pro-Hyp in the medium containing 10% FBS. After 144 h, cell number was estimated by the colorimetric method. The data are shown as the mean ± SE (n = 8). The different letters on the column indicate significant differences ($p < 0.05$). Refer to the paper by Shigemura et al. (2009).

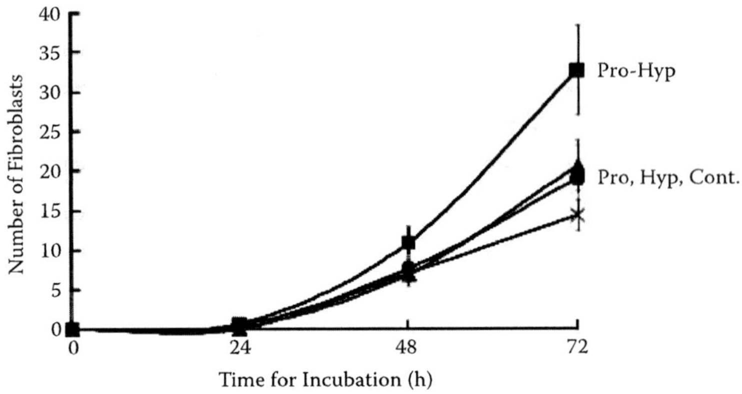

FIGURE 10.4 Effect of Pro-Hyp and its constituting amino acids on the number of fibroblasts migrated from a mouse skin disk. Refer the paper by Shigemura et al. (2009) for details.

fibroblasts on collagen and skin start to grow responding to Pro-Hyp. On the basis of these results, we proposed a hypothesis that Pro-Hyp can abolish the suppression of fibroblast growth by attaching collagen fibrils and consequently enhance wound healing in damaged tissues in the presence of growth factors. This hypothesis could at least partially explain the improvement of damaged skin. After our finding (Shigemura et al., 2009), two groups also demonstrated the effect of Pro-Hyp on the fibroblast and chondrocyte. Nakatani et al. (2009) demonstrated that Pro-Hyp increases glycosaminoglycan synthesis by chondrocytes and also suppresses calcification of chondrocytes in the cell culture system, and also confirmed that Pro-Hyp suppresses progress of high-phosphate diet-induced osteoarthritis in the rat model. Recently, Ohara et al. (2010) revealed that Pro-Hyp increases hyaluronic acid synthesis by fibroblasts in a culture system. Hyaluronic acid, a nonsulfated glycosaminoglycan, and proteoglycans that have a core protein with one or more covalently attached glycosaminoglycan are distributed in connective tissues, which contribute to tissue hydrodynamics and movement (Gandhi and Mancera, 2008). Hence these findings also could be linked to improvements of skin and joint conditions as suggested by human studies.

10.5 CONCLUSION

The major food-derived collagen peptide, Pro-Hyp, can affect growth of fibroblasts and synthesis of extracellular components *in vitro*, which might be linked to the suggested protective effect on connective tissues and increase of moisture on the skin. Little is known of the molecular mechanism behind Pro-Hyp-mediated stimulation of fibroblast growth and synthesis of connective tissue components. Pro-Hyp might act by binding the cell surface receptor or by passing into the target cell via the transporter. Alternatively, Pro-Hyp might affect the interaction between the extracellular matrix and cells.

According to our knowledge, it has not been reported that ingestion of collagen peptide or gelatin mediates pathological fibrosis and excess glycosaminoglycan synthesis. The underlying mechanism for control of excessive synthesis of connective tissue components in the presence of food-derived collagen peptides is unknown. Further studies on the molecular mechanism of the stimulatory effects of food-derived collagen peptides on cell growth and production of extracellular components and their control system are necessary.

REFERENCES

Aito-Inoue, M., Lackeyram, D., Fan, Z.M. Sato, K., and Mine, Y. 2007. Transport of a tripeptide, Gly-Pro-Hyp, across the porcine intestinal brush-border membrane. *J. Peptide Sci.* 13, 468–474.

Druml, W., Lochs, H., Roth, E., Hubl, W., Balcke, P., and Lenz, K. 1991. Utilization of tyrosine dipeptides and acetyltyrosine in normal and uremic humans. *Am. J. Physiol.* 260, E280–E285.

Foltz, M., Meynen, E.E., Bianco, V., Platerink C., van, Koning T.M.M.G., and Kloek, J. 2007. Angiotensin converting enzyme inhibitory peptides from a lactotripeptide-enriched milk beverage are absorbed intact into the circulation. *J. Nutr.* 137, 953–958.

Gandhi, N.S., and Mancera, R.L. 2008. The structure of glycosaminoglycans and their interactions with proteins. *Chem. Biol. Drug Des.* 72, 455–482.

Hata, S., Hayakawa, T., Okada, H., Hayashi, K., Akimoto, Y., and Yamamoto, H. 2008. Effect of oral administration of high advanced-collagen tripeptide (HACP) on bone healing process in rat. *J. Hard Tissue Biol.* 17, 17–22.

Heino, J. 2007. The collagen family members as cell adhesion proteins. *Bioessays* 29, 1001–1010.

Hubl, W., Druml, W., Langer, K., and Lochs, H. 1989. Influence of molecular structure and plasma hydrolysis on the metabolism of glutamine-containing dipeptides in humans. *Metabolism* 38, 59–62.

Ichikawa, S., Morifuji, M., Ohara, H., Matsumoto, H., Takeuchi, Y., and Sato, K. 2010. Hydroxyproline-containing dipeptides and tripeptides quantified at high concentration in human blood after oral administration of gelatin hydrolysate. *Int. J. Food Sci. Nutr.* 61, 52–60.

Iwai, K., Hasegawa, T., Taguchi, Y., Morimatsu, F., Sato, K., Nakamura, Y., Higashi, A., Kido, Y., Nakabo, Y., and Ohtsuki, K. 2005. Identification of food-derived collagen peptides in human blood after oral ingestion of gelatin hydrolysates. *J. Agric. Food Chem.* 53, 6531–6536.

Kelly, J., Tanaka, S., Hardt, T., Eikenberry, E.G., and Brodsky, B. 1988. Fibril-forming collagen in lamprey. *J. Biol. Chem.* 263, 980–987.

Kono, T., Tanii, T., Furukawa, M., Mizuno, N., Kitajima, J., Ishii, M., Hamada, T., and Yoshizato, K. 1990. Cell cycle analysis of human dermal fibroblasts cultured on or in hydrated type I collagen lattices. *Arch Dermatol. Res.* 282, 258–262.

Koyama, Y. 2009. Effect of collagen peptide ingestion on the skin. *Shokuhinto Kaihatsu* 44, 10–12.

Koyama, Y., Hirota, A., Mori, H., Takahara, H., Kuwaba, K., Kusubata, M., Matsubara, Y., Kasugai, S., Itoh, M., and Irie, S. 2001. Ingestion of gelatin has differential effect on bone mineral density and body weight in protein undernutrition. *J. Nutr. Sci. Vitaminol. (Tokyo)* 47, 84–86.

Laskin, D. L., Kimura, T., Sakakibara, S., Riley, D. J., and Berg, R.A. 1986. Chemotactic activity of collagen-like polypeptides for human peripheral blood nutrophils. *J. Leukocyte Biol.* 39, 255–266.

Martin, P., and Leibovich, S.J. 2005. Inflammatory cells during wound repair: The good, the bad and the ugly. *Trends Cell Biol.* 15, 599–607.

Matsuda, N., Koyama, Y., Hosaka, Y., Ueda, H., Watanabe, T., Araya, T., Irie, S., and Takehana, K. 2006. Effects of ingestion of collagen peptide on collagen fibrils and glycosaminoglycans in the dermis. *J. Nutr. Sci. Vitaminol.* 52, 211–215.

Matsumoto, H., Ohara, H., Itoh, K., Nakamura, Y., and Takahashi, S. 2006. Clinical effect of fish type I collagen hydrolysate on skin properties. *ITE Lett.* 7, 386–390.

Mazzi, G., Fioravanzo, F., and Burti, E. 1996. New marker of bone resorption: Hydroxyproline-containing peptide. High-performance liquid chromatographic assay without hydrolysis as an alternative to hydroxyproline determination: A preliminary report. *J. Chromatogr. B* 678, 165–172.

Moskowitz, R. W. 2000. Role of collagen hydrolysate in bone and joint disease. *Semin. Arthritis Rheum.* 30, 87–99.

Nakatani, S., Mano, H., Sampei, C., Shimizu, J., and Wada, S. 2009. Chondroprotective effect of the bioactive peptide prolyl-hydroxyproline in mouse articular cartilage *in vitro* and *in vivo*. *Osteoarthritis Carilage* 17, 1620–1627.

Nomura, Y., Oohashi, K., Watanabe, M., and Kasugai, S. 2005. Increase in bone mineral density through oral administration of shark gelatin to ovariectomized rats. *Nutrition* 21, 1120–1126.

Ohara, H., Ichikawa, S., Matsumoto, H., Akiyama, M., Fujimoto, N., Kobayashi, T., and Tajima, S. 2010. Collagen-derived dipeptide, proline-hydroxyproline, stimulates cell proliferation and hyaluronic acid synthesis in cultured human dermal fibroblasts. *J. Dermatol.* 37, 330–338.

Ohara, H., Ito, K., Iida, H., and Matsumoto, H. 2009. Improvement in the moisture content of the stratum corneum following 4 weeks of collagen hydrolysate ingestion. *Nippon Shokuhin Kogaku Kaishi* 56, 137–145.

Ohara, H., Matsumoto, H., Ito, K., Iwai, K., and Sato, K. 2007. Comparison of quantity and structures of hydroxyproline-containing peptides in human blood after oral ingestion of gelatin hydrolysates from different sources. *J. Agric. Food Chem.* 55, 1532–1535.

Poltlethwaite, A.E., and Kang, A. 1976. Collagen and collagen peptide induce chemotaxis of human blood monocytes. *J. Exp. Med.* 143, 1299–1307.

Poltlethwaite, A.E., Seyer, J.M., and Kang, A. 1978. Chemotactic attraction of human fibroblast to type I, II, and III collagens and collagen derived peptides. *Proc. Natl. Acad. Sci. USA* 75, 871–875.

Prockop, D.J., Keiser, H.R., and Sjoerdsma, A. 1962. Gastrointestinal absorption and renal excretion of hydroxyproline peptides. *Lancet* 2, 527–528.

Sato, K., Yoshinaka, R., Itoh, Y., and Sato, M. 1989. Molecular species of collagen in the intramusclar connective tissue of fish. *Comp. Biochem. Physiol.* 92B, 87–91.

Shigemura, Y., Iwai, K., Morimatsu, F., Iwamoto, T., Mori, T., Oda, C., Taira, T., Park, E.-Y, Nakamura, Y., and Sato, K. 2009. Effect of prolyl-hydroxyproline (Pro-Hyp), a food-derived collagen peptide in human blood, on growth of fibroblasts from mouse skin. *J. Agric. Food Chem.* 57, 444–449.

Shimizu, K., Sato, M., Zhang, Y., Kouguchi, T., Takahata, Y., Morimatsu, F., and Shimizu, M. 2010. Molecular size of collagen peptide reverses the permeability of Caco-2 cells. *Biosci. Biotechnol. Biochem.* 74, 1123–1125.

Tanaka, M., Koyama, Y., and Nomura Y. 2009. Effects of collagen peptide ingestion on UV-B induced skin damage. *Biosci. Biotechnol. Biochem.* 73, 930–932.

Trč, T., and Bohmová, J. 2010. Efficacy and tolerance of enzymatic hydrolysed collagen (EHC) vs. glucosamine sulphate (GS) in the treatment of knee osteoarthritis (KOA). *International Orthopaedics Online First*™, April 18.

Tsuruoka, N., Yamamoto, R., Sakai, Y., Yoshitake, Y., and Yonekura, H. 2007. Promotion by collagen tripeptide of type I collagen gene expression in human osteoblastic cell and fracture healing of rat femur. *Biosci. Biotechnol. Biochem.* 71, 2680–2687.

van der Rest, M., and Garrone, R. 1991. Collagen family of proteins. *FASEB J.* 5, 2814–2823.

Watanabe-Kamiyama, M., Shimizu, M., Kamiyama, S., Taguchi, Y., Sone, H., Morimatsu, M., Shirakawa, H., Furukawa, Y., and Komai, M. 2010. Absorption and effectiveness of orally administered low molecular weight collagen hydrolysate in rats. *J. Agric. Food Chem.* 58, 835–841.

Wu, J., Fujioka, M., Sugimoto, K., Mu, G., and Ishimi, Y. 2004. Assessment of effectiveness of oral administration of collagen peptide on bone metabolism in growing and mature rats. *J. Bone Miner. Metab.* 22, 547–553.

Yao, D.F., Zhang, Y.F., and Zhou, Y.F. 1989. Effect of Ejiao (colla coriiasini) on the hemodynamics, hemorheology, and microcirculation during endotoxin shock in dogs. *Zong guo Zhong Yao Za Zhi* 14, 44–46.

Zhuang, Y., Hou, H., Zhao, X., Zhang, Z., and Li, B. 2009. Effects of collagen and collagen hydrolysate from jellyfish (*Rhopilema esculentum*) on mice skin photoaging induced by UV irradiation. *J. Food Sci.* 74, H183–H188.

Zylberberg, L., Bonaventure, J., Cohen-Solal, L., Hartmann, D.J., and Bereiterhahn, J. 1992. Organization and characterization of fibrillar collagens in fish scales *in situ* and *in vitro*. *J. Cell Sci.* 103, 273–285.

11 Fish Collagen and Tissue Repair

Yoshihiko Hayashi, Shizuka Yamada, Takeshi Ikeda, and Kajiro Yanagiguchi

CONTENTS

11.1 INTRODUCTION

Collagen is a major extracellular matrix which accounts for about 30% of total proteins in the animal body. Bovine collagen has been widely used and supplied for medical and cosmetic purposes, and also as a healthy food for a long time. However, bovine spongiform encephalopathy (BSE,) which was found in England in the late 1980s, has made a significant drawback for manufacturing biomaterials based on this material. Hence the use of collagen from other species would serve as an alternative to this problem. This social phenomenon caused many researchers to pay attention to collagen from marine resources such as fish scales, skin, and bone sold, which are readily available in food waste, and the demand for collagen hydrolysates of marine origin has increased rapidly.

This chapter focuses on and reviews the characteristics and biomedical application of marine collagen. An understanding of the production processes and conformation of this material is indispensable for promoting theoretical and practical applications. Furthermore, marine collagen is also applied as a scaffold and carrier for molecular therapies, such as drug and gene delivery systems, and for regenerative medicine using stem cells.

11.2 THE CHARACTERISTICS OF FISH COLLAGEN

Collagen is the major component of the extracellular matrix and is useful for mechanical protection of tissues, organs, and physiological regulation of a cellular environment (Kielty et al., 1993). Among various types, type I collagen has been extensively applied as a biomaterial. Marine collagen, especially fish collagen, is evaluated for its potential application as an alternative to mammalian collagen. The characteristics of fish collagen must be summarized to understand its properties.

11.2.1 Differences among Species

Type I collagen is unique due to its extremely high solubility in dilute acid (Gallop and Seifter, 1963; Yamaguchi et al., 1976) compared to avian and mammalian collagens. Type I collagens derived from lower vertebrates such as bony fish and lamprey were found to exhibit a high degree of structural similarity between their α1 and α2 chains compared with type I collagen from calf.

Collagen substrates are known to influence the growth characteristics of cells and also modulate various aspects of cell behavior like cell-adhesion, proliferation, and differentiation (Kinner et al., 2001; Nehrer et al., 1997, 1998; Srivastava et al., 1990a, 1990b; Smith et al., 1992; Shortkroff et al., 1996; Fujisato et al., 1996; Lee et al., 2001a; Spector, 2002). The disadvantages of collagen as a biomaterial for tissue repair are its low biomechanical stiffness and rapid biodegradation (Koide et al., 1993).

Collagens are easily extracted and purified from fish skins and bones with high yield. Therefore, the practical use of fish collagens could contribute to the recycling of unutilized resources and to the development of value-added materials.

11.2.2 Significances

An IgE-reactive protein was clinically observed in surimi from walleye pollack, a fish food, by ELISA using one patient serum positive to the high molecular weight allergen, suggesting it to be collagen (Hamada et al., 2000). Studies have reported that the high molecular weight allergen recognized by plural patient sera is collagen (Hamada et al., 2001). Atelocollagen is a processed natural biomaterial produced from bovine type I collagen. It has useful biomaterial characteristics such as rare inflammatory responses, high biocompatibility, and high biodegradability (Miyata et al., 1992; Hanai et al., 2006). The parts of collagen attributed to its immunogenicity, namely, telopeptides, are eliminated in the process of atelocollagen production. Therefore, atelocollagen possesses little immunogenicity (Sano et al., 2003). Since substantial amounts of collagen can be obtained from fish wastes (scale, skin, and bone), they can be applied as an alternative to bovine collagen in food, cosmetics, and biomedical materials.

11.2.3 Biochemical Characteristics

A great amount of collagen could be obtained from many marine vertebrates and invertebrates (Nagai et al., 1999, 2000, 2002a, 2002b; Nagai and Suzuki, 2000a, 2000b, 2000c, 2002). Amino acid composition and denaturation temperature were summarized to understand the biomedical applications of acid-solubilized and pepsin-solubilized collagens.

11.2.3.1 Composition of Amino Acids

The biochemical composition of marine collagen is thought to be different from that of mammalian collagen. For biochemical analyses, strict conditions for sample preservation is important and indispensable before collagen extraction. This means that the hydroxyproline content in relation to collagen stability strongly depends on these sampling procedures (Swatschek et al., 2002). Several works showed that amino acid composition of fish collagens was almost similar to that of mammalian collagens (Kimura, 1983; Kimura et al., 1988; Nagai et al., 2001; Nagai, 2004; Bae, 2008). Glycine was the most abundant amino acid and accounted for more than 30% of all amino acids. Furthermore, the degree of hydroxylation of proline was calculated to be 40–48%, which was also a level similar to that of the mammalian (about 45%). The linear relationship between collagen stability and hydroxyproline content was demonstrated by the degree of hydroxylation of proline in the fish collagen peptides, which was calculated at about 35% (personal communication from the laboratory of Prof. M. Yamauchi, North Carolina Oral Health Institute). Furthermore, it is very interesting that the degree of hydroxylation of proline of fish in cold seas, for example, chum salmon, was reported at a low level (35–37%) (Kimura et al., 1988; Matsui et al., 1991) compared to that of fish in a relatively not so cold sea, which is related to the denaturation temperature of fish (see next section).

11.2.3.2 Denaturation Temperature

Fish collagen fibrillar gels have not been studied, with the exception of shark collagen (Nomura et al., 2000a, 2000b), probably due to their low denaturation temperature (Td), which renders these materials difficult to handle. The Td of shark collagen solution is approximately 30°C (Nomura et al., 1995), which results in the dissolution of the fibrillar gel of this collagen at 37°C (Nomura et al., 2000a). This indicates that the gel could not be practically used at the actual physical temperature in human medical applications. The Td of chum salmon is approximately 19°C (Kimura et al., 1988; Matsui et al., 1991), which is the main reason it is unstable at the actual physical temperature of the human body. As the denaturation temperature of fish collagen is lower than the mammalian body temperature, fish collagen melts when placed in contact with the human body for a clinical application. Recently, collagen extract from the skin of ray fish or the scales of a tropical fish (tilapia) has been reported to have a Td of 33–34°C (Bae, 2008) and 35°C (Ikoma et al., 2003), respectively. Furthermore, the improvement can be achieved by chemical cross-linking *in vitro* collagen fibrillogenesis. This method brings the Td of salmon collagen to 55°C, and its biocompatible properties have been demonstrated by several studies (Nagai et al., 2004, 2007).

11.3 BIOMEDICAL APPLICATION OF FISH COLLAGEN

Numerous attempts have been recently made to use type I collagen for biomaterials. The cross-linking methods for stabilization of collagen are divided into physical treatment, such as ultraviolet irradiation (Weadock et al., 1995), dehydrothermal treatment (Weadock et al., 1995; Gorham et al., 1992; Koide et al., 1993; Wang et al., 1994; Pieper et al., 1999), and treatments involving chemicals such as glutaraldehyde (White et al., 1973), carbodiimide (Pieper et al., 1999), and 1-ethyl-3-(3-dimethyl- aminopropyl)-carbodiimide (EDC) (Yunoki et al., 2004). Chemical treatments confer remarkably high strength and stability to the collagen matrix, but may result in potential cytotoxicity or poor biocompatibility (Huang-Lee et al., 1990), while physical treatments have no cytotoxicity and can provide sufficient stability (Koide et al., 1993; Yunoki et al., 2003).

Tissue engineering requires a porous, biodegradable scaffold to replicate the natural extracellular matrix, which serves to organize cells spatially, provide them with environmental signals, and direct site-specific cellular regulation (Wang et al., 2006). Pore size, pore number, surface area, and pore wall morphology are widely recognized as important parameters as scaffolds for cell seeding, migration, growth, and new tissue formation in tissue engineering (Chen et al., 2002; Yang et al., 2001).

11.3.1 Soft Tissue Repair

Bioactive materials including collagen are clinically used for chronic wound repair (Plotner and Mostow, 2010). When they are used in conjunction with standard therapy for venous leg ulcers and diabetic foot ulcers, bioactive materials may increase the possibility and rate of healing. Type I collagen from fish is an adequate culture substrate, in either serum-supplemented or serum-free conditions, to support the adhesion and spreading process of fish cell proliferation, although calf type I collagen is a substrate with limitations, depending on the fish cell species and the medium conditions (Kaneko et al., 1995).

The fish scale-derived scaffold without cross-linking can be developed as a promising material for tissue engineering of cornea (Lin et al., 2010). The regenerated tissue in the salmon milt DNA (sDNA)/SC sponge application group showed similar morphology to native dermis, while the SC sponge application group without sDNA showed epithelial overgrowth, and blood capillary formation was significantly enhanced in the sDNA/SC group compared to the SC only group (Shen et al., 2008).

The relationship between a composition of cross-linker and cell activity was investigated after application of cross-linked salmon atelocollagen (SAC) to the scaffold of human periodontal ligament (HPDL) cells. SAC was cross-linked by EDC or dehydrothermal treatment (DHT) (Nagai et al., 2004). There was no difference in the thermal stability, porous structure, and cell proliferation rate between EDC-SAC and EDC–bovine AC (BAC). In addition, the collagen helix of EDC-SAC was partially denaturated, and this structure resulted in the enhancement of alkaline phosphatase (ALP) activity of HPDL cells compared to that using EDC-BAC.

11.3.2 Hard Tissue Repair

Salmon atelocollagen-cultured samples were irradiated at doses of 10, 15, and 25 kGy. The γ-irradiation had a significant impact on cell proliferation rate and ALP activity of MC3T3-E1 cells with γ-irradiation increased significantly (Nagai et al., 2006).

Fish collagen peptides (FCPs; 0.1%) stimulate expression and production of bone-forming genes (osteocalcin, osteopontin, BMP-2, and integrin β3mRNAs) and their proteins. In addition, they strongly accelerate the bone cavity of newly formed bone after the direct application of FCP powder (Hayashi et al., 2007). Dietary hydrolyzed collagen including that of fish origin increases osteoblastic activity (ALP) as measured in primary mouse bone cell culture (from 0.2 to 1.0 mg/ml), which acts on bone remodeling and increases the external diameter of the cortical area of mouse femurs (25g/kg of hydrolyzed collagen) (Guillermint et al., 2010). Shark fin is commonly used in Chinese cuisine, but dermal collagen is not yet utilized despite being a promising source of collagen. Shark skin collagen has been reported as an important source of marine collagen (Nomura, 2004). In ovariectomized model rats, shark skin collagen would be useful as a dietary supplement in the form of gelatin for treating osteoporosis (Nomura et al., 2005).

11.3.3 Cosmetics

Hydrophilic polymers of high molecular weight are generally used to mask wrinkles. These products do not penetrate deep through the skin barrier, but they present a surface effect after a short period of time. They retain water and decrease the loss of moisture from the skin surface following deposition of a film hindering water evaporation and minimizing the aspect of fine wrinkling (Rawlings et al., 2004). Some of these compounds belong to glycosaminoglycans, collagen, elastin, and wheat protein.

Skin hydration and the skin firming were investigated through dose-response effects of cosmetic formulations enriched in compounds derived from algae (cream) and fish collagen (serum) (Xhauflaire-Uhoda et al., 2008). The serum enriched in marine compounds showed a better moisturizing effect in a short time. The cream appeared more active later, particularly following repeat applications. A sustained tensor (firming) effect was observed during treatment with both the lotion and the cream. Furthermore, FC derived from chum salmon is useful to increase the moisture content because of the low Td.

TABLE 11.1
Development of Collagen Scaffolds

Generation	Constitution	Effects	References
First	Collagen without cross-linking	Simple preparation from fish scale; higher thermal stability	Lin et al., 2010; Pati et al., 2010
Second	Collagen with cross-linking Chemical methods: WSC Physical methods: UV or γ- irradiation	Improvement of low Td fish collagen and improvement of stability	Yunoki et al., 2004; Nagai et al., 2004; Yunoki et al., 2003; Nagai et al., 2006
Third	Mixture with other components Collagen-glycosaminoglycan	Induction of organ regeneration	Yannas et al., 2010
	Collagen-calcium phosphate	High compressive modulus	Al-Munajjed and O'Brien, 2009
		In vitro osteogenic activity	Al-Munajjed, 2009; Lyons et al., 2010
	Collagen-hydroxyapatite	Improvement of mechanical properties, in vitro mineralization, and in vivo bone healing potential	Gleeson et al., 2010
	Collagen-chitosan	Improvement of mechanical properties, superior biocompatibility	Tan et al., 2001; Shi et al., 2008; Zhu et al., 2009a; Shi et al., 2009 ; Niu et al., 2009; Zhu et al., 2009b

The incidence of BSE has resulted in marine collagen hydrolysate (low molecular weight gelatin) being sold as supplements and cosmetics in Japan. Gelatin is a multifunctional ingredient used in foods, pharmaceuticals, cosmetics, photographic films, and other products (Gudmundsson, 2002; Karim and Bahat, 2009; Yang et al., 2007; Zhou and Regenstein, 2004). Gelatin gels have relatively lower melting temperatures than gels of other gelling agents (Williams, 2007). Gelatin gels generally have a melting temperature below 35°C, which is below human body temperature.

11.3.4 Regenerative Medicine

Regenerative medicine consists of three components: cell, nutrient, and scaffold. The scaffold functions to (1) provide structural integrity and to define a potential space for the engineered tissue, (2) guide the restructuring that occurs through the proliferation of the donor cells and the ingrowth of the host tissue, (3) maintain distances between the parenchymal cells that permit diffusion of gas and nutrients and, possibly, the ingrowth of vasculature from the host bed, and (4) transmit the tissue-specific mechanical forces to cue the behavior of the cells within it (Vacanti et al., 1998). Table 11.1 shows the classification according to processing steps of fish collagen scaffolds.

Polymer scaffolds are central to tissue engineering technology because they direct a variety of cellular processes based on the structural and biochemical properties of the scaffold (Yang et al., 2001; Seal et al., 2001; Chen, 2002). The materials used for scaffold fabrication not only determine such physical properties as biocompatibility, biodegradability, and mechanical stability, but also provide the appropriate signals for directing the cellular processes that lead to tissue formation (Dillow and Lowman, 2002; Yang et al., 2001). Collagen is an ideal scaffold or carrier for tissue engineering as it supports many kinds of connective tissues such as skin, tendon, bone, cartilage, blood vessel, and ligaments (Bateman et al., 1996; Brown and Timpl, 1995; Chevallay and Herbage,

2000; Eyre, 1980; Kadler, 1995; Kemp, 2000; Lee et al., 2001b; Patino et al., 2002; Prockop and Kivirikko, 1995; Ricard-Blum et al., 2000; Wong Po and Kaplan, 2002; Yang et al., 1997).

Jellyfish collagen with telopeptides enhanced not only the IgM production of human hybridoma (HB4C5) cells, but also the IgM and IgG production of human peripheral blood lymphocytes (Sugahara et al., 2006). In addition, collagen from jellyfish stimulated not only transcription activity, but also translation activity for enhanced immunoglobulin and cytokine production (Nishimoto et al., 2008). *In vivo* responses of jellyfish atelocollagen were investigated concerning the safety for biological applications in comparison with bovine collagen (Song et al., 2006). Its scaffolds had a highly porous and interconnected pore structure, which is useful for high-density cell seeding, an efficient nutrient and an oxygen supply to the cells cultured in the three-dimensional matrices. To determine whether this collagen evokes any specific inflammatory response compared to that induced by bovine collagen or gelatin, jellyfish atelocollagen was implanted onto the dorsal side of normal mice. It was found to induce immunological responses at least comparable to those caused by bovine collagen and gelatin. Further *in vitro* and *in vivo* studies are indispensable before the clinical application of fish collagens.

11.4 CONCLUSIONS

Fish collagen is useful in foods, cosmetics, and regeneration medicine because of its reduced potential for infectious diseases such as BSE, foot and mouth disease, and avian influenza, due to the large evolutionary distance between fish and mammals. Establishment of good manufacturing (GMP), good laboratory (GLP), and good clinical practices (GCP) are indispensable for clinical applications using fish collagens.

REFERENCES

Al-Munajjed, A. 2009. Development of novel collagen-calcium phosphate scaffolds for bone tissue engineering and bone repair. Dublin: Royal College of Surgeons in Ireland and Dublin University, Trinity College (PhD).

Al-Munajjed, A. A., and O'Brien, F. J. 2009. Influence of a novel calcium-phosphate coating on the mechanical properties of highly porous collagen scaffolds for bone repair. *J. Mech. Behav. Biomed. Mater.* 2:138–146.

Bae, I. 2008. Studies on biochemical utilization of acid-soluble collagen and gelatinolytic enzymes from ray species in Ariake Sound. Nagasaki University Graduate School of Science and Engineering (Doctor's thesis).

Bateman, J. F., Lamande, S. R., and Ramshaw, J. A. M. 1996. Collagen superfamily. In *Extracellular matrix*, ed. W. D. Comper, 22–67. Amsterdam: Harwood Academic Publishers.

Brown, J. C., and Timpl, R. 1995. The collagen superfamily. *Int. Arch. Allergy Immunol.* 104:484–490.

Chen, G., Ushida, T., and Tateishi, T. 2002. Scaffold design for tissue engineering. *Macromol. Biosci.* 2:67–77.

Chevallay, B., and Herbage, D. 2000. Collagen-based biomaterials as 3D scaffold for cell cultures: application for tissue engineering and gene therapy. *Med. Biol. Eng. Comput.* 38:211–218.

Dillow, A. K., and Lowman, A. M. 2002. *Biomimetic materials and design; biointerfacial strategies, tissue engineering, and targeted drug delivery*. New York: Marcel Dekker.

Eyre, D. R. 1980. Collagen: molecular diversity in the body's protein scaffold. *Science* 207:1315–1322.

Fujisato, T., Sajiki, T., Lin, Q., and Ikada, Y. 1996. Effect of basic fibroblast growth factor on cartilage regeneration in chondrocyte-seeded collagen sponge scaffold. *Biomaterials* 17:155–162.

Gallop, P., and Seifter, S. 1963. Preparation and properties of soluble collagens. In *Methods in enzymology*, ed. S. P. Colowick and N. O. Kaplan, 635–641. Vol. VI. New York: Academic Press.

Gleeson, J. P., Plunkett, N. A., and O'Brien, F. J. 2010. Addition of hydroxyapatite improves stiffness, interconnectivity and osteogenic potential of a highly porous collagen-based scaffold for bone tissue regeneration. *Eur. Cell. Mater.* 20:218–230.

Gorham, S. D., Light, N. D., Diamond, A. M., et al. 1992. Effect of chemical modifications on the susceptibility of collagen to proteolysis. II. Dehydrothermal crosslinking. *Int. J. Biol. Macromol.* 14:129–138.

Gudmundsson, M. 2002. Rheological properties of fish gelatins. *J. Food Sci.* 67:2172–2176.

Guillerminet, F., Beaupied, H., Fabien-Soule, V., et al. 2010. Hydrolyzed collagen improves bone metabolism and biomechanical parameters in ovariectomized mice: an *in vitro* and *in vivo* study. *Bone* 46:827–834.

Hamada, Y., Genka, E., Ohira, M., Nagashima, Y., and Shiomi, K. 2000. Allergenecity of fish paste products and surimi from walleye pollack. *Shokuhin Eiseigaku Zasshi* (in Japanese) 41:38–43.

Hamada, Y., Nagashima, Y., and Shiomi, K. 2001. Identification of collagen as a new fish allergen. *Biosci. Biotechnol. Biochem.* 65:285–291.

Hanai, K., Takeshita, F., Honma, K., et al. 2006. Atelocollagen-mediated systemic DDS for nucleic acid medicines. *Ann. N.Y. Acad. Sci.* 1082:9–17.

Hayashi, Y., Yamada, S., Ikeda, T., and Yanagiguchi, K. 2007. Bone regeneration agent and method. Patent application (Japan) 2007-27314.

Huang-Lee, L. L. H., Cheung, D. T., and Nimni, M. E. 1990. Biochemical changes and cytotoxicity associated with the degradation of polymeric glutaraldehyde derived crosslinks. *J. Biomed. Mater. Res.* 24:1185–1201.

Ikoma, T., Kobayashi, H., Tanaka, J., Walsh, D., and Mann, S. 2003. Physical properties of type I collagen extracted from fish scales of *Pagrus major* and *Oreochromis niloticas*. *Int. J. Biol. Macromol.* 32:199–204.

Kadler, K. 1995. Extracellular matrix 1: fibril-forming collagens. *Protein Profile* 2:491–619.

Kaneko, Y., Suzuki, K., Igarashi, M., et al. 1995. Effects of fish and calf type I collagens as culture substrate in the adhesion and spreading process of established fish cells. *In Vitro Cell. Dev. Biol. Anim.* 31:178–182.

Karim, A. A., and Bahat, R. 2009. Fish gelatin: properties, challenges, and prospect as an alternative to mammalian gelatins. *Food Hydrocolloid* 23:563–576.

Kemp, P. D. 2000. Tissue engineering and cell-populated collagen matrices. *Methods Mol. Biol.* 139:287–293.

Kielty, C. M., Hoplinson, I., and Grant, M. E. 1993. *Part I: connective tissue and its heritable disorder*, eds. P. M. Royce and B. Steinmann, 103–147. New York: Wiley-Liss.

Kimura, S. 1983. Vertebrate skin type I collagen: comparison of bony fishes with lamprey and calf. *Comp. Biochem. Physiol. B* 74:525–528.

Kimura, S., Zhu, X.-P., Matsui, R., Shinjoh, M., and Takamizawa, S. 1988. Characterization of fish muscle type I collagen. *J. Food Sci.* 53:1315–1318.

Kinner, B., Zaleskas, J., and Spector, M. 2001. Contractile behavior of human connective tissue and mesenchymal stem cells in collagen-GAG matrices for tissue engineering. *Tissue. Eng.* 7:617.

Koide, M., Osaki, K., Konishi, J., et al. 1993. A new type of biomaterial for artificial skin: dehydrothermally cross-linked composites of fibrillar and denatured collagen. *J. Biomed. Mater. Res.* 27:79–84.

Lee, C. H., Singla, A., and Lee, Y. 2001. Biomedical applications of collagen. *Int. J. Pharm.* 221:1–22.

Lee, C. R., Grodzinsky, A. J., and Spector, M. 2001. The effects of cross-linking of collagen-glycosaminoglycan scaffolds on compressive stiffness, chondrocyte-mediated contraction, proliferation and biosynthesis. *Biomaterials* 22:3145–3154.

Lin, C. C., Ritch, R., Lin, S. M., et al. 2010. A fish scale-derived scaffold for corneal regeneration. *Eur. Cell. Mater.* 19:50–57.

Lyons, F. G., Al-Munajjed, A. A., Kieran, S. M., et al. 2010. The healing of bony defects by cell-free collagen-based scaffolds compared to stem cell-seeded tissue engineered constructs. *Biomaterials* 31:9232–9243.

Matsui, R., Ishida, M., and Kimura, S. 1991. Characterization of an _3 chain from the skin type I collagen of chum salmon (*Oncorhynchus keta*). *Comp. Biochem. Physiol. B* 99:171–174.

Miyata, T., Taira, T., and Noishiki, Y. 1992. Collagen engineering for biomaterial use. *Clin. Mater.* 9:139–148.

Nagai, N., Anzawa, T., Satoh, Y., et al. 2006. Activities of MC3T-E1 cells cultured on γ-irradiated salmon atelocollagen scaffold. *J. Biosci. Bioeng.* 101:511–514.

Nagai, N., Mori, K., Satoh, Y., et al. 2007. *In vitro* growth and differentiated activities of human periodontal ligament fibroblasts cultured on salmon collagen gel. *J. Biomed. Mater. Res. A* 82:395–402.

Nagai, N., Yunoki, S., Suzuki, T., et al. 2004. Application of cross-linked salmon atelocollagen to the scaffold of human periodontal ligament cells. *J. Biosci. Bioeng.* 97:389–394.

Nagai, T. 2004. Collagen from diamondback squid (*Thysanoteuthis rhombus*) outer skin. *Z. Naturforsch. C* 59:271–275.

Nagai, T., Araki, Y., and Suzuki, N. 2002a. Collagen of the skin of ocellate phffer fish (*Takifugu rubripes*). *Food Chem.* 78:173–177.

Nagai, T., Nagamiori, K., Yamashita, E., and Suzuki, N. 2002b. Collagen of octopus *Callistoctopus arakawai* arm. *Int. J. Food Sci. Technol.* 37:285–289.

Nagai, T., Ogawa, T., Ito, T., et al. 1999. Collagen of edible jellyfish exumbrella. *J. Sci. Food. Agric.* 79:855–858.

Nagai, T., and Suzuki, N. 2000a. Isolation of collagen from fish waste material-skin, bone and fins. *Food Chem.* 68:277–281.

Nagai, T., and Suzuki, N. 2000b. Partial characterization of collagen from purple sea urchin (*Anthocidaris crassispina*) test. *Int. J. Food Sci. Technol.* 35:497–501.

Nagai, T., and Suzuki, N. 2000c. Preparation and characterization of several fish bone collagen. *J. Food Biochem.* 24:427–436.

Nagai, T., and Suzuki, N. 2002. Preparation and partial characterization of collagen from paper nautilus (*Argonauta argo*, Linnaeus) outer skin. *Food Chem.* 76:149–153.

Nagai, T., Worawattanamateekul, W., Suzuki, N., et al. 2000. Isolation and characterization of collagen from rhizostomous jellyfish (*Rhopilema asamushi*). *Food Chem.* 70:205–208.

Nagai, T., Yamashita, E., Taniguchi, K., Kanamori, N., and Suzuki, N. 2001. Isolation and characterization of collagen from the outer skin waste material of cuttlefish (*Sepia lycidas*). *Food Chem.* 72:425–429.

Nehrer, S., Breinan, H. A., Ramappa, A., et al. 1997. Matrix collagen type and pore size influence behavior of seeded canine chondrocytes. *Biomaterials* 18:769–776.

Nehrer, S., Breinan, H. A., Ramappa, A., et al. 1998. Chondrocyte-seeded collagen matrices implanted in a chondral defect in a canine model. *Biomaterials* 19:2313–2328.

Nishimoto, S., Goto, Y., Morishige, H., et al. 2008. Mode of action of the immunostimulatory effect of collagen from jelly fish. *Biosci. Biotechnol. Biochem.* 72:2806–2814.

Niu, X., Feng, Q., Wang, M., Guo, X., and Zheng, Q. 2009. Porous nano-HA/collagen/PLLA scaffold chitosan microspheres for controlled delivery of synthetic peptide derived from BMP-2. *J. Control. Release* 134:111–117.

Nomura, Y. 2004. Properties and utilization of shark collagen. In *More efficient utilization of fish and fisheries product*, 147–158. New York: Elsevier.

Nomura, Y., Oohashi, K., Watanabe, M., and Kasugai, S. 2005. Increase in bone mineral density through oral administration of shark gelatin to ovariectomized rats. *Nutrition* 21:1120–1126.

Nomura, Y., Toki, S., Ishii, Y., and Shirai, K. 2000a. The physicochemical property of shark type I collagen gel and membrane. *J. Agric. Food Chem.* 48:2028–2032.

Nomura, Y., Toki, S., Ishii, Y., and Shirai, K. 2000b. Improvement of the material property of shark type I collagen by composing with pig type I collagen. *J. Agric. Food Chem.* 48:6332–6336.

Nomura, Y., Yamano, M., and Shirai, K. 1995. Renaturation of α1 chains from shark skin collagen type I. *J. Food Sci.* 60:1233–1236.

Pati, F., Adhikari, B., and Dhara, S. 2010. Isolation and characterization of fish scale collagen of higher thermal stability. *Bioresource Technol.* 101:3737–3742.

Patino, M. G., Neiders, M. E., Andreana, S., et al. 2002. Collagen as an implantable material in medicine and dentistry. *J. Oral Implantol.* 28:220–225.

Pieper, J. S., Oosterhof, A., Dijkstra, P. J., Veerkamp, J. H., and van Kuppevelt, T. H. 1999. Preparation and characterization of porous crosslinked collagenous matrices containing bioavailable chondroitin sulphate. *Biomaterials* 20:847–858.

Plotner, A. N., and Mostow, E. N. 2010. A review of bioactive materials and chronic wounds. *Cutis* 85:259–266.

Prockop, D., and Kivirikko, K. 1995. Collagens: molecular biology, diseases, and potentials for therapy. *Annu. Rev. Biochem.* 64:403–434.

Rawlings, A. V., Cansestrari, D. A., and Dobkowski, B. 2004. Moisturizer technology versus clinical performance. *Dermatol. Ther.* 17:49–56.

Ricard-Blum, S., Dublet, B., and van der Rest, M. 2000. *Unconventional collagens*. New York: Oxford University Press.

Sano, A., Maeda, M., Nagahara, S., et al. 2003. Atelocollagen for proteinand gene delivery. *Adv. Drug Deliv. Rev.* 55:1651–1677.

Seal, B. L., Otero, T. C., and Panitch, A. 2001. Polymeric biomaterials for tissue and organ regeneration. *Mater. Sci. Eng. Res. Rep.* 34:147–230.

Shen, X., Nagai, N., Murata, M., et al. 2008. Development of salmon milt DNA/salmon collagen composite for wound dressing. *J. Mater. Sci. Mater. Med.* 19:3473–3479.

Shi, H., Han, C., Mao, Z., Ma, Lie., and Gao, C. 2008. Enhanced angiogenesis in porous collagen-chitosan scaffolds loaded with angiogenin. *Tissue Eng.* 14:1775–1785.

Shi, S., Cheng, X., Wang, J. et al. 2009. RhBMP-2 microspheres-loaded chitosan/collagen scaffold enhanced osteointegration: an experiment in dog. *J. Biomater. Appl.* 23:331–344.

Shorkroff, S., Barone, L., Hsu, H. P. et al. 1996. Healing of chondral and osteochondral defects in a canine model: the role of cultured chondrocytes in regeneration of articular cartilage. *Biomaterials* 17:147–154.

Smith, M. D., Shearer, M. G., Srivastava, S. et al. 1992. Quantitative evaluation of the growth of established cell lines on the surface of collagen, collagen composite and reconstituted basement membrane. *Urol. Res.* 20:285–288.

Song, E., Kim, S. Y., Chum, T., Byun, H.-J., and Lee, Y. M. 2006. Collagen-scaffold derived from a marine source and their biocompatibility. *Biomaterials* 27:2951–2961.

Spector, M. 2002. Novel cell-scaffold interactions encountered in tissue engineering contractile behavior of musculoskeletal connective tissue cells. *Tissue Eng.* 8:351–357.

Srivastava, S., Gorham, S. D., and Courtney, J. M. 1990a. The attachment and growth of an established cell line on collagen, chemically modified collagen, and collagen composite surface. *Biomaterials* 11:162–168.

Srivastava, S., Gorham, S. D., French, D. A., Shivas, A. A., and Courtrey, J. M. 1990b. *In vivo* evaluation and comparison of collagen, acetylated collagen and collagen/glycosaminoglycan composite films and sponges as candidate biomaterials. *Biomaterials* 11:155–161.

Sugahara, T., Ueno, M., Goto, Y., et al. 2006. Immmunostimulation effect of the jellyfish collagen. *Biosci. Biotechnol. Biochem.* 70:2132–2137.

Swatschek, D., Schatton, W., Kellermann, J., Mullaer, W. E. G., and Kreuter, J. 2002. Marine sponge collagen: isolation, characterization and effects on the skin parameters surface-pH, moisture and sebum. *Eur. J. Pharmaceut. Biopharmaceut.* 53:107–113.

Tan, W., Krishnaraj, R., and Desal, T. A. 2001. Evaluation of nanostructured composite collagen-chitosan matrices for tissue engineering. *Tissue Eng.* 7:203–210.

Vacanti, J. P., Langer, R., Upton, J., and Murler, J. J., 1998. Transplantation of cells in matrices for tissue generation. *Adv. Drug Deliv. Rev.* 33:165–182.

Wang, J.-H., Hung, C.-H., and Young, T.-H. 2006. Proliferation and differentiation of neural stem cells on lysine-alanine sequential polymer substrates. *Biomaterials* 27:3441–3450.

Wang, M. C., Pins, G. D., and Silver, F. H. 1994. Collagen fibers with improved strength for the repair of soft tissue injuries. *Biomaterials* 15:507–512.

Weadock, K. S., Miller, F. J., Bellincampi, L. D., Zawadsky, J. P., and Dunn, M. G. 1995. Physical crosslinking of collagen fibers: comparison of ultraviolet irradiation and dehydrothermal treatment. *J. Biomed. Mater. Res.* 29:1373–1379.

White, M. J., Kohno, I., Rubin, A. I., Stenzel, K. H., and Miyata, T. 1973. Collagen films: effect of cross-linking on physical and biological properties. *Biomater. Med. Dev. Art. Org.* 1:703–715.

Williams. P. A. 2007. *Handbook of industrial water soluble polymers*, 5–76. New York: Wiley-Blackwell.

Wong Po, F. C., and Kaplan, D. L. 2002. Genetic engineering of fibrous proteins: spider dragline silk and collagen. *Adv. Drug Deliv. Rev.* 54:1131–1143.

Xhauflaire-Uhoda, E., Fontaine, K., and Pierard, G. E. 2008. Kinetics of moisturizing and firming effects of cosmetic formulations. *Int. J. Cosmet. Sci.* 30:131–138.

Yamaguchi, K., Lavety, J., and Love, R. M. 1976. The connective tissues of fish. VIII. Comparative studies on hake, cod and catfish collagens. *J. Food Technol.* 11:389–399.

Yannas, I. V., Tzeranis, D. S., Harley, B. A., and So, P. T. C. 2010. Biologically active collagen-based scaffolds: advances in processing and characterization. *Phil. Trans. R. Soc. A* 368:2123–2139.

Yang, C., Li, S. W., Helmine, D. L., et al. 1997. Apoptosis of chondrocytes in transgenic mice lacking collagen II. *Exp. Cell Res.* 235:370–373.

Yang, H., Wang, Y., Jiang, M., et al. 2007. 2-step optimization of the extraction and subsequent physical properties of channel cat fish (*Ictalurus punctatus*) skin gelatin. *J. Food Sci.* 72:C188–C195.

Yang, S., Leong, K. F., Du, Z. H., and Chus, C. K. 2001. The design of scaffolds for use in tissue engineering. Part 1. Traditional factors. *Tissue Eng.* 7:679–689.

Yunoki, S., Nagai, N., Suzuki, T., and Munekata, M. 2004. Novel biomaterial from reinforced salmon collagen gel prepared by fibril formation and cross-linking. *J. Biosci. Bioeng.* 98:40–47.

Yunoki, S., Suzuki, T., and Takai, M. 2003. Stabilization of low denaturation temperature collagen from fish by physical cross-linking methods. *J. Biosci. Bioeng.* 96:575–577.

Zhou, P., and Regenstein, J. M. 2004. Optimization of extraction conditions for pollock skin gelatin. *J. Food Sci.* 69:393–398.

Zhu, C., Fan, D., Duan, Z., Xue, W., et al. 2009. Initial investigation of novel human-like collagen/chitosan scaffold for vascular tissue engineering. *J. Biomed. Mater. Res. A* 89:829–840.

Zhu, Y., Liu, T., Song, K., et al. 2009. Collagen-chitosan polymer as a scaffold for the proliferation of human adipose tissue-derived stem cell. *J. Mater. Sci. Mater. Med.* 20:799–808.

12 Red Algal Mycosporine-Like Amino Acids (MAAs) as Potential Cosmeceuticals

Yvonne V. Yuan and Yasantha Athukorala

CONTENTS

12.1 INTRODUCTION

Historically, edible marine macroalgae or seaweeds (sometimes referred to as sea vegetables) from the Protista taxonomies Phaeophyceae (brown (B)), Chlorophyceae (green (G)), and Rhodophyceae (red (R)) have been commonplace in Pacific (e.g., Indonesia, Philippines, New Zealand, Hawaii) and Asian (e.g., Japan, China, Korea) diets, and albeit to a lesser extent, in those of France, Iceland, Ireland, Norway, Wales, as well as the Atlantic Canadian provinces and Maine in the United States. A more recent, growing area of inquiry has focused on marine macroalgae and constituents as functional foods and nutraceuticals for potential beneficial health effects as sources of antioxidants and bioactives in reducing the risk for various diet-related chronic diseases such as atherosclerosis and hyperlipoproteinemia in cardiovascular disease (CVD), as well as breast and colon carcinogenesis. Coincident with the growing body of evidence elucidating the efficacy of edible macroalgae and their constituents in reducing the risk of chronic disease in animal models, such as mammary (Yamamoto et al., 1987; Teas et al., 1984), intestinal (Lee and Sung, 2003; Yamamoto and Maruyama, 1985), and skin carcinogenesis (Yamamoto et al., 1986; Higashi-Okai et al., 1999), is the rapidly expanding literature elucidating the endogenous antioxidant activities of marine macroalgae (reviewed by Yuan, 2007). Macroalgal antioxidant constituents include cellular enzymatic defenses (e.g., superoxide dismutase, glutathione reductase); aqueous and lipophilic molecules including L-ascorbic acid, glutathione (GSH), carotenoids, tocopherols, and chlorophyll derivatives (e.g., chlorophyll-a, pheophytin a); as well as secondary small molecule metabolites such as flavan-3-ols, phenolic acids, lignans, phlorotannins (phloroglucinol polymers) in brown macroalgae, and the UV-absorbing sunscreen mycosporine-like amino acids (e.g., mycosporine-glycine and shinorine; MAAs) in red marine macroalgae.

Mycosporines, which were first identified as fungal metabolites, contain an aminocyclohexenone ring system with UV absorption at 310 or 320 nm (Bandaranayake, 1998; Leach and Trione, 1965). Later, mycosporine-like amino acids, characterized by a cyclohexenone or cyclohexenimine chromophore conjugated with the nitrogen moiety of an amino acid (Table 12.1), were identified in cyanobacteria, phytoplankton, marine macroalgae, and marine animals spanning the entire globe, including tropical, subtropical, temperate, Arctic, and Antarctic waters (Shick and Dunlap, 2002; Conde et al., 2000; Bandaranayake, 1998; Karsten et al., 1998a, 1998b). These aqueous or water-soluble secondary metabolites, with conjugated double bonds in their structure, are noted to have absorbance maxima between 310 and 360 nm, and are thus thought to function as UV-absorbing sunscreens in marine species, including various red marine macroalgae (*Palmaria palmata*, *Porphyra tenera*, *Devaleraea ramentacea*), corals (*Palythoa tuberculosa*), and small animals such as sea squirts (*Lissoclinum stellatus*) (Aguilera et al., 2002; Karsten and Wiencke, 1999; Dunlap and Yamamoto, 1995; Takano et al., 1979). Similar to the flavonoids and phenolic acids, MAAs are derived from the Shikimic acid pathway via 3-dehydroquinic acid and 4-deoxygadusol (4-DG) (Shick and Dunlap, 2002; Dunlap et al., 1997). The resonance tautomers of 4-DG exist in equilibrium with each other prior to conjugation with an amino acid such as glycine or taurine to form the oxo-carbonyl MAA mycosporine-glycine or -taurine. Mycosporine-glycine can then undergo conversion to form one of several imino-MAAs, such as shinorine or porphyra-334 (Table 12.1). A protective UV screening function for these compounds can be inferred from the fact that their absorbance maxima overlap not only with UVB (280–320 nm) and UVA irradiation (320–400 nm), but also the absorbance maxima for DNA (260 and 280 nm). Moreover, 4-DG as a Shikimic acid pathway intermediate has been noted to possess strong antioxidant activity (Dunlap and Yamamoto, 1995).

Skin photodamage is manifested in alterations to the dermal extracellular matrix (ECM) (Pinnell, 2003); this damage is exhibited in the development of wrinkles, loss of elasticity, coarseness, mottled pigmentation as hyper- or hypomelanosis, increased epidermal thickness, as well as changes to the connective tissue (collagen, proteoglycans, and glycoproteins) (Pallela et al., 2010; Ryu et al., 2009; Pinnell, 2003). Moreover, degraded collagen is due to the inhibition of collagen biosynthesis mediated by matrix metalloproteinases (MMPs), an array of secreted or transmembrane Zn-endopeptidases capable of breaking down dermal ECM. MMP expression is induced by cell signaling mechanisms such as the presence of growth factors, cytokines, tumor promoters, as well as UV irradiation (Pallela et al., 2010; Ryu et al., 2009). It is well known that exposure to solar, and thereby, UV irradiation greatly enhances normal age-related skin damage. Moreover, not only are UVB wavelengths known to be absorbed by the epidermis causing sunburn, but they are also noted to be responsible for tumor initiation and squamous cell carcinomas (Schmid et al., 2006; Pinnell, 2003). Conversely, while UVA wavelengths are not as energetic as those of the shorter UVB range, they are noted to penetrate much more deeply into the epidermis and dermis layers and are thereby associated with premature aging and tumor promotion. It is noteworthy that in sunlight, UVA irradiation is 10-fold more efficient than UVB at causing lipid peroxidation. It is also known

TABLE 12.1
The Major Mycosporine-Like Amino Acids (MAAs) Identified in Red Marine Macroalgae and Their ESI/MS Structural Information[a]

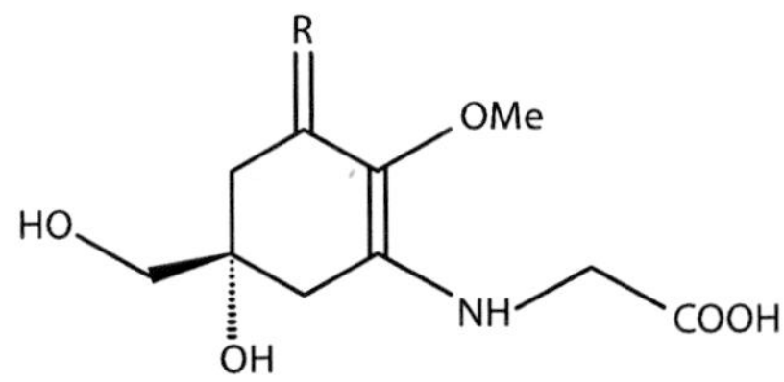

MAA	R	λ_{max}	ε ($M^{-1}cm^{-1}$)	Mass	$[M+H]^+$	ESI/MS Fragmentation Pattern Values
Palythine	HN	320	36,200	244.11	245.00	137, 150, 155, 162, 184, 186, 199, 209, 230, 245
Mycosporine-glycine	O	310	28,888	243.00	ND	ND
Shinorine	N-CH(CO_2H)CH_2OH	332	44,668	332.12	333.10	137, 168, 185, 186, 197, 211, 230, 241, 255, 333
Asterina-330	N-CH_2CH_2OH	330	4,800	288.13	289.10	137, 150, 168, 186, 197, 199, 213, 230, 243, 273, 289
Palythinol	N-CH(CH_3)CH_2OH	330	43,500	302.15	303.10	137, 150, 168, 186, 197, 199, 231, 243, 288, 303
Porphyra-334	N-CH(CO_2H)CH(OH)CH_3	332	42,300	346.14	347.10	137, 151, 168, 185, 186, 197, 200, 243, 303, 347
Usujirene	N-CH=CHCH_3 (cis)	356	50,000	284.00	285.10	137, 149, 166, 179, 193, 195, 197, 202, 223, 241, 285
Palythenic acid	N-C(CO_2H)=CHCH_3	337	29,200	328.00	329.00	138, 150, 175, 182, 193, 197, 225, 237, 241, 251, 268, 283, 296, 314, 329
Palythene	N-CH=CHCH_3 (trans)	360	50,000	284.00	285.00	137, 149, 166, 179, 193, 195, 197, 205, 223, 241, 285

[a] ESI/MS = electrospray ionization/mass spectroscopy, ND = not detected, ε = molar absorptivity. References: Rastogi et al., 2010; Yuan et al., 2009; Cardozo et al., 2006; Schick and Dunlap, 2002; Whitehead et al., 2001; Bandaranayake, 1998.

that UVA wavelengths are responsible for DNA strand scission and oxidation of nucleic acids, as well as inducing MMP synthesis (Pinnell, 2003). Thus, promotion of the use of topical sunscreens with sun protection factors (SPFs) of 15 or greater has been encouraged by health care providers as well as in the lay media. However, compliance rates and studies of sunscreen use have indicated that instead of application at the recommended 2 mg/cm^2, actual sunscreen use approaches levels of only 0.5 mg/cm^2 or less. Sunscreen use may be limited by the opacity of current sun block creams containing titanium dioxide or concerns about the safety of the active ingredients; thus development of "natural" plant-derived cosmeceutical products may improve the use and compliance of sunscreen products and thereby increase protection against photodamage and potentially skin carcinogenesis.

Taken together, the lines of evidence outlined above support the investigation of the potential utilization of red marine macroalgal MAAs as cosmeceuticals. For example, recent studies have evaluated the cosmeceutical potential of MAAs such as porphyra-334 incorporated into a commercial *Aloe vera* gel (Bhatia et al., 2010), as well as an MAA-containing extract derived from *Porphyra umbilicalis* commercialized as Helioguard-365 (Schmid et al., 2003, 2006), which will be discussed further below. Thus, this chapter reviews the extraction, identification, and quantitation of red marine macroalgal MAAs, as well as the evidence underlying the mechanisms of action and bioactivities of these small molecules as antioxidants, and thereby, potentially cosmeceuticals.

12.2 BACKGROUND AND OCCURRENCE OF MYCOSPORINE-LIKE AMINO ACIDS (MAAS)

MAAs are known to be produced *de novo* via the Shikimic acid pathway, which occurs in fresh- and seawater plants as well as certain bacteria, such as *Escherichia coli* (Volkmann and Gorbushina, 2006; Bandaranayake, 1998; Bentley, 1990). It is noteworthy that MAAs have not been reported in higher terrestrial plants (Tevini and Teramura, 1989; Karentz et al., 1991) where morphological adaptations, pigments, and other UV-protecting compounds (i.e., furanocoumarins, flavonoids) are synthesized to provide protection from solar irradiation. More recently, Balskus and Walsh (2010) identified an MAA biosynthetic gene cluster encoding four enzymes in the cyanobacterium *Anabaena variabilis* (ATCC 29413). These enzymes comprised a dehydroquinate synthase homologue (DHQS) adjacent to an *O*-methyltransferase (O-MT), a gene with homology to the adenosine triphosphate (ATP)-grasp superfamily as well as a putative amide bond-forming enzyme such as an ATP-grasp or nonribosomal peptide synthase (NRPS) homologue for synthesis of cyanobacterial MAAs. These workers hypothesized that the DHQS homologue identified as Ava_3858 and O-MT Ava_3857 could synthesize 4-DG, and that ATP-grasp homologue Ava_3856 and NRPS-like enzyme Ava_3855 might attach glycine and serine to 4-DG via imine bonds. It is known that MAAs can occur in cyanobacteria grown under a wide variety of environmental stresses, i.e., elevated temperatures, cold, drought, salinity, osmotic and UV stress (Sinha et al., 1998). Interestingly, very high intracellular MAA concentrations (98 mM; 3% of the cell wall weight) have been reported in unicellular cyanobacteria studied in Eilat, Israel (Oren, 1997). As well, terrestrial lichens (genera *Collema*, *Gonohymenia*, and *Peltula*), which are symbionts of fungi, algae, or cyanobacteria, have been reported to contain MAAs (Budel et al., 1997; Torres et al., 2004).

While various microalgae (phytoplankton, e.g., Bacillariophyceae, Chlorarachniophyceae, Chlorophyceae, Chrysophyceae, Cryptophyceae, Cyanophyceae, Dinophyceae, Euglenophyceae, Eustigmatophyceae, Prasinophyceae, Prymnesiophyceae, Raphidophyceae, Haptophyceae, and Phodophyceae) have been reported to contain MAAs or related compounds (Llewellyn and Airs, 2010; Jeffrey et al., 1999; Shick and Dunlap, 2002; Singh et al., 2008), surface bloom-forming dinoflagellates, cryptomonads, and raphidophytes accumulate much higher concentrations of MAAs than other microalgae (Carreto et al., 2001; Jeffrey et al., 1999).

Recently, considerable progress has been made in determining the MAA profiles of marine macroalgae. For example, MAAs have been reported to occur in Rhodophyceae, Chlorophyceae, and Phaeophyceae marine macroalgae; however, in the latter two taxonomies, MAAs are usually found in low and trace concentrations, respectively. MAAs are not only abundant in red marine macroalgae, but are also often present with great diversity (Rastogi et al., 2010); the most common MAAs identified in red marine macroalgae are summarized in Table 12.1. There are several comprehensive surveys published on the MAA composition of polar to warm-temperate region macroalgae (Karsten et al., 1998a), tropical macroalgae (Karsten et al., 1998b), phytoplankton as well as green, brown, and red marine macroalgae (Gröniger and Häder, 2000). Red marine macroalgal species belonging to *Acanthophora*, *Bangia*, *Bostrychia*, *Caloglossa*, *Catenella*, *Ceramium*, *Chondrus*, *Corallina*, *Devaleraea*, *Gelidiella*, *Gelidium*, *Gracilaria*, *Iridea*, *Palmaria*, *Phyllophora*, *Polysiphonia*, *Porphyra*, and *Stictosiphonia* have been reported to be rich in MAAs (Rastogi et al., 2010; Sinha et al., 2007). For example, in *Porphyra umbilicalis*, the most widely consumed red marine macroalgal species in the world (known as Nori in Japan and elsewhere), MAAs comprise 1.4% of the dry mass. On the other hand, the mass of MAAs in other red marine macroalgae ranges between 0.16 and 0.84% (Cockell and Knowland, 1999). Eight MAAs were identified in Dulse (*Palmaria palmata*) harvested from Spitsbergen, Norway: mycosporine-glycine, shinorine, porphyra-334, palythine, asterina-330, palythinol, palythene, and usujirene (Table 12.1) (Karsten and Wiencke, 1999). The greatest amount of MAAs (2.74 mg/g dry wt) was isolated from Dulse samples collected from 1.5 m water depths, whereas samples collected from 3 m depths contained substantially reduced amounts (0.68 mg/g dry wt). In the shallow water Dulse samples, palythine was the predominant MAA at approximately 1.4 mg/g dry wt compared to porphyra-334, and shinorine at 0.5 and 0.4 mg/g dry wt, respectively (Karsten and Wiencke, 1999). On the other hand, in Dulse samples collected between 3 and 10 m depths, shinorine, prophyra-334, and palythine were present in very similar amounts of approximately 0.22 mg/g dry wt each. Further evidence for a UV-protective role for MAAs was demonstrated by the regional tissue distribution of these compounds in the Dulse plant; the young, growing apical tips of this alga contained between six- and eightfold greater MAAs than the self-shaded older basal regions of the same plants. Similarly, when Aguilera and co-workers (2002) examined the MAA contents of *P. palmata* and *D. ramentacea* samples during a period ranging from sea-ice cover to ice breakup in Spitsbergen, Norway, total MAA tissue concentrations ranged from lows of 0.55 and 0.75 mg/g dry wt, respectively, under sea-ice cover in June, to highs of 1.0 and 1.3 mg/g dry wt, respectively, following ice breakup in July and August. The predominant MAA in *D. ramentacea* was palythine, whereas that for *P. palmata* was shinorine. While *P. palmata* contained the same eight MAAs as above, *D. ramentacea* samples contained all except for usujirene. More recently, Yuan and co-workers (2009) confirmed by high-performance liquid chromatography–mass spectrometry (HPLC/MS) that Dulse samples harvested from Grand Manan Island, New Brunswick, Canada, contained palythine, shinorine, asterina-330, palythinol, and porphyra-334, but that only the sample grown in high-UV exposed area contained usujirene, supporting the protective roles of MAAs in algal tissue against PAR and UV irradiation. Other workers have described three classification groupings of red marine macroalgae according to their MAA content: (1) species with no capability of producing MAAs (*Neuroglossum ligulatum*, *Plocamium cartilagineum*, *Polyides rotundus*, and *Gelidium amansii*); (2) species with a basic MAA content that is variable with alterations in environmental irradiation (*Kallymonia antarctica*, *Gymnogongrus antarcticus*, *Palmaria decipiens*, *Porphyra plocamiestris*, *Chondrus crispus*, and *Ptercladiella capillacea*), and (3) species with a high amount of MAAs irrespective of environmental conditions (*Lemanea fluviatilis* (freshwater red macroalga), *Porphyra endiviifolium*, *Gymnogongrus turquetti*, *Porphyra umbilicalis*, *Mastocarpus stellatus*, *Ceramium rubrum* and *Gracilaria cornea*) (Lee and Shiu, 2009; Singh et al., 2008).

MAAs have been reported in a broad spectrum of marine organisms, including *Porifera* (sponge), *Coelenterata* (sea anemone), *Ctenophora* (ctenophore), *Platyhelminthes* (planarian), *Nemertinea* (ribbon worm), *Annelida* (polychaete), *Mollusca* (pteropod, snail, nudibranch, chiton, and clam), *Arthropoda* (sea spider, copepods, isopod, krill, and amphipod), *Bryozoa* (bryozoans), *Echinodermata* (sea star, sea cucumber), and *Chaetognatha* (brittle star and chaetognath) (Karentz et al., 1991). Therefore, MAAs are virtually ubiquitous in the marine animal world. MAAs have been reported in the eggs (extraembryonic structures) or larvae of mollusks, polychaetes, fish (*Aspasmogaster costatus*), urchins, corals, ascidians, and gastropods (Przeslawski et al., 2005; Karentz, 2001). In addition, MAAs have been reported in the ocular tissue and skin of fish, eggs and ovaries of sea urchins, and embryos of some shrimp and certain fish species (Bandaranayake, 1998). However, biosynthesis of MAAs has not been demonstrated in marine animals, in that their concentration remains constant in Antarctic krill after 35 days of fasting (Newman et al., 2000); on the other hand, certain MAAs found in marine animals have not been detected in their macroalgal diet (Karentz et al., 1991).

12.3 MAA SOLUBILITY AND EXTRACTION METHODS

MAAs are polyfunctional compounds with several ionizable groups (i.e., carboxyl, sulfate, imine, secondary amine); therefore the solubility and the absorption maxima (λ_{max}) properties of MAAs will vary with the polarity and pH of the extraction solvent system. MAAs solubilize completely in polar solvents such as water, methanol, and ethanol (pH 7). Interestingly, slightly different spectral properties were observed for MAAs dissolved in dimethylsulfoxide (DMSO) (Gröniger and Häder, 2000); for example, porphyra-334 characterized by λ_{max} at 332–334 nm in aqueous media has λ_{max} of 338 nm in DMSO; similarly, mycosporine-glycine with λ_{max} of 305–308 nm in aqueous media has λ_{max} of 298 nm in DMSO. On the other hand, when porphyra-334 was solubilized in 100, 70, 50, and 30% ethanol, the λ_{max} were similar to those with water as the solvent, albeit with a shift to λ_{max} of 336 nm with increasing ethanol concentration (Zhang et al., 2005). As expected, MAAs do not solubilize completely in nonpolar solvents, e.g., hexane, benzene, and ether (Zhang et al., 2005; Gröniger and Häder, 2000). Taking advantage of these solvent characteristics, Sekikawa and co-workers (1986) precipitated a crude isolate of usujirene from an ethanolic solution of *P. palmata* using ether.

Polar solvents including water (Misonou et al., 2003; Schmid et al., 2003), ethanol (Leach and Trione, 1965), methanol (Dunlap et al., 1989), acetonitrile (Llewellyn and Airs, 2010), and phosphate buffer (20 mM, pH 7.0) (Yoshiki et al., 2009) have been successfully used to extract MAAs from red marine macroalgae. However, aqueous mixtures of polar solvents, most commonly aqueous methanol (with or without ~0.1% acetic acid), have been used in many studies for the separation of MAAs by HPLC (Huovinen et al., 2004; Schmid et al., 2003; Dunlap, et al., 1989). Methods used for the extraction of MAAs from algal biomass have been fairly simple; in the majority of studies, ground/milled red marine macroalgal samples were extracted sequentially with the desired polar solvent for 1 to 2.5 h at 45°C (Lee and Shiu, 2009; Huovinen et al., 2004; Schmid et al., 2003; Gröniger et al., 1999). The red marine macroalgal samples are usually either lyophilized, commercially available sun-dried specimens, or washed in clean seawater followed by freezer storage at –20 or –70°C. After collecting algal samples, it is essential to wash and clean them well before grinding or milling (Carefoot et al., 2000), because in some cases, microscopic algae, bacteria, or associated living organisms (epiphytic diatoms or other microalgae on the thalli that produce MAAs) will contribute to the total MAA composition of the sample.

Carreto and co-workers (2001, 2005) raised concerns about MAA stability during the serial extraction of macroalgal samples using 25% aqueous methanol over 2 h in a 45°C water bath, due to the possibility for hydrolysis, transformation, or degradation of MAAs during the extraction process. As an alternative, these authors proposed that lyophilized macroalgal tissues be rehydrated with a small volume of water overnight, followed by sonication in the presence of aqueous methanol to improve the MAA extraction efficacy (Carreto et al., 2005). More recently, we have used this protocol with some modifications for the fractionation and separation of MAAs from red marine macroalgae, including *P. palmata*, etc. (Figure 12.1) (Yuan et al., 2009). Rehydrating the algal sample (sample:water ratio was 9:1) in the dark prior to MAA extraction may hydrate the polar groups of the sample and improve MAA extraction.

Sample extraction with 0.2% aqueous acetic acid containing 0.5% methanol (v/v) for 12 h with shaking, 4°C, followed by centrifugation and filtration (0.2 μ) has been proposed by Volkmann and Gorbushina (2006) as a broadly applicable method for not only red marine macroalgae, but also terrestrial and freshwater organisms. These workers were able to demonstrate an improved extraction recovery and data quantification due to the absence of an extract "redissolving" step in the above protocol. In addition, Llewellyn and Airs (2010) have recently reported that 75% acetonitrile is as

Fresh algae sample (cleaned and washed of all epiphytes) or
commercially available sun-dried algae sample

↓

Cut into pieces and freeze dry/lyophilize

↓

Grind/mill the sample to pass through a 40-mesh screen

↓

Soak sample (0.25 g) in water overnight at 4°C (sample: water ratio, 1:9)

↓

Add 14 mL methanol (HPLC grade) and sonicate, 5 min. in ice

↓

Centrifuge mixture at 5000 rpm, 15 min., 4°C and collect the supernatant

↓

Re-extract the pellet with another 17 mL methanol, 2× and pool the supernatants

↓

Concentrate the extract using a rotary evaporator (40°C)

↓

Dissolve extract in 400 μL of distilled water

↓

Filter the extract using a 0.20 μ syringe filter

↓

Analyze using HPLC/PDAD or LC/MS to determine MAA profile and content

FIGURE 12.1 Scheme for the extraction of crude MAA mixture from red marine macroalgae.

good as 25 or 100% methanol for the extraction of MAAs from microalgae. These workers also reported that longer extraction times and increased temperatures did not improve extraction efficacy when using an ultrasonic probe vs. homogenization; however, pretreatment of algae by freezing or freeze-drying did facilitate MAA extraction efficiency. An exhaustive study by Tartarotti and Sommaruga (2002) with different species of algae and zooplankton demonstrated that both methanol concentration and extraction temperature can influence MAA extraction efficiency markedly; for example, the mean total MAA concentrations in *Porphyra* sp. and freshwater phytoplankton were approximately 13- and 3-fold greater, respectively, when extracted with 25% aqueous methanol, 45°C vs. 100% methanol, 4°C. Additionally, the extraction efficiency appeared to depend on the sample type studied, and it was suggested that there may be two compartments of MAAs in algal tissue. Indeed, Ingalls and co-workers (2010) described two distinct pools of MAAs in biological samples: the first, easily extractable, and the second, requiring extraction overnight using warmed methanol; however, further studies are needed to confirm this hypothesis. Thus, there currently remains considerable variability in the polar solvents used within the literature for MAA extraction. While the volume of extraction solvent also varied, sample concentrations ranged between 0.005 and 0.1 g/ml; thus, when a lower solvent volume is used, multiple extractions may be required to improve extraction efficiency.

Further methodological considerations include the paucity of information available on the effect of particle size or mean particle diameter of the ground or milled algae samples in MAA extraction. Also, due to lack of information about the localization of MAAs in marine macroalgal thalli tissue and cells, the quantitation of extracted MAAs is expressed on a dry weight basis from the whole thallus. Thus, not only have the MAA concentrations in macroalgae been demonstrated to be quite low (Karsten et al., 1998b), but also additional techniques or steps are required to remove the coextractives. Extraction of MAAs from different botanical components of the thallus (i.e., vegetative, reproductive, apical, or central portions of the algal thallus) have been suggested by some scientists

(Ye and Xiaojun, 2007; Huovinen et al., 2004; Karsten and Wiencke, 1999) and may prove useful to overcome the difficulties discussed above.

12.4 MAA IDENTIFICATION AND QUANTITATION

Identification of MAAs in a sample usually involves high-performance liquid chromatography with a photodiode array detector (HPLC-PDAD) or LC/MS instrumentation including a combination of parameters such as retention time, absorption maxima (λ_{max}), spectral properties, molecular weight or electrospray ionization–mass spectrometry (ESI/MS) positive ion spectra, fragmentation patterns (Yuan et al., 2009), or rarely cochromatography with purified standards (Karsten et al., 1998a). MAA concentrations are then determined via the molar extinction coefficient (ε, molar absorptivity in $M^{-1}cm^{-1}$) using the Beer-Lambert law: $A = \varepsilon \times I \times c$, where A = sample absorbance, ε = sample molar extinction coefficient ($M^{-1}cm^{-1}$), I = path length of the sample cuvette (typically 1 cm), and c = sample concentration (M). Since the typical path length is 1 cm, it follows that $c = A \div \varepsilon$. For optimal results, it is recommended to know the absorption maxima of the solvents employed and also to determine whether the sample compounds obey the Beer-Lambert law for linearity of response. In some cases when the ε was not available for certain MAAs, researchers have used the ε of a closely related compound. Generally, the MAA content of red marine macroalgal samples is quite low; therefore it is quite a challenging task to purify them (costly and time consuming). Therefore powerful analytical tools are required to isolate and identify MAAs; for example, UV/Vis, FTIR, HPLC, MS (LC/MS, MS/MS^n and FTIR-MS), NMR, and GC techniques have been used for identification and characterization of MAAs. However, few studies and less information are available evaluating the latter two techniques.

12.4.1 ULTRAVIOLET-VISIBLE SPECTROMETRY (UV/VIS)

The presence of MAAs in a sample is easily detected using UV/Vis spectrometry given that these compounds demonstrate a prominent absorption peak between 310 and 360 nm. Frequently, crude macroalgal extracts dissolved in 100% methanol are used to generate UV spectra using single- or double-beam spectrometry scanning from 250 to 700 nm. The intensity of an absorption peak at 324 nm is considered to be an indicator of MAA abundance. When algal samples were treated with different UV irradiation exposures, the induction of MAA synthesis was calculated using this technique, in which the chlorophyll-a absorption peak (665 nm) was considered an internal standard (the chlorophyll-a content did not vary with short-term UV irradiation); for example, the optical density, O.D. 324 nm ÷ O.D. 665 nm, was noted to increase with MAA biosynthesis (Gröniger and Häder, 2002). Whitehead and Hedges (2005) reported the degree of degradation, as loss of absorbance, of palythine solubilized in water after different treatments (i.e., irradiation and photosensitization) using UV/Vis absorption in which they observed slight increases in absorbance at lower UV wavelengths (approximately 285 nm) and losses of 4% at longer wavelengths (320 nm) characteristic of palythine, compared to the nonirradiated control. Furthermore, UV/Vis spectrometry has been widely utilized in evaluating the stability of MAAs in terms of pH (Zhang et al., 2005) and heat (Sinha et al., 2000). When purifying MAAs from a crude algal extract, researchers have often combined and concentrated to dryness the fractions containing UV-absorbing substances, followed by a second chromatography (Sekikawa et al., 1986; Misonou et al., 2003). This protocol can yield better results by eliminating coextractives and thereby increasing the purity of the isolated MAAs.

12.4.2 FOURIER TRANSFORM INFRARED SPECTROSCOPY (FTIR)

Infrared spectroscopy has a history of being used for qualitative and quantitative research on various types of natural products. Using a KBr disk technique (Shimadzu multipurpose recording spectrophotometer, model MPS-50L) Yoshida and Sivalingam (1970) studied the IR spectrum of a UV-absorbing substance at 337 nm, identified as porphyra-334, isolated and purified from *Porphyra yezoensis* (R) and observed bands at 3,840, 2,640, 1,720, 1,672, 1,690, and 750 cm^{-1}. Later, Sivalingam and co-workers (1976) reported that the IR spectra of porphyra-334, using the KBr disk technique with a Hitachi 215 type IR spectrophotometer, were distinguished by major absorption bands at 3,450 cm^{-1} (OH), 2,920 cm^{-1} (C-C), 1,600 cm^{-1} (COO-Na), 1,540 cm^{-1} (amide), and 1,368 cm^{-1}. More recently, Bhatia and co-workers (2010) reported that porphyra-334 isolated from *P. vietnamensis* (R) also exhibited similar IR absorption frequencies (3,100, 1,520, 1,670, 1,330, and 1,100 cm^{-1}). In other work, porphyra-334 isolated from the cyanobacterium *Aphanizomenon flos-aquae* exhibited absorbance bands at 3,300, 1,600, 1,540, 1,380, and 1,080 cm^{-1} using a KBr pellet technique (Perkin-Elmer 2000, FTIR instrument) (Torres et al., 2006). Thus, IR analyses elucidate spectral properties that are useful in identifying MAAs; however, limited information is available on the ratio between the sample and KBr, sample purity, and the sample moisture content employed. Despite the fact that IR techniques have not been used much in the literature, it is a simple and easy technique to detect the presence of MAAs and identify them according to their spectral properties.

12.4.3 High-Performance Liquid Chromatography (HPLC)

The most common analytical method used for the isolation and quantitation of MAAs from red marine macroalgal extracts has been HPLC coupled with photodiode array detection (PDAD). However, mobile phase composition and column parameters need to be carefully chosen to optimize the resolution of MAAs using HPLC-PDAD. In most studies, prior to analysis by HPLC, crude algal extracts are cleaned up using centrifugation and filtration (Figure 12.1) (Yuan et al., 2009; Helbling et al., 2004); however, in others a Sephadex LH-20/G10 column (Cardozo et al., 2008; Zhang et al., 2005), ultrafiltration techniques (Carreto et al., 2005) or solid phase extraction cartridges (Obermuller et al., 2005; Mason et al., 1998) have been incorporated. Normal phase (Karentz et al., 1991), reverse phase (RP), i.e., RP-8 (Przeslawski et al., 2005; Volkmann et al., 2003; Mason et al., 1998) and RP-18 (Yuan et al., 2009), and amino-modified silica gel columns (Dehorter and Bernillon, 1983) have been utilized in the separation and quantitation of MAAs. However, the RP-18 column (5 μ, 250 mm × 4 mm ID) has been used most commonly for MAA analyses with a flow rate ranging between 0.7 and 1.00 ml/min (Yuan et al., 2009; Carreto et al., 2005; Helbling et al., 2004; Klisch and Häder, 2002; Conde et al., 2000; Karsten et al., 1998a, 1998b). A few studies have included a guard column before the main column to prevent undesirable compounds from entering the main column (Carreto et al., 2005; Whitehead and Hedges, 2003; Mason et al., 1998). Water and methanol (75:25) containing 0.1% acetic acid have been used frequently at flow rates between 0.7 and 1.00 ml/min in most studies of the resolution of MAAs (Volkmann and Gorbushina, 2006; Karentz et al., 1991). Aqueous methanol is suitable for the separation of the more polar MAAs from macroalgal samples; however, to separate the less polar MAAs, namely, usujirene and palythene, approximately 0.1% acetic acid needs to be added to the aqueous methanol (Nakamura et al., 1982).

MAAs have been identified by using HPLC chromatograms combined with PDAD λ_{max} values and retention times in concert with known values from the existing literature. The λ_{max} values of the most common MAAs in red marine macroalgae are summarized in Table 12.1. However, great care should be taken when identifying compounds using HPLC-PDAD as MAAs with similar parameters, i.e., retention time, extinction coefficient, and λ_{max}, have been reported recently (Bandaranayake, 1998; Wu Won et al., 1997). For example, an abundant MAA in the coral *Pocillopora capitata*, palythine-threonine, was not conclusively identifiable using HPLC-PDAD alone, due to its λ_{max} of 320 nm (identical to palythine; Table 12.1) and an *m/z* value of 289 (identical to mycosporine-methyl-amine-serine), but was identified via high-resolution MS with a time-of-flight (TOF) mass spectrometer (Carignan et al., 2009). Moreover, HPLC-PDAD analyses of mycosporine-glycine-valine revealed similar characteristics to palythenic acid (the difference between the λ_{max} of these MAAs is 2 nm), making identification of the former MAA difficult (Whitehead et al., 2001).

However, due to the lack of commercially available MAA standards, relatively few researchers have been able to report use of their own or borrowed purified MAA standards for compound identification and quantitation. Thus, while HPLC-PDAD is very useful for the identification of known MAAs, it lacks the ability to identify related MAAs or their intermediate products; therefore other techniques to identify MAAs have been developed and optimized below.

12.4.4 Liquid Chromatography–Mass Spectroscopy (LC/MS)

Due to the limitations with HPLC-PDAD identification of MAAs discussed above, researchers have more recently combined HPLC with mass spectroscopy (LC/MS) to study MAAs. With LC/MS, the molecular fragmentation pattern produced when an MAA is bombarded by an ion source is key to the identification of MAAs in the absence of commercial standards (Yuan et al., 2009). Also, MS data can be utilized to predict the structure of the compound without complete resolution or with a very small amount of crude sample; in addition, a very small amount of sample is needed for the analyses. For example, if MAAs are solubilized in deionized water at nanomolar concentrations, LC/MS can detect the compound without the need for further concentration (Whitehead and Hedges, 2002). The MS fragmentation pattern of an MAA will vary with respect to its respective R group (Table 12.1). Even if some losses of molecular ions were observed from MAAs during MS analyses, fragmentation patterns with 137, 168, 186, 197, or 199 ions have been identified as diagnostic MAA fragments (Whitehead and Hedges, 2002, 2003). The characteristic ESI/MS positive ion spectra for the most common MAAs are shown in Table 12.1. The ion at 197 *m/z* is common for most of the MAAs, except for palythine, where it is not prominent. These workers also reported that MAAs varied in sensitivity to ionization and fragmentation of $[M+H]^+$ ions related to their relative polarities; at a cone voltage of 40 V, palythine (less polar) exhibited virtually no fragmentation but was most easily ionized, whereas the more polar MAAs, shinorine and porphyra-334, exhibited a decrease in sensitivity to ionization with substantial fragmentation (Whitehead and Hedges, 2002). Thus, due to the variable sensitivity of MAAs to ionization, quantitation of the target compound without a known standard is a challenging task. Whitehead and co-workers (2002) suggested using tyramine hydrochloride as an internal standard to assist in the quantitation of MAAs and to eliminate errors associated with different MS instrumentation.

For ionization sources in MS, electrospray ionization (ESI) (Yuan et al., 2009; Volkmann and Gorbushina, 2006), atmospheric pressure chemical ionization (API-ES) (Whitehead and Hedges, 2002; Carreto et al., 2001), and atmospheric pressure photoionization (APPI) have been utilized to detect and identify MAAs. Among these techniques, ESI ionization has proven to be the technique of choice, possibly due to easier access to this instrumentation. It is noteworthy that MAAs have been studied using a variety of MS instrumentation, including single-quadrupole MS (Whitehead et al., 2001), reflection time-of-flight (TOF) MS, triple-quadrupole MS (MS/MS) (Volkmann and Gorbushina, 2006; Whitehead et al., 2001), ion trap (MS^n) (Ingalls et al., 2010), and hybrid mass spectrometry (Cardozo et al., 2006). In most of those studies, fairly good data to support MAA identification were obtained. Based on recently published data, ESI/MS^n, that is ESI/MS/MS coupled with deuterium labeling and accurate-mass sequential MS (FTICR-MS) provided more information about MAAs than LC-MS (Cardozo et al., 2006, 2008; Carignan et al., 2009), particularly with respect to groups associated with the basic skeleton structure.

12.4.5 Nuclear Magnetic Resonance (NMR) Spectroscopy of MAAs

NMR spectroscopy is a desirable technique for the identification of MAA structure because it is a nondestructive and selective technique in which samples can be scanned multiple times. Sample preparation for NMR spectroscopy is simpler and less time consuming; however, its sensitivity is much lower than that of MS, and therefore requires a larger amount of sample for an assay

(Silva Elipe, 2003). In addition, in order to obtain clear chemical shifts in NMR, the purity of the compounds of interest is very important, thereby limiting the application of this technique in MAA analysis (Silva Elipe, 2003); as a result, NMR data on the red marine macroalgal MAAs are scarce and far from complete. Nevertheless, the chemical shifts data of some MAA compounds from a few red marine macroalgae using ^{1}H and ^{13}C-NMR have been reported to date: a 337 nm-absorbing MAA from *Porphyra yezoensis* Ueda, subsequently identified as porphyra-334 (Yoshida and Sivalingam, 1970; Sivalingam et al., 1976); palythene and porphyra-334 from *Porphyra tenera* Kjellman (Takano et al., 1979); palythene and usujirene from *P. palmata* (L.) O. Kuntze (Sekikawa et al., 1986); and usujirene from *P. yezoensis* (Nakayama et al., 1999). As HPLC and MS techniques have some limitations in identifying new MAAs as discussed above, the chemical shifts data derived from ^{1}H and ^{13}C-NMR together with amino acid analyses are very useful in identifying and confirming the structure of new MAAs (Wu Won et al., 1995, 1997). The limitations in NMR for the analysis of MAAs may be able to be overcome by LC/NMR and LC/MS/NMR; however, few developments have been published in this arena in the elucidation of the structures of new and complex MAAs.

12.4.6 Gas Chromatography and Mass Spectroscopy (GC/MS) of MAAs

Similar to amino acids, MAAs are relatively polar in nature and are less volatile for direct analysis by GC/MS. Therefore they need to be derivatized to allow elution at high temperatures without decomposition or molecular structure alterations during GC/MS (Whitehead and Hedges, 2003; Yeung et al., 1986). However, the variety of functional groups within MAAs results in considerable challenges during derivatization, due the formation of multiple derivatives (Whitehead and Hedges, 2003; Molnar-Perl and Katona, 2000). For example, the multiple functional groups characteristic of not only amino acids (i.e., $-NH_2$, $-COOH$, $-OH$, and $-SH$), but also MAAs have different sensitivities to silylating reagents, rendering the quantitative preparation of dominant derivatives difficult, even under optimized conditions (Whitehead and Hedges, 2003; Molnar-Perl and Katona, 2000). Moreover, Whitehead and Hedges (2003) described additional challenges when analyzing MAAs by GC/MS, including compound losses, decomposition, and complicated GC spectra. However, with purified MAA standards, ESI/MS of trimethylsilylated MAAs will be useful in elucidating the structure of MAAs (Whitehead and Hedges, 2003). Thus quantitative preparation of uniform MAA derivatives remains a challenge for GC/MS; however, hopefully, further studies can optimize the application of GC/MS for MAA identification.

12.5 ANTIOXIDANT ACTIVITIES OF MAAS

Support for the hypothesis that MAAs play a role in the endogenous antioxidant activity of red marine macroalgae, such as *P. palmata*, is derived from an earlier study by Dunlap and Yamamoto (1995) demonstrating that mycosporine-glycine was consumed in a reaction with the hydrophilic stable free radical 2,2′-azo-bis(2-amidinopropane) dihydrochloride (AAPH), whereas the imino-carbonyl MAAs such as shinorine, porphyra-334, palythine, asterina-330, and palythinol remained at initial levels in a *P. tuberculosa* extract. These workers reported that extracts containing an 8:1 ratio of asterina-330:palythine exhibited little inhibition of AAPH-induced phosphatidylcholine peroxidation (0.9 mM total MAAs) (Table 12.2), whereas those containing a 9.6:1 ratio of porphyra-334:shinorine (0.9 mM) or a 2.4:1 ratio of shinorine:mycosporine-glycine (1.2 mM) exhibited moderate inhibition, and that containing a 27.9:9:4.3:1 ratio of mycosporine-glycine:palythine:palythiol:asterina-330 (1.6 mM total MAAs) exhibited the greatest inhibition of AAPH-induced phosphatidylcholine peroxidation *in vitro*. When the UV absorbance and HPLC profiles of these MAA-containing extracts were further investigated from 0–10 h incubation, asterina-330, porphyra-334, shinorine, palythine, and palythinol exhibited little to no decreases, whereas mycosporine-glycine decreased steadily over time (Dunlap and Yamamoto, 1995). More recently, when de la Coba and co-workers (2009) evaluated the

TABLE 12.2
Antioxidant Activity of Selected Mycosporine-Like Amino Acids (MAAs) and Mixtures of MAAs from Red Marine Macroalgal Extracts

MAA or MAA Mixture and Ratio	Free Radical or Oxidation Model[a]	[MAA]	Antioxidant Activity[b]	References
Mycosporine-glycine	$^1O_2$ from eosin Y	15–60 μM	25–80% inhibition, dose dependent	Suh et al., 2003
	$O_2^{-\bullet}$ from pyrogallol, pH 8.2	50–1,000 μM	Nonreactive	de la Coba et al., 2009
	ß-carotene bleaching in linoleic acid emulsion	10–200 μM	4–28% inhibition vs. 10 μM α-tocopherol, dose dependent	
	$ABTS^+$ decolorization			
	pH 6	2.5–10 μM	12–36% inhibition	
	pH 7.5	2.5–10 μM	38–76% inhibition	
	pH 8.5	2.5–10 μM	46–82% inhibition, dose dependent	
	AAPH induced PC-liposome, 37°C	10–30 μM	Strong	Dunlap et al., 1998
Shinorine	AAPH induced PC-liposome, 37°C	10–30 μM	Weak	Dunlap et al., 1998
	$O_2^{-\bullet}$ from pyrogallol, pH 8.2	100–1,000 μM	13–46% inhibition, dose dependent	de la Coba et al., 2009
	ß-carotene bleaching in linoleic acid emulsion	10–200 μM	4–69% inhibition vs. 10 μM α-tocopherol, dose dependent	
	$ABTS^+$ decolorization			
	pH 6	2.5–10 μM	N/A	
	pH 7.5	2.5–10 μM	N/A at 2.5 μM, 2–6% inhibition at 5, 10 μM	
	pH 8.5	2.5–10 μM	2–9% inhibition, dose dependent	
Usujirene	Linoleic acid emulsion, 40°C, 15 days, TBARS	2 mg	Strong inhibition	Nakayama et al., 1999
	Ferric thiocyanate method, 40°C, 15 days	2 mg	Strong inhibition	
4-Deoxygadusol	AAPH induced PC-liposome, 37°C	2, 10–30 μM	Strong	Dunlap et al., 1998
MAA Mixtures				
Mycosporine-glycine + palythine + palythinol + asterina-330, 27.9:9:4.3:1	AAPH-induced PC-peroxidation	1.6 mM	Strong inhibition	Dunlap and Yamamoto, 1995

(continued)

TABLE 12.2 (continued)
Antioxidant Activity of Selected Mycosporine-Like Amino Acids (MAAs) and Mixtures of MAAs from Red Marine Macroalgal Extracts

MAA or MAA Mixture and Ratio	Free Radical or Oxidation Model[a]	[MAA]	Antioxidant Activity[b]	References
Asterina-330 + palythine, 86:14	$O_2^{-\bullet}$ from pyrogallol, pH 8.2	50–1,000 μM	15–85% inhibition, dose dependent	de la Coba et al., 2009
	ß-Carotene bleaching in linoleic acid emulsion	10–200 μM	48–81% inhibition vs. 10 μM α-tocopherol	
	$ABTS^+$ decolorization			
	pH 6	2.5–10 μM	N/A at 2.5, 5 μM, 2.76% inhibition at 10 μM	
	pH 7.5	2.5–10 μM	5–16% inhibition, dose dependent	
	pH 8.5	2.5–10 μM	12–47% inhibition, dose dependent	
Asterina-330 + palythine, 8:1	AAPH-induced PC-peroxidation	0.9 mM	Weak inihibition	Dunlap and Yamamoto, 1995
Porphyra-334 + shinorine, 88:12	$O_2^{-\bullet}$ from pyrogallol, pH 8.2	200–1,000 μM	16–46% inhibition, dose dependent	de la Coba et al., 2009
	ß-Carotene bleaching in linoleic acid emulsion	10–200 μM	11–71% inhibition vs. 10 μM α-tocopherol, dose dependent	
	$ABTS^+$ decolorization			
	pH 6	2.5–10 μM	N/A	
	pH 7.5	2.5–10 μM	N/A at 2.5, 5 μM, 1.8% inhibition at 10 μM	
	pH 8.5	2.5–10 μM	3–15% inhibition, dose dependent	
Porphyra-334 + shinorine, 9.6:1	AAPH-induced PC-peroxidation	0.9 mM	Moderately strong inhibition	Dunlap and Yamamoto, 1995
Shinorine + mycosporine-glycine, 2.4:1	AAPH-induced PC-peroxidation	1.2 mM	Moderately strong inhibition	

[a] $ABTS^+$ = 2,2´-azinobis (3-ethylbenzothiazoline-6-sulfonate), AAPH = 2,2´-azo-bis(2-amidinopropane) dihydrochloride, TBARS = 2-thiobarbituric acid reactive substances, PC = phosphatidylcholine.
[b] N/A = no activity detected.

antioxidant efficacies of MAA extracts from red marine macroalgae (*P. rosengurtii*, *Gelidium corneum*, *Ahnfeltiopsis devoniensis*) and a lichen, *Lichina pygmaea*, all harvested from Malaga, Spain, the greatest superoxide radical ($O_2^{-\bullet}$) scavenging activities, at pH 8.2, were observed with an 86:14 mixture of asterina-330:palythine (from *G. corneum*) resulting in 50% inhibition at 200 μM and 85% at 1,000 μM (Table 12.2), whereas lower scavenging efficacies were observed with an 88:12 mixture of porphyra-334:shinorine (from *P. rosengurtii*) providing approximately 50% inhibition only at 1,000 μM. Interestingly, purified shinorine (from *A. devoniensis*) also exhibited approximately 50% inhibition of $O_2^{-\bullet}$ at 1,000 μM, whereas purified mycosporine-glycine (from *L. pygmaea*) was nonreactive against $O_2^{-\bullet}$ between 50 and 1,000 μM. On the other hand, when these workers evaluated the inhibition of lipid peroxidation by the above MAA mixtures, these imino-mycosporines exhibited strong antioxidant activity against β-carotene bleaching in a linoleic acid emulsion system: 86:14 asterina-330:palythine provided maximal ~80% inhibition at 100 μM (albeit there was no increase in inhibition at 200 μM) compared to 10 μM α-tocopherol; 88:12 porphyra-334:shinorine exhibited ~60% inhibition of bleaching at 100 μM, which increased to 70% at 200 μM; whereas purified shinorine exhibited 55% inhibition at 100 μM and 70% at 200 μM; and purified mycosporine-glycine provided 28% inhibition of β-carotene bleaching at 200 μM (Table 12.2). Other studies with purified MAAs have indicated weak, but dose-dependent inhibition of AAPH-induced phosphatidylcholine peroxidation by shinorine, and strong, dose-dependent inhibition by mycosporine-glycine and the Shikimic acid pathway MAA precursor, 4-DG (Table 12.2) (Dunlap et al., 1998). Others have determined that usujirene strongly inhibits oxidation of a linoleic acid emulsion over several days at 40°C (Table 12.2) (Nakayama et al., 1999).

Mechanistic studies have hypothesized that the greater susceptibility of mycosporine-glycine to oxidation, and thereby its efficacy as an antioxidant, is related to its lower redox potential and greater ability to act as a reducing agent in donating an electron to stabilize a free radical (Dunlap and Yamamoto, 1995). More recently, mycosporine-glycine was reported to inhibit singlet oxygen ($^1O_2$)-mediated type II photosensitization in not only eosin Y-mediated red blood cell hemolysis, but also methylene blue-mediated lipid peroxidation of soybean hypocotyl microsomes (Suh et al., 2003) (Tables 12.2 and 12.3). The rate constant for $^1O_2$ quenching by mycosporine-glycine was reported as 5.6×10^7 M^{-1} s^{-1}, lower than that of α-tocopherol (15.4×10^7 M^{-1} s^{-1}) (Suh et al., 2003). Moreover, these workers reported that direct incubation of mycosporine-glycine with eosin Y under visible light irradiation degraded the MAA into at least three major and several minor unidentified products. The $^1O_2$ quenching efficacy of mycosporine-glycine may conceivably be associated with the carbonyl group of the cyclohexenone chromophore and some resonance stabilization by the carbon ring double-bond structure. On the other hand, the inhibition of AAPH-derived peroxyl radical-induced lipid peroxidation by mycosporine-glycine may be associated with hydrogen abstraction from the cyclohexenone ring at C-4 or C-6 with resonance stabilization provided as described above. In contrast, the relatively weak inhibitory effect of mycosporine-glycine against lipid peroxidation and β-carotene bleaching may demonstrate the polar paradox, whereby polar compounds exhibit weak antioxidant activity in emulsions, particularly at the droplet interface, due to dilution in the aqueous phase (Yuan et al., 2005). The nonreactivity of mycosporine-glycine against the $O_2^{-\bullet}$ may relate to the alkaline pH 8.2 used in this assay, which would be expected to be greater than the pK_a for this MAA (de la Coba et al., 2009), resulting in the loss of the proton from the carboxyl group, and thereby potentially influence its ability to dismutate $O_2^{-\bullet}$. For usujirene, Nakayama and co-workers (1999) hypothesized that the strong antioxidant activity of this MAA may be associated with hydrogen abstraction from the cycloheximine ring at C-4, C-6, or the methylene group at C-9 of the glycine residue with resonance stabilization provided by the conjugated double bonds of the *cis*-unsaturated chain at C-11 on the double-bonded nitrogen in series with the carbon ring double-bond structure. Likewise, it is plausible that the weak antioxidant activities of shinorine against AAPH-derived peroxyl radical lipid oxidation as well as $O_2^{-\bullet}$ are associated with hydrogen abstraction from the cycloheximine ring at C-4, C-6, or the methylene group at C-9 of the glycine residue with some resonance stabilization from the carbon ring double-bond structure.

A factor that may play a role in whether an MAA exhibits antioxidant activity is the relative acidity of these compounds, and thereby the degree of dissociation under experimental conditions. For example, mycosporine-glycine, palythine, palythinol, asterina-330, and usujirene are only weakly acidic, while both shinorine and porphyra-334 (pK_a = 5.7) are strong acids. Thus antioxidant activity, or the absence thereof, observed at pH 8.0 in the presence of 0.1 mM EDTA (Dunlap and Yamamoto, 1995), or pH 8.2 (de la Coba et al., 2009) may not be comparable to studies conducted at pH 7.0 (Nakayama et al., 1999; Suh et al., 2003) or pH 6.0 (de la Coba et al., 2009). For example, mycosporine-glycine exhibited variable efficacy in decolorizing

TABLE 12.3
Photoreactivity of Selected Mycosporine-Like Amino Acids (MAAs) and Mixtures of MAAs from Red Marine Macroalgal Extracts

MAA or MAA Mixture and Ratio	Photosensitization Model[a]	[MAA]	Reactivity/ Inhibition of Photosensitization	References
Mycosporine-glycine	Growth inhibition of *E. coli* by eosin Y with white light illumination	90 nM	Almost complete inhibition of photooxidation	Suh et al., 2003
	Red blood cell hemolysis induced by eosin Y, 37°C with continuous illumination	0.15 mM	21% inhibition of photohemolysis	
		0.3 mM	50% inhibition	
	Respiration of submitochondrial particles irradiated with visible light in the presence of eosin Y	0.1–0.3 mM	Dose-dependent stabilization of microsomal membranes and O_2 consumption	
Palythenic acid	[MAA] remaining after exposure to riboflavin (10:1, photosensitizer: MAA) and natural sunlight > 290 nm	0.1 μM	30–0% remaining at doses from 200–1,150 kJ/m²	Whitehead and Hedges, 2005
Shinorine		0.1 μM	20–5% remaining at doses from 200–600 kJ/m²	
Asterina-330		0.1 μM	20–0% remaining at doses from 200–600 kJ/m²	
Palythinol		0.1 μM	30–0% remaining at doses from 200–1,150 kJ/m²	
Porphyra-334		0.1 μM	30–0% remaining at doses from 200–1,150 kJ/m²	
	[MAA] remaining after exposure to riboflavin (10:1, photosensitizer:MAA) with ultrafiltered deep seawater as solvent medium and source of photosensitizer	0.1 μM	67–0% remaining at doses from 200–6,000 kJ/m²	Whitehead and Hedges, 2005
	Absorbance spectra over time, 21°C, constant stirring, photolysis with light source, λ > 325 nm	20 μM	94–88% remaining after 130 and 300 min UVA irradiation	Conde et al., 2000
	Crude extract of *G. cornea* exposed to UVB, UVA, and PAR simultaneously	Not given	Absorption peak at 334 nm not reduced after 6 h irradiation; but decreased after 24 and 72 h irradiation	Sinha et al., 2000

(continued)

TABLE 12.3 (Continued)
Photoreactivity of Selected Mycosporine-Like Amino Acids (MAAs) and Mixtures of MAAs from Red Marine Macroalgal Extracts

MAA or MAA Mixture and Ratio	Photosensitization Model[a]	[MAA]	Reactivity/ Inhibition of Photosensitization	References
Palythene	[MAA] remaining after exposure to riboflavin (10:1, photosensitizer:MAA) and natural sunlight > 290 nm	0.1 μM	0% remaining at doses from 200–1,150 kJ/m^2	Whitehead and Hedges, 2005
Palythine	[MAA] remaining after exposure to riboflavin (10:1, photosensitizer:MAA)	0.1 μM	10–0.2% remaining at doses from 500–3,000 kJ/m^2	Whitehead and Hedges, 2005
	[MAA] remaining after exposure to rose bengal (10:1, photosensitizer:MAA)	0.1 μM	55–50% remaining at doses from 500–3,000 kJ/m^2	
	[MAA] remaining after exposure to light irradiation with ultrafiltered deep seawater as solvent medium and source of photosensitizer	0.1 μM	40–0% remaining at doses from 4,000–37,000 kJ/m^2	Whitehead and Hedges, 2005
MAA Mixtures				
Porphyra-334 + shinorine (*Porphyra umbilicalis* extract), 11.5:1	*P. umbilicalis* extract in 3.3% lecithin liposome preparation, stored at 4°C, RT, 37°C	0.1%	1 month storage, no reductions; 3 months storage, no reductions for 4°C and RT, 20% reduction at 37°C	Schmid et al., 2006
	As above, but with exposure to UVA irradiation, 320 nm	0.1%	1 month storage, no reductions; 3 months storage, no reductions	

[a] RT = ambient, room temperature, UVB = 280–320 nm, UVA = 320–400 nm, PAR = photosynthetically active radiation, 400–700 nm.

or scavenging the 2,2′-azinobis (3-ethylbenzothiazoline-6-sulfonate) ($ABTS^+$) radical cation, depending on the system pH, with the greatest antioxidant activity at pH 8.5 with 84% inhibition at 10 μM MAA (Table 12.2), whereas shinorine exhibited very weak antioxidant activity ranging from 0–9% inhibition between 2.5 and 10 μM at pH 6, 7.5, or 8.5 (de la Coba et al., 2009). On the other hand, when MAA mixtures were evaluated for $ABTS^+$ scavenging activity, an 86:14 ratio of asterina-330:palythine, 2.5–10 μM, provided very little to weak inhibition at pH 6 and 7.5, respectively, and 47% inhibition at 10 μM, pH 8.2; an 88:12 ratio of porphyra-334:shinorine, 2.5–10 μM, exhibited no inhibition at pH 6, and very little to weak inhibition at pH 7.5 and 8.2, respectively (Table 12.2) (de la Coba et al., 2009). Thus, the purified MAAs and MAA mixtures exhibited ionization characteristics in concert with their acidities: shinorine, a strong acid with two carboxyl groups (Table 12.1) would have a net negative charge at pH 8.5, potentially decreasing its ability to scavenge $ABTS^+$, whereas mycosporine-glycine, a weak acid, would be less ionized at pH 7.5 and 8.5 and thereby retain strong $ABTS^+$ scavenging activity. Similarly, asterina-330 and palythine are both weak acids with moderate $ABTS^+$ scavenging at pH 8.5, while porphyra-334 and shinorine are both strong acids with very weak $ABTS^+$ scavenging at pH 7.5 and 8.5 (Table 12.2).

12.6 STABILITY OF MAAS

The antioxidant activity of MAAs clearly plays a role in their stability, as discussed above, whereby mycosporine-glycine was consumed in the reaction with the stable free radical AAPH, compared to the robustness of the imino-carbonyl MAAs, shinorine, porphyra-334, palythine, asterina-330, and palythinol, which were retained at initial concentrations (Dunlap and Yamamoto, 1995). MAAs have also been evaluated for their stability in terms of storage time, temperature, and UV exposure. This latter characteristic comprises the photostability or photoreactivity of MAAs related to their UV sunscreen function *in situ*.

12.6.1 Thermal Stability of MAAs

MAAs were reported to be quite stable and did not degrade during the storage of dried macroalgae when Karsten and co-workers (1998a) stored sun-dried algae packed in airtight bags under cool, dry, and dark conditions. Similarly, Przeslawski and co-workers (2005) did not observe any degradation of MAAs in the egg masses from 46 species of mollusks stored at −80°C over 18 months. An ethanolic MAA extract isolated from *P. umbilicalis* (Heliogard 365), containing an 11.5:1 ratio of porphyra-334:shinorine incorporated at 0.1% in a lecithin liposome preparation, was stable at 4°C and room temperature after 1 and 3 months storage; MAA content was reduced by 20% after 3 months at 37°C (Table 12.3) (Schmid et al., 2006).

When Sinha and co-workers (2000) studied the heat sensitivity of MAA-containing crude extracts from the red marine macroalga, *G. cornea*, that were incubated in a 75°C water bath, the MAA absorption spectra from 250–500 nm remained constant even after 6 h, demonstrating the heat tolerance of MAAs. Purified MAAs such as shinorine, porphyra-334, and mycosporine-glycine were stable for up to 19 h, 70°C (Gröniger and Häder, 2000). Differential scanning calorimetry (DSC) analyses conducted by Bhatia and co-workers (2010) indicated that porphyra-334 decomposed around 233°C, with a melting point between 164 and 165°C. Only a slight decrease was observed in the λ_{max} of porphyra-334 even after 3 months at room temperature in slightly alkaline and acidic solutions (Yoshiki et al., 2009). It is noteworthy that when porphyra-334 ($C_{14}H_{21}O_8N_2$) was solubilized in phosphate buffer and heated at 120°C, 30 min, a dehydrated derivative, $C_{14}H_{20}O_7N_2$, resulted that exhibited a high antioxidant activity (Yoshiki et al., 2009).

12.6.2 pH Stability of MAAs

With aqueous solutions of porphyra-334, UV/Vis spectrometry at pH 1 and 3 exhibited hypsochromic shifts for λ_{max} to 330 and 332 nm, respectively (Zhang et al., 2005). Interestingly, when the pH values were adjusted back to initial values, the λ_{max} for porphyra-334 returned to 334 nm. Sivalingam and co-workers (1976) attributed these λ_{max} shifts to the resonance delocalization within the conjugated double-bond structure of the porphyra-334 molecule core. In contrast, when porphyra-334 was solubilized in highly alkaline conditions (pH 12 and 13), the λ_{max} and extinction coefficient were not altered (Zhang et al., 2004, 2005). The absorbance of porphyra-334 scanned from 200–400 nm at pH 12 and 13 decreased substantially between 20, 40, 60, and 80°C, whereas those at pH 1–11 were constant at 20 and 40°C, with small decreases at 60°C and marked decreases at 80°C (Zhang et al., 2004). Purified porphyra-334 incorporated into a commercial *Aloe vera* gel exhibited similar UV/Vis spectra when scanned between 200 and 400 nm at pH 1, 3, and 6; however, at pH 12 the MAA exhibited a dramatic increase in absorption at low wavelengths, 200–250 nm, albeit the spectral properties between 250 and 400 nm were not different (Bhatia et al., 2010).

12.6.3 Photostability and Photoreactivity of MAAs

Table 12.3 summarizes the photoreactivity of several purified MAAs as well as one red marine macroalgal mixture when exposed to various photosensitization models *in vitro*. The UV/Vis absorption peak of porphyra-334 in a crude extract of *G. cornea* (R), when scanned from 250 to 700 nm, remained constant after 6 h continuous combined UVB, UVA, and PAR irradiation (Table 12.3) (Sinha et al., 2000). A significant decrease in the peak intensity was observed only after 24 h irradiation, although complete elimination of the porphyra-334 absorbance was not observed even after 72 h. When porphyra-334 and shinorine, extracted and purified from *Porphyra* sp. and solubilized in aqueous solutions, were studied for their photodecomposition rate, the absorption spectra were relatively constant after continued photolysis (Table 12.3) (Conde et al., 2000, 2004). Moreover, when the porphyra-334 solutions from the photolysis experiments above were analyzed by HPLC-PDAD, at 330 nm, the absence of MAA derivative products indicated high photostability (Conde et al., 2000). In other studies of photochemical and photophysical characteristics, porphyra-334 and shinorine exhibited high photostability and a poor fluorescence quantum yield, respectively; approximately 96–98% of absorbed photon energy was dissipated as heat to the medium, thus protecting live cells in culture (Conde et al., 2000, 2004).

When an aqueous solution of shinorine was exposed to UVA and UVB (1.0 W m^{-2}) irradiation and observed at 0, 20, 40, 80, and 160 min, up to 24 h, there were no differences in the MAA concentration compared to the zero time control (Adams and Shick, 1996). Similarly, palythine, shinorine, and porphyra-334 exhibited low photodegradation when solubilized in deep seawater (with photodegradation rate constants $k = 0.018$ to 0.26×10^{-3} m^2 kJ^{-1}); however, the presence of photosensitizers such as riboflavin and rose bengal enhanced photodegradation rates considerably ($k = 0.77$ to 1.19×10^{-3} m^2 kJ^{-1}) in the latter two samples (Table 12.3) (Whitehead and Hedges, 2005). These workers also reported that degradation products were observed via LC/MS analyses, suggesting that MAA degradation originated from the amino moiety; thus, differences in MAA photostability may relate to the chemistry of the R group. Light, oxygen, type of photosensitizer, UV irradiation, and organic compounds dissolved in the deep seawater all affected the rate of MAA photodegradation above. It is clear that red marine macroalgal MAAs are stable and do not produce breakdown derivatives under normal UVB, UVA, and PAR exposure; thus, they absorb UV irradiation and dissipate energy thermally without showing fluorescence or generating oxygen radicals.

12.7 BIOLOGICAL ACTIVITIES OF MAAS

Studies evaluating the biological activities of MAAs have documented UV-protective roles for MAAs in not only micro- and macroalgae, but also coral reefs, as well as in the eyes of teleost Japanese medaka fish (*Oryzias latipes*) and in the eggs of sea urchins (*Strongylocentrotus droebachiensis*) (Shick and Dunlap, 2002; Mason et al., 1998).

12.7.1 Biological Activities of MAAs in Cell Culture: UV Protection

Oyamada and co-workers (2007) demonstrated a protective effect of MAAs against UV-induced human fibroblast cell (WI-38) death. The cells were grown in the presence of MAAs: 0–0.02 mM mycosporine-glycine, 0–0.5 mM shinorine, and 0–2 mM porphyra-334 for 24 h and then exposed to UV irradiation (302 nm; 576 mJ/cm^2) for 1 h. Twenty-four hours after UV irradiation, fibroblast cell viability was dose dependent with EC_{50} values as follows: mycosporine-glycine, 24 μM; shinorine, 64 μM; and porphyra-334, 294 μM. The greater efficacy of mycosporine-glycine was thought to be a function of its $\lambda_{max} = 310$ nm, close to the UV source peak output of 302 nm (Oyamada et al., 2007), but also likely reflected the efficacy of this MAA as an antioxidant scavenging $^1O_2$ as well as aqueous stable free radicals such as AAPH, as discussed above. On the other hand, the lower efficacies of shinorine and porphyra-334 may be a reflection of the weaker antioxidant activity of the former

MAA previously observed with AAPH (Table 12.2, as above), as well as the fact that both these MAAs are strongly acidic (vs. the weak acidity of mycosporine-glycine) and may have been ionized in culture. Thus, MAAs present in the cell culture media exhibited sunscreen protection against UV irradiation, which was lost when the medium was replaced with fresh media devoid of MAAs, as evidenced by the decline in cell viability to control levels of approximately 45% (Oyamada et al., 2007). It is noteworthy that human skin carcinoma (A431) cells grown in media containing shinorine for 48 h demonstrated an apparently saturable, active transport of MAAs into the cells, reaching a maximum of ~2.5 mM (Mason et al., 1998). Thus, MAAs are transported into human skin carcinoma cells without any difficulties within 48 h incubation.

When a mixture of MAAs isolated from *P. umbilicalis*, known as Helioguard-365, containing porphyra-334 and shinorine at a ratio of 11.5:1, was added to the media of mouse fibroblasts (3T3) in culture at concentrations of 0, 0.1, 0.5, 1, and 5 μg/ml followed by UVA irradiation for 1 min and a further 48 h incubation, a dose-dependent protection against the growth-limiting effects of UVA irradiation was observed. Moreover, the nonirradiated control cells did not show any deleterious effects of the MAAs at the concentration levels studied (Schmid et al., 2003). Similarly, when Helioguard-365 was added to the culture medium of human keratinocytes (HaCaT) at 0.125 and 0.25% for 24 h, followed by exposure (10 min) to UVA irradiation, the MAAs improved cell viability in a dose-dependent manner almost back to control levels with the 0.25% treatment (Schmid et al., 2006). When these workers treated fibroblasts (IMR-90) with Helioguard-365 at 0, 3, and 5%, followed by exposure to UVA irradiation, the presence of MAAs reduced DNA damage as determined by comet tail length in a dose-dependent manner. The DNA scission damage in the UVA-irradiated control was reduced by >50% with 5% Helioguard-365; thus MAA treatment protected against cellular DNA damage *in vitro*.

In other work with HaCaT cells treated with a novel mycosporine, collemin A ($C_{19}H_{32}N_2O_{13}$, MW 496) isolated from the lichenized ascomycete *Collema cristatum* for 24 h and exposed to UVB irradiation at 3.4 mW × s^{-1} with a dose of 200 mJ × cm^{-2} (to induce cell death), cell survival was increased from 20% with the untreated control to 35 and 75% with 0.6 and 3 mg/cm^2 MAA, respectively (Torres et al., 2004). Moreover, when HaCaT cells incubated with collemin A were exposed to 3.4 mW × s^{-1} UVB irradiation with a dose of 60 mJ × cm^{-2} (sufficient to induce DNA damage), a very low pyrimidine dimer content (~20%) resulted when compared to the untreated irradiated sample (100%), thus demonstrating the ability of this MAA to block pyrimidine dimer formation in cells exposed to UVB irradiation (Torres et al., 2004). Therefore it is possible that MAAs may have the ability to protect DNA in HaCaT cells and thereby protect cells from UVB-induced carcinogenesis and tumor initiation. For example, when a UV-absorbing extract fractionated from *P. yezoensis*, containing palythine, shinorine, and porphyra-334, was combined with 1 mM thymine at different concentrations or inserted as a filter between the light source and thymine solution prior to irradiation at 6.7 kJ/m^2 by UV-Vis light from a Xe lamp at –20°C, the production of the 6-4 thymine photodimer and 5-6 dimer was reduced to 40 and 25% in the samples containing thymine mixed with MAAs (Misonou et al., 2003). In contrast, when the MAA extract was employed as a filter in a cuvette inserted between the thymine solution and the light source, 6-4 and 5-6 dimer formation was inhibited by only 15 and 31%, respectively (Misonou et al., 2003). Taken together, these reports suggest that MAAs are able to not only directly quench excited thymine residues (via a direct molecule-to-molecule energy transfer) and protect cellular DNA from UV irradiation damage, but also provide some measure of UV sunscreen protection to fibroblast and keratinocytes. These efficacies likely reflect the photostability of MAAs, whereby they are able to dissipate photon energy as heat into the surrounding medium, as discussed above.

12.7.2 Biological Activities of MAAs as Topical Sunscreens

In addition to the cell culture studies with Helioguard-365 above, Schmid and co-workers (2003, 2006) also conducted human topical sunscreen application trials incorporating a 15% ethanolic extract containing an 11.5:1 ratio of porphyra-334:shinorine into a cream liposome formulation containing 3.3% lecithin (phosphatidylcholine) and 0.4% phenoxyethanol; the final MAA concentration was 0.005% in the cream. When 20 women (36–54 years of age) applied the cream twice daily to the inner forearm and the face, followed by UVA irradiation (10 J/cm^2) twice weekly over a period of 4 weeks, forearm skin firmness was improved over protection provided by a synthetic UVB (4%) + UVA (1%) sunscreen combination and an untreated control cream after 2 and 4 weeks; forearm skin smoothness was not different from the synthetic UVB + UVA combination after 2 weeks, but was improved after 4 weeks; depth of facial wrinkles in the "crow's feet" area was decreased by the MAA-containing cream compared to an untreated control after 2 and 4 weeks treatment (Schmid et al., 2003, 2006). Forearm skin levels of lipid peroxidation (measured as squalene hydroperoxide by HPLC) were inhibited by both the MAA-containing cream and the synthetic UVB + UVA sunscreen combinations compared to the untreated controls, albeit there was no difference between the two treatment groups. More recently, Ryu and co-workers (2009) demonstrated that crude methanol extracts of the red marine macroalga *Corallina pilulifera* not only exhibited potent antioxidant activities in the ferric thiocyanate assay and against the 1,1-diphenyl-2-picrylhydrazyl (DPPH$^•$) stable free radical and hydroxyl ($^•$OH) radicals, but also conferred protection on human dermal fibroblast (HDF) cells exposed to UVA irradiation, 100 J/cm^2, 10 s, followed by treatment with the *C. pilulifera* extract at 10, 100, and 200 μg/ml for 4 h. HDF cell viability was increased in a dose-dependent manner with the algal extract, but did not recover to nonirradiated control levels. Interestingly, gelatin zymography of human fibrosarcoma (HT-1080) cells exposed to the *C. pilulifera* extract for 1 h at 10, 50, and 100 μg/ml prior to stimulation with phorbol 12-myristate 13-acetate and culturing for 3 days demonstrated inhibition of specific matrix metalloproteinases (MMPs): expression of MMP-9 and MMP-2 were both inhibited dose dependently. MMP-2 and -9 are both known to degrade the ECM, thereby influencing skin thickness and wrinkling (Ryu et al., 2009). Thus, while these workers could not attribute these effects to the presence of MAAs in the *C. pilulifera* extract used in these studies, the extraction procedure they used was equivalent to that used for MAA extraction and isolation, as discussed above. Taken together, these studies suggest that the potential positive cosmeceutical effects of MAAs in topical sunscreen applications likely involve a combination of direct UVB and UVA sunscreen absorption activities as well as cell signaling effects on MMP expression, particularly if absorption of MAAs by these cells can be confirmed.

12.7.3 Bioavailability and Absorption of MAAs

Few *in vivo* feeding trials have been conducted to date in order to determine the uptake, absorption, tissue deposition, or potential bioactivity of dietary MAAs *in situ*. When *Mastocarpus stellatus* (R; rich in shinorine) was fed at 5% w/w to SKH-1 hairless mice (*Mus musculus*), dietary MAAs were absent from the eyes, skin, and liver tissues, but were present in the proximal, middle, and distal small intestine and large intestinal tissues as well as in the feces (Mason et al., 1998). While the albino hairless, euthymic, and immunocompetent SKH-1 mouse is an animal model used in many studies evaluating the effects of phytochemicals on carcinogen- or UV-induced skin cancer, this rodent model has also been demonstrated to exhibit unique regional distribution and frequency of gastrointestinal endocrine cells that may influence the physiological and absorptive functions of the digestive tract (Ku et al., 2002). Therefore, it remains unclear whether mammals absorb dietary MAAs or if particular MAAs are better absorbed than others depending on their polarity or charge.

12.8 CONCLUSIONS AND FUTURE WORK

The evidence presented above does provide credence for the further evaluation and potential commercial application of red marine macroalgal mycosporine-like amino acids and mixtures of MAAs as effective UVB and UVA sunscreens to protect the skin from photoaging and the associated effects on dermal extracellular matrices. The efficacy of MAAs as topical UV sunscreens likely comprises not only their efficacy as antioxidants, but also their photostability against UVA and UVB irradiation as well as cell signaling with respect to MMP expression. The variability in the stability of MAAs against oxidative stress is no doubt a reflection of the differences in their polarity and acidity. Due to the fact that standards of the main MAAs are not yet commercially available, researchers have not been able to ascertain whether there are synergistic effects between MAAs or within extract mixtures; hopefully this latter work can take place in the near future in concert with studies to determine MAA uptake or absorption by cells in culture to confirm their efficacy as cosmeceuticals.

ACKNOWLEDGMENTS

The work of the authors was supported by a Natural Sciences and Engineering Research Council of Canada (NSERC) discovery grant to Y.V.Y., Ryerson University postdoctoral fellowship to Y.A., and Ontario Workstudy Program funds, and was conducted with the help of postdoctoral fellows, technicians, and numerous student research assistants.

REFERENCES

Adams, N.L., and J.M. Shick. 1996. Mycosporine-like amino acids provide protection against ultraviolet radiation in eggs of the green sea urchin *Strongylocentrotus droebachiensis*. *Photochem. Photobiol.* 64:149–158.

Aguilera, J., A. Dummermuth, U. Karsten, R. Schriek, and C. Wiencke. 2002. Enzymatic defences against photooxidative stress induced by ultraviolet radiation in Arctic marine macroalgae. *Polar Biol.* 25:432–441.

Balskus, E.P., and C.T. Walsh. 2010. The genetic and molecular basis for sunscreen biosynthesis in cyanobacteria. *Science* 329:1653–1656.

Bandaranayake, W.M. 1998. Mycosporines: Are they nature's sunscreens? *Nat. Prod. Rep.* 15:159–172.

Bentley, R. 1990. The shikimate pathway—A metabolic tree with many branches. *Crit. Rev. Biochem. Mol. Biol.* 25:307–384.

Bhatia, S., K. Sharma, A.G. Namdeo, B.B. Chaugule, M. Kavale, and S. Nanda. 2010. Broad-spectrum sun-protective action of porphyra-334 derived from *Porphyra vietnamensis*. *Pharmacognosy Res.* 2:45–49.

Budel, B., U. Karsten, and F. Garcia-Pichel. 1997. Ultraviolet-absorbing scytonemin and mycosporine-like amino acid derivatives in exposed, rock-inhabiting cyanobacterial lichens. *Oecologia* 112:165–172.

Cardozo, K.H.M., V.M. Carvalho, E. Pinto, and P. Colepicolo. 2006. Fragmentation of mycosporine-like amino acids by hydrogen/deuterium exchange and electrospray ionisation tandem mass spectrometry. *Rapid Commun. Mass Spectrom.* 20:253–258.

Cardozo K.H.M., R. Vessecchi, V.M. Carvalho, et al. 2008. A theoretical and mass spectrometry study of the fragmentation of mycosporine-like amino acids. *Int. J. Mass Spectrom.* 273:11–19.

Carefoot, T.H., D. Karentz, S.C. Pennings, and C.L. Young. 2000. Distribution of mycosporine-like amino acids in the sea hare *Aplysia dactylomela*: Effect of diet on amounts and types sequestered over time in tissues and spawn. *Comp. Biochem. Physiol. C Toxicol. Pharmacol.* 126:91–104.

Carignan, M.O., K.H.M. Cardozo, D. Oliveira-Silva, P. Colepicolo, and J.I. Carreto. 2009. Palythine-threonine, a major novel mycosporine-like amino acid (MAA) isolated from the hermatypic coral *Pocillopora capitata*. *J. Photochem. Photobiol. B Biol.* 94:191–200.

Carreto, J.I., M.O. Carignan, and N.G. Montoya. 2001. Comparative studies on mycosporine-like amino acids, paralytic shellfish toxins and pigment profiles of the toxic dinofagellates *Alexandrium tamarense*, *A. catenella* and *A. minutum*. *Mar. Ecol. Prog. Ser.* 223:49–60.

Carreto, J.I., M.O. Carignan, and N.G. Montoya. 2005. A high-resolution reverse phase liquid chromatography method for the analysis of mycosporine-like amino acids (MAAs) in marine organisms. *Marine Biol.* 146:237–252.

Cockell, C.S., and J. Knowland. 1999. Ultraviolet radiation screening compounds. *Biol. Rev.* 74: 311–345.

Conde, F.R., M.S. Churio, and C.M. Previtali. 2000. The photoprotector mechanism of mycosporine-like amino acids. Excited-state properties and photostability of porphyra-334 in aqueous solution. *J. Photochem. Photobiol. B Biol.* 56:139–144.

Conde, F.R., M.S. Churio, and C.M. Previtali. 2004. The deactivation pathways of the excited-states of the mycosporine-like amino acids shinorine and porphyra-334 in aqueous solution. *Photochem. Photobiol. Sci.* 3:960–967.

Dehorter, B., and J. Bernillon. 1983. Photoinduction des périthèces du Nectria galligena: Production et activité photomorphogéne des mycosporines. *Can. J. Bot.* 61:1435–1442.

de la Coba, F., J. Aguilera, F.L. Figueroa, M.V. de Gálvez, and E. Herrera. 2009. Antioxidant activity of mycosporine-like amino acids isolated from three red macroalgae and one marine lichen. *J. Appl. Phycol.* 21:161–169.

Dunlap, W.C., K. Masaki, Y. Yamamoto, R.M. Larsen, and I. Karube. 1997. A novel antioxidant derived from seaweed. Presented at 4th International Marine Biotechnology Conference Sorrento, Italy.

Dunlap, W.C., K. Masaki, Y. Yamamoto, R.M. Larsen, and I. Karube. 1998. A novel antioxidant derived from seaweed. In *New developments in marine biotechnology*, ed. Y. LeGal and H. Halvorson, 33–35. New York: Plenum Press.

Dunlap, W.C., D.M. Williams, B.E. Chalker, and A.T. Banaszak. 1989. Biochemical photoadaptions in vision: UV-absorbing pigments in fish-eye tissues. *Comp. Biochem. Physiol. B Biochem. Mol. Biol. B* 93:601–607.

Dunlap, W.C., and Y. Yamamoto. 1995. Small-molecule antioxidants in marine organisms: Antioxidant activity of mycosporine-glycine. *Comp. Biochem. Physiol.* 112B:105–114.

Gröniger, A., and D.P. Häder. 2002. Induction of the synthesis of an UV-absorbing substance in the green alga *Prasiola stipitata. J. Photochem. Photobiol. B Biol.* 66:54–59.

Gröniger, A., and D.P. Häder. 2000. Stability of mycosporine-like amino acids. *Recent Res. Dev. Photochem. Photobiol.* 4:247–252.

Gröniger, A., C. Hallier, and D.P. Häder. 1999. Influence of UV radiation and visible light on *Porphyra umbilicalis*: Photoinhibition and MAA concentration. *J. Appl. Phycol.* 11:437–445.

Helbling, E.W., E.S. Barbieri, R.P. Sinha, V.E. Villafañe, and D.P. Häder. 2004. Dynamics of potentially protective compounds in *Rhodophyta* species from Patagonia (Argentina) exposed to solar radiation. *J. Photochem. Photobiol. B Biol.* 75:63–71.

Higashi-Okai, K., S. Otani, and Y. Okai. 1999. Potent suppressive effect of a Japanese edible seaweed, *Enteromorpha prolifera* (Sujiao-nori) on initiation and promotion phases of chemically induced mouse skin tumorigenesis. *Cancer Lett.* 140:21–25.

Huovinen, P., I. Gòmez, F.L. Figueroa, N. Ulloa, V. Morales, and C. Lovengreen. 2004. Ultraviolet-absorbing mycosporine-like amino acids in red macroalgae from Chile. *Bot. Mar.* 47:21–29.

Ingalls, A.E., K. Whitehead, and M.C. Bridoux. 2010. Tinted windows: The presence of the UV absorbing compounds called mycosporine-like amino acids embedded in the frustules of marine diatoms. *Geochim. Cosmochim. Acta* 74:104–115.

Jeffrey, S.W., H.S. MacTavish, W.C. Dunlap, M. Vesk, and K. Groenewoud. 1999. Occurrence of UVA- and UVB-absorbing compounds in 152 species (206 strains) of marine microalgae. *Mar. Ecol. Prog. Ser.* 189:35–51.

Karentz, D. 2001. Chemical defences of marine organisms against solar radiation exposure: UV-absorbing mycosporine-like amino acids and scytonemin. In *Marine chemical ecology*, ed. J. B. McClintock and B.J. Baker, 481–520. Boca Raton, FL: CRC Press.

Karentz, D., F.S. McEuen, M.C. Land, and W.C. Dunlap. 1991. Survey of mycosporine-like amino acid compounds in Antarctic marine organisms: Potential protection from ultraviolet exposure. *Mar. Biol.* 108:157–166.

Karsten, U., T. Sawall, D. Hanelt, et al. 1998a. An inventory of UV-absorbing mycosporine-like amino acids in macroalgae from polar to warm-temperate regions. *Bot. Mar.* 41:443–453.

Karsten, U., T. Sawall, and D. Wiencke. 1998b. A survey of distribution of UV-absorbing substances in tropical macroalgae. *Phycol. Res.* 46:271–279.

Karsten, U., and C. Wiencke. 1999. Factors controlling the formation of UV-absorbing mycosporine-like amino acids in the marine red alga *Palmaria palmata* from Spitsbergen (Norway). *J. Plant Physiol.* 155:407–415.

Klisch, M., and D.P. Häder. 2002. Wavelength dependence of mycosporine-like amino acid synthesis in *Gyrodinium dorsum. J. Photochem. Photobiol. B Biol.* 66:60–66.

Ku, S.-K., J.-H. Lee, H.-S. Lee, and K.-D. Park. 2002. The regional distribution and relative frequency of gastrointestinal endocrine cells in SKH-1 hairless mice: An immunohistochemical study. *Anat. Histol. Embryol.* 31:78–84.

Leach C.M., and E.J. Trione. 1965. An action spectrum for light induced sporulation in the fungus *Ascochyta pisi. Plant Physiol.* 40:808–812.

Lee, E.-J., and M.-K. Sung. 2003. Chemoprevention of azoxymethane-induced rat colon carcinogenesis by seatangle, a fiber-rich seaweed. *Plant Foods Hum. Nutr.* 58:1–8.

Lee, T.M., and C.T. Shiu. 2009. Implications of mycosporine-like amino acid and antioxidant defenses in UV-B radiation tolerance for the algae species *Ptercladiella capillacea* and *Gelidium amansii*. *Mar. Environ. Res.* 67:8–16.

Llewellyn, C.A., and R.L. Airs. 2010. Distribution and abundance of MAAs in 33 species of microalgae. *Mar. Drugs* 8:1273–1292.

Mason, D.S., F. Schafer, J.M. Shick, and W.C. Dunlap. 1998. Ultraviolet radiation-absorbing mycosporine-like amino acids (MAAs) are acquired from their diet by Medaka fish (*Oryzias latipes*) but not by SKH-1 hairless mice. *Comp. Biochem. Physiol.* 120:587–598.

Misonou, T., J. Saitoh, S. Oshiba, et al. 2003. UV-absorbing substance in the red alga *Porphyra yezoensis* (Bangiales, Rhodophyta) block thymine photodimer production. *J. Mar. Biotechnol.* 5:194–200.

Molnar-Perl, I., and Z.F. Katona. 2000. GC-MS of amino acids as their trimethylsilyl and t-butyldimethylsilyl derivatives: III. In model solutions. *Chromatographia Suppl.* 51:228–236.

Nakamura, H., J. Kobayashi, and Y. Hirata. 1982. Separation of mycosporine-like amino acids in marine organisms using reverse phase high performance chromatography. *J. Chromatogr.* 250:113–118.

Nakayama, R., Y. Tamura, H. Kikuzaki, and N. Nakatani. 1999. Antioxidant effect of the constituents of Susabinori (*Porphyra yezoensis*). *J. Am. Oil Chem. Soc.* 76:649–653.

Newman, S.J., W.C. Dunlap, S. Nicol, and D. Ritz. 2000. Antarctic krill (*Euphausia superba*) acquire a UV-absorbing mycosporine-like amino acid from dietary algae. *J. Exp. Mar. Biol. Ecol.* 255:93–110.

Obermuller, B., U. Karsten, and D. Abele. 2005. Response of oxidative stress parameters and sun screening compounds in Arctic amphipods during experimental exposure to maximal natural UVB radiation. *J. Exp. Mar. Biol. Ecol.* 323:100–117.

Oren, A. 1997. Mycosporine-like amino acids as osmotic solutes in a community of halophilic cyanobacteria. *J. Geomicrobiol.* 14:231–240.

Oyamada, C., M. Kaneniwa, K. Ebitani, M. Murata, and K. Ishihara. 2007. Mycosporine-like amino acids extracted from scallop (*Patinopecten yessoensis*) ovaries: UV protection and growth stimulation activities on human cells. *Mar. Biotechnol.* 10:141–150.

Pallela, R., Y. Na-Young, and S.-K. Kim. 2010. Antiphotoaging and photoprotective compounds derived from marine organisms. *Mar. Drugs* 8:1189–1202.

Pinnell, S.R. 2003. Cutaneous photodamage, oxidative stress, and topical antioxidant protection. *J. Am. Acad. Dermatol.* 48:1–19.

Przeslawski, R., K. Benkendorff, and A.R. Davis. 2005. A quantitative survey of mycosporine-like amino acids (MAAs) in intertidal egg masses from temperature rocky shores. *J. Chem. Ecol.* 31:2417–2438.

Rastogi, R.P., Richa, R.P. Sinha, S.P. Singh, and D.P. Häder. 2010. Photoprotecive compounds from marine organisms. *J. Ind. Microbiol. Biotechnol.* 37:537–558.

Ryu, B., Z.-J. Qian, M.-M. Kim, K.W. Nam, and S.-K. Kim. 2009. Anti-photoaging activity and inhibition of matrix metalloproteinase (MMP) by marine red alga, *Corallina pilulifera* methanol extract. *Radia. Phys. Chem.* 78:98–105.

Schmid, D., C. Schürch, and F. Zülli. 2006. Mycosporine-like amino acids from red algae protect against premature skin-aging. *Euro. Cosmetics* 9:1–4.

Schmid, D., C. Schürch, F. Zülli, H.-P. Nissen, and H. Prieur. 2003. Mycosporine-like amino acids: Natural UV-screening compounds from red algae to protect the skin against photoaging. *SOFW-Journal* 129:2–5.

Sekikawa, I., C. Kubota, T. Hiraoki, and I. Tsujino. 1986. Isolation and structure of a 357 nm UV-absorbing substance, usujirene, from red alga *Palmaria palmata* (L.) O. Kuntze. *Jap. J. Phycol.* 34:185–188.

Shick, J.M., and W.C. Dunlap. 2002. Mycosporine-like amino acids and related gadusols: Biosynthesis, accumulation, and UV-protective functions in aquatic organisms. *Annu. Rev. Physiol.* 64:223–262.

Silva Elipe, M.V. 2003. Advantages and disadvantages of nuclear magnetic resonance spectroscopy as a hyphenated technique. *Anal. Chim. Acta* 497:1–25.

Singh, S.P., S. Kumari, R.P. Rastogi, K.L. Singh, and R.P. Sinha. 2008. Mycosporine-like amino acids (MAAs): Chemical structure, biosynthesis and significance as UV-absorbing/screening compounds. *Ind. J. Exp. Biol.* 46:7–17.

Sinha, R.P., M. Klisch, A. Gröniger, and D.P. Häder. 1998. Ultraviolet-absorbing/screening substances in cyanobacteria, phytoplankton and macroalgae. *J. Photochem. Photobiol. B Biol.* 47:83–94.

Sinha R.P., M. Klisch, A. Gröniger, and D.P. Häder. 2000. Mycosporine-like amino acids in the marine red alga *Gracilaria cornea*—Effects of UV and heat. *Environ. Exp. Bot.* 43:33–43.

Sinha, R.P., S.P. Singh, and D.P. Häder. 2007. Database on mycosporines and mycosporine-like amino acids (MAAs) in fungi, cyanobacteria, macroalgae, phytoplankton and animals. *J. Photochem. Photobiol. B Biol.* 66:47–53.

Sivalingam, P.M., T. Ikawa, and K. Nisizawa. 1976. Isolation and physico-chemical properties of a substance 334 from the red alga, *Porphyra yezoensis* Ueda. *Bot. Mar.* 19:1–7.

Suh, H.-J., H.-W. Lee, and J. Jung. 2003. Mycosporine glycine protects biological systems against photodynamic damage by quenching singlet oxygen with a high efficiency. *Photochem. Photobiol.* 78:109–113.

Takano, S., A. Nakanishi, D. Uemura, and Y. Hirata. 1979. Isolation and structure of a 334 nm UV-absorbing substance, porphyra-334 from the red alga *Porphyra tenera* Kjellman. *Chem. Lett.* 26:419–420.

Tartarotti, B., and R. Sommaruga. 2002. The effect of methanol concentrations and temperature on the extraction of mycosporine-like amino acids (MAAs) in algae and zooplankton. *Arch. Hydrobiol.* 154:691–703.

Teas, J., M.L. Harbison, and R.S. Gelman. 1984. Dietary seaweed (*Laminaria*) and mammary carcinogenesis in rats. *Cancer Res.* 44:2758–2761.

Tevini, M., and A.H.Teramura. 1989. UV-B effects on terrestrial plants. *Photochem. Photobiol.* 50:479–487.

Torres, A., C.D. Enk, M. Hochberg, and M. Serbnik. 2006. Porphyra-334, a natural source for UVA protective sun screens. *Photochem. Photobiol. Sci.* 5:432–435.

Torres, A., M. Hochberg, I. Pergament, et al. 2004. A new UV-B absorbing mycosporine with photoprotective activity from the lichenized ascomycete *Collema cristatum*. *Eur. J. Biochem.* 271:780–784.

Volkmann, M., and A.A. Gorbushina. 2006. A broadly applicable method for extraction and characterization of mycosporines and mycosporine-like amino acids of terrestrial, marine and freshwater origin. *FEMS. Microbiol. Lett.* 255:286–295.

Volkmann, M., K. Whitehead, H. Rütters, J. Rullkötter, and A.A. Gorbushina. 2003. Mycosporine-glutamicoglucoside: A natural UV-absorbing secondary metabolite of rock inhabiting microcolonial fungi. *Rapid Commun. Mass Spectrom.* 17:897–902.

Whitehead, K., and J.I. Hedges. 2002. Analysis of mycosporine-like amino acids in plankton by liquid chromatography electrospray ionization mass spectrometry. *Mar. Chem.* 80:27–39.

Whitehead, K., and J.I. Hedges. 2003. Electrospray ionization tandem mass spectrometric and electron impact mass spectrometric characterization of mycosporine-like amino acids. *Rapid Commun. Mass Spectrom.* 17:2133–2138.

Whitehead, K., and J.I. Hedges. 2005. Photodegradation and photosensitization of mycosporine-like amino acids. *J. Photochem. Photobiol. B Biol.* 80:115–121.

Whitehead, K., D. Karentz, and J.I. Hedges. 2001. Mycosporine-like amino acids (MAAs) in phytoplankton, a herbivorous pteropod (*Limacina helicina*), and its pteropod predator (*Clione antarctica*) in McMurdo Bay, Antarctica. *Mar. Biol.* 139:1013–1019.

Wu Won, J.J., B.E. Chalker, and J.A. Rideout. 1997. Two new UV-absorbing compounds from *Stylophorapi sulphate* esters of mycosporine-like amino acids. *Tetrahedron Lett.* 3:2525–2526.

Wu Won, J.J., J.A. Rideout, and B.E. Chalker. 1995. Isolation and structure of a novel micosporine-like amino acid from the reef-building corals *Pocillopora damicornis* and *Stylophora pistillata*. *Tetrahedron Lett.* 36:5255–5256.

Yamamoto, I., and H. Maruyama. 1985. Effect of dietary seaweed preparations on 1,2-dimethylhydrazine-induced intestinal carcinogenesis in rats. *Cancer Lett.* 26:241–251.

Yamamoto, I., H. Maruyama, and M. Moriguchi. 1987. The effect of dietary seaweeds on 7,12-dimethylbenz[α] anthracene-induced mammary tumorigenesis in rats. *Cancer Lett.* 35:109–118.

Yamamoto, I., H. Maruyama, M. Takahashi, and K. Komiyama. 1986. The effect of dietary or intraperitoneally injected seaweed preparations on the growth of sarcoma-180 cells subcutaneously implanted into mice. *Cancer Lett.* 30:125–131.

Ye, C., and Y. Xiaojun. 2007. The UV-absorbing compounds in gametophytic blades of *Porphyra haitanensis* and the effect on formation of sporangial branchlets of conchocelis. *Chin. J. Oceanogr. Limnogr.* 25:343–346.

Yeung, J.M., G.B. Baker, and R.T. Coutts. 1986. Sample automated gas chromatographic analysis of amino acids and its application to brain tissue and urine. *J. Chromatogr.* 378:293–304.

Yoshida, T., and P.M. Sivalingam. 1970. Isolation and characterization of the 337 mμ UV-absorbing substance in red alga *Porphyra yezoensis* Ueda. *Cell Physiol.* 11:427–434.

Yoshiki, M., K. Tsuge, Y. Tsuruta, et al. 2009. Production of new antioxidant compound from mycosporine-like amino acid, porphyra-334 by heat treatment. *Food Chem.* 113:1127–1132.

Yuan, Y.V. 2007. Antioxidants from edible seaweeds. In *Antioxidant measurement and applications*, ed. F. Shahidi and C.-T. Ho, 268–301. Washington, DC: Oxford University Press.

Yuan, Y.V., D.E. Bone, and M.F. Carrington. 2005. Antioxidant activity of dulse (*Palmaria palmata*) extract evaluated *in vitro*. *Food Chem.* 91:485–494.

Yuan, Y.V., N.D. Westcott, C. Hu, and D.D. Kitts. 2009. Mycosporine-like amino acid composition of the edible red alga, *Palmaria palmata* (dulse) harvested from the west and east coasts of Grand Manan Island, New Brunswick. *Food Chem.* 112:321–328.

Zhang, Z., X. Gao, Y. Tashiro, S. Matsukawa, and H. Ogawa. 2004. Researches on the stability of porphyra-334 solution and its influence factors. *J. Ocean Univ. China* 3:166–170.

Zhang, Z., X. Gao, Y. Tashiro, S. Matsukawa, and H. Ogawa. 2005. The isolation of porphyra-334 from marine algae and its UV-absorption behaviour. *Chin. J. Oceanogr. Limnogr.* 23:400–405.

13 Cosmeceutical Applications of Chitosan and Its Derivatives

Mahinda Senevirathne, Chang-Bum Ahn, Se-Kwon Kim, and Jae-Young Je

CONTENTS

13.1 INTRODUCTION

Cosmeceutical products fill the gap between cosmetic products that simply cleanse and improve beauty, and pharmaceuticals that cure and heal problems. They provide a health benefit in addition to traditional cosmetic qualities. Healthy growth in the cosmeceutical market worldwide is primarily attributed to innovative technological knowledge. Technological innovations will continue to drive growth of cosmeceuticals in the cosmetic industry. In the cosmetic industry, chitosan is used as a base material (Minami et al., 1996), and its safety has been demonstrated in mice and rat experiments as a food additive and cosmetic base material (Mita, 1987; Seo, 1990). For cosmeceutical applications of chitosan and its derivatives, they have shown antioxidant, antiaging, matrix metalloproteinase (MMP) inhibitory, fungicidal, and fungistatic properties, offering good ingredient characteristics for use in the cosmetics industry. Chitosan is the only natural cationic gum that becomes viscous on being neutralized with acid. These biomaterials are therefore used as ingredients in sun creams, lotions, and permanent waving lotions in the cosmetics industry (Mark et al., 1985).

In the cosmetics industry, many novel families of biological macromolecules such as chitin, chitosan, and their derivatives are increasingly becoming of interest to many research groups. Their potential and useful applications in diverse fields are estimated to be more than 200 (Kumar, 2000). Those applications include biomedical, veterinary medicine, biotechnology, food, agriculture, and cosmetic fields (Domard et al., 1997; Chen and Chen, 1999; Mita, 1987; Seo, 1990; Pospieszny et al., 1994; Doares et al., 1995; Liu et al., 2007; Shi et al., 2006; Imeri and Knorr, 1988; Soto-Perlata et al., 1989; Chen and Li, 1996; Rwan and Wu, 1996; Spagna et al., 1996; Khor, 2011). In relation to personnel care, performance cosmetics, functional cosmetics, dermoceuticals, and active cosmetics have become rapidly popularized words among scientists and the general public. Healthful cosmetics provide protective, healing, and rejuvenative characteristics. Therefore, cosmeceutical materials are ingredients with medicinal properties that demonstrate beneficial localized actions and provide protection against degenerative skin diseases. Among thousands of bioactive materials, chitin, chitosan, and their derivatives have received much attention in the cosmetics field due to their characteristic features.

Chitosan is a linear polymer of d-glucosamine by β-(1,4) linkages and is a cellulose-like biopolymer present in the shells of crustaceans and in the cell walls of fungi and insects (Pae et al., 2001). Chitosan contains three types of functional groups, an amino/acetamido group as well as both primary and secondary hydroxyl groups at C-2 and C-3, respectively. The amino contents of chitosan molecules are mainly responsible for differences between their molecular structures and physicochemical properties, which are correlated with their various biological activities (Xia, 2003). Furthermore, it is a biocompatible, biodegradable, and nontoxic biopolymer (Muzzarelli, 1997). Chitosan and its derivatives have been of key interest among many research groups due to their potential biological activities, such as antibacterial (Allan and Hadwiger, 1979; Feng and Xia, 2011), antioxidant (Ngo et al., 2008; Park et al., 2004), matrix metalloproteinase inhibitory (Kim and Kim, 2006), immunity enhancing and antitumor (Suzuki et al., 1986), hypocholesterolemic (Kim and Rajapakshe, 2005), calcium binding (Jeon et al., 2000; Bravo-Osuna et al., 2007), as well as drug delivery capabilities (Bravo-Osuna et al., 2007; Saboktakin et al., 2010). However, to use these compounds in different applications, they must be converted or modified into absorbable forms by utilizing various grafting or synthesizing methods.

Chitosan oligosaccharides (low molecular weight COSs) are depolymerized products of chitosan and can be obtained by chemical and enzymatic hydrolysis of chitosan (Dou et al., 2007). Also, fermentation processes have been studied for decades for various crustacean shells (Jung et al., 2007; Oh et al., 2003), shrimp waste (Cira et al., 2002), and prawn waste (Fagbenro, 1996). Furthermore, the enzymatic hydrolysis of chitin and chitosan has been proposed as a preferred method for the production of bioactive chitin/chitosan oligomers. The hydrolytic enzymes in this process include protease, lipase, chitinase, lysozyme, and chitosanase, while variously modified chitosans and derivatives are widely used in the cosmetics industry.

13.2 COSMECEUTICAL APPLICATIONS OF CHITOSAN AND ITS DERIVATIVES

13.2.1 Chitosan and Its Derivatives as UV Protective Molecules

A high dose of UV irradiation is often known to be one of the most ubiquitous environmental hazards that impact on every living creature. Skin is the largest human organ that is directly exposed to UV irradiation (Isoir et al., 2006; Assefa et al., 2005). A growing body of evidence suggests that reactive oxygen species are generated by UV radiation, resulting in damage to both epidermal and dermal cells oxidatively (Assefa et al., 2005; Fisher et al., 2002; Rabe et al., 2006). Due to the increased public awareness and recognition of the adverse effects of UV radiation, the use of sun protection products is growing tremendously (Lautenschlager et al., 2007). However, there are some in-use criteria requirements of such products that include mainly effective protection against acute and long-term UV-induced skin damage, high skin accumulation of UV absorbers with minimal absorption, photostability of sunscreen actives, low skin irritation potential, user-friendly application requirements such as desirable aesthetic attributes, and reduced frequency of application (Forestier, 2008; Tanner, 2006).

Chitosan is another biopolymer widely investigated as a versatile carrier matrix for the delivery of drugs and cosmetic materials (Paolicelli et al., 2009). Incorporation of organic UV filters in gelatin (Patel et al., 2006), polymethylmethacrylate (Gogna et al., 2007), hyaluronic acid benzyl ester (Anselmi et al., 2002), ethylcellulose, and PLGA (Perugini et al., 2002) delivery systems greatly enhanced their activities. Renbutsu et al. (2007) have produced some UV-curable chitosan derivatives (3-methoxy-4-(2-hydroxy-3-methacryloyloxypropoxy) benzyl, 3,4-bis(2-hydroxy-3- metha cryloyl oxy propoxy) benzyl, 3-methoxy-4-methacryloyloxybenzyl, and 3,5-dimethacryloyloxybenzyl). Those compounds have shown better solubility in several organic solvents and were transformed

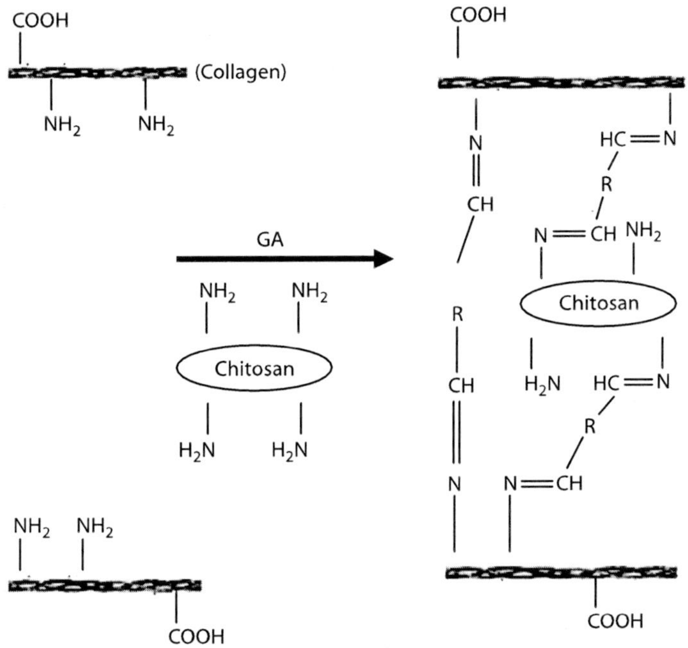

FIGURE 13.1 Schematic presentation of collagen cross-linked with glutaraldehyde in the presence of chitosan. (Adapted from Ma et al., *Collagen/chitosan porous scaffolds with improved biostability for skin tissue engineering*, Elsevier 4833–4841, 2003.)

into gels (Renbutsu et al., 2007) and showed strong results. Phenylbenzimidazole sulfonic acid (PBSA) is a hydrophilic relatively photostable UV-B filter (Couteau et al., 2007) commonly used in cosmetic products at a maximum concentration of 8% (w/w) with no skin irritation or photosensitization potential (COLIPA, 2006). Due to its water solubility, it is added to the water phase in emulsion systems and used in clear gels, resulting in aesthetically pleasing moisturizing sunscreen formulations with light skin feeling. PBSA was reported to boost the photoprotective effect of other organic and inorganic sunscreens (Johncock, 1999). Further, chitosan shows skin compatibility and adhesive properties, water resistance enhanced by cross-linking, and UV absorption (Shimoda et al., 2001; Lim and Wan, 1995; Anumansirikul et al., 2008). Ma et al. (2003) produced a porous chitosan/collagen scaffold by their cross-linking with glutaraldehyde and later freeze-dried to improve biostability and good biocompatibility. Further, they also confirmed that the potential cytotoxicity of glutaraldehyde might be decreased through the presence of chitosan. Hence chitosan can function as a bridge to increase the cross-linking efficiency of glutaraldehyde in collagen-based scaffolds due to the presence of a large number of amino groups in its chain (Figure 13.1).

Chitosans potentially satisfy the requirements of a good polymer matrix for the delivery of sunscreens especially in hydrophilic nature. Therefore PBSA-loaded chitosan was used as a sunscreen compound and showed highly positive results (Gomaa et al., 2010). *In vitro* investigation of the UV effect of PBSA-chitosan microparticle (MP) gel using Transpore™ tape as a substrate is indicated in Figure 13.2. Results showed that plain chitosan gel did screen UV radiation at the wavelength of 290–400 nm with a sun protection factor (SPF) of 0.89. However, cross-linking of chitosan slightly increased the UV screening effect, as shown by the slightly higher SPF of plain chitosan MP gel (0.93). Further, incorporation of 2% free PBSA in the gel increased the SPF from 0.89 to 1.14, while incorporation of the drug at the same concentration level loaded in chitosan MPs further raised the SPF to 1.61.

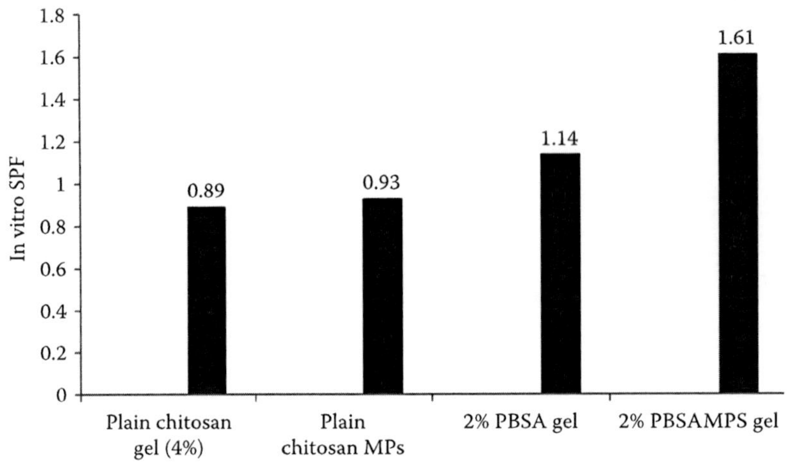

FIGURE 13.2 *In vitro* sun protection factors (SPFs) of tested microparticle (MP) formulations. (Adapted from Gomaa et al., Y.A., *Carbohyd. Polym.*, 81, 234, 2010.)

13.2.2 Chitosan and Its Derivatives as Skin Cleansing and Moisturizing Agents

Alpha-hydroxy acids (AHAs) have becoming important substances as cosmetic active ingredients. AHAs act as a mild exfoliator, reducing excessive epidermal keratinization. AHAs have the ability to reverse photodamage and reduce the formation of brown spots on the skin (Ditre et al., 1996). However, the application of high concentrations of AHAs generally causes bad effects such as skin irritation (Perugini et al., 2000), but a sustained delivery system containing AHAs may provide an advantage by lowering the side effects and increasing the efficacy of AHAs in skin applications. Liposomes have been shown to possess great potential as a delivery system for a variety of drugs (Agarwal et al., 2001; Betz et al., 2001; Verma et al., 2003). An increase in the rigidity of the shells of liposome by some means such as interacting/coating of liposomes with polymers may reduce this problem (Cho et al., 2007), and this property can easily be obtained by coating the liposomes with a suitable polymer like chitosan. The sustained delivery of the AHAs is the result of the above process. Phetdee et al. (2008) developed chitosan-coated liposomes for sustained delivery of tamarind fruit pulp's extracts in the skin and showed prominent results.

Further, Boriwanwattanarak et al. (2008) have done a study to develop a curcuminoid hydrogel patch using chitosan as a controlled release matrix. The hydrogel is designed for a great advantage over conventional cosmetics by delivering the active ingredients into skin at a desired amount and time. The volume of water in a hydrogel also plays an important role in skin moisturizing. Moreover, it can remove dead skin cells during peel-off because of its mild adhesion property. The effects of the addition of water-soluble chitosans on skin hydration, physicochemical properties, and the vitamin E release ratio of vital moisture creams were studied by Chen and Heh (2000), and the results showed that vitamin E creams containing different molecular weights or concentrations of water-soluble chitosans are pseudoplastic fluids. Further, chitinous materials have very good film formation properties (Alexander, 1991; Izume, 1989; Muzzarelli et al., 1989; Sakurai et al., 1983). Sakurai et al. (1983) reported that application of 0.3% hydroxypropyl chitosan to skin will form a smooth and pliable film that has good water-holding properties, while Gross et al. (1982) reported that chitosan film was stable in high humidity environments.

13.2.3 Chitosan and Its Derivatives as Hair Care Ingredients

Chitosan is able to form electrostatic complexes under acidic conditions as it is a polyelectrolyte compound. Here, two different types of complexes are considered. They are electrostatic

TABLE 13.1
Properties of Chitin, Chitosan, and Their Derivatives in Hair Care

Characteristics of the Compounds	Uses
Aqueous solution interacting with negatively charged hair (electrostatic interaction)	Shampoos
Antistatic effect (hydrophilic character), maintains moisture in low humidity and hair style in high humidity	Hair tonics
	Rinses
	Permanent wave agents
	Hair colorants
	Hair sprays
	Time release delivery (chitosan beads, gels, or granules)
Removing sebum and oils from hairs (due to hydrophobic character)	
Antibacterial and antifungal activity	
Act as thickening polymer	
Role in surfactant stability; stabilize emulsion	
Make hair softer; increase mechanical strength of hair	
Protect elastic film on hairs	

Source: Adapted from Rinaudo, M., *Prog. Polym. Sci.*, 31, 624, 2006.

complexes with an oppositely charged surfactant and polyelectrolyte complexes. Electrostatic complexes are formed in the presence of a low DA chitosan involving cooperative stacking of surfactant alkyl chains. A general behavior of polyelectrolytes is demonstrated with chitosan and sodium dodecyl sulfate. At physiological pH, chitosan could bind itself to the negatively charged surfaces of most of the cells and living tissues due to its positive charges, and hence can be used as a bioadhesive, which increases the retention of the therapeutic materials at the site of application (Calvo et al., 1997; He et al., 1998). Due to this behavior of chitosan, it is used in the cosmeceutical industry as a hair care ingredient for shampoo, hair gel, hair colorants, hair sprays, etc. (Table 13.1). Hair care products containing chitosan show superior film formation ability, compatibility, stiffness and curl retention to synthetic polymers (Alexander, 1991; Wachter and Stenberg, 1996; Horner et al., 1997).

New types of depilatories and means for curling and doing hair have been prepared by some researchers of Sonat Co. USA using a composition based on chitosan and other hydrocolloids containing antioxidant, antiallergic, and anti-inflammatory materials of vegetable origin. Chitin, chitosan, and their derivatives can be used in three different areas of the cosmetics industry as skin care, oral care, and hair care (www.meronbiopolymers.com).

Chitosan and hair are complementary to each other as they possess opposite electrical charges. Chitosan has positive charges while hair shows negative charges. A clear solution that contains chitosan forms a clear, elastic film on hair that causes it to increase its softness, smoothness, and mechanical strength. Further, chitosan can form a gel when added to mixtures of alcohol and water, while it can be used in shampoos, rinses, permanent wave agents, hair colorants, styling lotions, hair sprays, and hair tonics. Several derivatives of chitosan and chitin have potential applications in hair care. Glyceryl chitosan, an adduct of an oligomer of hydrolyzed chitosan, n-hydroxypropyl chitosan, quaternary hydroxyprophyl-substituted chitosan, and polyoxyallylene chitosan are some of them. Some derivatives of chitosan can form foam and create emulsifying action, and chitin powder can be used directly in shampoo (www.meronbiopolymers.com).

13.2.4 Chitosan and Its Derivatives as Agents for Oral Hygiene

Oral hygiene is important among the hygienic precursors that the general public follows. Being social animals, people always tend to keep and improve their beauty in the presence of large gatherings. Hence they may upgrade their oral and facial hygiene at any cost, and many people use various ways and means to meet their personal hygiene, among which oral hygiene may be most prominent. Natural compounds have attracted more attention due to fewer side effects, and hence global attention is now focused on natural compounds from underutilized marine-based resources, of which chitin, chitosan, and their derivatives are prominent. The combined effect of drugs with chitosan and its derivatives on oral diseases is given in Table 13.2.

Giunchedi et al. (2001) have formulated a buccal tablet using drug-loaded chitosan microspores. This study deals with the development of buccal chlorhexidine in chitosan microspheres, which dissolves more quickly *in vitro* than does chlorhexidine powder. The antibacterial activity of the microparticles was evaluated as a minimum inhibitory concentration, minimum bactericidal concentration, and killing time. The loading of chlorhexidine into chitosan is able to maintain or improve the antibacterial potential of the drug, and it showed the highest activity against *Candida albicans*. Therefore its potential application against buccal infections is very important. Drug-empty microparticles have an antimicrobial activity due to the polymer itself. Buccal tablets were prepared by direct compression of the microparticles with mannitol alone or with sodium alginate. After their *in vivo* administration the determination of chlorhexidine in saliva showed the capacity of these formulations to give a prolonged release of the drug in the buccal cavity.

Aksungur et al. (2004) have developed an occlusive bioadhesive system for prophylaxis or treatment of oral mucositis. In that study, gel and film formulations were prepared using chitosans at different molecular weights and in different solvents. Nystatin was incorporated into the formulations, as it is considered a prophylactic agent for oral mucositis. The *in vitro* release of nystatin from the formulations was decreased with the increasing molecular weight of chitosan. The effect of the formulations was investigated *in vivo* in hamsters with chemotherapy-induced mucositis. Mucositis scores in groups treated with nystatin incorporated into gel and suspension formulations were significantly lower ($p < 0.05$) than those treated with the chitosan gel alone. Survival of animals in the treated groups was higher than that in the control group. The retention time and distribution of the gels in the oral cavity were also investigated in this study and showed significant results.

Gum chewing is believed to be a physically and mechanically effective method for cleaning teeth, second only to tooth brushing. Natural biomaterials have recently been investigated as potential agents to prevent oral diseases such as dental caries. Chitosan is positively charged and combines with the bacterial cell wall and membrane with bacteriostatic and bactericidal results (Fujiwara et al., 2004; Oh et al., 2001; Helander et al., 2003; Kumar et al., 2005; Voet and Voet, 1995). Water-soluble, reduced chitosan was used as a mouth rinse reagent, and it displayed an antibacterial and plaque-reducing action (Sano et al., 2003; Bae et al., 2006). Also, the effect of phosphorylated chitosan rinsing on plaque development on the buffering capacity of plaque suspension was evaluated (Sano et al., 2001). A mouth rinse containing 0.5% phosphorylated chitosan significantly suppressed both newly formed plaque areas and old plaques areas compared to placebo rinsing. Furthermore, recent studies have demonstrated that chewing

TABLE 13.2
Effect of Chitosan and Its Derivatives on Oral Problems

Nature of Chitosan	Drug	Disease	Reference
Chitosan gel/film	Nystatin	Oral mucosistis	Aksungur et al., 2004
Chitosan microspheres	Chlorhexidine	Plaque formation	Giunchedi et al., 2002
Chitosan microspheres	Amphicillin	Control microbial growth	Anal et al., 2006
Chitosan gel/film	Chlorhexidine gluconate	Periodontal diseases	Ikinci et al., 2002

chitosan-containing gum effectively inhibited the growth of cariogenic bacteria in saliva. Hayashi et al. (2007) have used chitosan-containing gum to combat bacterial activity and to increase salivary secretion. The amount of oral bacteria in subjects who chewed chitosan-containing gum significantly decreased during all three sampling times except at 60 min for total bacteria in comparison to those in the rinse group. Chitosan-containing gum chewing also significantly increased the secretion of saliva. Therefore, chitosan-containing gum chewing has a greater antibacterial effect and it also increases salivary secretion; hence it can be used as a potential natural material in the formulations of oral hygiene.

13.3 CONCLUSIONS

Chitosan and its derivatives have gained much attention due to their promising characteristics and biological activities leading to their extensive use over a wide range of applications. Due to their specific characteristics they can be included in cosmetic formulations. Chitosan and its derivatives showed sunscreening power, while their potential can be increased by incorporating them with compounds such as PBSA. Hence both chitosan and PBSA may contribute to formulations for the improvement of sunscreen products. Further, chitosan and its derivatives are potential agents that can be used in many cosmeceutical products relating to hair care, oral care and skin cleansing and moisturizing due to their various bioactivities.

REFERENCES

Agarwal, R., Katare, O.P., and Vyas, S.P. 2001. Preparation and *in vitro* evaluation of liposomal/niosomal delivery systems for antipsoriatic drug dithranol. *Int. J. Pharm.* 228:43–52.

Aksungur, P., Sungur, A., Unal, S., Iskit, A.B., Squier, C.A., and Senel, S. 2004. Chitosan delivery systems for the treatment of oral mucositis: *in vitro* and *in vivo* studies. *J. Controlled Release* 98:269–279.

Alexander, P. 1991. The foundation of beauty. *Manuf. Chemist*, May, pp. 27–33.

Allan, C.R., and Hadwiger, L.A. 1979. The fungicidal effect of chitosan on fungi of varying cell wall composition. *Exp. Mycol.* 3:285–287.

Anal, A.K., Stevens, W.F., and Remuñán-López, C. 2006. Ionotropic cross-linked chitosan microspheres for controlled release of ampicillin. *Int. J. Pharm.* 312:166–173.

Anselmi, C., Centini, M., Rossi, C., Ricci, M., Rastrelli, A., Andreassi, M., et al. 2002. New microencapsulated sunscreens: technology and comparative evaluation. *Int. J. Pharm.* 242:207–211.

Anumansirikul, N., Wittayasuporn, M., Klinubol, P., Tachaprutinum, A., and Wanichwe Charun-gruang, S. 2008. UV screening chitosan nanocontainers: increasing the photo stability of encapsulated materials and controlled release. *Nanotechnology* 19:1–9.

Assefa, Z., Laethem, A.V., Garmyn, M., and Agostinis, P. 2005. Ultraviolet radiation-induced apoptosis in keratinocytes: on the role of cytosolic factors. *Biochim. Biophys. Acta* 1755:90–106.

Bae, K., Jun, E.J., Lee, S.M., Paik, D.I., and Kim, J.B. 2006. Effect of water-soluble reduced chitosan on *Streptococcus mutans*, plaque regrowth and biofilm vitality. *Clin. Oral Invest.* 10:102–107.

Betz, G., Imboden, R., and Imanidis, G. 2001. Interaction of liposome formulations with human skin *in vitro*. *Int. J. Pharm.* 229:117–129.

Boriwanwattanarak, P., Ingkaninan, K., Khorana, N., and Viyoch, J. 2008. Development of curcuminoids hydrogel patch using chitosan from various sources as controlled-release matrix. *Int. J. Cosmetic Sci.* 30:205–218.

Bravo-Osuna, I., Millotti, G., Vauthier C., and Ponchel, G. 2007. *In vitro* evaluation of calcium binding capacity of chitosan and thiolated chitosan poly(isobutyl cyanoacrylate) core-shell nanoparticles. *Int. J. Pharm.* 338:284–290.

Calvo, P., Vila-Jato, J.L., and Alonso, M.J. 1997. Evaluation of cationic polymer coated nanocapsules as ocular drug delivery. *Int. J. Pharm.* 153:41–50.

Chen, R.H., and Chen, H.C. 1999. *Advances in chitin science*. RITA Advertising Co. Ltd., Taipei.

Chen, R.H., and Heh, R.S. 2000. Skin hydration effects, physico-chemical properties and vitamin E release ratio of vital moisture creams containing water-soluble chitosans. *Int. J. Cosmetic Sci.* 22:349–360.

Chen, Y., and Li, C. 1996. Studies on the application of chitosan to clarification of grapefruit juice. *Food Sci. Taiwan* 23:617–628.

Cho, E.-C., Lim, H.-J., Shim, J., Kim, J., and Chang, I.-S. 2007. Improved stability of liposome in oil/water emulsion by association of amphiphilic polymer with liposome and its effect on bioactive skin permeation. *Colloid Surface A* 299:160–168.

Cira, L.A., Huerta, S., Hall, G.M., and Shirai, K. 2002. Pilot scale lactic acid fermentation of shrimp wastes for chitin recovery. *Process Biochem.* 37:1359–1366.

COLIPA. 2006. Opinion on phenylbenzimidazole sulfonic acid and its salts. Opinion adopted by the SCCP, S45.

Couteau, C., Faure, A., Fortin, J., Paparis, E., and Coiffard, L. 2007. Study of the photostability of 18 sunscreens in creams by measuring the SPF *in vitro*. *J. Pharmaceut. Biomed.* 44:270–273.

Ditre, C.M., Griffin, T.D., Murphy, G.F., et al. 1996. Effect of alpha hydroxyl acids on photoaged skin: a pilot clinical, histologic and ultrastructural study. *J. Am. Acad. Dermatol.* 34:187–195.

Doares, S.H, Syrovets, T., Weiler, E.W., and Ryan, C.A. 1995. Oligogalacturonides and chitosan activate plant defensive genes through the octadecanoid pathway. *Proc. Natl. Acad. Sci. USA* 92:4095–4098.

Domard, A., Roberts, G.A.F., and Varum, K.M. 1997. *Advances in chitin science*. Jacques Andre Publishers, Lyon.

Dou, J.L., Tan, C.Y., Du, Y.G., Bai, X.F., Wang K.Y., and Ma, X.J. 2007. Effect of chitooligosaccharides on rabbit neutrophils *in vitro*. *Carbohydr. Polym.* 69:209–213.

Fagbenro, O.A. 1996. Preparation, properties and preservation of lactic acid fermented shrimp heads. *Food Res. Int.* 29:595–599.

Feng, Y., and Xia, W. 2011. Preparation, characterization and antibacterial activity of water-soluble O-fumaryl-chitosan. *Carbohrdr. Polym.* 83:1169–1173.

Fisher, G.J., Kang, S.-W., Varani, J., Bata-Csorgo, Z., Wan, Y.S., Datta S., and Voorhees, J.J. 2002. Mechanisms of photoaging and chronological skin aging. *Arch. Dermatol.* 138:1462–1470.

Forestier, S. 2008. Rationale for sunscreen development. *J. Am. Acad. Dermatol.* 58:134S–138S.

Fujiwara, M., Hayashi, Y., and Ohara, N. 2004. Inhibitory effect of water soluble chitosan on growth of *Streprococcus mutans*. *New Microbiol.* 27:83–86.

Giunchedi, P., Juliano, C., Gavini, E., Cossu, M., and Sorrenti, M. 2001. Formulation and *in vivo* evaluation of chlorhexidine buccal tablets prepared using drug-loaded chitosan microspheres. *Eur. J. Pharmacol. Biopharm.* 53:233–239.

Gogna, D., Jain, S., Yadav, A., and Agrawal, G. 2007. Microsphere based improved sunscreen formulation of ethylhexyl methoxycinnamate. *Curr. Drug Delivery* 14:153–159.

Gomaa, Y.A., El-Khordagui, L.K., Boraei, N.A., and Darwish, I.A. 2010. Chitosan microparticles incorporating a hydrophilic sunscreen agent. *Carbohydr. Polym.* 81:234–242.

Gross, F., Konrad, E., and Mager, H. 1982. Investigation on chitosan as a natural film forming ingredient in hair cosmetic products under the consideration of ecological aspects. In *Chitin and chitosan*, Proceedings of the Second International Conference on Chitin and Chitosan, Hirano, S., and Tokura, S. (Eds). Japanese Society of Chitin and Chitosan, Tottori, pp. 205–209.

Hayashi, Y., Ohara, N. Ganno, T., Ishizaki, H., and Yanagiguchi, K. 2007. Chitosan-containing gum chewing accelerates antibacterial effect with an increase in salivary secretion. *J. Dentistry* 35:871–874.

He, P., Davis, S.S., and Illum L. 1998. *In vitro* evaluation of the mucoadhesive properties of chitosan microspheres. *Int. J. Pharm.* 166:75–88.

Helander, I.M., Latva-Kala, K., and Lounatmaa, K. 2003. Permeabilizing action of olyethyleneimine on *Salmonella typhimurium* involves disruption of the outer membrane and interactions with lipopolysaccharide. *Microbiology* 344:385–390.

Horner, V., Pittermann, W., and Wachter, R. 1997. Efficiency of high molecular weight chitosan in skin care application. In *Advances in chitin science*, Domard, A., Roberts, G.A.F., and Varum, K.M. (Eds.). Jacques Andre, Lyon, pp. 671–677.

Ikinci, G., Senel, S., Akincibay, H., Kas, S., Ercis, S., Wilson, C.G., and Hincal, A.A. 2002. Effect of chitosan on a periodontal pathogen *Porphyromonas gingivalis*. *Int. J. Pharm.* 235:121–127.

Imeri, A.G., and Knorr, D. 1988. Effect of chitosan on yield and compositional data of carrot and apple juice. *J. Food Sci.* 53:1707–1709.

Isoir, M., Buard, V., Gasser, P., Voisin, P., Lati E., and Benderitter, M. 2006. Human keratinocyte radio sensitivity is linked to redox modulation. *J. Dermatol. Sci.* 41:55–65.

Izume, M. 1989. Application of chitin chitosan in toilet products. *Biol. Ind.* 6:75–83.

Jeon, Y.J., Shahidi, F., and Kim, S.K. 2000. Preparation of chitin and chitosan logomers and their applications in physiological functional foods. *Food Rev. Int.* 16:159–176.

Johncock, W. 1999. Sunscreen interactions in formulations. *Cosmetics Toiletries* 114:75–82.

Jung, W.-J., Jo, G.-Y., Kuk, J.-H. et al. 2007. Production of chitin from red crab shell waste by successive fermentation with *Lactobacillus paracasei* KCTC-3074 and *Serratia marcescens* FS-3. *Carbohydr. Polym.* 68:746–750.

Khor, E. 2011. In *Chitin, chitosan and their derivatives: biological activities and applications*, Kim, S.K (Ed.). CRC Press, Boca Raton, FL, pp. 405–412.

Kim, M.-M., and Kim, S.-K. 2006. Chitooligosaccharides inhibit activation and expression of matrix metalloproteinase-2 in human dermal fibroblasts. *FEBS Lett.* 580:2661–2666.

Kim, S.-K., and Rajapakse, N. 2005. Enzymatic production and biological activities of chitosan oligosaccharides (COS): a review. *Carbohydr. Polym.* 62:357–368.

Kumar, A.B.V., Varadaraj, M.C., Gowda, L.R., and Tharnathan, R.N. 2005. Characterization of chito-oligosaccharides prepared by chitosanolysis with the aid of papain and pronase, and their bactericidal action against *Bacillus cereus* and *Escherichia coli. Biochem. J.* 391:167–175.

Kumar, M.N.V.R. 2000. A review of chitin and chitosan applications. *Reaction Functional Polym.* 46:1–27.

Lautenschlager, S., Wulf, H., and Pittelkow, M. 2007. Photoprotection. *Lancet* 370:528–537.

Lim, L.-Y., and Wan, L.-S. 1995. Heat treatment of chitosan films. *Drug Dev. Ind. Pharm.* 21:839–846.

Liu, Z., Jiao, Y., and Zhang, Z. 2007. Calcium-carboxymethyl chitosan hydrogel beads for protein drug delivery system. *J. Appl. Polym. Sci.* 103:3164–3168.

Ma et al. 2003. *Collagen/chitosan porous scaffolds with improved biostability for skin tissue engineering*, Elsevier 4833–4841.

Mark, H.F., Bikales, N.M., Overberger, C.G., and Menges, G. 1985. *Encyclopedia of polymer science and engineering*. Wiley, New York, p. 20.

Minami, S., Oh-oka, M., Okamoto, Y., Miyatake, K., Matsuhashi, A., Shigemasa, Y., and Fukumoto, Y. 1996. Chitosan-inducing hemorrhagic pneumonia in dogs. *Carbohydr. Polym.* 29:241–246.

Mita, Y. 1987. Character of chitosan and its application to cosmetics. In *Development and application of chitin and chitosan*. Kogyo-gijyutsu-kai, Tokyo, pp. 248–269 (in Japanese).

Muzzarelli, R.A.A. 1997. Human enzymatic activities related to the therapeutic administration of chitin derivatives. *Cell. Mol. Life Sci.* 53:131–140.

Muzzarelli, R.A.A., Weckx, M., Filippini, O., and Lough, C. 1989. Characteristic properties of N-carboxybutyl chitosan. *Carbohydr. Polym.* 11:307–320.

Ngo, D.N., Kim, M.-M., and Kim, S.-K. 2008. Chitin oligosaccharides inhibit oxidative stress in live cells. *Carbohydr. Polym.* 74:228–234.

Oh, H.I., Kim, Y.J., Chang, E.J., and Kim, J.Y. 2001. Antimicrobial characteristics of chitosan against food spoilage microorganisms in liquid media and mayonnaise. *Biosci. Biotechnol. Biochem.* 65:2378–2383.

Oh, S.-H., Kang, S.-G., Kim, E.-S., Cho, S.-H., and Lee, J.-H. 2003. Fabrication and characterization of hydrophilic poly(lactic-co-glycolic acid)/poly(vinyl alcohol) blend cell scaffolds by melt-molding particulate leaching method. *Biomaterials* 24:4011–4021.

Pae, H.-O., Seo, W.-G., Kim, N.-Y., Oh, G.-S., Kim G.-E., and Kim, Y.-H. et al. 2001. Induction of granulocytic differentiation in acute promyelocytic leukemia cells (HL-60) by water-soluble chitosan oligomer. *Leukemia Res.* 25:339–346.

Paolicelli, P., de la Fuente, M., Sánchez, A., Begoña, S., and Alonso, M. 2009. Chitosan nanoparticles for drug delivery to the eye. *Expert Opin. Drug Deliv.* 6:239–253.

Park, P.-J., Je, J.-Y., and Kim, S.-K. 2004. Free radical scavenging activities of differently deacetylated chitosans using an ESR spectrometer. *Carbohydr. Polym.* 55:17–22.

Patel, M., Jain, S., Yadav, A., Gogna D., and Agrawal, G. 2006. Preparation and characterization of oxybenzone-loaded gelatin microspheres for enhancement of sun screening efficacy. *Drug Deliv.* 13:323–330.

Perugini, P., Genta, I., Pavanetto, F., Conti, B., Scalia, S., and Baruffini, A. 2000. Study on glycolic acid delivery by liposomes and microspheres. *Int. J. Pharm.* 196:51–61.

Perugini, P., Simeoni, S., Scalia, S., Genta, I., Modena, I., Conti, B., et al. 2002. Effect of nanoparticles encapsulation on the photostability of the sunscreen agent, 2-ethylhexyl-p-methoxycinnamate. *Int. J. Pharm.* 246:37–45.

Phetdee, M., Polnok, A., and Viyoch, J. 2008. Development of chitosan-coated sustained delivery of tamarind to the skin. *Int. J. Cosmetic Sci.* 30:285–295.

Pospieszny, H., Struszczyk, H., Chirkov, S.N., and Atabekov, J.G. 1994. New applications of chitosan in agriculture. In *Chitin world*, Karniki, S.Z. (Ed.). Irtschaftsverlag NW, Bremerhaven, Germany, pp. 246–254.

Rabe, J.H., Mamelak, A.J., McElgunn, P.J.S., Morison W.L., and Sauder, D.N. 2006. Photoaging: mechanisms and repair. *J. Am. Acad. Dermatol.* 55:1–19.

Renbutsu, E., Okabe, S., Omura, Y., Nakatsubo, F., Minami, S., Saimoto, H., and Shigemasa, Y. 2007. Synthesis of UV-curable chitosan derivatives and palladium (II) adsorption behavior on their UV-exposed films. *Carbohydr. Polym.* 69:697–706.

Rinaudo, M. 2006. Chitin and chitosan: properties and applications. *Prog. Polym. Sci.* 31:603–632.

Rwan, J., and Wu, J. 1996. Deacidification of grapefruit juice with chitosan. *Food Sci. Taiwan* 23:509–519.

Saboktakin, M.R., Tabatabaie, R. Maharramov, A., and Ramazanov, M.A. 2010. Synthesis and characterization of superparamagnetic chitosan–dextran sulfate hydrogels as nano carriers for colon-specific drug delivery. *Carbohydr. Polym.* 81:372–376.

Sakurai, K., Fujimoto, J., Shibano, T., and Takahashi, T. 1983. Gas permeation through chitosan membranes. *Sen-I Gakkaishi* 39:493–498.

Sano, H., Shibasaki, K., Matsukubo, T., and Takaesu, Y. 2003. Effect of chitosan rinsing on reduction of dental plaque formation. *Bull. Tokyo Dental College* 44:9–16.

Seo, H. 1990. Processing and utilization of chitin and chitosan. *Sen-i Gakkaishi* 46:564–569.

Shi, X., Du, Y., Yang, J., Zhang, B., and Sun, L. 2006. Effect of degree of substitution and molecular weight of carboxymethyl chitosan nanoparticles on doxorubicin delivery. *J. Appl. Polym. Sci.* 100:4689–4696.

Shimoda, J., Onishi, H., and Machida, Y. 2001. Bioadhesive characteristics of chitosan microspheres to the mucosa of rat small intestine. *Drug Dev. Ind. Pharm.* 27:567–576.

Soto-Perlata, N.V., Muller, H., and Knorr, D. 1989. Effect of chitosan treatments on the clarity and color of apple juice. *J. Food Sci.* 54:495–496.

Spagna, G., Pifferi, P.G., Rangoni, C., Mattivi, F., Nicolini, G., and Palmonari, R. 1996. The stabilization of white wines by adsorption of phenolic compounds on chitin and chitosan. *Food Res. Int.* 29:241–248.

Suzuki, K., Mikami, T., Okawa Y., Tokoro, A., Suzuki S., and Suzuki, M. 1986. Antitumor effect of hexa-N-acetylchitohexaose and chitohexaose. *Carbohyd. Res.* 151:403–408.

Tanner, P. 2006. Sunscreen product formulation. *Dermatol. Clin.* 24:53–62.

Verma, D.D., Verma, S., Blume, G., and Fahr, A. 2003. Liposomes increase skin penetration of entrapped and non-entrapped hydrophilic substances into human skin: a skin penetration and confocal laser scanning microscopy study. *Eur. J. Pharm. Biopharm.* 55:271–277.

Voet, J., and Voet, D. 1995. *Biochemistry*. 2nd ed. Wiley, New York, pp. 360–362.

Wachter, R., and Stenberg, E. 1996. Hydagen® CMF in cosmetic applications efficacy in different in vitro and in vivo measurements. In *Advances in chitin science*, Domard, A., Jeuniaux, C., Muzzarelli, R., and Roberts, G. (Eds.). Jacques Andre, Lyon, pp. 381–388.

Xia, W.S. 2003. Physiological activities of chitosan and its application in functional foods. *J. Chin. Inst. Food Sci. Technol.* 3:77–81.

14 Brown Algae-Derived Compounds as Potential Cosmeceuticals

Mahinda Senevirathne and Se-Kwon Kim

CONTENTS

14.1 INTRODUCTION

Cosmeceuticals are cosmetic products with biologically active ingredients that have medical or drug-like benefits. A vast number of cosmetics and toiletries exist to care for and positively alter the appearance of the skin, hair, and nails. Skin care products, in the form of cleansers, toners, exfoliants, masks, moisturizers, specialty creams, and antiperspirants/deodorants, may conflict with or augment the treatment of cutaneous dermatoses. Since seaweeds have been proven to have compounds with various biological activities, they can be used as potential agents in cosmeceutical formulations.

In Asian cultures seaweeds have been of great interest as food sources as well as ingredients for cosmeceutical and pharmaceuticals (Fleurence, 1999; Heo et al., 2010). They possess lipids, proteins, vitamins, and minerals (Ito and Hori, 1989), and also a large array of nutraceutical components, including antioxidants such as polyphenols (Heo et al., 2009; Ahn et al., 2007), pigments such as carotenoids (Heo and Jeon, 2009), and bioactive polysaccharides such as fucoidan and alginate (Plaza et al., 2008). Although marine algae are exposed to adverse environmental conditions such as high salinity and temperature, these conditions do not affect any serious photodynamic damage or kind of sunburns. This fact suggests that photosynthesizing plants like brown algae have metabolites that can act as antioxidative and photoprotective agents (Heo and Jeon, 2009; Ahn et al., 2007; Heo et al., 2009; Groniger et al., 2000). Further, it is well established that brown algae possess phenolic compounds known as phlorotannins, which have only been found to exist within brown algae, and have shown various biological activities, including antioxidant, photoprotection, antiplasmin inhibitors, antiallergic, skin whitening, anti-HIV-1, antibacterial, and anticancer activities (Heo et al., 2010; Sugiura et al., 2006, 2007; Le et al., 2009; Artan et al., 2008).

Polysaccharides are omnipresent ingredients of cosmetics other than phlorotannins and they do an absolutely reliable task. Since polysaccharides are inexpensive and easily available, they have been used in cosmetics for a long time. Polysaccharides have been shown to act as antioxidant, anticancer, and immunomodulating agents (Rocha de Souza et al., 2007; Wang et al., 2008; Ye et al., 2008; Athukorala et al., 2009; Leiro et al., 2007). Carotenoids are pigments that are available in brown seaweeds and show a variety of biological activities, such as antioxidant, anticancer, photoprotective, and anti-inflammatory activities (Kotake-Nara, 2005; Heo and Jeon, 2009; Kim et al., 2010; Heo et al., 2010). Fucoxanthin has been isolated as a major carotenoid from several brown seaweeds and has shown photoprotective and antioxidant activities (Heo and Jeon, 2009; Kim et al., 2010). Micosporine-like amino acids (MAAs) isolated from some brown algae show strong UV-absorbing activities.

A number of classical methods have been used for centuries to purify biologically active compounds from brown algae. Polyphenolic compounds are obtained using organic solvents such as methanol, ethanol, ethyl acetate, and n-hexane, and the amount of bioactive compounds may depend on the concentration and type of the solvent used (Heo et al., 2010; Kim et al., 2010). Enzymatic hydrolysis has also been used, but the amounts of compounds have been decreased significantly (Heo et al., 2005; Je et al., 2009). Recently, new methods such as supercritical CO_2 extraction and pressurized liquid extraction have been deployed to obtain bioactive compounds with minimal waste (Herrero et al., 2006; Zhang et al., 2011). However, conventional methods of isolating pigments, which usually require grinding the tissue with some organic solvents such as acetone or methanol (Murakami et al., 2010; Mori et al., 2004), are not suitable for fucoxanthin extraction, because they will be associated with a lot of Chl *a* and Chl *c*. Thus a specific solvent that targets fucoxanthin rather than chlorophyll is needed. Isolation of the pigments from brown algae as initially investigated using dimethyl sulfoxide (DMSO) by Seely et al. (1972) and Wheeler (1980). Wang et al. (2005) also have used DMSO to extract fucoxanthin from the rhizoid of *Laminaria japonica*. However, fucoxanthin has been extracted and isolated successfully using methanol (Heo et al., 2010). Further, pressurized liquid extraction was a successful method to get fucoxanthin from *Eisenia bicyclis* since plant carotenoids are sensitive to degradation during extraction (Zhang et al., 2011). Hot water extraction is the most common method for the extraction of polysaccharides (Zhang et al., 2010). However, some researchers have used both water and enzymatic extractions to obtain polysaccharides (Athukorala et al., 2007; Heo et al., 2010), and MAAs are also obtained by water extraction (Schmid et al., 2003).

14.2 POTENTIAL COSMECEUTICAL EFFECTS OF BROWN ALGAE

14.2.1 Photoprotective Compounds

Human skin is frequently exposed to ultraviolet (UV) radiation from sunlight, and hence suffers from deleterious effects due to its actinic nature, manifested genetically and physiologically. UV-B radiation is a mutagen, and epidemiological evidence has indicated that UV-B radiation is responsible for the induction of melanoma and nonmelanoma skin cancers (Urbach, 1978). The skin represents the first barrier that protects the body from external environmental pollutants, including UV radiation and environmental chemicals (Katiyar, 2007). The reactive oxygen species (ROSs) formed under these conditions are presumed to play an important role in cutaneous disease such as carcinogenesis, skin aging, and causing oxidative stress to DNA (Fuchs et al., 1989; Emerit, 1992).

Therefore regular intake of dietary antioxidants or treatment of the skin with cosmeceutical products containing antioxidant ingredients may be a useful strategy for preventing UV-induced skin damage. Since UV-protective compounds themselves may become reactive radicals when activated by UV radiation, and few of them can provide full spectral protection against UV light, the use of formulations containing both sunscreens and naturally occurring antioxidants may be highly beneficial for effective prevention of skin damage by UV radiation (Tomaino et al., 2006).

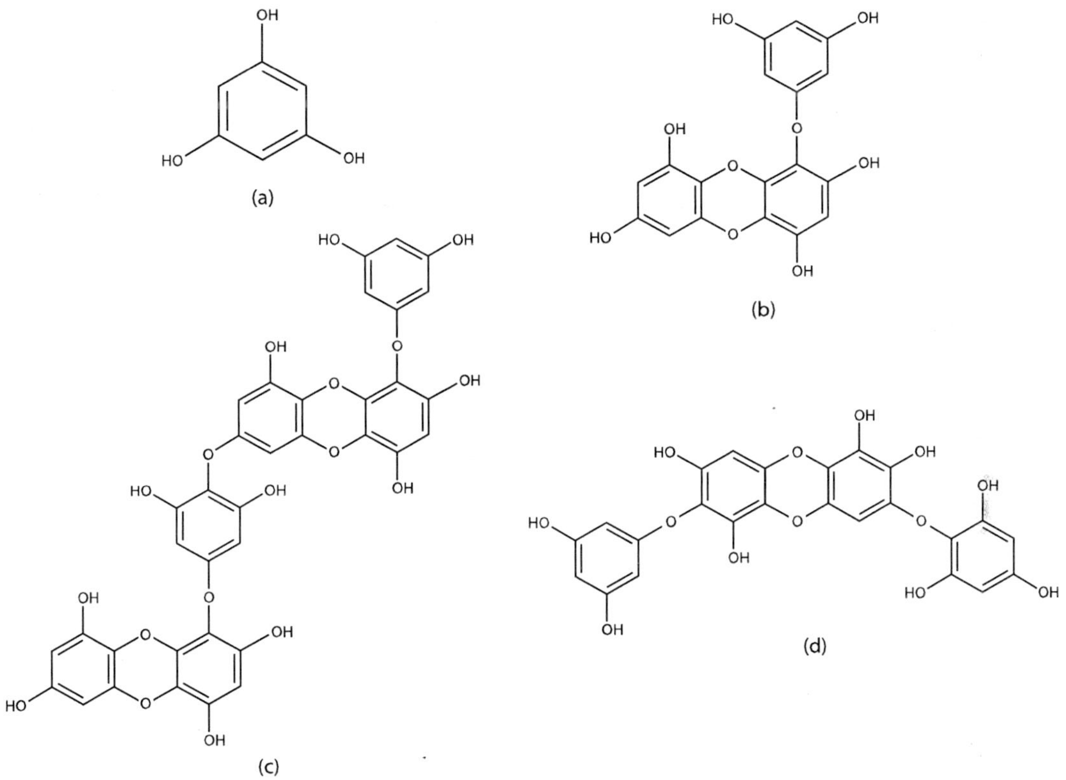

FIGURE 14.1 Chemical structures of some phlorotannins: (a) phloroglucinol, (b) eckol, (c) dieckol, (d) and diphlorethohydroxycarmalol.

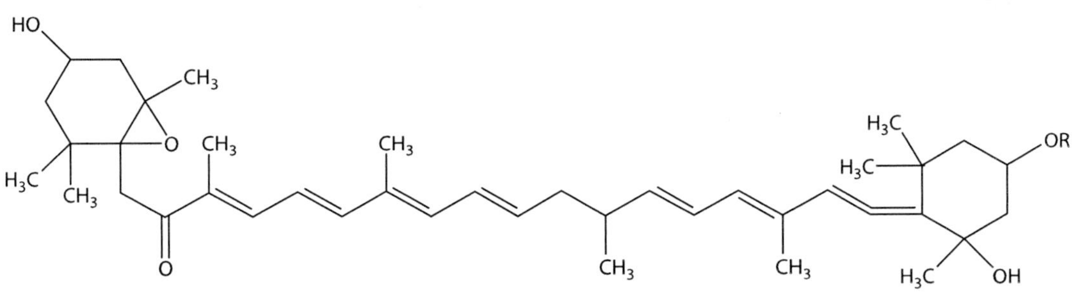

FIGURE 14.2 Chemical structure of fucoxanthin.

Polyphenolic compounds (Figure 14.1) such as eckol, dieckol, phloroglucinol, diphlorethohydroxy carmalol (DPHC), and carotenoids such as fucoxanthin (Figure 14.2) are known to act as photoprotective compounds (Heo et al., 2009, 2010; Heo and Jeon, 2009). Few reports are available on photoprotective activities of the active compounds from brown seaweeds (Table 14.1). DPHC, a phlorotannin isolated from *Ishige okamurae*, has been shown to be active against UV-B radiation-induced cell damage (Heo et al., 2010). Further, some other phlorotannin compounds such as eckol, dieckol, and phloroglucinol extracted from *Ecklonia cava* have also shown strong photoprotective abilities (Heo et al., 2009).

Fucoidan inhibits UV-B-induced MMP-1 enzyme activity in human skin fibroblasts (Moon et al., 2008). It mainly inhibits UV-B induced MMP-1 expression by inhibiting the extracellular signal-regulated kinase (ERK) pathways. Fucoidan obtained from *Costaria costata* is also active against

TABLE 14.1
Photoprotective Compounds Isolated from Brown Algae

Compounds	Species	Reference
	Phlorotanins	
Eckol	*Ecklonia cava*	Heo et al., 2009
Dieckol	*Ecklonia cava*	Heo et al., 2009
Phloroglucinol	*Ecklonia cava*	Heo et al., 2009
Diphlorethohydroxycarmalol	*Ecklonia cava*	Heo et al., 2009
	Carotinoid	
Fucoxanthin	*Sargassum siliquastrum*	Heo and Jeon, 2009
	Polysaccharide	
Fucoidan	*Costaria costata*	Moon et al., 2009
	MAAs	
Porphyra-334, palythine, shinorine	*Sargassum oligocystun*	Karsten et al., 1998
Palythine, palythinol, porphyra-334, shinorine	*Padina crassa*	Nakamura et al., 1982
Porphyra-334, shinorine	*Desmarestia aculeatia*	Groniger et al., 2000
Porphyra-334, palythine	*Desmarestia menzies*	Karentz et al., 1991

UVB-induced MMP-1 promoter, mRNA, and protein expression *in vitro* in human keratinocyte (HaCaT) cells (Moon et al., 2009). Pretreatment with fucoidan significantly inhibited MMP-1 protein expression. MMPs are responsible for the degradation or inhibition of synthesis of collagenous extracellular matrix in connective tissues.

Carotenoids are important natural lipid-soluble pigments that directly provide photoprotection against UV light-induced photooxidation in the skin (Sies and Stahl, 2004; Aust et al., 2005), while β-carotene is also shown to modulate UVA-induced gene expression in human keratinocytes (Wertz et al., 2004, 2005). Fucoxanthin, a major carotenoid of edible brown seaweeds, has a unique structure including allenic, conjugated carbonyl, epoxide, and acetyl group in its molecule. Fucoxanthin isolated from *Sargasum siliquastrum* has revealed that it has strong activities against UV-B-induced cell damage (Heo et al., 2009).

MAAs are another type of UV-absorbing metabolites in some brown algae that contain aminocyclohexenimine ring systems. They are low molecular weight water-soluble molecules with UV absorption maxima between 310 and 360 nm. Glycine is the most common amino acid present in all MAAs (Bandaranayake, 1998). Further, various MAAs in brown algae have been demonstrated to act as photoprotective compounds (Groniger et al., 2000). Some MAAs such as porphyra-334, shinorine, palythinol, asterina-330, palathene, and palythine (Figure 14.3) have been isolated from brown algae varieties such as *Sargassum oligocystun*, *Padina crassa*, *Desmarestia aculeatia*, *Halopteris scoparia*, and *Desmarestia menzies* (Groniger et al., 2000).

14.2.2 Skin Care and Whitening Compounds

Melanin is an essential pigment needed for the protection of human skin against radiation, but the accumulation of abnormal melanin induces pigmentation disorders, such as melasma, freckles, ephelide, and senile lentigines (Baumann, 2001; Solano, 2006). In an effort to inhibit the activity of tyrosinase, many different types of tyrosinase inhibitors have been developed via either synthesis or isolation from natural sources (Kim and Uyama, 2005; Heo et al., 2010). Thus inhibition of tyrosinase activity or its production can prevent melanogenesis, and also those compounds can be used

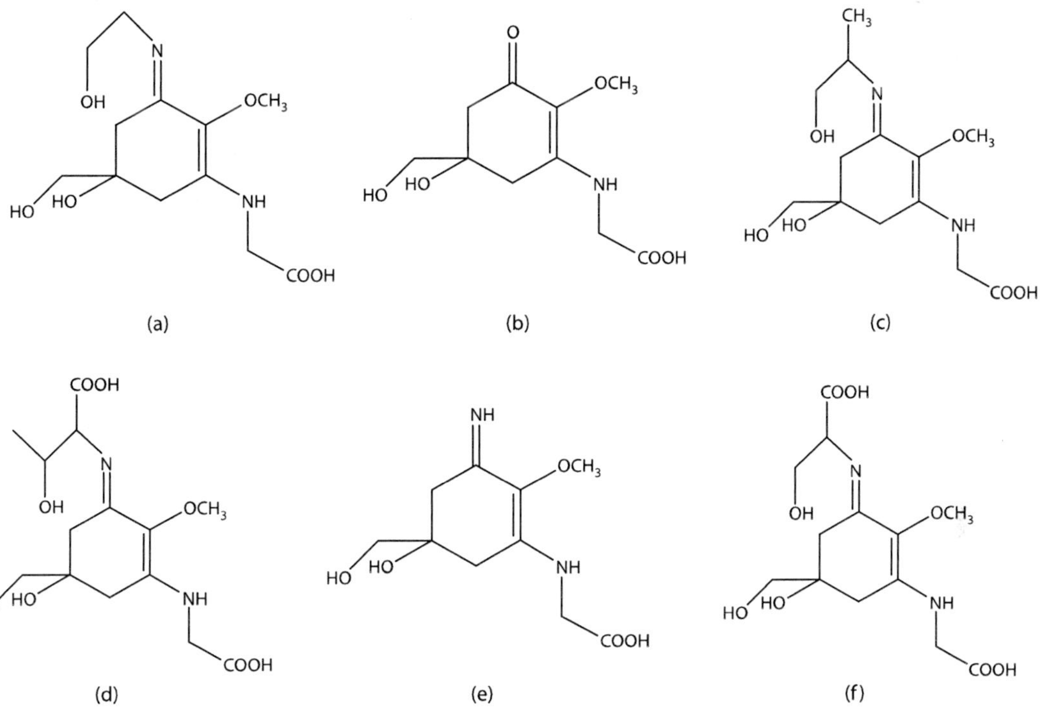

FIGURE 14.3 Chemical structures of micosporine-like amino acids: (a) asterina-330, (b) micosporine glycine, (c) palythinol, (d) porphyra-334, (e) palythine, (f) shinorine.

in the cosmetics industry as skin whitening agents and as agents in the treatment of pigmentation disturbances (Kim and Uyama, 2005).

DPHC has been investigated for its inhibitory effect against melanogenesis and cosmetic-related experiments (Heo et al., 2010). It shows both antioxidant and cosmeceutical activities, including inhibitory effects against tyrosinase and melanin synthesis. Three phlorotannin compounds, 7-phloroeckol, phloroglucinol, and dioxinodehydroeckol, were isolated from *E. cava*, and 7-phloroeckol showed strong tyrosinase inhibitory activity out of those three compounds (Yoon et al., 2009). Further, they were evaluated against 3-isobutyl-1-methylxanthin-induced (IBMX) melanin formation in B16F10 melanoma cells, and treatment with 7-phloroeckol resulted in a significant inhibition of melanin synthesis. In addition, some other phlorotanins, such as eckol and dieckol isolated from *E. cava*, other than phloroglucinol, showed strong activity against melanogenesis (Heo et al., 2009). The effect of melanogenesis was investigated via inhibitory effect on tyrosinase and melanin synthesis. The value of dieckol in a tyrosinase inhibitory assay was relatively higher than that of the commercial cosmeceutical compound, kojic acid. Hence a number of depigmenting compounds isolated from various brown algae showed strong activities in relation to undesirable skin discoloration as well as aging, and they may be potential compounds in the cosmetics industry (Table 14.2).

Some 29 seaweed extracts have been studied on transcriptional activities of NF-κB and Ap-1, and matrix metalloproteinase-1 (MMP-1) expression, with the view of developing an antiphotoaging agent (Joe et al., 2006). Among them, *Eisenia bicyclis*, *Ecklonia cava*, and *Ecklonia stolonifera* have shown strong inhibition on both NF-κB and Ap-1 reporter activity, which were well correlated with their abilities to inhibit MMP-1 expression. In addition, MMP-1 expression was dramatically attenuated by treatment with eckol or dieckol, which was purely isolated from *E. stolonifera*, indicating that those compounds are active principles to inhibit MMP-1 expression in human dermal fibroblasts. Acute exposure of human skin to UV radiation may upregulate the synthesis of several

TABLE 14.2
Skin Whitening and Skin Care Compounds from Brown Algae

Compounds	Species	References
	Phlorotanins	
Eckol	*Ecklonia cava, E. stolonifera*	Heo et al., 2009; Joe et al., 2006
Dieckol	*Ecklonia cava, E. stolonifera*	Heo et al., 2009; Joe et al., 2006
Phloroglucinol	*Ecklonia cava*	Yoon et al., 2009; Heo et al., 2009
Dioxinodehydroeckol	*Ecklonia cava*	Yoon et al., 2009
7-Phloroeckol	*Ecklonia cava*	Yoon et al., 2009
Diphlorethohydroxy Carmalol	*Ishige okamurae*	Heo et al., 2010

MMPs that degrade skin collagen (Fisher et al., 1996, 1997). Recent studies showed that some MMPs are highly induced in inflamed and photoaged skin and break down the dermal matrix proteins such as elastin and collagen that possibly lead to prolonged skin damage and wrinkle formation (Fisher and Voorhees, 1998). Therefore the agents that inhibit collagenase activity may have beneficial effects for maintaining healthy skin by preventing degradation of the dermal matrix. MMPs are necessary for tissue remodeling and the healing cascade under normal physiological conditions. Clinically, naturally aged skin is smooth, pale, and finely wrinkled; in contrast, photoaged skin is coarsely wrinkled (Gilchrest, 1989). Therefore agents such as phlorotannin that inhibit collagenase activity and MMP activity may have beneficial effects for maintaining healthy skin by preventing degradation of the dermal matrix. Phlorotannins from *E. cava* have been reported to inhibit MMP activity with strong activities (Kim et al., 2006). Both fucoidan and phloroglucinol have demonstrated cosmetic benefits, including the direct inhibition of matrix-degrading enzymes such as elastase and collagenase. Elastase inhibition is an especially important aspect of effective cosmetic active ingredients, as UV-induced elastase activity is a key activator of wrinkles.

Glycation, the nonenzymatic attachment of sugars to proteins, is also a marker for aging skin. Advanced glycation end products (AGEs) are nonenzymatic glycosylated proteins that accumulate in aging or damaged tissues (Lohwasser et al., 2006). Fucoidan can block some receptors for AGEs and reduce their accumulation (Tamura et al., 2003; Kuniyasu et al., 2003). Polyphenols from some seaweeds have also been demonstrated to inhibit glycation (Okada et al., 2004). Thus seaweed-derived ingredients can potentially inhibit glycation formation and accumulation processes as they are good antioxidative compounds.

A simple fucoidan isolated from *Fucus vesiculosus* showed beneficial effects as a skin active, decreasing skin thickness and increasing elasticity (Fujimura et al., 2002). It has been assessed as nonphototoxic and not a primary irritant ingredient. *In vitro* screening indicated that the product inhibited a variety of matrix-degrading enzymes, including elastase and collagenase, and the pigment-forming enzyme tyrosinase. Alginates are linear unbranched polymers containing β-(1-4)-linked D-mannuronic acid and α-(1-4)-linked L-guluronic acid residues. Sodium alginate is a natural water-soluble polysaccharide that produces a gel consistency when hydrated. Sodium alginate has been used in the medical field, in the food industry, and more recently in cosmetics as a skin care ingredient due to its thickening, gelling, and water-absorbing abilities (Onsoyen, 1996). Further, alginates act as humectants and increase the water holding capacity of the skin, preventing it from drying. Further, by definition all humectants are also moisturizers.

Tetrahydropyridine derivatives with an λ_{max} of 305–311 nm, with extremely high absorptivity (ε 32,000), were prepared from 3-alkanoyl-1,4,5,6-tetrahydropyridines (Dunlap et al., 1998; Bird et al., 1990), sufficiently stable for commercial considerations as skin care products. This class of compounds with the 4,4-dimethyl structure had the most promising candidates. The *gem*-dimethyl

TABLE 14.3
Antioxidant and Antiaging Compounds from Brown Algae

Compounds	Species	References
	Phlorotanins	
Eckol	*Ecklonia cava, E. stolonifera*	Heo et al., 2009; Joe et al., 2006
Dieckol	*Ecklonia cava, E. stolonifera*	Heo et al., 2009; Joe et al., 2006
Phloroglucinol	*Ecklonia cava*	Yoon et al., 2009; Heo et al., 2009
Dioxinodehydroeckol	*Ecklonia cava*	Yoon et al., 2009
7-Phloroeckol	*Ecklonia cava*	Yoon et al., 2009
Diphlorethohydroxy Carmalol	*Ishige okamurae*	Heo et al., 2010

group (also present in vitamin A and retinal) prevents radical processes at the α-ring position of the chromophore, thereby preventing oxidative decomposition of the compound, and also restricts free rotation of the alkanoyl group, hence increasing the λ_{max} of the derivatives.

14.2.3 Antioxidant and Antiaging Compounds

Since cosmetic products have to protect from oxidation, essentially antioxidant compounds can play an additional role as preservatives to protect the active ingredients. Polyphenolic, polysaccharide, and carotenoid compounds present in brown algae show potent antioxidant activities (Table 14.3). Some phlorotannins such as eckol, dieckol, phloroglucinol, and diphlorethohydroxycarmalol have been isolated from brown algae such as *Ecklonia cava*, *Ecklonia stolonifera*, *Ishige okamurae*, and *Sargassum saliquastrum* and showed potent antioxidant activities in both chemical and cell-based systems (Heo and Jeon, 2009; Ahn et al., 2007; Heo et al., 2009; Kang et al., 2003). Further, phloroglucinol isolated from brown seaweeds has demonstrated both potential antioxidant and anti-inflammatory activities (Kim and Kim, 2010). Fucoxanthin obtained from brown alga, *Sargassum saliquastrum*, showed strong antioxidant activities against hydrogen peroxide-induced cell damage (Heo et al., 2008). Further, fucoxanthin from *Hijikia fusiformis* also showed potent antioxidant activity against DPPH radical scavenging (Yan et al., 1999). Polysaccharides made up of many monosaccharides joined together by glycosidic bonds possess marked immunological properties ranging from nonspecific stimulation of host immune, antitumor, antimutagenic, and antioxidant activities (Bohn and BeMiller, 1995; Kennedy, 1989). Further, sulfated polysaccharides extracted from *Laminaria japonica* and *Terbinaria ornata* are excellent antioxidants (Zhang et al., 2010; Ananthi et. al., 2010). In addition, the fucoidan fraction obtained from *Laminaria japonica* has been reported to have a positive correlation for sulfate content and superoxide radical scavenging activity (Wang et al., 2008). One alginic acid and a natural glucan were isolated from *Turbinaria conoides* and they were evaluated for the antioxidant activity by FRAP and DPPH assays (Chattopadhyay et al., 2010). The fucoidan fraction showed the highest activity, followed by the alginate and glucan.

The sargachromanols, meroterpenoids of the chromene class, were isolated from *S. siliquastrum*, and all the isolated compounds exhibited significant activity against DPPH radical scavenging (Jang et al., 2005). The known plastiquinones were isolated from *S. micracanthum* and one compound out of them displayed significant activity.

14.2.4 Antibacterial and Antifungal Compounds

Preservatives are added to cosmetics to prevent the growth of microorganisms (e.g., bacteria and fungi), which can spoil the product and possibly harm the user. Preservatives used in cosmetics can include parabens, benzyl alcohol, and tetrasodium EDTA (ethylenediaminetetraacetic acid). However, due to the side effects they cause, consumers are very interested in natural-based compounds in their products. Several pharmacological activities have been reported for phlorotannins, including enzyme inhibition and antibacterial activity (Shibata et al., 2003; Nagayama et al., 2002). Compounds isolated from *Fucus vesiculosus* had strong antibacterial activities (Sandsdalen et al., 2003). A 1,4-napthaquinone derivative (deoxy lapachol) was isolated from a New Zealand brown alga *Landsburgia quercifolia* by the bioactivity-directed isolation method, and it showed antifungal activity (Perry et al., 1991).

14.2.5 Other Potential Effects of Compounds

In order to create a moist environment for rapid wound healing, a hydrogel sheet composed of a blended powder of alginate, chitin/chitosan, and fucoidan (ACF-HS; 60:20:2:4 w/w) has been developed as a functional wound dressing (Murakami et al., 2010). Sometimes the ingredients in cosmetics can have unintended side effects. For example, skin allergies (allergic dermatitis) to specific ingredients can be a problem. Allergies to cosmetic products can be due to chemicals such as added fragrances and preservatives. This can lead to a skin rash where the product is applied. However, antiallergy compounds may be the answer for that kind of problem. Some phlorotannin derivatives such as deikol, phloroglucinol, 6,6′-bieckol, and 1-(3′,5′-dihydroxyphenoxy)-7-(2″,4″,6-trihydroxyphenoxy)-2,4,9-trihydroxydibenzo-1,4-dioxin isolated from *Ecklonia cava* showed potential antiallergy activities (Le et al., 2009). Also, fucoxanthin isolated from brown algae showed anti-inflammatory activity (Heo et al., 2010).

14.3 CONCLUSION

Cosmetics and other toiletries are used by men, women, and children on a daily basis and for various reasons. Recently interest has increased toward natural materials-based cosmetic products since people are very concerned about their personal health and are willing to reduce the side effects caused by some cosmetics produced with synthetic compounds. Hence many researchers have focused their attention on bioactive materials from the ocean due to various reasons, such as their availability, inexpensiveness, and richness with a large number of secondary metabolites. They have been reported to show a series of biological activities *in vitro* and *in vivo*, and hence they have potential for inclusion in cosmetic products. Many *in vivo* and clinical experiments have clearly shown that bioactive compounds from brown algae prevent the progression of oxidative stress, skin photoaging, and collagen degradation. Hence, based on the recently published literature, we conclude that bioactive compounds derived from brown algae are potential agents in cosmeceutical formulations.

REFERENCES

Ahn, G. N., Kim, K. N., Cha, S. H., Song, C. B., Lee, J. H., et al. 2007. Antioxidant activities of phlorotannins purified from *Ecklonia cava* on free radical scavenging using ESR and H_2O_2-mediated DNA damage. *Eur. Food Res. Technol.* 226: 71–79.

Ananthi, S., Raghavendran, H. R. B., Sunil, A. G., Gayathri, V., Ramakrishnan, G., and Vasanthi, H. R. 2010. *In vitro* antioxidant and *in vivo* anti-inflammatory potential of crude polysaccharide from *Turbinaria ornata* (marine brown alga). *Food Chem. Toxicol.* 48: 187–192.

Artan, M., Li, Y., Karadeniz, F., Lee, S. H., Kim, M. M., et al. 2008. Anti-HIV-1 activity of phloroglucinol derivative, 6,6-bieckol, from *Ecklonia cava*. *Bioorg. Med. Chem.* 16: 7921–7926.

Athukorala, A., Lee, K. W., Kim, S. K., and Jeon, Y. J. 2007. Anticoagulant activity of marine green and brown algae collected from Jeju Island in Korea. *Bioresource Technol.* 98: 1711–1716.

Athukorala, Y., Ahn, G. N., Jee, Y. H., Kim, G. Y., Kim, S. H., et al. 2009. Antiproliferative activity of sulfated polysaccharide isolated from an enzymatic digest of *Ecklonia cava* on the U-937 cell line. *J. Appl. Phycol.* 21: 307–314.

Aust, O., Stahl, W., Sies, H., Tronnier, H., and Heinrich, U. 2005. Supplementation with tomato-based products increases lycopene, phytofluene, and phytoene levels in human serum and protects against UV-light-induced erythema. *Int. J. Vitam. Nutr. Res.* 75: 54–60.

Bandaranayake, W. M. 1998. Mycosporines: are they nature's sunscreens? *Nat. Pro. Rep.* 15: 159–172.

Baumann, L. 2001. *Cosmetic dermatology*. McGraw-Hill, New York, p. 99.

Bird, G., Fitzmaurice, N., Dunlap, W. C., Chalker, B. E., and Bandaranayake, W. M. 1990. Patent Abridgment 80728/87.

Bohn, J. A., and BeMiller, J. N. 1995. (1→3)-β-D-Glucans as biological response modifiers: a review of structure-functional activity relationships. *Carbohyd. Polym.* 28: 3–14.

Chattopadhyay, N., Ghosh, T., Sinha, S., Chattopadhyay, K., Karmakar, P., et al. 2010. Polysaccharides from *Turbinaria conoides*: structural features and antioxidant capacity. *Food Chem.* 118: 823–829.

Dunlap, W. C., Chalker, B. E., and Bandaranayake, W. M. 1998. Nature's sunscreen from the Great Barrier Reef, Australia. *Int. J. Cosmet. Sci.* 20: 41–51.

Emerit, I. 1992. Free radicals and aging of the skin. *EXS* 62: 328–341.

Fisher, G. J., Datta, S. C., Talwar, H. S., Wang, Z. Q., Varani, J., Kang, S., and Vorhees, J. J. 1996. Molecular basis of sun-induced premature skin aging and retinoid antagonism. *Nature* 379: 335–339.

Fisher, G. J., and Voorhees, J. J. 1998. Molecular mechanism of photoaging and its prevention by retinoic acid: ultraviolet irradiation induces MAP kinase signaling transduction cascade that induces AP-1 regulated matrix metalloproteinases that degrade human skin *in vivo*. *J. Invest. Dermatol.* 3: 61–68.

Fisher, G. J., Wang, Z. Q., Datta, S. C., Varani, J., Kang, S., and Voorhees, J. J. 1997. Pathophysiology of premature skin aging induced by ultraviolet light. *N. Engl. J. Med.* 337: 1419–1428.

Fleurence, J. 1999. Seaweed proteins: biochemical, nutritional aspects and potentials uses. *Trends Food Sci. Technol.* 10: 25–28.

Fuchs, J., Huflejt, M. E., Rothfuss, L. M., Wilson, D. S., Carcamo, G., and Packer, L. 1989. Impairment of enzymic and nonenzymic antioxidants in skin by UVB irradiation. *J. Invest. Dermatol.* 93: 769–773.

Fujimura, T., Tsukahara, K., Moriwaki, S., Kitahara, T., Sano, T., and Takema, Y. 2002. Treatment of human skin with an extract of *Fucus vesiculosus* changes its thickness and mechanical properties. *J. Cosmet. Sci.* 53: 1–9.

Gilchrest, B. A. 1989. Skin aging and photoaging: an overview. *J. Am. Acad. Dermatol.* 21: 610–613.

Groniger, A., Sinha, R. P., Klisch, M., and Hader, D. P. 2000. Photoprotective compounds in cyanobacteria, phytoplankton and macroalgae—a database. *J. Photochem. Photobiol. B Biol.* 58: 115–122.

Heo, S. J., and Jeon, Y. J. 2009. Protective effect of fucoxanthin isolated from *Sargassum siliquastrum* on UV-B induced cell damage. *J. Photochem. Photobiol. B:* 95: 101–107.

Heo, S. J., Ko, S. C., Cha, S. H., Kang, D. H., Park, H. S., et al. 2009. Effect of phlorotannins isolated from *Ecklonia cava* on melanogenesis and their protective effect against photo-oxidative stress induced by UV-B radiation. *Toxicol. In Vitro* 23: 1123–1130.

Heo, S. J., Ko, S. C., Kang, S. M., Kang, H. S., Kim, J. P., et al. 2008. Cytoprotective effect of fucoxanthin isolated from brown algae *Sargassum siliquastrum* against H_2O_2-induced cell damage. *Eur. Food Res. Technol.* 228: 145–151.

Heo, S. J., Park, E. J., Lee, K. W., and Jeon, Y. J. 2005. Antioxidant activities of enzymatic extracts from brown seaweeds. *Bioresource Technol.* 96: 1613–1623.

Heo, S. J., Yoon, W. J., Kim, K. N., Ahn, G. N., Kang, S. M., et al. 2010. Evaluation of anti-inflammatory effect of fucoxanthin isolated from brown algae in lipopolysaccharide-stimulated RAW 264.7 macrophages. *Food Chem. Toxicol.* 48: 2045–2051.

Herrero, M., Cifuentes, A., and Ibanez, E. 2006. Sub- and supercritical fluid extraction of functional ingredients from different natural sources: plants, food-by-products, algae and microalgae. A review. *Food Chem.* 98: 136–148.

Ito, K., and Hori, K. 1989. Seaweed: chemical composition and potential food uses. *Food Rev. Int.* 5: 101–144.

Jang, K. H., Lee, H. B., Choi, B. W., Lee, H. S., and Shin, J. 2005. Chromenes from the brown alga *Sargassum siliquastrum*. *J. Nat. Prod.* 68: 716–723.

Je, J. Y., Park, P. J., Kim, E. K., Park, J. S., and Yoon, H. D. 2009. Antioxidant activity of enzymatic extracts from the brown seaweed *Undaria pinnatifida* by electron spin resonance spectroscopy. *LWT—Food Sci. Technol.* 42: 874–878.

Joe, M. J., Kim, S. N., Choi, H. Y., Shin, W. S., Park, G. M., et al. 2006. The inhibitory effects of eckol and diekol from *Ecklonia stolonifera* on the expression of matrix metalloprotenase-1 in human dermal fibroblasts. *Biol. Pharm. Bull.* 29: 1735–1744.

Kang, H. S., Chung, H. Y., Jung, J. H., Son, B. W., and Cho, J. S. 2003. A new phlorotannin from the brown alga *Ecklonia stolonifera*. *Chem. Pharm. Bull.* 51: 1012–1014.

Karentz, D., McEuen, F. S., Land, M. C., and Dunlap, W. C. 1991. Survey of mycosporine-like amino acid compounds in Antarctic marine organisms: potential protection from ultraviolet exposure. *Mar. Biol.* 108, 157–166.

Karsten, U., Sawall, T., Hanelt, D., Bischof, K., Figueroa, F. L., Flores-Moya, A., and Wiencke, C. (1998). An inventory of UV-absorbing mycosporine-like amino acids in macroalgae from polar to warm-temperate regions. *Bot. Mar.* 41, 443–454.

Katiyar, S. K. 2007. UV-induced immune suppression and photocarcinogenesis: chemoprevention by dietary botanical agents. *Cancer Lett.* 255: 1–11.

Kennedy, J. F. 1989. *In carbohydrates*. Oxford University Press, Oxford.

Kim, K. N., Heo, S. J., Kang, S., M., Ahn, G. N., and Jeon, Y. J. 2010. Fucoxanthin induces apoptosis in human leukemia HL-60 cells through a ROS-mediated Bcl-xL pathway. *Toxicol. In Vitro* 24: 1648–1654.

Kim, M. M., and Kim, S. K. 2010. Effect of phloroglucinol on oxidative stress and inflammation. *Food Chem. Toxicol.* 48: 2925–2933.

Kim, M. M., Ta, Q. V., Mendis, E., Rajapakse, N., Jung, W. K., et al. 2006. Phlorotannins in *Ecklonia cava* extract inhibit matrix metalloproteinase activity. *Life Sci.* 79: 1436–1443.

Kim, Y. J., and Uyama, H. 2005. Tyrosinase inhibitors from natural and synthetic sources: structure, inhibition mechanism and perspective for the future. *Cell. Mol. Life Sci.* 62: 1707–1723.

Kotake-Nara, E., Asai, A., and Nagao, A. 2005. Neoxanthin and fucoxanthin induce apoptosis in PC-3 human prostate cancer cells. *Cancer Lett.* 220: 75–84.

Kuniyasu, A., Ohgami, N., Hayashi, S., Miyazaki, A., Horiuchi, S., and Nakayama, H. 2003. CD36-mediated endocytic uptake of advanced glycation end products (AGE) in mouse 3T3-L1 and human subcutaneous adipocytes. *FEBS Lett.* 537: 85–90.

Le, Q. T., Li, Y., Qian, Z. J., Kim, M. M., and Kim, S. W. 2009. Inhibitory effects of polyphenols isolated from marine alga *Ecklonia cava* on histamine release. *Process Biochem.* 44: 168–176.

Leiro, J. M., Castro, R., Arranz, J. A., and Lamas, J. 2007. Immunomodulating activities of acidic sulfated polysaccharides obtained from the seaweed *Ulva rigida* C. Agardh. *Int. Immunopharmacol.* 7: 879–888.

Lohwasser, C., Neureiter, D., Weigle, B., Kirchner, T., and Schuppan, D. 2006. The receptor for advanced glycation end products is highly expressed in the skin and up-regulated by advanced glycation end products and tumor necrosis factor-alpha. *J. Invest. Dermatol.* 126: 291–299.

Moon, H. J., Lee, S. R., Shim, S. N., Jeong, S. H., Stonik, V. A., et al. 2008. Fucoidan inhibits UVB-induced MMP-1 expression in human skin fibroblasts. *Biol. Pharm. Bull.* 31: 284–289.

Moon, H. J., Park, K. S., Ku, M. J., Lee, M. S., Jeong, S. H., et al. 2009. Effect of *Costaria costata* fucoidan on expression of matrix metalloproteinase-1 promoter, mRNA, and protein. *J. Nat. Prod. 72*: 1731–1734.

Mori, K., Ooi, T., Hiraoka, M., Oka, N., Hamada, H., Tamura, M., and Kusumi, T. 2004. Fucoxanthin and its metabolites in edible brown algae cultivated in deep seawater. *Mar. Drugs* 2: 63–72.

Murakami, K., Aoki, H., Nakamura, S., Nakamura, S. I., Takikawa, M., et al. 2010. Hydrogel blends of chitin/chitosan, fucoidan and alginate as healing-impaired wound dressings. *Biomaterials* 31: 83–90.

Nagayama, K., Iwamura, Y., Shibata, T., Hirayama, I., and Nakamura, T. 2002. Bactericidal activity of phlorotannins from the brown alga *Ecklonia kurome*. *J. Antimicrob. Chemother.* 50: 889–893.

Nakamura, H., Kobayashi, J., and Hirata, Y. 1982. Separation of mycosporine-like amino acids in marine organisms using reverse-phase high-performance liquid chromatography. *J. Chromatogr.* 250, 113–118.

Okada, Y., Ishimaru, A., Suzuki, R., and Okuyama, T. 2004. A new phloroglucinol derivative from the brown alga *Eisenia bicyclis*: potential for the effective treatment of diabetic complications. *J. Nat. Prod.* 67: 103–105.

Onsoyen, E. 1996. Commercial application of aliginates. *Carbohydr. Eur.* 14: 26–31.

Perry, N. B., Bluent, J. W., and Munro, M. H. G. 1991. A cytotoxic and antifungal 1,4-naphthoquinone and related compounds from a New Zealand brown alga *Landsburgia Quercifolia*. *J. Nat. Prod.* 54: 978–985.

Plaza, M, Cifuentes, A., and Ibanez, E. 2008. In the search of new functional food ingredients from algae. *Trends Food Sci. Technol.* 19: 31–39.

Rocha de Souza, M. C., Marques, C. T., Dore, C. M. G., Ferreirada Silva, F. R., Rocha, H. A. O., et al. 2007. Antioxidant activities of sulfated polysaccharides from brown and red seaweeds. *J. Appl. Phycol.* 19: 153–160.

Sandsdalen, E., Haug, T., Stensvag, K., and Styrvold, O. B. 2003. The antibacterial effect of a polyhydroxylated fucophlorethol from the marine brown alga, *Fucus vesiculosus*. *World J. Microbiol. Biotechnol.* 19: 777–782.

Schmid, D., Schurch, C., Zulli, F., Nissen, H. P., and Prieur, H. 2003. Mycosporine-like amino acids: natural UV-screening compounds from red algae to protect the skin against photoaging. www.mib-bio.com.

Seely, G. R., Duncan, M. J., and Vidaver, W. E. 1972. Preparative and analytical extraction of pigments from brown algae with dimethyl sulfoxide. *Mar. Biol.* 12: 184–188.

Shibata, T., Nagayama, K., Tanaka, R., Yamaguchi, K., and Nakamura, T. 2003. Inhibitory effects of brown algal phlorotannins on secretory phospholipase A_2s, lipoxygenases and cyclooxygenases. *J. Appl. Phycol.* 15: 61–66.

Sies, H., and Stahl, W. 2004. Nutritional protection against skin damage from sunlight. *Annu. Rev. Nutr.* 24: 173–200.

Solano, F., Briganti, S., Picardo, M., and Ghanem, G. H. 2006. Hypopigmenting agents: an updated review on biological, chemical and clinical aspects. *Pigm. Cell Melanoma Res.* 19: 550–571.

Sugiura, Y., Matsuda, K., Yamada, Y., Nishikawa, M., Shioya, K., et al. 2006. Isolation of a new anti-allergic phlorotannin, phlorofucofuroeckol-B, from an edible brown alga, *Eisenia arborea. Biosci. Biotech. Biochem.* 70: 2807–2811.

Sugiura, Y., Matsuda, K., Yamada, Y., Nishikawa, M., Shioya, K., et al. 2007. Anti-allergic phlorotannins from the edible brown alga, *Eisenia arborea. Food Sci. Technol. Res.* 13: 54–60.

Tamura, Y., Adachi, H., Osuga, J., Ohashi, K., Yahagi, N., et al. 2003. FEEL-1 and FEEL-2 are endocytic receptors for advanced glycation end products. *J. Biol. Chem.* 278: 12613–12617.

Tomaino, A., Cristani, M., Cimino, F., Speciale, A., Trombetta, D., Bonina, F., and Saija, A. 2006. *In vitro* protective effect of a Jacquez grapes wine extract on UVB-induced skin damage. *Toxicol. In Vitro* 20: 1395–1402.

Urbach, F. 1978. Evidence of epidemiology of UV-induced carcinogenesis in man. *Natl. Cancer Inst. Monogr.* 50: 5–10.

Wang, J., Zhang, Q., Zhang, Z., and Li, Z. 2008. Antioxidant activity of sulfated polysaccharide fractions extracted from *Laminaria japonica. Int. J. Biol. Macromol.* 42: 127–132.

Wang, W. J., Wang, G. C., Zhang, M., and Tseng, C. K. 2005. Isolation of fucoxanthin from the rhizoid of *Laminaria japonica* Aresch. *J. Integr. Plant Biol.* 47: 1009–1015.

Wertz, K., Hunziker-Buchwald, P., Seifert, N., Riss, G., Neeb, M., Steiner, G., and Goralczyk, R. 2005. Beta-carotene interferes with ultraviolet light A-induced gene expression by multiple pathways. *J. Invest. Dermatol.* 124: 428–434.

Wertz, K., Seifert, N., Hunziker-Buchwald, P., Riss, G., Wyss, A., et al. 2004. Beta-carotene inhibits UVA-induced matrix metalloprotease 1 and 10 expression in keratinocytes by a singlet oxygen-dependent mechanism. *Free Radical Biol. Med.* 37: 654–670.

Wheeler, W. N. 1980. Pigment content and photosynthetic rate of the fronds of *Macrocystis pyrifera. Mar. Biol.* 56: 97–102.

Yan, X., Chuda, Y., Suzuki, M., and Nagata, T. 1999. Focoxanthin as the major antioxidant in *Hijikia fusiformis*, a major edible seaweed. *Biosci. Biotechnol. Biochem.* 63: 605–607.

Ye, H., Wang, K., Zhou, C., Liu, J., and Zeng, X. 2008. Purification, antitumor and antioxidant activities *in vitro* of polysaccharides from the brown seaweed *Sargassum pallidum. Food Chem.* 111: 428–432.

Yoon, N. Y., Eom, T. K., Kim, M. M., and Kim, S. W. 2009. Inhibitory effect of phlorotannins isolated from *Ecklonia cava* on mushroom tyrosinase activity and melanin formation in mouse B16F10 melanoma cells. *J. Agric. Food Chem.* 57: 4124–4129.

Zhang, Y. F., Kim, S. M., Lee, W. J., and Um, B. H. 2011. Pressurized liquid method for fucoxanthin extraction from *Eisenia bicyclis* (Kjellman) Setchell. *J. Biosci. Bioeng.* 111: 237–241.

Zhang, Z., Wang, F., Wang, X., Liu, X., Hou, Y., et al. 2010. Extraction of the polysaccharides from five algae and their potential antioxidant activity *in vitro. Carbohydr. Polym.* 82: 118–121.

15 Biological Properties of Cosmeceuticals Derived from Marine Algae

Se-Kwon Kim and Ratih Pangestuti

CONTENTS

15.1 INTRODUCTION

Physical appearance has been a great focus in human life, and for that reason cosmetics have been used since the oldest known civilization (Cox and Glick, 1986; Kohl et al., 2000). The word *cosmetics* derived from the Latin word *cosmetae*, which is actually used to describe slaves in the Roman era whose function was to bathe men and women in perfume (Chaudhri and Jain, 2009). In Egypt, as early as 10,000 BC, humans used scented oils and ointments to clean and soften their skin and mask body odor. Two thousand years ago, Chinese ancient documents mentioned blackened eyebrows (Chen, 2009). Furthermore, a flowering plant called henna has been used to dye skin, hair, and fingernails worldwide since ancient times. It has been proven that the earliest civilizations used henna, including the Babylonians, Assyrians, Sumerians, Semites, Ugaritics and Canaanites. Henna has also been used extensively since the fourth and fifth centuries in the Deccan, west India (Chaudhri and Jain, 2009).

In this era, cosmetics have become an inherent part of our daily life and cosmetics usage has increased significantly each year (Azeem et al., 2008). There are far more consumers for the beautification of the human body than for the treatment of illness. However, the continuous use of cosmetics over time may result in various undesirable effects, and recently much attention has been paid by consumers to beauty as well as maintaining health and a youthful appearance. Hence it can be suggested that cosmeceuticals are an excellent alternative novel cosmetics that can contribute to the consumer's well-being.

Cosmeceuticals, or cosmetic pharmaceuticals, may be defined as cosmetic products having drug-like benefits that enhance or protect the appearance of the human body (Kim et al., 2008).

Cosmeceuticals contain active ingredients such as vitamins, phytochemicals, enzymes, antioxidants, and essential oils and can be applied to products such as creams, lotions, and ointments. Cosmeceuticals have recently attracted increased attention due to their beneficial effects on human health and beautification. Therefore, the search for and development of innovative and economical cosmeceuticals has attracted great attention. Many research groups have searched for novel cosmeceuticals from natural products, microbial sources, chemical synthesis, and so forth.

As more than 70% of the world's surface is covered by oceans, the wide diversity of marine organisms offers a rich source of natural products and bioactive compounds. These bioactive compounds have diverse functional roles as secondary metabolites and thus can be applied to the development of novel cosmeceuticals. The use of marine animals, however, has been prohibited due to ecological reasons. Accordingly, marine botanicals such as marine algae, which is sometimes referred to as seaweed, will open various ways for the development of marine organisms as cosmeceutical agents.

15.2 DEVELOPMENT OF MARINE ALGAE IN COSMECEUTICAL INDUSTRIES

The total biomass of marine algae is estimated to be more than twice the total mass of all other plants and animals combined. Despite the fact that marine algae have long been used for human beautification, they are still identified as an underexploited plant resource. Most people have unknowingly utilized marine algae daily in some of their cosmetic products. Three major phycocolloids extracted from marine algae, including carrageenan, alginate, and agar, are used in cosmetics as thickeners, gelling agents, emulsifiers, and so forth. Marine algae have become key ingredients in cosmetic products such as soaps, shampoos, creams, powders, sprays, and so forth. A large number of cosmetics derived from marine algae are claiming to have pharmaceutical and health benefits. However, some of these products are not labeled as cosmeceuticals due to regulatory difficulties, e.g., FDA regulations in the United States. Brown and red algae are the common algal types used in cosmeticeutical products.

The development of cosmeticeuticals derived from marine algae is rapidly advancing. The largest market for marine algae cosmeticeuticals is France, with an estimated 5,000 tons of wet algae used to cater to the demand. Meanwhile, several manufacturers of cosmetic products in the UK have started to use small quantities of marine algae. It has been reported that the cosmeceutical industry is one of the fastest growing markets in the United States. Therefore the use of marine algae in cosmeceutical industries is believed to be a high potential industry, as marine algae has long been believed to enhance human beauty, health, and fitness.

Marine algae have cosmeceutical potential since they contain high amounts of minerals and biologically active ingredients. Fucoidan and fucoxanthin, which are derived from marine brown algae, are the most common marine algal products on the market. Furthermore, diverse groups of ingredients such as vitamins, minerals, amino acids, sugars, lipids, and other biologically active compounds are extracted from the algae, and these compounds are used for varieties of cosmeceutical products. Some components of algal extracts react with diverse skin proteins and form a protective gel on the surface, and thereby reduce moisture loss. Kelp, brown algae, found in offshore areas, contains many essential vitamins, minerals, and fatty acids, like omega-3 and -6, which are known to facilitate cell regeneration and skin health. Kelp extracts also can be used as thickeners in many cosmetics and hair products, especially conditioners.

Bioactive compounds derived from marine algae are potential candidates as functional ingredients for cosmeceuticals. Great interest has developed to isolate and identify bioactive compounds derived from marine algae, which can be used for cosmeceuticals. This chapter presents an overview of the development and biological activity of marine algal bioactive compounds and their potential application as cosmeceutical agents.

15.3 BIOLOGICAL ACTIVITIES OF MARINE ALGAE

15.3.1 Skin Whitening Effects

Skin whitening has been in practice around the world, with Asia as the largest market. As many Asian females prefer a fairer skin tone, skin whiteners have become and continue to be the best-selling skin care product in Asia (Wang et al., 1997). *Skin whitening* is a term used for lightening the complexion through artificial means, such as creams, lotions, soaps, and injections. It can be achieved by several mechanisms of action, such as (1) microphthalmia-associated transcription factor inhibition, (2) downregulation of melanocortin receptor 1 activity, (3) interference with melanosome maturation and transfer, (4) melanocyte loss, and (5) tyrosinase inhibition.

Tyrosinase (EC 1.14.18.1), which is also known as polyphenol oxidase (PPO), is a copper-containing monooxygenase that is considered a key enzyme in melanin synthesis. Tyrosinase catalyzes two oxidative reactions: the hydroxylation of L-tyrosine to L-dihydroxyphenylalanine (DOPA), which is followed by the oxidation of DOPA to DOPA quinone (EC 1.10.3.1, catechol oxidase). Quinones, highly reactive compounds, can polymerize spontaneously to form high-molecular-weight compounds or brown pigments (i.e., eumelanin or pheomelanin) or react with amino acids and proteins that enhance the brown color produced. As tyrosinase catalyzes the rate-limiting step of pigmentation, tyrosinase inhibition has become the most common and is increasingly important in the cosmeceutical products used in the prevention of hyperpigmention and in skin whitening.

In an effort to inhibit the activity of tyrosinase, many different types of tyrosinase inhibitors have been developed via either synthesis or isolation from natural sources. These compounds are applicable for cosmetics as skin whitening agents and also as drugs for use in the treatment of pigmentation disturbances. Despite the enormous amount of research conducted thus far into the development of new whitening agents, the use of synthetic tyrosinase inhibitors is rather limited, owing to high toxicity, low stability, poor skin penetration, and insufficient activity. Hence development of novel tyrosinase inhibitors from natural resources continues to arouse great attention, and in recent years marine algae have attracted great attention in search of natural tyrosinase inhibitor agents (Solano et al., 2006).

Even though studies of tyrosinase inhibitory activities of marine algae are still in their infancy, a number of studies have reported tyrosinase inhibitory activity of phlorotannins derived from marine algae. Various tyrosinase inhibitors from marine algae have been isolated and characterized. A list of phlorotannins reported to have significant tyrosinase inhibitory activity is presented in Table 15.1. The tyrosinase inhibitory activity of the above phlorotannins has been determined by inhibition against oxidation of L-tyrosine catalyzed by mushroom tyrosinase assays. Preincubation of mushroom tyrosinase with phlorotannins in the absence of L-tyrosine did not diminish mushroom tyrosinase activity, suggesting that phlorotannins are inhibitors rather than inactivators of the mushroom tyrosinase. In several studies, the potencies of phlorotannins as tyrosinase inhibitors are expressed as an IC_{50} value, which is the tyrosinase inhibitor concentration leading to 50% inhibition of tyrosinase activity. Some phlorotannins derived from marine brown algae, such as 7-phloroeckol and dioxinodehydroeckol, have been described to inhibit tyrosinase activity more than arbutin and kojic acid (Yoon et al., 2009). Phlorotannins derived from marine algae showed two types of tyrosinase inhibition: competitive and noncompetitive. A competitive inhibitor is a substance that combines with free enzyme in a manner that prevents substrate binding. In this case, phlorotannins as inhibitors and L-tyrosine as substrate are mutually exclusive, because of true competition for the same site. Furthermore, a noncompetitive tyrosinase inhibitor caused a change in the structure and shape of the enzyme, and the modified enzyme was no longer capable of binding correctly with the substrate. A relationship between structure and types of tyrosinase inhibition activity of some phlorotannins has been reported. Nerya et al. suggested that one of the structural requirements for noncompetitive tyrosinase inhibitors is the presence of a resorcinol moiety, and that most of the competitive inhibitors

TABLE 15.1
Tyrosinase Inhibitory Activities of Several Marine Algae

Marine Algae	Compounds	Type of Inhibition	IC_{50}	References
Ecklonia stolonifera	Phloroglucinol	Competitive	92.8 μg ml^{-1}	Kang et al., 2004
Ecklonia stolonifera	Eckstolonol	Competitive	126 μg ml^{-1}	Kang et al., 2004
Ecklonia stolonifera	Eckol	Noncompetitive	33.2 μg ml^{-1}	Kang et al., 2004
Ecklonia stolonifera	Phlorofucofuroeckol A	Noncompetitive	177 μg ml^{-1}	Kang et al., 2004
Ecklonia stolonifera	Dieckol	Noncompetitive	2.16 μg ml^{-1}	Kang et al., 2004
Ecklonia cava	7-Phloroeckol	Noncompetitive	0.85 μM	Yoon et al., 2009
Ecklonia cava	Dioxinodehydroeckol	—	222.94 μM	Yoon et al., 2009
Ishige okamurae	Diphlorethohydroxycarmalol	—	142.20 μM	Heo et al., 2010
—	Arbutin[a]	—	243.16 μg ml^{-1}	Yoon et al., 2009
—	Kojic acid[a]	—	40. μg ml^{-1}	Yoon et al., 2009

[a] Positive control.

were able to chelate copper in this enzyme, implying a possible inhibitory mechanism (Nerya et al., 2003). Additionally, a competitive inhibitor might be a copper chelator, nonmetabolizable analogs, or derivatives of the true substrate (Chang, 2009).

Exposure of the skin to ultraviolet (UV) radiation has been continuously reported to increase melanin production, and thus increase pigmentation (Moyal, 2004). Fucoxanthin isolated from *Laminaria japonica* has been demonstrated to suppress tyrosinase activity in UVB-irradiated guinea pigs and melanogenesis in UVB-irradiated mice. Topical and oral treatment of fucoxanthin significantly suppressed skin mRNA expression related to melanogenesis, suggesting that fucoxanthin negatively regulated melanogenesis factor at the transcriptional level (Shimoda et al., 2010). It also has been consistently demonstrated that phlorotannins and fucoxanthin derived from marine brown algae are able to inhibit melanogenesis in mouse B16F10 melanoma cells (Heo et al., 2009, 2010; Yoon et al., 2009; Shimoda et al., 2010). Collectively, this evidence suggests that bioactive compounds derived from marine algae have promising potential to be used as skin whitening agents in the cosmeceutical industry. There are numerous advantages over other classes of skin whitening agents, such as relatively low production costs, broad spectrum of skin whitening properties, low cytotoxicity, safety, wide acceptability, and novel modes of action, suggesting marine algal bioactive compounds as promising skin whitening candidates in the near future. Further studies are needed with clinical trials for these skin whitening effects.

15.3.2 Photoprotective Effects

The sun emits a wide spectrum of electromagnetical radiation, including infrared, visible, ultraviolet A radiation (UVA; 320 to 400 nm), ultraviolet B radiation (UVB; 290 to 320 nm), and ultraviolet C radiation (UVC; 10 to 290 nm). Large amounts of UVB and UVC are screened out by the ozone; hence solar ultraviolet (UV) radiation that reaches the earth, as well as our skin, is composed of 5–10% highly energetic UVB (290–315 nm) and 90–95% UVA (315–400 nm), which is less energetic but penetrates the skin more deeply due to its longer wavelength. Despite the skin protecting our body from UV radiation, high levels of UV exposure have been demonstrated to induce numerous skin pathologies. Therefore it is important to offer our skin adequate photoprotection to prevent UV-related disorders. A large number of photoprotective agents have been identified from different organisms. Compared to other marine organisms, marine algae have been demonstrated to be a good source of photoprotective agents. Photoprotective compounds such as mycosporine-like

TABLE 15.2
Summary of Some Selected Marine Algae Compounds/Extracts and Their Molecular Targets or Mechanism of Action in Skin Photoprotection

Marine Algae	Compounds/Extracts	Molecular Targets/ Mechanisms	References
Ecklonia cava	Phloroglucinol, eckol, dieckol, triphlorethol, eckstolonol	Inhibit ROS, inhibit ion of DNA damage	Heo et al., 2009; Ko et al., 2010
Ishige okamurae	Diphlorethohydroxycarmalol	Inhibit ROS, inhibit ion of DNA damage	Heo et al., 2010
—	Brown algae polyphenol	Inhibit oxidative stress, inflammation	Hwang et al., 2006
Sargassum siliquastrum	Fucoxanthin	Inhibit ROS, inhibit ion of DNA damage	Heo and Jeon, 2009
Algal extract AST-101 (SPI-Switzerland)	Astaxanthin	Inhibit DNA damage, enhance antioxidant defense enzymes	Lyons and O'Brien, 2002
Corallina pilulifera	Methanolic extracts	Inhibit DNA damage, inhibit MMP, enhance antioxidant defense enzymes	Ryu et al., 2009
Porphyra rosengurttii	Porphyra-334, shinorine, asterina, palythine	Enhance antioxidant defense enzymes	de la Coba et al., 2009
Ahnfeltiopsis devoniensis	Shinorine	Enhance antioxidant defense enzyme	de la Coba et al., 2009
Gelidium corneum	Shinorine, asterina-330, palythine	Enhance antioxidant defense enzyme	de la Coba et al., 2009
Porphyra umbilicalis	Porphyra-334	Enhance antioxidant defense enzyme, increase skin firmness and smoothness	Daniel et al., 2004

amino acids (MAAs), carotenoids, and polyphenols have been identified from different classes of marine algae. They are briefly discussed below as well as summarized in Table 15.2.

Extract from *Corallina pilulifera*, which belong to red algae, has been reported to possess a photoprotective effect on UVA-induced oxidative stress in human dermal fibroblasts (HDF) (Ryu et al., 2009). Several studies have reported that red algae contain high amounts of MAAs, and since some of them present absorption maximally in UVA, it can be assumed that the photoprotective properties of *C. pilulifera* may correlate to the MAA content in this alga. MAAs are intracellular, low molecular weight (<400 Da), water-soluble molecules, absorbing UV radiation with the maximum absorbance between 310 and 365 nm (Oren and Gunde-Cimerman, 2007). Several lines of studies demonstrated that MAAs are the most common sunscreen compounds in marine algae (Gröniger et al., 2000; Rastogi et al., 2010). A study of the competition between two red algae, *Chondrus crispus* and *Mastocarpus stellatus*, in the intertidal and upper sublittoral zone in Germany showed a much higher MAA content in *M. stellatus*, which live in more light-exposed sites (Bischof et al., 2000). Those results suggest that MAAs are a UV-absorbing compound that protects marine algae from UV radiation damage.

Recently, porphyra-334 MAAs isolated from *Porphyra umbilicalis* were encapsulated in liposomes and used as sunscreen in a study on UVA-induced skin aging. Comparing skin lipid oxidation and skin aging parameters such as elasticity, wrinkle depth, and roughness, porphyra-334 liposomal MAAs performed as well as a cream with a synthetic UVA sunscreen (Daniel et al., 2004). Upon irradiation, reactive intermediates were not produced by porphyra-334 MAAs, suggesting that porphyra-334 MAAs were able to transform absorbed UV into harmless thermal energy with no subsequent loss of protective power effectively and rapidly (Conde et al., 2000). Hence porphyra-334 MAAs may be considered potential qualified photoprotectors.

In recent years, there has been accumulating evidence that MAAs play additional roles as antioxidants. Some MAAs may protect the skin against UV radiation not only by absorbing high-energy photons and dissipating the energy as heat, but also by scavenging ROSs. Several MAAs, such as shinorine, asterina, and palythine, have been isolated from marine algae and demonstrated potential antioxidant activity. Combinations of photoprotection and photostability with antioxidant substances may be introduced in order to improve the photoprotection efficacy of topical sunscreens. Hence MAAs derived from marine algae showed great potential to develop as a photoprotective agent against UV deleterious effects due to their wide spectral range of protection and photostability as well as their antioxidant properties.

In contrast to the wide abundance of MAAs in red algae, the occurrence of MAAs in green and brown algae is limited. Within the green and brown algae, most species do not contain MAAs, or sometimes only in trace concentrations. Since the occurrence of MAAs is restricted in green and brown algae, it may be assumed that certain UV-absorbing compounds might be present in order to protect them against UV radiation damage.

The occurrence of a polyphenolic compound, phlorotannin, as a UV-absorbing compound (absorbing between 280 and 320 nm) from the brown alga *Ascophyllum nodosum* was reported more than a decade ago (Pavia and Brock, 1997). Recently, dietary feeding and topical treatment of brown algae phlorotannin have been shown to decrease UVB-induced skin tumor development in SKH mice, which are associated, in part, with suppression of prostaglandin E_2 (PGE_2) levels, cyclooxygenase (COX)-2 expression, and cell proliferation. It is well documented that exposure to UVB is initially associated with an inflammatory response characterized by the induction of pro-inflammatory mediators; there is an increasing body of evidence demonstrating that COX-2 and PGE_2 are elevated in human tumors compared to adjacent normal tissue. Induction of COX-2 elevates levels of PGE_2, which acts as a promoter in carcinogenesis, inflammatory cell infiltration and activation, and the development of skin tumors. Therefore, brown algae phlorotannins are clearly able to protect skin cells from UVB-induced mutation and prevent chronic inflammation (Hwang et al., 2006). In Jeju Island (Korea), the leaves of *E. cava* have traditionally been used to heal sunburned skin, skin inflammation caused by overexposure to UV (Hwang, 2010). Recently, protective effects of *E. cava* on UVB-induced photooxidative stress have been investigated (Ko et al., 2010; Heo et al., 2009). *E. cava* phlorotannins and extracts provide a strong protective effect against UVB radiation-induced cell damage in HaCaT cells and strong protective properties against photooxidative stress induced by UVB radiation in human fibroblast cells. Furthermore, diphlorethohydroxycarmalol, a carmalol derivative from *Ishige okamurae*, demonstrated strong protective properties against UVB radiation via damaged DNA tail length and morphological changes in fibroblasts (Heo et al., 2010). Fucoxanthin and astaxanthin, two carotenoids isolated from brown algae, have been demonstrated to possess photoprotective properties in human fibroblast cells via inhibition of DNA damage and enhance antioxidant activity (Lyons and O'Brien, 2002; Heo and Jeon, 2009).

The photoprotective effects of phlorotannin and carotenoids are mainly through antioxidant activity, inhibition of apoptosis, and inhibition of DNA damage. In addition, it is important to highlight the potency of phlorotannin and carotenoids in inhibiting DNA damage, since inhibiting apoptosis without inhibiting DNA damage allows UVB-mutated cells to survive and results in the development and ultimate increase in tumor formation. Taken together, it may be assumed that MAAs derived from marine red algae and polyphenols and carotenoids from marine brown algae have potential to be used as photoprotective agents.

15.3.3 Restore Skin Elasticity

It is believed that skin elasticity is a hallmark of young and healthy skin, yet the loss of skin elasticity progresses over the entire life span. Collagen, a component of the skin that increases and enhances its elasticity, is estimated to lose an average of 1% every year once we enter our twenties. Skin that is significantly exposed to UV has been shown to lose collagen at an even faster rate. Once collagen is initially cleaved by matrix metalloproteinase (MMP)-1, MMP-3, and other MMPs, its breakdown is further promoted. The enzyme mainly responsible for collagen breakdown in skin is MMP-1 (fibroblast collagenase), which cleaves type I, III, VII, VIII, and X collagens.

Even though development of marine algal bioactive compounds with MMP-1 inhibition activities is still in its infancy, a number of studies have reported MMP-1 inhibitory activity of several marine algae species. Fucoidan, a sulfated fucosylated polymer derived from marine brown algae cell walls, has been shown to inhibit UVB-induced MMP-1 expression in human skin fibroblasts (Moon et al., 2008, 2009a, 2009b). Furthermore, *Eisenia bicyclis*, *Ecklonia cava*, and *Ecklonia stolonifera* also possess strong abilities to inhibit MMP-1 expression (Joe et al., 2006).

It has been shown that several natural antioxidant compounds directly inhibit the expression of MMP-1, and thus reduce losses of collagen. The inhibitory effect of these antioxidant compounds on MMP-1 activity is associated with their antioxidant activities. A link between ROSs and MMP-1 has been identified previously. ROSs regulate MMP-1 gene expression and activation of MMP-1 proenzymes. MMP-1 is activated by ROSs through interactions with thiol groups, in which the thiol residue is converted by ROSs to sulfinic acid (Touyz, 2006). Marine algae have long been reported to exhibit antioxidant activity in several cell lines and in animal models. Recently, Kang et al. confirmed the protective effect of triphlorethol-A, which is isolated from *E. cava* on MMP-1 induced by oxidative stress in human keratinocyte cells (Kang et al., 2008).

Collagen is indeed important for the skin owing to the principal structural protein holding the skin together. The elastin network in the dermis, which is crucial for elastic skin recovery from gravity or other mechanical forces, is also important. Elastin is responsible for the ability of tissues to resume their original shape after being stretched; in several pathological conditions it is often associated with inflammation and oxidative stress. Under chronic inflammation or oxidative stress, an enzyme called elastase, which hydrolyzes elastin in the dermis, is overactivated to excessively break down elastin fibers, leading to loss of dermal elasticity. Several studies reported that both anti-skin-aging and antiwrinkle effects are significantly correlated with decreased elastase activity. Two marine algae species, *Eisenia arborea* and *Ecklonia cava*, have been demonstrated as potential elastase inhibitors (Tsukui et al., 2006; Bu et al., 2006). Furthermore, Xhauflaire Uhoda et al. found that cosmeticeutical formulations enriched with marine green algae (*Ulva lactuca* and *Lola implexa*) have been reported to increase skin hydrating and firming effects (Xhauflaire Uhoda et al., 2008).

This evidence suggests that marine algae is one of the useful candidates in the search for effective, nontoxic substances with potential application in cosmeceuticals to regulate both upstream and downstream risk factors that contribute to the loss of skin elasticity. Therefore marine algae could be used as a rich source of natural compounds with potential application to restore skin elasticity by being a part of cosmeceuticals.

15.3.4 Antiacne Effects

Acne vulgaris (acne) is the most common skin disease treated by dermatologists. It is characterized by open and closed comedones, papules, pustules, and cysts (Fredricks, 2001). Acne treatments consist of reducing sebum production, removing dead skin cells, and killing bacteria with topical drugs and oral medications. Marine algae have recently been shown to be effective in the treatment of acne. Sargafuran, an active ingredient from brown algae, *Sargassum macrocarpum*, has potential to be used as an effective acne treatment. Sargafuran has low cytotoxicity, and the minimum inhibitory concentration (MIC) against *P. acnes* was 15 μg/ml, showing a broad antibacterial activity against Gram-positive bacteria. A time-kill study showed that sargafuran was bactericidal and completely killed *P. acnes* at 4 × MIC by lysing bacterial cells (Kamei et al., 2009). Moreover, bladderwrack (*Fucus*) extracts have been used as integral ingredients of acne care systems called OXY range and Clearskin Max, which are manufactured by the Mentholatum Company and Roduve, respectively. Both acne care systems are currently available in the market and seem to be safe and effective in the treatment of acne. Therefore it may be assumed that marine algae have great potential for acne management.

15.3.5 Other Biological Activities

Marine algal products have long been used by humans for spa and thalassotherapy products. In addition, private spas and salons offer cellulite massage treatments that include body scrubs, pressure point massages, essential oils, and herbal products using marine algal extracts. Serum enriched with *Ulva lactuca* has been shown to be absorbed across the human skin and thus produces a better moisturizing effect and diverse biological effects. Moreover, a large number of marine algal pigments such as chlorophyll *a*, zeaxanthin, luthein, fucoxanthin, phycoerythrin, phycocyanobillin, etc., have been identified and characterized. These marine algal pigments have potential as natural colorants to be used in cosmeceuticals, particularly as substitutes for synthetic dyes. In addition, biological activities of algal pigments *in vitro* and *in vivo* are well documented.

15.4 CONCLUSION

Much attention has been given by consumers to their quest to stay healthier and look younger later in life, and hence it can be suggested that cosmeceuticals are excellent alternative sources of cosmetics that can contribute to enhancing human physical appearance. The possibilities of designing novel cosmeceutical products derived from marine algae are promising. Various metabolites derived from marine algae such as polysaccharides, carotenes, phenolic compounds, MAAs, and sargafuran have been investigated for their potential applications as cosmeceutical agents due to valuable biological functions with beneficial health effects. Bioactive compounds derived from marine algae have great potential as active ingredients for preparation of various cosmeceutical products.

REFERENCES

Azeem, A, M Rizwan, FJ Ahmad, et al. 2008. Emerging role of microemulsions in cosmetics. *Recent Patents Drug Deliv Formulation* 2 (3):275–289.

Bischof, K, G Kräbs, D Hanelt, and C Wiencke. 2000. Photosynthetic characteristics and mycosporine-like amino acids under UV radiation: a competitive advantage of *Mastocarpus stellatus* over *Chondrus crispus* at the Helgoland shoreline? *Helgoland Marine Res* 54 (1):47–52.

Bu, HJ, YM Ham, JM Kim, SJ Lee, JW Hyun, and NH Lee. 2006. Elastase and hyaluronidase inhibition activities of phlorotannins isolated from *Ecklonia cava*. *Korean J Pharmacognosy* 37 (2):92–96.

Chang, TS. 2009. An updated review of tyrosinase inhibitors. *Int J Mol Sci* 10 (6):2440–2475.

Chaudhri, SK, and NK Jain. 2009. History of cosmetics. *Asian J Pharm* 3 (3):164–167.

Chen, Q. 2009. Evaluate the effectiveness of the natural cosmetic product compared to chemical-based products. *Int J Chem* 1 (2):57–59.

Conde, FR, MS Churio, and CM Previtali. 2000. The photoprotector mechanism of mycosporine-like amino acids. Excited-state properties and photostability of porphyra-334 in aqueous solution. *J Photochem Photobiol B Biol* 56 (2–3):139–144.

Cox, CL, and WH Glick. 1986. Resume evaluations and cosmetics use: when more is not better. *Sex Roles* 14 (1):51–58.

Daniel, S, S Cornelia, and Z Fred. 2004. UV-A sunscreen from red algae for protection against premature skin aging. *Cosmet Toiletries Manufacture Worldwide*, 139–143.

de la Coba, F, J Aguilera, F Figueroa, M de Gálvez, and E Herrera. 2009. Antioxidant activity of mycosporine-like amino acids isolated from three red macroalgae and one marine lichen. *J Appl Phycol* 21 (2):161–169.

Fredricks, DN. 2001. Microbial ecology of human skin in health and disease. *J Invest Dermatol Symp Proc* 6 (3):167–169.

Gröniger, A, RP Sinha, M Klisch, and DP Häder. 2000. Photoprotective compounds in cyanobacteria, phytoplankton and macroalgae—a database. *J Photochem Photobiol B Biol* 58 (2–3):115–122.

Heo, SJ, and YJ Jeon. 2009. Protective effect of fucoxanthin isolated from *Sargassum siliquastrum* on UV-B induced cell damage. *J Photochem Photobiol B Biol* 95 (2):101–107.

Heo, SJ, SC Ko, SH Cha, et al. 2009. Effect of phlorotannins isolated from *Ecklonia cava* on melanogenesis and their protective effect against photo-oxidative stress induced by UV-B radiation. *Toxicol In Vitro* 23 (6):1123–1130.

Heo, SJ, SC Ko, SM Kang, et al. 2010. Inhibitory effect of diphlorethohydroxycarmalol on melanogenesis and its protective effect against UV-B radiation-induced cell damage. *Food Chem Toxicol* 48 (5):1355–1361.

Hwang, H, T Chen, RG Nines, HC Shin, and GD Stoner. 2006. Photochemoprevention of UVB induced skin carcinogenesis in SKH 1 mice by brown algae polyphenols. *Int J Cancer* 119 (12):2742–2749.

Hwang, HJ. 2010. Skin elasticity and sea polyphenols. *Seanol Sci Center Rev* 1 (1):1–10.

Joe, MJ, SN Kim, HY Choi, et al. 2006. The inhibitory effects of eckol and dieckol from *Ecklonia stolonifera* on the expression of matrix metalloproteinase-1 in human dermal fibroblasts. *Biol Pharm Bull* 29 (8):1735–1739.

Kamei, Y, M Sueyoshi, K Hayashi, R Terada, and H Nozaki. 2009. ChemInform abstract: the novel anti-*Propionibacterium acnes* compound, sargafuran, found in the marine brown alga *Sargassum macrocarpum*. *ChemInform* 40 (40).

Kang, HS, HR Kim, DS Byun, BW Son, TJ Nam, and JS Choi. 2004. Tyrosinase inhibitors isolated from the edible brown alga *Ecklonia stolonifera*. *Arch Pharmacol Res* 27 (12):1226–1232.

Kang, KA, R Zhang, MJ Piao, et al. 2008. Inhibitory effects of triphlorethol-A on MMP-1 induced by oxidative stress in human keratinocytes via ERK and AP-1 inhibition. *J Toxicol Environ Health A* 71 (15):992–999.

Kim, SK, YD Ravichandran, SB Khan, and YT Kim. 2008. Prospective of the cosmeceuticals derived from marine organisms. *Biotechnol Bioprocess Eng* 13 (5):511–523.

Ko, SC, SH Cha, SJ Heo, SH Lee, SM Kang, and YJ Jeon. 2010. Protective effect of *Ecklonia cava* on UVB-induced oxidative stress: *in vitro* and *in vivo* zebrafish model. *J Appl Phycol* 23 (4):697–708.

Kohl, L, A Blondeel, and M Song. 2000. Allergic contact dermatitis from cosmetics. *Dermatology* 204 (4):334–337.

Lyons, NM, and NM O'Brien. 2002. Modulatory effects of an algal extract containing astaxanthin on UVA-irradiated cells in culture. *J Dermatol Sci* 30 (1):73–84.

Moon, HJ, SH Lee, MJ Ku, et al. 2009a. Fucoidan inhibits UVB-induced MMP-1 promoter expression and down regulation of type I procollagen synthesis in human skin fibroblasts. *Eur J Dermatol* 19 (2):129–134.

Moon, HJ, SR Lee, SN Shim, et al. 2008. Fucoidan inhibits UVB-induced MMP-1 expression in human skin fibroblasts. *Biol Pharm Bull* 31 (2):284–289.

Moon, HJ, KS Park, MJ Ku, et al. 2009b. Effect of *Costaria costata* fucoidan on expression of matrix metalloproteinase-1 promoter, mRNA, and protein. *J Nat Prod* 72 (10):1731–1734.

Moyal, D. 2004. Prevention of ultraviolet-induced skin pigmentation. *Photodermatol Photoimmunol Photomed* 20 (5):243–247.

Nerya, O, J Vaya, R Musa, S Izrael, R Ben-Arie, and S Tamir. 2003. Glabrene and isoliquiritigenin as tyrosinase inhibitors from licorice roots. *J Agric Food Chem* 51 (5):1201–1207.

Oren, A, and N Gunde-Cimerman. 2007. Mycosporines and mycosporine-like amino acids: UV protectants or multipurpose secondary metabolites? *FEMS Microbiol Lett* 269 (1):1–10.

Pavia, H, and E Brock. 1997. Ascophyllum nodosum. *Marine Ecol Progr Ser* 157:139–146.

Rastogi, RP, RP Richa, SP Sinha, SP Singh, and D-P Häder. 2010. Photoprotective compounds from marine organisms. *J Indust Microbiol Biotechnol* 37 (6):537–558.

Ryu, BM, ZJ Qian, MM Kim, KW Nam, and SK Kim. 2009. Anti-photoaging activity and inhibition of matrix metalloproteinase (MMP) by marine red alga, *Corallina pilulifera* methanol extract. *Radiation Physics Chem* 78 (2):98–105.

Shimoda, H, J Tanaka, SJ Shan, and T Maoka. 2010. Anti-pigmentary activity of fucoxanthin and its influence on skin mRNA expression of melanogenic molecules. *J Pharmacy Pharmacol* 62 (9):1137–1145.

Solano, F, S Briganti, M Picardo, and G Ghanem. 2006. Hypopigmenting agents: an updated review on biological, chemical and clinical aspects. *Pigment Cell Res* 19 (6):550–571.

Touyz, RM. 2006. Mitochondrial redox control of matrix metalloproteinase signaling in resistance arteries. *Arterioscler Thromb Vasc Biol* 26 (4):685–688.

Tsukui, K, M Kakinuma, H Amano, et al. 2006. Identification of elastase inhibitors from brown alga, *Eisenia arborea*. *ITE Lett Batter New Technol Med* 7 (6):616–619.

Wang, CM, CL Huang, CTS Hu, and HL Chan. 1997. The effect of glycolic acid on the treatment of acne in Asian skin. *Dermatol Surg* 23 (1):23–29.

Xhauflaire Uhoda, E, K Fontaine, and GE Piérard. 2008. Kinetics of moisturizing and firming effects of cosmetic formulations. *Int J Cosmet Sci* 30 (2):131–138.

Yoon, NY, TK Eom, MM Kim, and SK Kim. 2009. Inhibitory effect of phlorotannins isolated from *Ecklonia cava* on mushroom tyrosinase activity and melanin formation in mouse B16F10 melanoma cells. *J Agric Food Chem* 57 (10):4124–4129.

16 Hair Biology and Care Product Ingredients from Marine Organisms

Se-Kwon Kim and Soon-Sun Bak

CONTENTS

16.1 INTRODUCTION

Hair is a filamentous biomaterial that grows from hair follicles found in the dermis, which is one of the defining characteristics of the mammalian class. A hair follicle is a very complex structure consisting of the hair shaft, dermal papilla cells, matrix, inner root sheath cells, outer root sheath cells, etc. Hair provides thermal regulation for many animals, and for some animals hair provides protection from environmental damage. Hair has a great social and sexual role in human beings (Randall 2008). Human hair follows a specific growth cycle with three distinct and concurrent phases: anagen, catagen, and telogen phases that occur simultaneously and are regulated by different genes and hormones (Stenn and Paus, 2001). Recent studies have reported the regulation of proliferation and function in hair follicle cells on the human hair cycle, aberrant action of androgen and testosterone, phsyical stress, and environment pollution related to human baldness (Elliott et al., 1990; Kaufman, 2002; Botchkarev, 2003; Winiarska et al., 2006).

The alopecia population is increasing worldwide due to the change of food habits, lifestyle, and environment. Baldness, sometimes referred to as alopecia, is a problem that has been limited to middle-aged people and older men in the past. However, it has been occurring more frequently among younger men and women lately. In Korea, the population of people with hair loss was increased by 5 to 10 million during 2005 to 2009. According to the census of the Korea National Health Insurance Corporation, the rate of the hair loss increased around 30% in people ages 20–30 from 2001 to 2008.

According to documented hair loss trends, not only hair growth treatment but also hair products related to loss protection, hair care, and hair management are in the forefront lately. Hence there are many kinds of hair cosmeceutical products. Ingredients of hair cosmeceuticals were limited to chemicals in the past, but have been extended to natural materials and development of natural, environmentally friendly product applications. While natural herbs have been evidenced to have hair care and growth effects when applied to hair cosmeceutical products, only a few studies have reported application products derived from marine natural sources (Jadhav et al., 2009; Abdullah and Rashid, 2010). As more than 70% of the earth's surface is covered by the sea, and marine species comprise about half of the total global biodiversity, this wide diversity of organisms is recognized as a reservoir of potent bioactive molecules. For this reason, marine natural sources are explored and exploited for cosmeceuticals as future alternative resources (Kim et al., 2008).

This chapter focuses on human hair structure, growth cycle, gene expression profile of hair papilla cells, and the relationship between hormones and hair alopecia for understanding hair biology. The main subject is materials from marine organisms as hair cosmeceutical products, including hair growth promoters and scalp care products.

16.2 FUNCTION AND STRUCTURE OF HUMAN HAIR

16.2.1 Function of Human Hair

The human body surface involves approximately 5 million hairs in an adult, and the scalp contains around 100,000 to 110,000 hairs at 175–300 hairs/cm^2 density (Szabo, 1958). Hair function is classified as both protection and decoration. The hair maintains body temperature and protects against environmental damage, including mechanical injuries or ultraviolet irradiation from sunlight. Protection is the main function of hair in other mammals, whereas few parts, such as eyelashes and vibrissa of hair, are required in humans. The decoration pattern of human hair is the important function as correlated with the indication of social and sexual communication, including the enhancement of sexual development and appeal (Stenn and Paus, 2001). The hair of the scalp and eyebrows might be shaped to influence social intercourse in humans. Recent studies have reported that human hair is becoming the center of dermatological science via the development of biogenetic experimental methods. Human hair biology might be applied to various parts, including histopathological formulation (Elder et al., 1997), keratinization (Bertolino and O'Guin, 2003), hair follicle stem cells (Akiyama et al., 2000), and hair immunology (Paus et al., 1999).

16.2.2 Structure of the Human Hair Follicle

Hair follicles have a very complicated structure and are composed of the hair shaft, dermal papilla cell (DPC), hair matrix, inner root sheath (IRS), outer root sheath (ORS), and so forth (Jahoda et al., 1984; Millar, 2002). A hair shaft penetrates deep into the scalp, formed and maintained via the interaction of dermal and epidermal components. The bulb of the hair follicle is the origin point of the hair and dermal papilla cell, which is found at the base of the follicle. DPCs are one of the specialized fibroblast cells that participate in morphogenesis of the hair follicle and are related to hair growth both normal and pathogenic conditions (Jahoda et al., 1984). The DPCs are surrounded by a capsule that is composed of epidermal cells, called matrix cells, and interconnected by nerve fibers and blood vessels. The hair fiber is surrounded by root sheath layers and divided into IRS and ORS. IRS is the innermost layer of hair fiber, while ORS is the outermost layer. The muscle and sebaceous gland are located in the ORS, which is related to hair standing erect and producing oils, respectively.

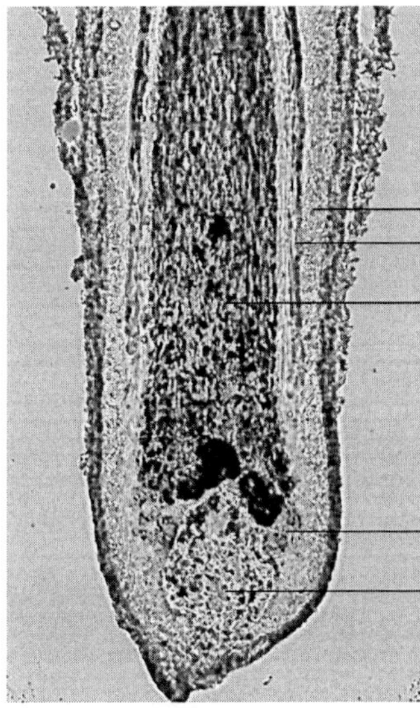

FIGURE 16.1 Structure of human hair follicle.

16.3 HUMAN HAIR GROWTH

Human hair follows a specific growth cycle that occurs simultaneously with anagen, catagen, and telogen phases. The mechanism of each phase is regulated by various factors involved in different signaling pathways. Human hair growth is also related to hormones, including androgen, testosterone, and dihydrotestosterone, and these factors affect hair loss (alopecia).

16.3.1 Hair Growth Cycle

As mentioned earlier, the hair growth cycle is classified into three main phases: anagen, catagen, and telogen. Epithelial cells and matrix cells differentiate and become specialized to each function in each follicular layer. During the anagen phase, matrix cells and epithelial cells proliferate and differentiate to form the IRS and hair shaft elongation (Stenn and Paus, 2001). Melanocyte also acts in melanogenesis for pigmented hair follicles. In the human hair cycle, the anagen phase is the longest among all the phases, lasting about 5–10 years, and normally 90% of the follicles remain. Catagen, a regression or transitional phase, lasts for 2–3 weeks and only 1–2% of the follicles exist. During the telogen phase, the follicles stay in a resting condition for a longer period, and around 10% of telogen follicles are maintained for 1–3 months (Saitoh et al., 1970; Millar, 2002).

16.3.2 Gene Expression Profile: Mechanism of Hair Growth Modulators

The mechanism of each phase of hair cycle is controlled by the regulation of genes involved in various signaling pathways. An initiation stage of anagen is induced by epithelial and mesenchymal interaction (EMI), which plays an important role in the organogenesis in many organs, including the hair follicles. EMI also affects the developmental stage and the hair follicle cycling process, particularly during anagen induction. The hair cycle of the follicles is regulated by the cascade reaction of the signaling pathway between follicular keratinocyte stem cells in the bulge linked with hair follicle-specific differentiation and DPC (Stenn and Paus, 2001; Hardy, 1992). During the anagen induction, stem cells near the follicle degenerate into DPC. This existing condition suggests that there are specific interactions between these two cells (Costsarelis et al., 1990). It has been reported that the anagen phase signifies the initiating signals secreted by follicular stem cells or degenerated DPC at the bulge. Moreover, versican (VCAN) in the proteoglycan of the extracellular matrix (ECM) acts to induce and maintain the anagen phase (Choocheep et al., 2010). Although the signaling pathway and mechanism are not yet clear, recent studies have demonstrated that several morphogenesis-related genes were expressed on other hair organization (Soma et al., 2006). Factors implicated in anagen follicles include insulin growth factor (IGF)-1, keratinocyte growth factor (KGF), hepatocyte growth factor (HGF), fibroblast growth factor (FGF), sonic hedgehog (shh), transforming growth factor (TGF-β), noggin and bone morphogenic protein (BMP), Wnt, and stem cell factor (SCF) (Millar, 2002). Wnt signals are needed for the induction of hair follicle development and regeneration after wounding (Andl et al., 2002). SCF is an essential molecule for proliferation, differentiation, and melanin production in the DPC (Hibberts et al., 1996). Several studies have revealed that Erk and Akt signaling pathways participate in various aspects of cells proliferation in hair follicles, and the intracellular levels of p53, caspase-3, Bcl-2, and Bax regulate apoptosis (Raj et al., 2006). In the anagen phase, many genes in the hair follicles influence each other and regulate the activity on different levels.

After the anagen phase, hair follicles enter the regressing phase. During the catagen phase, follicular epithelial cells stop their proliferation and induce apoptosis. Hence some of the endothelial cells are degenerated by apoptosis from the hair follicles and apoptotic cells are increased at terminated catagen (Ebling, 1987). Recent studies demonstrated that TGF-β is a strong factor that terminates anagen to early catagen in the normal condition of mammalian hair cycling. The TGF-β is classified into TGF-β1, TGF-β2, and TGF-β3, based on capacity. At the late stage of anagen phase, the TGF-β1 principally exists in the ORS (Foitzik et al., 2000). During the anagen to catagen transition, expression of TGF-β significantly increased but growth factors such as IGF-1 that simulate follicular epithelial cells decreased (Weger and Schlake, 2005). In the matrix and DPC, accumulation of TGF-β2 was characteristic of initiating the catagen phase in humans. These studies suggested that TGF-β regulated the hair growth phase in follicular epithelial cells. Moreover, termination of melanogenesis, proteolysis of molecular ECM, and remodeling of blood vessels are processed in the catagen phase (Ebling, 1987).

In the telogen phase, the resting phase of the hair cycle, biological activities of a variety of genes diminish, compared to the anagen stage (Kishimoto et al., 1999). Some specific factors such as plasminogen activator inhibitor type 2 (PAI-2) and desmoglein-3 are localized in telogen hair. PAI-2, a serine protease inhibitor, works to prevent premature hair shedding, and desmoglein-3 anchors telogen hair in the follicle (Lavker et al., 1998; Wu et al., 2003). The reentry of the hair follicles into the anagen phase with cellular growth and gene expression occurs in human terminal hairs.

Many studies also suggest that angiogenesis regulates hair growth via the activation of angiogenesis-related factors, including vascular endothelial growth factor-A (VEGF-A) and thrombospondin-1 (TSP-1). VEGF-A-mediated angiogenesis regulates the perifollicular vessel size, which is related to follicular size and endothelial cell proliferation by increasing in the anagen and decreasing in the catagen and telogen phases (Yano et al., 2001). In the catagen phase, TSP-1, the endogenous angiogenesis inhibitor, affects induction of the hair follicle and vascular degeneration (Yano et al., 2003). In the animal test model, the mRNA and protein expression of TSP-1 are increased in DPC and the follicular keratinocyte. Moreover, apoptotic vascular regression was also induced by TSP-1 during the catagen and telogen phases (Krause and Foitzik, 2006).

16.3.3 Androgen

There are various kinds of hormones related to hair growth, including androgen, testosterone, estrogen, and so forth. Androgens regulate human hair growth in normal healthy conditions and with normal thyroid function. Hypothyroidism in humans is also correlated with hair loss. Androgen acts on the scalp and body hair in humans, which is related to androgenic alopecia (Randall, 1994). Testosterone is the major androgenic hormone in humans and is changed to dihydrotestosterone (DHT), a more potent androgen, by the steroid 5α-reductase (5αR). There are two types of 5αR in humans: type I 5αR is usually expressed in the liver, skin, and sebaceous gland, whereas type II 5αR is present in the prostate, foreskin beard, and DPC of the frontal scalp (Andersson et al., 1991; Thigpen et al., 1993). Prolactin and 17β-estradiol receptors also exist in human follicles and reduce human follicular growth *in vitro* (Foitzik et al., 2006; Thornton et al., 2006).

16.4 ALOPECIA

Alopecia is a medical condition of hair loss from the head or body, caused by various biological, genetic, environmental, and social factors. A recent study found that alopecia was related to suppression of proliferation or function on DPC in the human hair cycle (Elliott et al., 1990), aberrant action of androgen and testosterone (Kaufman, 2002; Winiarska et al., 2006), physical stress, pollution of the environment (Botchkarev, 2003), etc. Despite all the studies on alopecia, the mechanism and pathology of alopecia were still not clearly defined. Recent studies suggest that a bald scalp in men with androgenetic alopecia maintains hair follicle stem cells as normal haired scalp tissue (Garza et al., 2011).

Alopecia is classified with an underlying characteristic and cause, including alopecia areata, androgenic alopecia, anagen effluvium, telogen effluvium, and self-induced hair loss (Soma et al., 2006). Alopecia areata is the immunological and genetic complication of hair loss from all over the body, particularly from the scalp. Androgenic alopecia is the most common cause of hair loss in both females and males. It also named male-pattern baldness, and often a rim of hair around the sides and rear of the head is left in the male. In females, the hair becomes thinner all over the head and the hairline does not recede. Anagen effluvium is pathologic hair loss by systemic and radiation chemotherapy (Piraccini et al., 2006). Telogen effluvium is caused by emotional or physiological stress.

16.5 MATERIALS OF HAIR COSMECEUTICAL PRODUCTS

There are various kinds of hair cosmeceuticals, including shampoo, conditioner, special care agents, and hair growth promoter. Hair cosmeceuticals should be harmless to the scalp and hair, and also should not cause any toxic effect when used frequently. Recently, the main component of hair cosmetic products has been changed from chemicals to natural resources. For the development of natural hair cosmeceutical products, biological active ingredients are extracted and isolated from natural resources. As the ocean covers more than 70% of the earth's surface, there are abundant, multifarious marine species. Marine organisms are excellent reservoirs for identifying and extracting biologically active substances (Kim et al., 2008). For these reasons, extracts of marine organisms and their metabolites were recently used as potential materials for hair cosmeceutical products. There have been a number of patents for cosmeceutical products and ingredients for hair treatment from marine organisms, including marine algae and fish.

16.5.1 Marine Algae

In this era, many natural and organic cosmeceuticals have been isolated from marine algae. Marine algae, sometimes referred to as seaweed, are classified based on their pigmentation: red (Rhodophyceae), brown (Phaeophyceae), or green (Chlorophyceae). They contains abundant metabolites and nutrients, including minerals, vitamins, and amino acids. Marine algae are rich in iodine, selenium, calcium, iron, potassium, magnesium, and zinc, and these trace minerals are needed by the thyroid gland for producing hair. It also involves B group vitamins such as biotin and vitamin B12, which are known to strengthen hair. Many reports show that marine algae and their metabolites have proficient bioactive efficacies, including antiphotoaging, anticoagulant, antioxidant, anti-inflammatory, matrix metalloproteinase (MMP) inhibitory, and antiallergic activities (Kim et al., 2008; Hwang, 2010). For these reasons, marine algae are believed to promote hair growth.

There are many patents on hair cosmeceutical products that are involved with marine algae, including *Palmaria palmata* (dulse), *Ecklonia cava*, *Undaria pinnatifida* (wakame), *Eucheuma cottonii*, *Chondrus crispus* (Irish moss, carrageen), *Laminaria japonica* (oarweed), *Porphyra tenera*, and *Sargassum fulvellum* (Kim, 2005; Han, 2007; Min and Jeong, 2007; Hirata and Sugawara, 2008; Burlando et al., 2010). Moreover, recent studies suggested that extracts and metabolites of marine algae have hair growth promotion effects.

P. palmate is an edible species of red algae in Northern Europe, Scotland, and Ireland that contributes to the regeneration of hair as containing the high content of the vitamin B group (Yuan et al., 2005). Extracts of *E. cava* have been demonstrated to suggest a potential hair growth-promoting effect via the increased proliferation of immortalized vibrissa DPC and hair elongation of rat vibrissa follicles. Moreover, *E. cava* extract decreased the activity of 5α-reductase, which is an essential enzyme in the transformation of testosterone to DHT. *E. cava* has a variety of metabolites, including fucoxanthin, astaxanthin, and other phenolic compounds. Diekol, the sea polyphenol isolated from *E. cava*, promoted cell proliferation and downregulated the activity of 5α-reductase on immortalized vibrissa DPC, and anagen progression of hair shaft was induced in C57BL/6 mice (Kang et al., 2010). *U. pinnatifida*, one of the brown algae, is abundant on the seashore of Korea, Japan, and China, and also has been spread worldwide through marine farming (Stuart, 2003). It is beneficial for healthy hair because of a synergistic action of folic acid, iodine, and fucoidan (Hirata and Sugawara, 2008). Another algae, *E. cottonii*, native to the tropical region, possesses a variety of compounds, including phycoerythrin, phycocyanin, flavonoid, and sulfated polysaccharides (Chandini et al., 2008, Chew et al., 2008). In a recent study, it has been reported that ethanolic extract of *E. cottonii* showed hair growth promoting effects and wound healing as tested in the animal model Sprague-Dawley rats (Fard et al., 2011). A species of red algae, *C. crispus*, may contribute to human skin and hair follicle health due to the richness of nutrients it contains, such as iodine, calcium, magnesium, and potassium. Therefore it has been applied to hair care products for emollient action via its specific binding to keratin fibers (Roh and Shin, 2006; Assaad, 2011). *L. japonica*, large-size brown algae, is composed of abundant amounts of natural vitamins and minerals to supply various nutrients for the hair. By possessing high amounts of polysaccharides, extracts of *L. japonica* are used for care products because of their emollient and hydrating actions on the epidermis and hair (D'Amelio, 1999; Park, 2010).

16.5.2 Fish

Fish contain various kinds of essential nutrients, including protein, lipids, minerals, and vitamins. Certain proteins extracted from marine fish have been reported to promote hair growth in females, particularly cartilaginous fish extract (Lassus et al., 1991; Oikawa et al., 1990). A number of studies have reported novel angiogeic inhibitor derived from shark cartilage.

Dupont reported that shark cartilage extract protects against excessive hair loss caused by antineoplastic activity, reduces the incidence of nausea typically associated with administration of an antineoplastic agent, and improves nail and hair health into a population that is being administered by shark cartilage extract, compared to a similar population that is not being administered the shark cartilage extract (Dupont, 2002). Shark cartilage has been used as an ingredient in hair growth promoters such as viviscal and nourkrin. Viviscal has been shown effective in the treatment of androgenic alopecia in both young and middle-aged males (Lassus et al., 1992, 1994). Moreover, Thom demonstrated that nourkrin may provide an alternative to pharmacotherapy for the treatment of hair loss problems because of its good tolerability by individuals with androgenetic alopecia and its lack of side effects (Thom, 2006). Essential fatty acids (EFAs), including omega-3 and omega-6, are necessary for skin and hair growth, and it has been demonstrated that a deficiency of these oils causes abnormality of skin and hair. However, omega-3 and omega-6 cannot be synthesized by our bodies. Therefore it seems reasonable to propose marine resources as a potential source of EFAs. The main source of preformed dietary EFAs, including eicosahexaenoic acid (EPA) and docosahexaenoic acid (DHA), is fish and shellfish (Skolinik et al., 1997; Mahaffey, 2004). Therefore some hair care products contain fish protein and oil.

16.6 SUMMARY

The the alopecia and baldness population is increasing worldwide due to changing of food habits, lifestyle, and environmental conditions. Because of this, cosmeceuticals for hair growth promotion and scalp care products are getting a great deal of attention nowadays. Marine organisms and their metabolites contain varieties of nutrients, including polysaccharides, amino acids, fatty acids, vitamins, and minerals. Therefore marine organisms have great potential application in the cosmeceuticals area. There are numerous advantages to marine organisms being used in cosmeceuticals, e.g., relatively low production cost, low cytotoxicity, occurrence of novel molecules, and so forth. In addition, the wide chemistry of marine organisms and a great number of undiscovered materials are an interesting source for use in the manufacture of novel hair cosmeceutical products. However, further studies in human subjects with clinical trials are needed.

REFERENCES

Abdullah F, Rashid RM. 2010. Alopecia: botanical approaches in review. *J Drugs Dermatol* 9(5), 537–541.

Akiyama M, Smith LT, Shimizu H. 2000. Changing patterns of localization of putative stem cells in developing human hair follicles. *J Invest Dermatol* 114, 321–327.

Andersson S, Berman DM, Jenkins EP, Russell DW. 1991. Deletion of steroid 5 alpha-reductase 2 gene in male pseudohermaphroditism. *Nature* 354, 159–161.

Andl T, Reddy ST, Gaddapara T, Millar SE. 2002. WNT signals are required for the initiation of hair follicle development. *Dev Cell* 2(5), 643–653.

Assaad P. 2011. Hand and body cream. U.S. Patent 7887853.

Bertolino AP, O'Guin WM. 2003. Differentiation of the hair shaft. In Olsen EA, ed., *Disorders of hair growth: diagnosis and treatment*. 2nd ed. New York: McGraw-Hill Medical Publishing Division, 23–48.

Botchkarev VA. 2003. Stress and the hair follicle: exploring the connections. *Am J Pathol* 162, 709–712.

Burlando B, Verotta L, Comara L, Bottini-Massa E. 2010. *Herbal principles in cosmetics: properties and mechanisms of action*. Boca Raton, FL: CRC Press.

Chandini SK, Ganesan P, Bhaskar N. 2008. *In vitro* antioxidant activities of three selected brown seaweeds of India. *Food Chem* 107, 707–713.

Chew YL, Lim YY, Omar M, Khoo KS. 2008. Antioxidant activity of three edible seaweeds from two areas in South East Asia. *LWT* 41, 1067–1072.

Choocheep K, Hatano S, Takagi H, Watanabe H, Kimata K, Kongtawelert P, Watanabe H. 2010.Versican facilitates chondrocyte differentiation and regulates joint morphogenesis. *J Biol Chem* 285(27), 21114–21125.

Costsarelis G, Sun TT, Lavker RM. 1990. Label-retaining cells reside in the bulge area of pilosebaceous unit: implications for follicular stem cells, hair cycle, and skin carcinogenesis. *Cell* 61, 1329–1337.

D'Amelio FS. 1999. *Botanicals—a phytocosmetic desk reference*. Boca Raton, FL: CRC Press.

Dupont E. 2002. Toxicity reduced composition containing an anti-neoplastic agent and a shark cartilage extract. U.S. Patent 6383522.

Ebling FJ. 1987. The biology of hair. *Dermatol Clin* 5, 467–481.

Elder D, Elenitsas R, Jaworky C, Johnson B Jr. 1997. *Lever's histopathology of the skin*. 8th ed. Philadelphia: Lippincott-Raven, 20–24.

Elliott K, Stephenson TJ, Messenger AG. 1990. Differences in hair follicle dermal papilla volume are due to extracellular matrix volume and cell number: implications for the control of hair follicle size and androgen responses. *J Invest Dermatol* 113, 873–877.

Fard SG, Shamsabadi FT, Emadi M, Meng GY, Muhammad K, Mohamed S. 2011. Ethanolic extract of *Eucheuma cottonii* promotes *in vivo* hair growth and wound healing. *J Anim Vet Adv* 10(5), 601–605.

Foitzik K, Krause K, Conrad F, Nakamura M, Funk W, Paus R. 2006. Human scalp hair follicles are both a target and a source of prolactin, which serves as an autocrine and/or paracrine promoter of apoptosis-driven hair follicle regression. *Am J Pathol* 168(3), 748–756.

Foitzik K, Lindner G, Mueller-Roever S, Maurer M, Botchkareva N, Botchkarev V, Handjiski B, Metz M, Hibino T, Soma T, Dotto GP, Paus R. 2000. Control of murine hair follicle regression (catagen) by TGF-beta1 *in vivo*. *FASEB J* 14(5), 752–760.

Garza LA, Yang CC, Zhao T, Blatt HB, Lee M, He H, Stanton DC, Carrasco L, Spiegel JH, Tobias JW, Cotsarelis G. 2011. Bald scalp in men with androgenetic alopecia retains hair follicle stem cells but lacks CD200-rich and CD34-positive hair follicle progenitor cells. *J Clin Invest* 121(2), 613–622.

Han WS. 2007. Composition for promoting hair-restoration which generates hair roots and promotes the metabolism. Korean Patent 100704754.

Hardy MH. 1992. The secret life of the hair follicle. *Trend Genet* 8, 55–61.

Hibberts NA, Messenger AG, Randall VA. 1996. Dermal papilla cells derived from beard hair follicles secrete more stem cell factor (SCF) in culture than scalp cells or dermal fibroblasts. *Biochem Biophys Res Commun* 222(2), 401–405.

Hirata T, Sugawara T. 2008. Neovascularization inhibitors containing fucoxanthin and/or fucoxanthinol and their uses as pharmaceuticals and cosmetics. JP2008001623A 20080110.

Hwang HJ. 2010. Skin elasticity and sea polyphenols. *Seanol Sci Center Rev* 1, 1–10.

Jadhav VM, Thorat RM, Kadam VJ, Gholve SB. Kesharaja. 2009. Hair vitalizing herbs. *Int J Pharm Tech Res* 1, 454–467.

Jahoda CA, Horne KA, Oliver RF. 1984. Induction of hair growth by implantation of cultured dermal papilla cells. *Nature* 311, 560–562.

Kang JI, Kim SC, Boo HJ, Jeon YJ, Yoo ES, Kang HK. 2010. The hair growth promoting effect of diekol, a component from Ecklonia cava. In *6th KSMB Annual Meeting and International Mini-Symposium*, Korean Society of Marine Biotechnology, 81.

Kaufman KD. 2002. Androgens and alopecia. *Mol Cell Endocrinol* 198, 89–95.

Kim MY. 2005. Process for producing hair regrowth agent in the forms of topical solution and pill. Korean Patent 100476731.

Kim SK, Ravichandran YD, Khan SB, Kim YT. 2008. Prospective of the cosmeceuticals derived from marine organisms. *Biotechnol Bioprocess Eng* 13, 511–523.

Kishimoto J, Ehama R, Wu L, Jiang S, Jiang N, Burgeson RE. 1999. Selective activation of the versican promoter by epithelial-mesenchymal interactions during hair follicle development. *Proc Natl Acad Sci USA* 96(13), 7336–7341.

Krause K, Foitzik K. 2006. Biology of the hair follicle: the basics. *Semin Cutan Med Surg* 25(1), 2–10.

Lassus A, Eskelinen E. 1992. A comparative study of a new food supplement, viviscal, with fish extract for the treatment of hereditary androgenic alopecia in young males. *J Int Med Res* 20, 445–453.

Lassus A, Jeskanen L, Happonen HP Santalahti J. 1991. Imedeen for the treatment of degenerated skin in females. *J Int Med Res* 19, 147–152.

Lassus A, Santalahti J, Sellmann. 1994. Treatment of hereditary androgenic alopecia in middle-aged males by combined oral and topical administration of special marine extract-compound. *J Int Med Res* 13, 254–255.

Lavker RM, Risse B, Brown H, Ginsburg D, Pearson J, Baker MS, Jensen PJ. 1998. Localization of plasminogen activator inhibitor type 2 (PAI-2) in hair and nail: implications for terminal differentiation. *J Invest Dermatol* 110(6), 917–922.

Mahaffey KR. 2004. Fish and shellfish as dietary sources of methylmercury and the ω-3 fatty acids, eicosahexaenoic acid and docosahexaenoic acid: risk and benefits. *Environ Res* 95(3), 441–428.

Millar SE. 2002. Molecular mechanisms regulating hair follicle development. *J Invest Dermatol* 118, 216–225.

Min YK, Jeong JY. 2007. Safe composition comprising medicinal herbs capable of promoting production of hair root and blood circulation in the scalp. Korean Patent 100775668.

Oikawa T, Ashino-Fuse H, Shimamura M, Koide U, Iwaguchi T. 1990. A novel angiogenic inhibitor derived from Japanese shark cartilage (I). Extraction and estimation of inhibitory activities toward tumor and embryonic angiogenesis. *Cancer Lett* 51, 181–186.

Park YJ. 2010. Concentrated beverage composition for hair health care, methods of manufacturing the concentrated beverage composition and natural tea comprising the same. U.S. Patent 7740887 B2.

Paus R, Christoph T, Muller-Rover S. 1999. Immunology of the hair follicle: a short journey into terra incognita. *J Invest Dermatol Symp Proc* 4, 226–234.

Piraccini BM, Iorizzo M, Rech G, Tosti A. 2006. Drug-induced hair disorders. *Curr Drug Saf* 1(3), 301–305.

Raj D, Brash DE, Grossman D. 2006. Keratinocyte apoptosis in epidermal development and disease. *J Invest Dermatol* 126(2), 243–257.

Randall VA. 1994. Androgens and human hair growth. *Clin Endocrinol* (Oxf) 40, 439–457.

Randall VA. 2008. Androgens and hair growth. *Dermatol Ther* 21, 314–328.

Roh YH, Shin CS. 2006. Preparation and characterization of alginate-carrageenan complex films. *J Appl Polym Sci* 99, 3483–3490.

Saitoh M, Uzuka M, Sakamoto M. 1970. Human hair cycle. *J Invest Dermatol* 54, 65–81.

Skolinik P, Eaglstein WH, Ziboh VA. 1977. Human essential fatty acid deficiency: treatment by topical application of linoleic acid. *Arch Dermatol* 113, 939–941.

Soma T, Yano K, Kishimoto J. 2006. Biotechnology in hair care (II): growth modulation. In Lad R, ed., *Biotechnology in personal care*. 1st ed. New York: Taylor & Francis Group, 267–297.

Stenn KS, Paus R. 2001. Controls of hair follicle cycling. *Physiol Rev* 81, 449–494.

Stuart MD. 2003. Review of research on *Undaria pinnatifida* in New Zealand and its potential impacts on the eastern coast of the South Island. DOC Science Internal Series 166. Wellington: Department of Conservation.

Szabo G. 1958. The regional frequency and distribution of hair follicles in human skin. In Mantagna W, Ellis RA, eds., *The biology of hair growth*. New York: Academic Press, 33–38.

Thigpen AE, Silver RI, Guileyardo JM, Casey ML, McConnell JD, Russell DW. 1993. Tissue distribution and ontogeny of steroid 5 alpha-reductase isozyme expression. *J Clin Invest* 92, 903–910.

Thom E. 2006. Nourkrin: objective and subjective effects and tolerability in persons with hair loss. *J Int Med Res* 34, 514–519.

Thornton MJ, Nelson LD, Taylor AH, Birch MP, Laing I, Messenger AG. 2006. The modulation of aromatase and estrogen receptor alpha in cultured human dermal papilla cells by dexamethasone: a novel mechanism for selective action of estrogen via estrogen receptor beta? *J Invest Dermatol* 126(9), 2010–2018.

Weger N, Schlake T. 2005. Igf-I signalling controls the hair growth cycle and the differentiation of hair shafts. *J Invest Dermatol* 125(5), 873–882.

Winiarska A, Mandt N, Kamp H, Hossini A, Seltmann H, Zouboulis CC, Blume-Peytavi U. 2006. Effect of 5α-dihydrotestosterone and testosterone on apoptosis in human dermal papilla cells. *Skin Pharmacol Physiol* 19, 311–321.

Wu H, Stanley JR, Cotsarelis G. 2003. Desmoglein isotype expression in the hair follicle and its cysts correlates with type of keratinization and degree of differentiation. *J Invest Dermatol* 120(6), 1052–1057.

Yano K, Brown LF, Detmar M. 2001. Control of hair growth and follicle size by VEGF-mediated angiogenesis. *J Clin Invest* 107, 409–417.

Yano K, Brown LF, Lawler J, Miyakawa T, Detmar M. 2003. Thrombospondin-1 plays a critical role in the induction of hair follicle involution and vascular regression during the catagen phase. *J Invest Dermatol* 120(1), 14–19.

Yuan YV, Carrington MF, Walsh NA. 2005. Extracts from dulse (*Palmaria palmata*) are effective antioxidants and inhibitors of cell proliferation *in vitro*. *Food Chem Toxicol* 43(7), 1073–1081.

17 Cosmeceuticals from Marine Fish and Shellfish

Vazhiyil Venugopal

CONTENTS

17.1 INTRODUCTION

Recent years have witnessed increasing public interest in health, well-being, and physical appearance, which have been reflected in demand not only for nutritional foods but also for cosmetics. Of the various cosmetic products, skin cosmetics have important roles, since protecting and preserving the skin is essential for good health and appearance. Mature skin suffers from essential nutrient deficiencies (minerals, vitamins, and amino acids), causing slackening, thinning, dull complexion, irregular surface, and dehydration. These effects are attributed to functional alterations in skin, such as increase in transepidermal water loss and decreased production of collagen, elastin, and the extracellular matrix. Environmental pollution and ultraviolet light (280–320 nm) from the sun have a profound influence on the aging process (Lamy et al., 2006). Protection of the skin requires a supply of bioactive components, including antioxidants and immunomodulant compounds. The use of cosmetics containing these compounds can reduce undesirable changes in skin (although they may have limitations in reversing the aging process). Traditional chemical sunscreens, based on para-amino benzoic acid and its derivatives, cinnamates, various salicylates and benzophenones, dibenzoylmethanes, anthraline derivatives, etc., act primarily by binding to skin protein and absorbing ultraviolet rays. Apart from skin care, hair color and style also play important roles in physical appearance and self-perception of an individual. Since many synthetic cosmeceuticals can have side effects, consumers are increasingly preferring cosmetic products that contain ingredients derived from natural sources, because of their known safety, biodegradability, and negligible toxicity (Santhanam, 2010).

The marine environment is unique in its support of diverse living organisms, which are rich in a variety of different nutrients and bioactive compounds. The contents of these compounds vary according to the organisms, and also environmental factors, such as temperatures and pressure regimes in the oceanic locations and their depth levels. Interests in bioactive properties of oceanic ingredients have encouraged their "biomining," which has resulted in the isolation and characterization of more than 3,000 compounds from marine fishery products, seaweed, corals, and microorganisms having therapeutic and nutritive values (Volkman, 1999; Faulkner, 2000). The nutraceutical industry makes use of these compounds to enhance the functional properties of food products, whereas the pharmaceutical industry is interested in the development of newer low-cost drugs that can be cures for diverse diseases. Recent developments in the field of food fortification and encapsulation have contributed significantly to enhanced nutritional values of processed food products. Cosmeceuticals, derived from the words *cosmetic* and *pharmaceutical*, have drug-like benefits and contain active ingredients such as proteins, peptides, vitamins, phytochemicals, enzymes, antioxidants, and essential oils. Cosmeceuticals have attracted research interest because of their beneficial effects on human health and appearance. In the last few years, with the recognition of the role of certain nutrients in protection of skin and aging, a novel discipline of "nutricosmetics" has emerged, which relates to bioactive compounds that can have both nutraceutical and cosmeceutical values. Several bioactive compounds derived from fish and shellfish can have nutricosmetic functions. Regular intake of these compounds can be useful to maintain a youthful appearance (Kim et al., 2008). A number of these marine ingredients can also have significant bioactivities and functional roles with respect to cosmetic products (Dillingham, 2010). However, interest in harnessing these ingredients for cosmetic formulations is rather nascent. This chapter discusses major classes of ingredients from marine fishery products and their functional properties and applications as cosmeceuticals.

17.2 MARINE FISHERY PRODUCTS

Marine fish and shellfish comprise more than 7,000 species, out of which a few thousand are commercially important as food. The global availability of seafood has been generally rising as a consequence of increasing fishing fleets and sophistication of fishing operations assisted by advance in electronics and engineering. Global fish landings increased from 39.2 mt in 1961 to 122.1 mt in 1997 (at an annual growth rate of 3.6%), while food fish supply grew from 27.6 mt to nearly 94 mt during the same period. In addition, mariculture of fish and shellfish is supplementing available marine fishery products. The captures (in mt) of the top 10 marine fish in 2004 were anchovita (10.7), Alaska pollock (2.7), blue whiting (2.4), skipjack tuna (2.1), Atlantic herring (2.0), chub mackerel (2.0), Japanese anchovy (1.8), Chilian jack mackerel (1.8), large head hairtail (1.6), and yellowfin tuna (1.4). However, the annual global marine catch is showing saturation. In 2006 the total catch declined to about 82 mt, with a global export of fish and fishery products at US$85.9 billion. The year also witnessed a culture of about 20 mt of marine fish and shellfish, mostly in Asian countries, including China, India, Malaysia, and Thailand (FAO, 2008).

17.3 BY-CATCH AND SEAFOOD PROCESSING WASTE

Although the supply of several commercially important fish species is dwindling due to overfishing, a significant amount of the available fish remains underutilized. These fish species consist of by-catch of fishing operations of targeted species such as shrimp. This includes several varieties of pelagic, demersal, and unconventional fish species, which are not fully utilized as human food due to reasons such as poor appearance, texture, and flavor. Global discard of by-catch is currently estimated at more than 20 mt per year (Delgado et al., 2003). Currently, most of the by-catch is used for fish meal or dumped into the sea. For instance, 23.3 mt of fish were used for nonfood purposes, such as fish meal and oil in the year 2006 (FAO, 2008). Because seafood is essentially a foreign exchange earner for most developing countries, industrial plants for commercial seafood processing generally operate near landing centers. Processing operations in these plants are intended to separate edible parts from whole fish and shellfish employing techniques such as beheading, skinning, gutting, removal of scales, filleting, etc. The separated edible portions are further processed by techniques such as chilling, curing, freezing, canning, smoking, etc. Mollusk and crustacean processing activities essentially are deshelling and deveining; the meat portions are subjected to washing, cooking, cooling, and packaging of the product. The processing operations lead to large amounts of organic waste consisting of shell, head, intestines, fin, skin, viscera (liver, roe, stomachs, etc.), backbones, cuts, and rejected fish from processing. On a wet weight basis, the wastes constitute as high as 40% of whole shrimp and krill, 50% of crab, and 24% in the case of squid. Finfish wastes may be 24–51% of the raw material, depending upon the species. Approximate annual estimates of crustacean waste generated from shrimp, squid, crabs, and krill on a global basis are 500,000, 100,000, 490,000, and 60,000 t, respectively (Subasinghe, 1999). Table 17.1 gives approximate global availability of crustacean waste. In the United Kingdom, about 7,000 t of crab shell wastes are generated annually, most of which are discarded as landfill, a practice that has caused serious environmental concern (Archer, 2001). Norwegian fisheries annually produce more than 615.000 t of waste; most of it is converted into fish silage and fish meal as animal feed, while about 180.000 t of by-catch from the fishing fleet is dumped into the sea (Gildberg, 2004). Global tuna catch is about 4 mt; processing the fish as loins/steaks or as a canned food generates 40–60% of the catch as discards, the canning industry alone annually generating about 450,000 t of discards (Sultanbawa and Aksnes, 2003). About 250,000 t of cod waste such as heads, backbones, viscera, and skin is generated annually, most of it either dis-

TABLE 17.1
Global Availability of Crustacean Waste

Resource	Total Landing	Waste Available[a]	Dry Waste[b]	Chitin Content[c]
Shrimp	1,292,476*	516,990	129,475	32,311
Squid	398,219	99,531	24,882	1,244
Crabs	943,826	482,744	144,823	28,964
Krill	150,000**	60,000	15,000	3,750

Source: Adapted from Subasinghe, *Infofish Int.*, 3, 58–62, 1999.
[a] Assuming 40% is waste.
[b] Multiplication factor, 0.25 for calculating dry waste from wet waste.
[c] Multiplication factor 0.25 for calculating chitin from dry waste.
* Total landing during 2002; ** average annual global krill landing.

carded at sea or sold at a low price for feed production. Cod head contains about 55% muscle, 20% bone, 15% gill, 5% skin, and 4% eyes, with an average protein content of 15% (Anonymous, 2007a).

17.4 INGREDIENTS FROM BY-CATCH AND PROCESSING WASTE

The voluminous amounts of by-catch and seafood wastes can be raw material for several important ingredients, namely, proteins including collagen and gelatin, protein hydrolyzates, peptides, enzymes, oil, carotenoids, bones, chitin, chitosan, and other compounds, which can have a variety of uses (Arvanitoyannis and Kassaveti, 2008; Kim and Mendis, 2006). Whereas the residual meat recovered from operations such as deboning and filleting can be used for production of fish protein hydrolyzate (FPH), the bone waste can be a good source of proteins including collagen and gelatin. The functional properties of seafood proteins include solubility, viscosity, water holding capacity, gelation, adhesion, elasticity, and emulsification. These properties are bestowed due to their characteristic compositional and structural features, leading to hydrophilic character, hydrodynamic size and shape, hydrogen and bonding and ionic hydration, water entrapment, network formation, sulfide cross-links, and film formation. Their functional properties useful for various food and cosmetic applications include water holding, thickening, colloid stabilization, crystallization control, film formation, whipping, and emulsification (Damodaran, 1996). Adequate calcium and vitamin D as part of a healthy diet is critical for the development and maintenance of healthy bones. Foods are considered the preferred source of calcium and vitamin D over supplements. Tuna bone powder is a potential value-added product that contains a proper balance of calcium and phosphorus; it can be used as a calcium food supplement and for treating osteoporosis (Sultanbawa and Aksnes, 2006). Shrimp and other crustaceans can be a source of carotenoids, particularly astaxanthin. Crustacean waste also serves as a good source of chitin and its deacetylated form, chitosan. The latter, because of its remarkable biocompatibility, biodegradability, nontoxicity, antibacterial and gel-forming properties, and hydrophilicity, has found numerous applications. Bioprocessing of waste not only helps to address seafood-related environmental issues but also provides an opportunity to isolate interesting drugs and nutraceuticals, thereby offering potentials for value addition (Arvanitoyannis, and Kassaveti, 2008). For instance, the National Institute of Water and Atmospheric Research, New Zealand, isolated 2,200 samples from 154 different fish species and observed about 150 samples having bioactivities. Of these, 10 UV-protective extracts looked particularly promising, which showed excellent potential as skin protectants or skin antiaging compounds (Allens and Jaspars, 2009; Blanco et al., 2007; Proffitt, 2007). The major ingredients from marine fishery sources having potential as cosmeceuticals are given in Table 17.2. A brief account of their characteristics, isolation procedures, and properties will be pointed out, followed by discussion indicating some specific examples of their applications in the cosmeceutical industry.

17.4.1 Collagen

Collagen, in different types, is found in tendons, skin, artery walls, myofibrils, cartilage, and the organic part of bones and teeth. Collagen consists of three identical or different peptide chains, each chain possessing a helical structure forming together a triple-stranded helix consisting of repeating triplets, $(\text{glycine-X-Y})_n$, where X and Y are often proline (Pro) or hydroxyproline (Hyp). This basic structural unit of collagen fiber is called tropocollagen. It has a molecular weight of approximately 30 kDa with a length of about 280 nm and diameter of 1.4–1.5 nm. During maturation and aging, tropocollagen fibers strengthen and are stabilized by covalent cross-linkages giving mechanical strength to the fibers (through enzymatic oxidation of lysine and hydroxyl lysine to aldols and aldimines). The number and nature of the cross-links have a profound effect on the stabilization of the collagen and its physical properties. Type I collagen accounts for 90–95% of the extracellular matrix of cod with other noncollagenous proteins (Morimura et al., 1999). Collagen swells in water but is not soluble. It is amenable to hydrolysis by the enzyme collagenase. Heating of collagen

TABLE 17.2
Ingredients from Marine Fish and Shellfish Having Potential as Cosmeceuticals

Ingredient	Source	Purported Action
Proteins including collagen and gelatin	Various fish and shellfish	Provides softness and prevents skin irritation Collagen is antistatic, film forming, and humectant moisturizer
Peptides	Protein hydrolyzates from various fish and shellfish	Enhance collagen synthesis and protect skin from effects of UV radiation
Chitin, chitosan, and their derivatives, and glucosamine	Crustacean shellfish, cephalopods	Offer moisturizing, film forming, and ability to control flow characteristics Provide hair conditioning Increase skin flexibility/elasticity, smoothness, softness, and suppleness
Essential fatty acids (linoleic, linolenic, and arachidonic acids, omega-3 fatty acids)	Fish oils	Emollient, skin conditioning Omega-3 fatty acids provide significant skin care benefits in a broad spectrum of topical applications Hydrogenated fish oil glycerides soften the skin
Vitamins (A, C, and E), nicotinamide, retinoic acid, coenzyme Q	Various fish/fish oils Vitamin A found in three forms: retinol (vitamin A1), 3,4-didehydroretinol (vitamin A2), and 3-hydroxyretinol (vitamin A3).	Antioxidant activities Control lipid oxidation
Carotenoids including β-carotene, astaxanthin, and zeaxanthin	Shellfish	Antioxidant activities Control lipid oxidation β-Carotene is a precursor of the antioxidant vitamin A
Hyaluronic acid	Fish bone	Moisturizer, lubricant, antistatic, humectant, and skin conditioner
Squalene	Shark cartilage	Antistatic, emollient, hair conditioning, and skin conditioning agent
Chondroitin sulfate	Shark cartilage	Antistatic hair conditioning and skin conditioning agent
Guanine	Fish scales	Provides shining effect
Minerals (calcium and phosphate)	Fish bones	Enhance calcium supply

results in its shrinking, the shrinkage temperature (T_s) being different for different collagens. For fish collagen, T_s is 45°C, and for mammals, the value ranges between 60 and 70°C. Heating destroys the triple-helix structure to random coils, which are soluble in water, and gives gelatin (see below). Frames discarded from fish filleting operations and also fish scales are good sources of collagen. Head waste in cod fisheries, yielding about 20% of the fish weight, forms an excellent source of collagen and gelatin (Anonymous, 2007a). The process of extraction of collagen from bones of Baltic cod and other fish (and also pig skin) involves initial removal of lipid using acetone for 12 h at room temperature followed by drying. The treated bone is decalcified using a 10-fold 0.6 M hydrochloric acid at room temperature for a day to yield collagen. Collagen from acetic or citric acid extracts of skins of fish such as cod could be precipitated using 0.1% κ-carrageenan (Hofman et al., 2007). Acid-soluble collagen (ASC) and pepsin-solubilized collagen (PSC) were isolated from the bones and scales of two fish: black drum and sheephead sea bream caught in the Gulf of Mexico. The molecular weights of the collagen subunits were about 130 kDa for α_1 and 110 kDa for α_2, respectively. The amino acid composition of the PSCs was closer to that of calf skin ASC than to that of cod skin ASC. The melting temperatures of ASC and PSC were above 34°C. Intrinsic viscosity of the PSC was similar to that of collagen from fish species such as hake, cod, and catfish. The transition temperature of black drum ASC was 36.3°C, similar to that of calf hide ASC. The collagen from black drum skin was comparable to typical type I collagens from land-based animals and has potential therapeutic effects in terms of anti-inflammatory activity, inhibition of angiogenesis, and for tissue (Ogawa et al., 2004). Hake skin collagen had optimum protein solubility, apparent viscosity and water-binding capacity at a pH between 2 and 4, and these values were higher than with collagen from trout (Moneto et al., 1999). Seafood wastes including skin and frame by-products from surimi manufacturing are other good resources of collagen (Arvanitoyannis and Kassaveti, 2008; Kim and Park, 2004). Extraction of collagen from fish scales consists of decalcification and disaggregation followed by treatment with pepsin digestion. The yields of collagen from different fish scales are sardine, 51%; red sea bream, 37.5%; and Japanese sea bass, 41%. The fish scale collagens have a characteristic amino acid pattern similar to that of calf collagen; however, there are differences in the contents of imino acids and amino acids with hydroxyl groups in the side chain. Collagen can be hydrolyzed to various extents using the enzyme collagenase or with *Bacillus subtilis* protease at pH 8.0 for 1 h at 60°C. Fractionation of the hydrolyzate by ultrafiltration gives collagen peptides (Nagai et al., 2004).

17.4.2 Gelatin

Gelatin is a gelling protein obtained by hydrolysis of collagen, and has wide industrial applications. Gelatin is primarily known for its thermoreversible gelation behavior. Most of the current production of gelatin is from porcine and bovine hides, bones, and hooves. However, frequent occurrences of bovine spongiform encephalopathy (BSE) and foot and mouth diseases have caused serious concerns of human health safety that have hampered uses of mammalian gelatin (Baziwane and He, 2003). Gelatin was extracted from yellowfin tuna, employing aqueous NaOH (1.9% w/w) treatment at 58°C for 4.7 h. The extracted gelatin had a bloom strength of 429, in comparison to 216 and 295 for bovine and porcine gelatins, respectively. Dynamic viscoelastic properties of yellowfin tuna skin gelatin did not change at 20°C, but increased at 10°C, similar to bovine and porcine gelatins (Cho et al., 2005). Tuna gelatin contains a higher percentage of molecules larger than 10,000 Da, and therefore has good gelling properties (Sultanbawa and Aksnes, 2003). The process of extraction of gelatin from fish waste such as cod head involves successive extraction of the raw material at room temperature in dilute NaOH and HCl. It has a low melting point of 10°C, compared with about 30°C for mammalian gelatin; hence it can be used for encapsulation of compounds that are sensitive to heat (Anonymous, 2007a). Muscle proteins from minced fish backbone wastes were hydrolyzed using a mild proteinase, followed by subsequent heating to recover gelatin and a protein-calcium fraction from the purified bone fraction. The recovered gelatin had a lower molecular weight (up to 30 kDa) than standard gelatin preparations (>100 kDa) (Anonymous, 2007a). North Sea horse mackerel skin was extracted using aqueous sodium hydrogen carbonate (0.125%), sodium hydroxide (0.2%), sulfuric acid (0.2%), and citric acid (0.715%). Its color, molecular weight (196 kDa), and bloom strength (230) compared well with those of commercial gelatins (Zhou et al., 2006).

Modified gelatins, such as cold-water-soluble gelatin, hydrolyzed gelatin-based bioactive peptides, and esterified gelatin, have also been prepared (Ward et al., 2001). Gel formation of fish gelatin, like mammalian gelatins, involves electrostatic interactions, which are important in the stabilization of the protein gel network. Microbial transglutaminase (MTGase) could also be useful for extraction of gelatin from the skin of big-eye snapper and brown stripe red snapper (Jongjareonrak et al., 2006). Gelatins from marine fish skins and bones could be plausible substitutes to mammalian gelatins, since these gelatins have a bloom strength (the gel strength equivalent referred to in the

TABLE 17.3
Comparison of Physicochemical Properties of Gelatin from Fsh and Land Animals

Property	Alaska Pollock	Tilapia	Pork
Hydroxyproline content (%)	7.3 ± 0.3	10.6 ± 0.2	13.1 ± 0.3
Bloom strength (bloom)	98 ± 2	273 ± 4	240 ± 5
Gel melting temperature (°C)	21.2 ± 0.2	25.4 ± 0.2	31.2 ± 0.1
Viscosity (mP)	120 ± 2	38 ± 1	47 ± 1

Source: Adapted from Zhou, P., Mulvaney, S. J., and Regenstein, J. M. (2006), Properties of Alaska pollock skin gelatin: a comparison with tilapia and pork skin gelatins, *J. Food Sci.*, 71, C313.

industry), viscosity, and solubility comparable with those of mammalian gelatin. However, fish gelatin gels generally have a considerably lower storage modulus and melting temperatures (12–13°C). The bloom strength varies from less than 100 or more than 300 for most commercial gelatins. These gelatins also have a viscosity between 15 and 75 mP. The melting point of gelatin gels is usually lower than the human body temperature. This melt-in-the-mouth property is one of the important characteristics responsible for wide applications of gelatins in food and pharmaceutical industries (Lu et al., 1997). Table 17.3 compares some physicochemical properties of gelatin from fish and land animals.

17.4.3 Bioactive Peptides

Bioactive peptides can be produced *in vitro* by enzymatic hydrolysis of food proteins. Whereas these peptides are inactive within the sequence of the parent protein, they become active upon hydrolysis. Usually, pancreatic enzymes, generally trypsin, and other combinations of enzymes, including protinases, alcalase, chymotrypsin, and pepsin, as well as enzymes from bacterial and fungal sources, have been utilized to generate bioactive peptides. After hydrolysis the peptides in the hydrolyzate are fractionated and enriched using chromotagraphic, isoelectric focusing, and other suitable methods. Ultrafiltration membranes have been successfully used to enrich specific peptide fractions. Apart from hydrolysis with proteolytic enzymes, physical removal, chemical modification, and heat denaturation are some of the other strategies to remove epitopes (Philanto and Korhonen, 2003). Proteins from by-catch fish and seafood processing waste are good sources of bioactive peptides with interesting functional properties (Aneiros and Garateox, 2004; Kwon-Kim et al., 2001). *In vitro* and *in vivo* studies have shown that certain peptide fractions in fish protein hydrolysates may stimulate the nonspecific immune defense system (Gildberg, 2004). Table 17.4 indicates important bioactivities of peptides. Zonase, an enzyme, was first discovered at the salmon hatcheries in Norway. The enzyme was found to help break the eggshell during hatching, and has been shown to have a protein structure similar to that of human skin proteins (Anonymous, 2006). Heat-stable enzymes from marine sources including deep-sea bacteria have been recognized to offer protection to skin against UV-induced free radicals (Buche et al., 2002; Haard, 1998).

17.4.4 Marine Fish Lipids

Marine fish generally contain high amounts of fat. According to the fat contents of their fillets, marine fish can be grouped as lean (<3% fat), medium (3–7%), and high fat (>7%). Lipids in fatty fish are mostly subcutaneous in nature, while in lean fish they are deposited in the liver, muscle tissue, and mature gonads. Marine lipids are composed of neutral lipids, including triacyl glycerols,

TABLE 17.4
Important Bioactivities of Peptides

Antihypertensive (ACE inhibitory) activity
Antioxidant activity
Antimicrobial activity
Antihypoallergenic activity
Obesity control
Cell immunity
Cellular growth
Secretion of digestive enzymes
Ripening of fermented foods
Influencer of food taste
Antifreeze proteins

phospholipids, sterols, wax esters, and some unusual lipids, such as glyceryl esters, glycolipids, sulfolipids, and hydrocarbons. Most of the variations in lipids are found in the triacyl glycerol fraction, while the phospholipids show few variations. The triacyl glycerol functions as a reserve of fatty acids that provide energy through oxidation. They are also converted into phospholipids, which are present in cell membranes. Fatty acids are straight-chain carboxylic acids, which may be short-chain fatty acids (with chain lengths of 10 carbon atoms or less), medium-chain fatty acids (up to 14 carbon atoms), and long-chain fatty acids (having more than 14 carbon atoms). Whereas the short-chain fatty acids are always saturated, others may be saturated or unsaturated. In unsaturated fatty acids, the position of the first double bond is given by the (n-x) notation, counting the number of carbon atoms from the methyl end, according to the international nomenclature. For example, omega-3 and omega-6 (also referred to as ω-3 and ω-6 or n-3 and n-6) fatty acids denote such acids, in which the first double bond starts at three and six carbons, respectively, from the methyl end. The symbol 18:4 ω-3 identifies a fatty acid with 18 carbon atoms and 4 double bonds, the first double bond occurring after the third carbon atom. The fatty acids most studied for their health benefits are the ω-3 fatty acids; the 20-carbon eicosa pentaenoic acid (EPA; $C_{20:5w3}$, *cis*-5.8.11.14.17-eicosapentaenoic acid), which contains 5 double bonds; and the 22-carbon docosahexaenoic acid (DHA; $C_{22:6w3}$, *cis*-4,7,10,13,16,19-docosahexaenoic acid), containing six double bonds. Marine oils are rich in EPA and DHA. Marine fish and shellfish accumulate these fatty acids from microalgae, which serves as primary food for oceanic life. For instance, mollusks having good access to phytoplankton accumulate a high proportion of EPA and DHA. Accumulation of the polyunsaturated fatty acid (PUFA) has also been observed in other bivalves, including oysters, patella, clams, and scallops. Major dietary n-6 PUFA include linoleic, $C_{18:2w6}$, γ-linolenic, $C_{18:3w6}$, and arachidonic, $C_{20:4w6}$ acids, whereas major dietary omega-3 PUFA include α-linolenic, $C_{18:3w6}$, EPA, and DHA (Venugopal, 2008).

Polyunsaturated n-3 and n-6 fatty acids are widely accepted as a part of modern nutrition because of their beneficial effects on metabolism. Marine fish oils, because of the rich presence of omega-3 fatty acids, have been well documented for beneficial health effects such as reducing the risk of cardiovascular disease, hypertension, and autoimmune and inflammatory diseases. They are beneficial in controlling blood vessel disorders and inflammatory diseases, and moderating an overactive immune function resulting in alleviation of autoimmune disease, such as arthritis and some types of dermatitis. Omega-3 and omega-6 fatty acids are also involved in the structure of cell membranes, development of the nervous system, and influence the synthesis of cell mediators (prostaglandins and leukotrienies), which play important roles in coagulation, inflammation, and proliferation of certain cells. Deficiency of these compounds causes disorders such as restrictive growth, abnormality of skin and hair, damage of the reproductive system, and abnormal composition of serum and tissue fatty acids (Gogus and Smith, 2010; Minihane and Lovegrove, 2006; Lands, 2005). The

TABLE 17.5
Production of Shark Liver Oil, Its Composition, and Uses

Process of Extraction	
Natural decomposition (incubation at 30°C)	
Acid ensilage	
Alkali digestion	
Steam rendering (90°C for 30 min)	
Composition	
Hydrocarbons, squalene, and pristine	65%
Diacetyl glyceryl ether, selacyl, chimyl, and butyl alcohols	26%
Triglycerides	6%
Uses	
Squalene and squalane are used as moisturizers in cosmetics	
Squalene also has therapeutic activities such as antitumor activity	
Diacayl glyceryl ethers have bacteriostatic action	
As surfactants and for wound dressings	

Source: Adapted from Buranudeen, F., and Rajadurai, P. N. R., Shark liver oil, *Infofish Marketing Dig.*, 1, 42, 1986.

health benefits of PUFA have encouraged regulatory agencies to recommend regular consumption of fatty fish such as herring, sardine, mackerel, or fish oil by the general public. The World Health Organization recommends consumption of one to two servings of fish, containing 200–500 mg of EPA and DHA, per week. The American Heart Association recommends that healthy individuals consume a 2–3 oz serving of fatty fish per week, and that persons diagnosed with cardiovascular diseases consume 1 g each of the fatty acids per day (Satterlee, 2004; Dallongeville et al., 2003; Kris-Etherton et al., 2002). Marine fish species including anchovies, capelin, Atlantic cod, Atlantic herring, Atlantic mackerel, Atlantic menhaden, salmonids, and sardines, and the crustacean Antarctic krill are commercial sources of oil. The oil contents in these fish vary with marine environment and season and can be as high as 21% in herring and 18% in sardines. Several industries specialize in their production and purification through cold pressing, further concentration, chilling, and other technologies. Table 17.5 gives production of shark liver oil, its composition, and uses. More than 70% of the all the omega-3 is used as a dietary supplement, 15% goes into functional foods, 5% goes to baby formulae, and finally 10% goes into pet food (Sultanbawa and Aksnes, 2003; Alasalvar and Taylor, 2002).

Fish oils, being rich in polyunsaturated acids, are extremely sensitive to autooxidation. Autooxidation leads to formation of several reactive oxygen species (ROSs), which include peroxide radical (ROO•), alkoxy (RO•), superoxide anion (O_2^-), hydroxyl radical (HO•), hydroperoxy (HOO•), and nitric oxide (NO•) derived radicals. Nonradical derivatives are hydrogen peroxide (H_2O_2), ozone (O_3), and singlet oxygen ($^1O_2$). Hydroxy radical is the most reactive ROS, followed by singlet oxygen. Unsaturated lipids are easily oxidized by ROSs, which results in low molecular volatile aldehydes, alcohols, and hydrocarbons. ROS-induced reactions are favored by activation of hemoproteins and increase in free iron (Clemans and Pressman, 2007). Reactive oxygen (ROS) and reactive nitrogen species (RNO) are continuously produced in the human body. These cause oxidative damage to DNA, block cellular signal transduction, and modulate gene expression and enzyme activities. ROSs also react with proteins, sugars, and vitamins, causing cross-linking or cleavage of proteins, causing loss of their functionality, and destroying essential fatty acids, amino acids, and vitamins. These reactions have roles in the acceleration of the process of aging. Besides, these interactions also lead to induction of serious human diseases, including atherosclerosis, rheumatoid arthritis, muscular dystrophy, cataracts, neurological disorders, and cancer (Cho and Min, 1996).

Antioxidants may be defined as substances that are capable of delaying, retarding, or preventing the development of rancidity. Antioxidants can inhibit or retard oxidation and loss of flavor in foods; the mechanisms involve binding of metal ions, scavenging of oxygen, converting hydroperoxides to nonradical species, deactivating singlet oxygen, and thereby suppressing the generation of free radicals. These reactions have a positive influence on health, reducing the risks of chronic diseases. Primary antioxidants are butyl hydroxyanisole (BHA), hutylhydroxyl toluene (BHT), tertiary butyl hydroxyquinone (TBHQ), and esters of gallic acid, such as propyl gallate, α-tocopherol, flavanoids, etc. They are used generally in concentrations as low as 0.02% of the fat or oil content, and sometimes, in combinations, for synergistic antioxidant effects. Antioxidants such as tocopherol (vitamin E), BHA, BHT, TBHQ, the gallate esters, and ascorbyl palmitate are lipid soluble, while sulfur dioxide, ascorbic acid, and cysteine are water soluble. Natural antioxidants include ascorbic acid, carotenoids, α-tocopherol, and flavanoids. Ascorbic acid has been demonstrated to be an effective radical scavenger of superoxide, hydrogen peroxide, hydrochlorite, peroxyl radical, and singlet oxygen. The common dietary antioxidants, which include vitamin E, vitamin C, polyphenols including flavanoids, and the carotenoids, β-carotene and lycopene, augment cellular defenses and protect components of the cell from oxidation damage. Natural antioxidants extracted from plants such as rosemary, sage, tea, soybean, citrus peel, sesame seed, olives, carob pod, and grapes can be used as alternatives to synthetic antioxidants because of their equivalent or greater effect on the inhibition of lipid oxidation. It has to be mentioned that while these compounds can offer protection against free radical-mediated diseases, they are unable to function as drugs to cure diseases (Lichtenstein et al., 2005; Willcox et al., 2004).

17.4.5 Squalene

Liver oils of some deep-sea sharks, mainly *Centrophorus* spp., found under depths varying between 300 and 3,000 m in the Pacific, North Atlantic, and Indian Oceans contain about 85–90% unsaponifiable matter, which is essentially the hydrocarbon squalene ($C_{30}H_{50}$, 2,6,10,15,19,23-hexamethyl 2,6,10,14,18,22-tetracosahexene). The shovenose dogfish liver oil contains 60% hydrocarbons, consisting mainly of squalene and pristane, and 25% diacyl glyceryl ether. Squalene is colorless but becomes pale yellow and thick on storage due to oxidation. Physical methods (refractive index, density, and viscosity) and chemical characteristics (unsaponifiable matter and iodine value) could be used to evaluate the presence of squalene contents in various shark species (Bimbo, 2007; Thankappan, 2003). Squalene can have a remarkable antioxidant activity (Ko et al., 2002). In addition, this compound has been reported to possess antilipidemic and membrane-stabilizing properties. Squalene is safe and tolerable, as shown by feeding studies on humans, who were administered 860 mg squalene daily for 20 weeks (Chan et al., 1996).

17.4.6 Chitin and Chitosan

Chitin represents the second most abundant natural biopolymer, which is present in the exoskeletons of crustaceans and also in the cell walls of fungi, insects, and marine diatoms (Tharanathan and Kuttur, 2003). While the estimated annual potential availability of chitin is about 100 bt, about 150,000 mt chitin is produced annually (Subasinghe, 1999). Marine crustaceans, which include shrimp, crabs, squid, cuttlefish, krill, and oysters, are richer in chitin, in comparison with terrestrial organisms such as insects, fungi, etc. In these organisms, chitin forms the outer protection coatings in a covalently bound network with proteins and dihydroxy phenylalanine together with some metals and carotenoids. The isolation process for chitin generally consists of three steps: demineralization, deproteinization, and bleaching, followed by extraction of chitin using acetone (Synowiecki and Al-Khateeb, 2003). Chitin is a very light, white or yellowish, powdery/flaky product. It is insoluble in water, almost all common organic solvents, and acidic and basic aqueous solutions. After a simple treatment in caustic soda, chitin may be solubilized with carbon disulfide and reprecipitated as a filament or film in the manner of viscose rayon. Chitin swells in cold alkali when some deacetylation takes place (Muffler and Ulber, 2005). Chitin is a cationic polysaccharide formed by units of N-acetyl-D-glucosamine, joined by β-bonds (1–4), viz., [β-(1–4)-N-acetyl-D-glucosamine], which is β-(1–4)-N-acetyl-2 amino-2-deoxy-D-glucose. It may be regarded as a derivative of cellulose, in which the C-2 hydroxyl group is substituted by an acetyl amino group. Chitin occurs in three polymorphic forms, α, β, and γ, which differ in three-dimensional arrangements of the molecular chains. Its most common form is α-chitin, where two units of N-acetyl-D-glucosamine are in an antiparallel arrangement (No and Meyers, 1999; Shahidi and Abuzaytoon, 2005).

Chitosan is deacetylated chitin. It is a polycationic, long-chain biopolymer with a natural affinity for the normally negatively charged biological membranes, because of the presence of one free amino group and two free hydroxyl groups for each glucose ring in chitosan. Structurally, chitosan is β-(1–4)-linked-D-glucosamine, i.e., poly[β-(1–4)-linked-2-amino-2-deoxy-D-glucose]. The difference between chitosan and chitin is only in the functional group situated at carbon-2 of the monomeric unit. Commercially chitosan is produced by deacetylation of crustacean chitin using 30–60% (w/v) sodium or potassium hydroxide at 80–140°C. The preparation is purified by dissolving it in dilute acetic acid, reprecipitation with alkali, followed by washing and drying to get flakes of chitosan. The characteristics of the product in terms of molecular weight and the extent of deacetylation depend upon the treatment conditions. Increasing temperature enhances deacetylation, but also results in fragmentation of the chitosan, which affects its functional properties and applications. The molecular weight of natural chitosan is generally higher than 10^6 Da. Chitosan is insoluble in pure water, but unlike chitin it is soluble in weakly acidic aqueous media.

A minimum deacetylation of 70% is required for chitosan to be acceptable for various purposes. The value of pKa for the positively charged ammonium group is about 6.2. When the pH is raised to about 6.5, chitosan precipitates in a gel form. Because of its cationic nature, chitosan is incompatible in solution with most anionic water-soluble gums such as alginates, pectate, sulfated carrageenan, as well as carboxymethyl cellulose. On the other hand, acidic solution of chitosan is compatible with nonionic water-soluble gums such as starch, dextrins, glucose, polyhydric alcohols, oils, fats, and nonionic emulsifiers. Chitosan can form films, which are tough, flexible, and transparent. Film can be extruded from the acidic solution of chitosan into a 70°C coagulating bath containing caustic soda and sulfonic acid esters of high molecular weight alcohols. Chitosan is also biodegradable by specific enzyme chitosanase. There can also be some digestion by nonspecific activity of digestive enzymes such as amylases and lipases. By enzymatic treatment, soluble chitosan can also be obtained in oligosaccharide form. Chitosan derivatives in the form of acetate, ascorbate, lactate, malate, and others are water soluble (No and Meyers, 1999; Shahidi and Abuzaytoon, 2005).

Chitosan in microcrystalline form has several advantages compared to standard chitosan. These include high stability, water retention, and biodegradability, higher film-forming capacity, and high chelating properties. The product is manufactured by the aggregation of glucosamine macromolecules from solution. It occurs as a gelatinous aqueous dispersion characterized by a high water retention value, ranging from 500 to 5,000%. In powder form it has lower water retention values ranging from 200 to 500%.

The application potential of chitosan is multidimensional and includes cosmetics (see below), food and nutrition, biotechnology, material science, drugs and pharmaceuticals, agriculture and environmental protection, and recently gene therapy too. Recent years have witnessed a marked growth in dietary chitosan products having therapeutic and health attributes (Subasinghe, 1999; Shahidi and Abuzavtoon, 2005). In India, a few entrepreneurs are producing chitin and chitosan on a commercial scale under the technical guidance of the Central Institute of Fisheries Technology, Cochin. Table 17.6 depicts specifications for chitin and chitosan.

TABLE 17.6
Specifications for Chitin and Chitosan

Properties	Chitin Food Grade	Chitosan Pharmceutical Grade	Liquid Chitosan Technical Grade
Appearance	White/yellow flake	White/yellow powder or flake	Clear yellow liquid
Odor and taste	Odorless, tasteless	Odorless, tasteless	Odorless, tasteless
Moisture	<10%	<10%	—
Ash	<2.5%	<0.2%	<0.5%
Protein	<1.0%	<0.3%	<0.5%
Deacetylation	Nil	70–100%	>90%
Viscosity (0.5% solution)	600 cps	<5 cps	50 cps
pH	7.9	7.9	<5.5
Heavy metals (arsenic and lead)	<10 ppm	<10 ppm	<10 ppm

Source: Adapted from Subasinghe, *Infofish Int.*, 3, 58–62, 1999.

17.4.7 Glucosamine

Glucosamine is a natural amino sugar found in large concentrations in certain foods such as milk, eggs, liver, yeast, molasses, and yeast. It is synthesized in the body from L-glutamine and glucose. Glucosamine is obtained by extensive hydrolysis of chitosan with a mineral acid such as hydrochloric acid. If it is produced from chitin, it has to be deacetylated. Glucosamine is absorbed easily into the human intestine (Tharanathan and Kuttur, 2003). Glucosamine has been reported to have positive effects on dry skin (Kajimoto et al., 2001).

17.4.8 Chondroitin Sulfate

Chondroitin sulfate (CS) is a heteropolysaccharide found in connective tissue, known as cartilage. CS is composed of an alternating sequence of sulfated N-acetyl-D-galactosamine (GalNAc) and D-glucuronic acid (GlcA) residues. In dermatan sulfate (DS), some GlcA residues are epimerized into L-iduronic acid (IdoA). Apart from CS and DS, cartilage consists of collagen, elastin, hyaluronic acid, fibronectin, laminin, and water, which form a complex extracellular matrix providing a cushion between bones and joints. There are two main types of cartilage, hyaline and elastic. Hyaline cartilage found in articular joint surfaces withstands repetitive loading and provides a low-friction interface, whereas elastic cartilage, found in rib, trachea, pharynx, ears, and nose, is flexible to maintain the shape of these organs. Arthritis indicates inflammation of the joints, while rheumatism means aches and pains in bones, muscles, and joints. Osteoarthritis impacts the joints, including hands, knees, hips, feet, and back. Osteoporosis affects the bones, leading to fracture due to an imbalance between bone resorption and formation. High incidences of obesity, sedentary lifestyle, and poor diet have been held responsible for the disease. Chondroitin sulfate, dermatan sulfate, and hyaluronic acid (collectively called glycosaminoglycans) are used by the body for the biosynthesis of proteoglycans. These compounds possess a very strong affinity for water molecules in the joints to give a lubricant that fills the space between the joints. While chondroitin provides cartilage with strength and resilience, glucosamine inhibits inflammation and stimulates cartilage cell growth (McAlindon et al., 2000). Shark cartilage has been found to possess several therapeutic effects against diseases such as arthritis and tumor. The endoskeletons of shark are rich sources of chondroitin sulfate. Sea mussel also has a high level of chondroitin. A process for preparation of chondroitin sulfate from skate (*Raja avirostris*) cartilage is available (Jo et al., 2004). Recently the cranial cartilages of some squid and cuttlefishes have been reported to be rich in chondroitin sulfate (Manjusha, 2011).

17.4.9 Hyaluronic Acid

Hyaluronic acid (HA) is a naturally occurring polysaccharide consisting of glucoronic acid and N-acetyl-D-glucosamine units. In the human body, it is found in high concentrations in the skin, connective tissues, joints, and eye fluids. HA functions as an effective moisturizer due to its strong water holding capacity, which binds water as high as 500 times its weight. With age, the levels of HA in the body decrease, reducing the moisture binding capacity of the skin, resulting in dry skin and wrinkles. Oral consumption of HA can help increase moisture on the surface and within the skin. A combination of gelatin, hyaluronic acid, and chondroitin can have potential in wound healing (Wang et al., 2006).

17.4.10 Carotenoids

Carotenoids are a family of fat-binding compounds, responsible for the red and yellow color of crustaceans and also many plants, algae, and cyanobacteria. The carotenoids found in nature can be classified as: hydrocarbons, such as β-carotene, and xanthophylls, the oxygenated derivatives of carotenes such as astacene, astaxanthin, canthaxanthin, cryptoxanthin, echionine, lutein, neoxanthin, violaxanthin, and zeaxanthin. The color of carotenoids is due to a chromophore consisting of a chain of conjugated double bonds. Because of this property, carotenoids are commonly used as natural colorants in foods. The naturally abundant β-carotene ($C_{40}H_{56}$) is a polyunsaturated hydrocarbon made up of two molecules of retinal. It possesses maximal provitamin A activity, while approximately 60 carotenoids possess varying levels of provitamin A activity. Carotenoids possess remarkable antioxidant properties; astaxanthin is about a 500 times more powerful antioxidant than β-carotene. Animals including humans do not synthesize carotenoids *de novo* and rely upon the diet for these compounds. Astaxanthin is widely used to pigment salmon and trout, although in recent times canthaxanthin also became very popular in fish pigmentation and also in poultry to give a red color to egg yolks (Shahidi et al., 1998). Oral supplementation of carotenoids appears to increase the photoprotective properties of the epidermis and dermis against environmental stress (e.g., UV radiation, pollution, smoke), and by quenching free radicals generated by oxidative stress. Studies have also shown the ability of carotenoids to reduce DNA damage and protect against depletion of Langerhans cells, a key component of immune function. The health benefits due to consumption of carotenoids increase with daily consumption, resulting in smooth and elastic skin (Huang et al., 2005).

Crustacean wastes are abundant sources of carotenoids such as astaxanthin and β-carotene (Sachindra et al., 2005; Chakraborti et al., 2002; Shahidi et al., 1998). Carotenoids can be extracted from shell waste employing proteases, which detach them from the bound proteins. The carotenoid-rich head wastes of four commercially important Indian shrimp species, namely, *Penaeus monodon*, *P. indicus*, *Metapenaeus monocerous*, and the aquacultured *P. monodon* (which constituted 53–67% of the whole shellfish), were treated with trypsin at an optimum temperature of 45–55°C for 2 h to extract the pigments. The extracts contained significant quantities of astaxanthin, although other carotenoids, such as β-carotene, canthxanthin, lutein, zeaxanthin, and crustacyanin, were present in minor quantities. Loss of β-carotene and its derivatives was noticed due to their sensitivity to oxygen, light, and hydroperoxide radicals of lipid peroxidation (Chakraborti et al., 2002; Madhu et al., 2008). Astaxanthin in the form of carotenoprotein was extracted with a yield of 49% by treating shrimp (*Xiphopenaeus kroyeri*) waste with proteolytic enzymes such as trypsin in the presence of ethylenediamine tetraacetic acid (EDTA) at a pH of 7.7 and 4°C (Holanda and Netto, 2006).

17.4.11 Vitamins and Minerals

Marine fatty fish species contain significant amounts of fat-soluble vitamins A and D. Halibut and cod liver oils are rich sources of vitamins A and D. Herring, mackerel, salmon, and lake trout contain varying amounts of vitamin D in their tissues. The popular fish salmon contains up to 12% lipids; a 3.5 oz portion provides 90% of the daily human need for vitamin D. Fish cannot synthesize vitamin E (α-tocopherol), a powerful antioxidant, and hence the concentration of this vitamin is related to feed (Batternfeind, 1977). Vitamin E is used in cosmetics and food product development. Seafood provides moderate amounts of thiamin. The average content of thiamin (vitamin B_1) in fish species can be as high as 430 mg per 100 g meat. Fish also contain modest amounts of biotin, folic acid, niacin, and panthothenic acid. The best sources of pyridoxine (vitamin B_6) are salmon and tuna, and to some extent shellfish. Modest amounts of riboflavin (vitamin B_2) are present particularly in the dark flesh of some species like canned herring, mackerel, and pilchard. Pyridoxin is present in fish and shellfish in reasonable amounts, with tuna and salmon being rich in this vitamin. In general, seafood contains up to 42 μg of vitamin B_{12} per serving (3 oz); clams offer the highest amount of the vitamin followed by crustaceans and flesh fish. Clams, oysters, crab, salmon, and rainbow trout provide 2–14 times the recommended daily allowance for the vitamin. Therefore adding the seafood to one's diet can be the ideal way to get the vitamin. Vitamins are considered the most susceptible to heat treatment, but the magnitude of loss depends on the specific vitamins and the conditions employed. Folate and vitamin B_6 are susceptible to destruction due to oxidation. Riboflavin is reasonably stable during cooking, but is sensitive to light as it decomposes on exposure to ultraviolet rays. The fat-soluble vitamin A and carotenes are relatively stable at normal cooking temperatures. Vitamin E is slowly destroyed during frying and is decomposed by light (Venugopal, 2006). Whereas some vitamins such as vitamin E can help protect skin cells and tissues because of their antioxidant activity, others can correct damage caused to skin, hair, and nails.

Bone comprises a significant part of seafood processing wastes. Filleting of fish generates a large amount of fish frames. For instance, backbone wastes from processing of Atlantic cod (*Gadus morhua*) account for approximately 15% of the wet weight of the fish. Fish bone contains about 40% crude protein and 6% collagen, on a dry weight basis. Inorganic minerals constitute approximately 65% of bones, the remaining the extracellular organic matrix. The major ash components of fish bones are calcium and phosphorus (Kim et al., 1998; Gildberg et al., 2002). The mineral constituents can be classified into three groups, composed mainly of either hydroxyapatite (HAP), tricalcium phosphate (TCP), or a mixture of HAP and TCP. Sea bream, horse mackerel, carp, and shark have hydroxyapatite type phosphate (as in the case of cattle, swine, and fowl), while Japanese anchovy has TCP type phosphate. HAP-TCP types are ashes of sardine, mackerel, tilefish, croaker, triggerfish, lizardfish, Spanish mackerel, flying fish, conger eel, and flat fish (Hamada et al., 1995). Fish bone is a potential source of bioavailable calcium. In order to make calcium bioavailable, the bone should be converted into an edible form by softening its structure, by either hot water or hot acetic acid treatment. Superheated steam has been used to reduce the loss of soluble components from fish tissue, which enables better recovery of bone within a shorter period. The treated bones are subjected to saponification, degreasing, and degumming. The preparation, which is a source of dietary calcium, may also contain some phosphorus (Ishikawa et al., 1990) The skeleton discarded from industrial processing of hoki (*Johnius belengerii*) was digested by a heterogeneous enzyme extracted from the intestine of bluefin tuna (*Thunnus thynnus*), which yielded a fish bone oligophosphopeptide containing 23.6% phosphorus. The peptide (molecular weight of 3.5 kDa) could be a nutraceutical with a potential calcium-binding activity (Jung et al., 2005). Peak bone stone is the bony structure situated near the vertebral column of the dorsal fin base obtained from large fishes like ghol (*Pygoplites diacanthus*), koth (*Otolithes biauratusa*), and dara (*Filamanus heptadactyla*). About 900 g of peak bone stone is obtained from the above fish of an average weight of 12 kg, with a yield of 6–7%. From large fishes, up to 13 peak bone stones can be obtained. There is potential to use the product as raw material for calcium powder (Venugopal, 2008). Selenium is a component of antioxidant enzymes. Plant foods like rice and wheat are the major dietary sources of selenium in most countries. However, fish and shellfish are also rich sources of the mineral (Venugopal, 2006). Regular consumption of either seafood or fish-based mineral supplements can be effective in keeping bones healthy by delaying osteoporosis.

17.4.12 Guanine and Pearl Essence

Guanine (2-amino-6-oxypurin), found as a constituent in scale membranes, is responsible for the gleaming effect of fish. Guanine combines with collagen and calcium phosphate, yielding a silvery white thinning substance. The lustrous material can be extracted from fish scales like those of sardine, herring, ribbonfish, carp, etc. For extraction, the scales are first agitated with water to free the shining crystals. After setting, the clear supernatant water is decanted and the settled fluid is treated with pepsin in the presence of acetic acid at room temperature for 2 days. The deproteinized guanine crystals are then separated and extracted using ether. The crystals are drained off, washed with dry ether to remove fat, and stored in ether or ethyl acetate. The yield of the recovered material, often called pearl essence because of its luster, is about 0.3% of the body weight in the case of ribbon fish. Another important source of pearl essence is herring (Dey, 2008).

17.5 USES OF SEAFOOD COMPONENTS AS COSMECEUTICAL INGREDIENTS

Some of the cosmeceutical applications of major classes of seafood components as well as examples of some commercial products containing these ingredients will be pointed out below.

17.5.1 Proteins and Peptides

Protein ingredients make formulations milder and reduce irritation generally associated with chemical surfactants such as sodium lauryl sulfate. The effect is largely due to the ability of protein to form a protective coating to skin and hair. Fish plasma protein is classified as a skin conditioner. Cosmeceutically active peptides enhance collagen synthesis, chemotaxis, and protect skin from the effects of UV radiation. These peptides belong to three main categories: signal peptides, neurotransmitter-affecting peptides, and carrier peptides. Their bioactivities can be substantiated with chemically modified peptide sequences. Long-chain fatty acid conjugates of peptides and proteins improve skin penetration, specific activity, and economic feasibility (Peschard and Lintner, 2000; Lintner, 2007). Foaming polypeptides are stable at high pH and have widespread use as hair relaxers in hair dye formulations. A skin cream known as luminescence gold caviar serum containing peptides and marine DNA has been claimed to rejuvenate mature skin through delivery of recombinant epidermal growth factor. This product enhances repair of damaged skin. Another commercial product, Caviar LS 9747, which contains ingredients extracted from fish roe (caviar) in combination with actin-derived oligopeptides and an extract of the seaweed *Fucus vesiculosus*, has been found useful to revitalize and regenerate skin. The product has been claimed to nourish skin cells by enhancing the synthesis of fibroblast proteins, thereby boosting cell viability. It is used in antiaging and rejuvenating products, face and body care formulations (Anonymous, 2007b). Peptide CPC® is a polylysine peptide containing amino acids used for antiaging and also against sun tanning. It is claimed to enhance collagen synthesis (Anonymous, 2010b). Another peptide, Biopeptide CL™ produced by Sederma, France, is an antiwrinkle cream having a collagen fiber repairing effect. Biopeptide EL™, produced by the same manufacturer, contains elastin with dermal reconstruction properties and possesses firming and antiwrinkle action, with the ability to protect skin against UV rays (http://www.sederma.fr/Market-trends/peptides.pdf). Sea beet peptide is used in skin whitening, while sea fennel peptide has been developed as an antiblemish ingredient. Another product, sea lavender peptide, has been recognized as useful in antiaging and sun care formulations (Mathews, 2010). Carboxyethyl aminobutyric acid is used in hair scalp treatments, sensitive skin, and antiaging products, and in skin care and eye care. Although the benefits of peptide cosmetics may not be as rigorously tested as most FDA-regulated drugs, the evidence to support their uses is growing (Lupo and Cole, 2007).

A number of collagen and elastin-based products are commercially available (http://www.nutreeplus.com/antiaging.html). The products are based on collagen, elastin, and peptides derived from them. Nutripeptin™ (90% protein and less than 0.1% fat), a marine peptide shown to reduce the glycemic effect of foods, is manufactured by enzymatic hydrolysis of cod fish meat using peptidases (http://www.nutrimarine.com/?id=33). Collactive™ is a marine ingredient composed of collagen and elastin polypeptides present in the same ratio that is naturally present in skin. They have a synergistic antiwrinkle action. The product is soluble when used in beverages and can be used as dietary supplements for beauty. A marine collagen cosmeceutical product is Foam-Coll 4 CM (potassium cocoyl hydrolyzed collagen) derived from collagen from fish such as tuna, cod, etc. It has foaming and protective properties and is used as a texturizer, conditioning agent, and facial cleanser. The product is prepared by covalently binding hydrolyzed collagen with a reactive fatty acid reagent (Arch Personal Care Products, Brooks, http://www.in-cosmetics.com/ExhibitorLibrary/3/FoamingPolypeptides.pdf). More than 75% of squid skin is made of crude protein, and half of this is collagen. Collagen combined with the moisturizing properties of chitosan is claimed to help restore the elasticity of skin (Anonymous, 2010a). A commercial fish collagen, Solu-Mar, is used in general skin lotions and creams, hair shampoos, conditioners, and hair treatment products. It is claimed to possess moisturizing, film-forming, conditioning, and strengthening properties. Procollagen is a precursor of collagen, useful for hair and skin conditioning. Employing the culture of murine keratinocytes, it has been shown that collagen has better physiological effects than gelatin and collagen hydrolysate as skin care cosmetic materials (Li et al., 2005). Piscum Cartilage Extract and Milt Extract are skin conditioners, extracted from fish cartilage and fish milt, respectively. A cream that combines squalane and marine collagen has been found to infuse moisture and provide UV protection and antiaging. Maricol CLR™ is a commercial collagen preparation of native soluble collagen of tuna fish skin, useful for skin care. It possesses film-forming, moisturizing, and protective properties and is classified as antistatic, film forming, hair conditioning, humectant, and skin conditioning (http://www.cosmeticanalysis.com/cosmetic-ingredients/piscum-cartilage-extract.html). Ichtyocollagene™ is a commercial fish collagen used in creams or gels that possesses film-making and hydrating properties (http://www.sederma.fr/Market-trends/peptides.pdf).

A fish-based skin care product has shown promising results in the treatment of psoariasis and eczema. The cream contains enzymes and gelatin from fish, the important component being the enzyme zonase, found in fish eggs, which is responsible for the emergence of juveniles from the eggs. The enzyme can break down skin cells without damaging live cells. When used in psoriasis treatment, the cream helps dead skin to peel off and stimulates growth of new skin cells (Anonymous, 2006). Collagen is also used in tissue engineering and as an implant (Lee et al., 2001).

17.5.2 Chitin, Chitosan, and Derivatives

Chitin, chitosan, and their derivatives in combination with antioxidants, hyaluronic acid, peptides, and other immunomodulant compounds can address problems indicative of aging. Chitosan is known for its film-forming and antimicrobial properties. Because of this, it produces a fine, highly elastic film that minimizes moisture loss and protects against sun. It also increases skin flexibility/elasticity, smoothness, softness, and suppleness. Use of low concentrations of chitosan in skin care products offers moisturizing, film forming, and the ability to control flow characteristics. Chitosan at 0.05–0.1% has the potential to improve the dermatological compatibility of lipsticks, providing long-lasting protection against drying, besides giving suppleness and gloss. Chitin and chitosan generate chitooligomers glucosamine and N-acetyl-glucosamine, which are incorporated into the extracellular matrix of skin through phosphorylation (Morganti et al., 2010). Derivatives of chitin and chitosan that have applications as cosmetic ingredients include O-carboxymethylchitin and O-hydroxypropylchitin, carboxymethyl chitosan, chitosan acetate, adipate, glycolate, hydrolyzed chitosan, pyrrolidone carboxylic acid (PCA), hydroxyl methyl, hydroxyl propyl, formate, ascorbate, and salicylate of chitosan, among others. Chitosan PCA possesses moisturizing, protective, and curl retention properties, offering improved manageability and smooth appearance and luster to hair. Kytamer™ PC is a chitosan-pyrrolidone carboxylic acid. Aquatrix™ is a versatile commercial cosmetic hydrogel that contains carboxymethyl chitosan and other ingredients. It has applications in toiletries, oral care, hair conditioning, and skin care (Subasinghe, 1999). Other examples of chitin and chitosan products include Hydagen® HCMF, Hydagen® DCMF, Chitosan MM 222, and Curasan™ Ritachitosan LVHydagen® (http://www.specialchem4cosmetics.com). Recent developments in nanotechnology have contributed successful applications of chitin nanocrystals in cosmetics (Morganti et al., 2010).

Glucosamine and its derivatives, including acetate, ascorbate, salicylate, and sulfate, are widely used for skin care and hair conditioning. Glucosasmine HCl is an antistatic hair conditioner, which is also highly effective in skin lightening. N-acetyl glucosamine promotes production of hyaluronic acid. Commercial products include Trichogen® and Olevasan™ GL, a combination of glucosamine and niacinamide (http://www.specialchem4cosmetics.com).

Sodium chondroitin sulfate is classified as an antistatic hair and skin conditioner. Commercial products include Cromoist CS®, Ursolisome®, Marine® Juvenium, Marine Filling Spheres®, and Completech MBAC-OS. Cromoist CS contains sodium chondroitin sulfate and hydrolyzed collagen, which has moisturizing properties imparting increased skin extensibility and elasticity. It is used in mature skin treatments and nutrient creams, eye, neck, and bust creams, body toning products, night creams and face masks, and other applications. Ursolisome contains sodium chondritin sulfate and other ingredients, lecithin, ursolic acid, and xanthan gum. Marine Juvenium contains sodium chondroitin sulfate and other ingredients, including butylene glycol, magnesium ascorbyl phosphate, and xanthan gum, among others. It has applications in facial and body care for babies as well as adults. Hyaluronic acid is classified as antistatic, humectant, moisturizing, and skin conditioning. Some of the commercial products include Biocare™ Polymer HA-24, Biocare Polymer BHA-10, and Cromoist HYA (http://www.specialchem4cosmetics.com).

17.5.3 Marine Oils

Lipids, particularly glycolipids, are widely used as surfactants in cosmetics for their softening and smoothening effects on the skin. Sophorolipids, rhamnolipids, and mannosylerythritol lipids are the most widely used glycolipid biosurfactants in cosmetics (Lourith and Kanlayavattanakul, 2009). Fish oils and their hydrogenated counterparts are classified as emollients. Essential fatty acids are common ingredients in nutricosmetic formulations. PUFA is used as an ingredient in hair care cosmetics (shampoos, conditioners), skin care (facial care, facial cleansing, body care, baby care), and slimming preparations. Omega-3 fatty acids are known to provide significant skin care health benefits in a broad spectrum of topical applications, as well as soft tissue repair. Among the reported effects of omega-3 on skin are the stimulation of tissue repair and enhancement of collagen production. PUFAs, including DHA and EPA, are also used in lipstick. Encapsulated and nanoencapsulated omega-3 fatty acids are used in cosmetic formulations. The encapsulation technology protects the fatty acids against oxidation and, together with an optimized antioxidant mixture, keeps the fatty acids fresh. The nanoencapsulation technology, in addition, functions as a vehicle for delivering the omega-3 lipids into the skin. The nanocapsules can withstand high temperatures, making them suitable for use in skin care formulations. The efficacy of dietary n-3 PUFA docosahexaenoic acid (DHA) in patients with atopic eczema has been tested. In a randomized, double-blind, controlled trial employing 53 patients suffering from atopic eczema aged 18–40 years the patients received either 4–5 g DHA daily or an isoenergetic control of saturated fatty acids for 8 weeks. At periodic intervals clinical outcome was assessed in terms of atopic dermatitis index, IgE production, and activation of peripheral blood mononuclear cells (PBMCs). It was observed that DHA resulted in a significant clinical improvement of atopic eczema within 8 weeks (Koch et al., 2008). Nevique™ is a skin care product that employs the natural health benefits of pharmaceutical grade fish oil containing omega-3 fatty acids from North Atlantic fish. Euphoryl™, an omega-3 concentrate together with extract of pink pepper, has been claimed to stimulate the production of dopamine, which is known to improve microcirculation, inducing a healthy rise of skin color. It can be incorporated into a variety of products, such as face moisturizers, body care applications, and fragrance-free formulations (http://www.specialchem4cosmetics.com).

Deodorized and refined oil from the skin of the orange roughy, a deep-sea fish found in the South Pacific Ocean, has been reported to possess moisture-retaining properties, and therefore could be used as an ingredient in skin ointments (Anonymous, 2010b). Squalene is classified as an antistatic, emollient, and hair and skin conditioner. Technical products of squalene include Phytosqualene, Presperse Incorporated—Dermolane, Squalene EX, Lipogard Fitoderm®, and Nikkol Olive Squalane, among others (http://www.specialchem4cosmetics.com).

17.5.4 Vitamins as Antioxidants

Antioxidants function as free radical scavengers and therefore can function as antiaging ingredients positively influencing the appearance of visible signs of aging. These compounds are important in sustaining normal cell processes and promote cell growth. Antioxidants also control the oxidation of ingredients in a formulation. Tocopherol, for example, controls oxidative damage by protection of cellular membranes and biomolecules such as lipids, proteins, and nucleic acids. Topical application is shown to reduce the appearance of fine facial lines and wrinkles as well as increasing stratum corneum hydration and enhancing its water-binding capacity. In addition, α-tocopherol can reduce the age-dependent increase of collagenase activity, potentially delaying the progression of skin aging. The most important vitamins for use in topical cosmetics and nutricosmetics are vitamin A, which enhances skin elasticity, normalizes dry, photodamaged skin, and reduces wrinkles; vitamin C, an antioxidant and essential cofactor in the formation of collagen; vitamin D, required for normal differentiation and growth of skin cells; and vitamin E, an antioxidant that quenches free radicals and blocks lipid peroxidation, while also moisturizing the skin from within. Use of fish oils that are rich in vitamins A and D as cosmetic ingredients can be beneficial to derive the antioxidant properties of the vitamins.

17.5.5 Minerals

Mineral supplements such as calcium from fish bone waste can be used in cosmetics and also for fortification of diverse food products. In addition to calcium supplements, pointed out earlier, Phoscalim® is a wild fish-derived mineral extract that provides bioavailable calcium and phosphorus to promote optimal bone health. Apart from marine fishery products, deep-sea water, with its sodium chloride removed, is of commercial interest in Japan. Off-label claims are being made for improved health through the consumption of deep-sea water, including improvement of the immune function, relief of atopic dermatitis, prevention of certain cataracts, etc. The health benefits are claimed to be related to the minerals in the sea water and to the purity and quality of the deep-sea water sources. Phytomer, a leading marine-based cosmetics brand, uses freeze-dried sea water in its skin care products (Mathews, 2010; Bailey, 2010).

Combinations of marine ingredients have been shown to enhance functional properties in some cosmetic products. A dietary supplement called oxyprolane, launched by Sephora in France (www.nutreeplus.com/antiaging.html), claims to improve skin regeneration during the healing process after aesthetic surgery, and can also be used for antiaging. The formula contains amino acids and micropeptides from hydrolyzed fish proteins, caroteno-protein fish and krill oils rich in omega-3 fatty acids, and natural tocopherols (vitamin E). Crystalline guanine extracted from fish as pearl essence is used in the cosmetic industry as an adhesive in various products such as shampoos, and it provides a pearly effect and shimmering luster in eye shadow and nail polish. Guanine is highly prized for its iridescent qualities. However, recently it has been replaced by bismuth oxychloride, a synthetic ingredient that has comparable visual properties as guanine (Dey, 2008). Table 17.7 provides some examples of cosmeceutical products from fish and shellfish.

17.6 COMMERCIAL STATUS

The world cosmeceutical market is about US$20–30 billion. In India, the current organized market for antiaging creams is considered at around US$35 million, and growing at over 25% annually

TABLE 17.7
Some Cosmeceutical Products from Fish and Shellfish

Product	Ingredients and Sources	Functions
Active wrinkle cream, skin care products	Purified extract from male sturgeon gonads	UV protection and antiaging effects
Skin cream	Zonase enzyme and gelatin from fish eggs	Useful for treatment of psoariasis and eczema
	Fish oil	Zonase breaks down skin cells without damaging live cells
Body lotions	Cod sperm as water binder	Makeup, stimulates immune system
Various creams	Shark cartilage containing chondroitin sulfate	Joint flexibility, prevents osteoporosis
Various skin care products, shampoo, toothpaste	Chitosan from crustaceans and collagen from fish and mollusks	Ingredients function as gelling agents, moisturizers, and stimulate cell regeneration
Nourishing creams	Squalene from shark liver oil	Moisturizer
Nail polish	Herring, ribbon fish, etc.	

Source: Adapted from Proffitt, *Infofish Int.*, 6, 44, 2007; Dey, *Infofish Int.*, 1, 46, 2008.

(Sharma, 2010). Marine resources can have significant potential to provide interesting ingredients for the industry, as pointed out in this chapter. Developments in the techniques for isolation of various ingredients employing the tools of biotechnology and nanotechnology have elicited interesting applications of marine ingredients in nutricosmetics. According to a report published by Global Industry Analysts, advance in scientific disciplines such as molecular biology, bioinformatics, and genomics have also contributed to market growth. As a result, commercial revenue from marine biotechnology is projected to reach about US$4 billion by 2012. Nevertheless, the industry is still at an early stage of development and accounts for only a small percentage of the overall biotechnology market (www.bccresearch.com).

As pointed out in this chapter, in recent years an increasing number of new products are appearing that incorporate ingredients from the sea. These products use sea water and marine ingredients, including collagen, astaxanthin, caviar, pearls, and algal extracts (Mathews, 2010). A number of entrepreneurs like Aquapreneur, Strategro Int., New Jersey (www.aquapreneur.com), Sederma, Paris, France (http://www.sederma.fr), Copalis Le Portel, Boulogne, France (http://www.copalis.fr/), and Frutarom, Israel (www.frutarom.com) have developed dietary supplements and nutricosmetic ingredients, including various marine ingredients such as collagen, elastin, various peptides, glycosaminoglycans, alkyl glycerols, shark liver oil, etc. It must be pointed out that although there may be some reservation in the use of animal-derived products in cosmetic products, this could be overcome by future developments in finding ways to manipulate these ingredients and combining them with other synthetic or natural compounds that can create targeted results, particularly within the antiaging and anticellulite markets (Bird, 2008).

ACKNOWLEDGMENTS

Information on some commercial products was received from SpecialChem4Cosmetics (http://www.specialchem4cosmetics.com).

REFERENCES

Alasalvar, C., and Taylor, T., (2002), Ω-3 fatty acid concentrates—a review of production technologies, in *Seafoods—quality, technology and nutraceutical applications*, Springer-Verlag, Heidelberg, 157.

Allens, M. J., and Jaspars, M., (2009), Realizing the potential of marine biotechnology: challenges and opportunities, *Ind. Biotechnol.*, 5, 77–83.

Aneiros, A., and Garateox, A., (2004), Bioactive peptides from marine sources: pharmacological properties and isolation procedures, *J. Chromat. B*, 803, 41–45.

Anonymous, (2006), Fish based skin cream, *Infofish Int.*, 2, 71.

Anonymous, (2007a), Gelatin from cod skin, *Infofish Int.*, 5, 70.

Anonymous, (2007b), Wellness solutions with tangible benefits Special Chem-4, April 20, http://www.specialchem4cosmetics.com/services/news.aspx?id=1824&q=fish.

Anonymous, (2008), Cosmetic products from squid, *Infofish Int.*, 2, 73.

Anonymous, (2010a), Special Chem-4 cosmetics, http://www.specialchem4cosmetics.com/tds/peptide-cpc/solabia/8824/index.aspx.

Anonymous, (2010b), Skin moisturizer from fish oil, *Infofish Int.*, 4, 72.

Archer, M., (2001), Fish waste production in the United Kingdom: the quantities produced and opportunities for better utilization, Seafish Report SR 537, Seafish Industry Authority, Edinburg, Scotland.

Arvanitoyannis, I. S., and Kassaveti, A., (2008), Fish industry waste: treatments, environmental impacts, current and potential uses, *Int. J. Food Sci. Technol.*, 43, 726–745.

Batternfeind, J. C., (1977), The tocopherol contents of food and influencing factors, *Crit. Rev. Food Sci. Nutr.*, 9, 337–341.

Bailey, R., Marine products in Japan, http://nutraceuticalsworld.com/contents/view/13208.

Baziwane, D., and He, Q., (2003), Gelatin: the paramount food additive, *Food Rev. Int.*, 19, 423.

Bimbo, A. P., (2007), Current and future sources of raw materials for the long-chain omega-3 fatty acid market, *Lipid Technol.*, 19, 176–191.

Bird, K., (2008), Cosmetics industry drives growth in marine biotech, http://www.cosmeticsdesign.com/Products-Markets/Cosmetics-industry-drives-growth-in-marine-biotech.

Blanco, M., et al., (2007), Towards sustainable and efficient use of fishery resources: present and future trends, *Trends Food Sci. Technol.*, 18, 29.

Buche, P., Girard, P., Lamy, F., Mas-Chamberlin, C., Mondon, P., Scocci, S., and Lintner, K., (2002), Heat-stable enzymes from deep sea bacteria: a key tool for skin protection against UV-A induced free radicals, *IFSCC Magazine*, 5, 3, 195–200.

Buranudeen, F., and Rajadurai, P. N. R., (1986), Shark liver oil, *Infofish Marketing Dig.*, 1, 42.

Chakraborti, R., (2002), Carotenoprotein from tropical brown shrimp shell waste by enzymatic process, *Food Biotechnology*, 16, 81–86.

Chan, P., (1996), Effectiveness and safety of low dose pravastatin and squalene, alone and in combination in elderly patients with hypercholesterolemia, *J. Clin. Pharmacol.*, 36, 422–426.

Cho, S. M., Gu, Y. S., and Kim, S. B., (2005), Extracting optimization and physical properties of yellowfin tuna (*Thunnus albacares*) skin gelatin compared to mammalian gelatins, *Food Hydrocoll.*, 19, 221–224.

Choe, E., and Min, D. B., (2006), Chemistry and reactions of reactive oxygen species in foods, *Crit. Rev. Food Sci. Nutr.*, 46, 1–24.

Clemans, R., and Pressman, P., (2007), Oxidative stress: defense or disease, *Food Technol.*, 11.07, 17, www.ift.org.

Dallongeville, J., et al., (2003), Fish consumption is associated with lower heart rates, *Circulation*, 108, 820–824.

Damodaran, S., (1996), Amino acids, peptides, and proteins, in *Food chemistry*, ed. O. R. Fennema, Marcel Dekker, New York, 321.

Delgado, C. L., Wada, N., Rosegrant, M. W., Meijer, S., and Ahmed, M., (2003), *Outlook for fish: meeting global demand*, Vision 2020, World Fish Center, International Food Policy Research Institute, Malaysia.

Dey, V. K., (2008). The making of pearl essence, *Infofish Int.*, 1, 46.

Dillingham, S., (2010), Marine ingredients poised for accelerated development, www.insidecosmeceuticals.com/articles/2010/05/marine.ingredients.

FAO, (2008), *State of world fisheries and aquaculture*, Food and Agriculture Organization of the United Nations, Rome.

Faulkner, D. J., (2000), Highlights of marine natural products chemistry, 1972–1999, *Natural Products Rep.*, 17, 1–11.

Gildberg, A., (2004), Enzymes and bioactive peptides from fish waste related to fish silage, fish feed and fish sauce preparation, *J. Aquatic Food Prod. Technol.*, 13, 3–11.

Gildberg, A., Arnesen, J. A., and Carlehog, M., (2002), Utilization of cod backbone by biochemical fractionation, *Proc. Biochem.*, 38, 475–480.

Gogus, U., and Smith, C., (2010), N-3 omega fatty acids—a review of current knowledge, *Int. J. Food Sci. Technol.*, 45, 417–436.

Haard, N. F., (1998), Specialty enzymes from marine organisms, *Food Technol.*, 52 (7), 64–78.

Hamada, M., et al., (1995), Inorganic constituents of bone of fish, *Fish. Sci.* (Tokyo), 61, 517–521.

Hofman, K., Hall, B., and Avery, N., (2007), Fish and mammalian skin collagen—similarities and differences, paper presented at 29th Annual Seafood Science and Technology Society of the Americas Conference, Norfolk, VA, November 7–9.

Holanda, D. H., and Netto, F. M., (2006), Recovery of components from shrimp (*Xiphopenaeus kroyeri*) processing waste by enzymatic hydrolysis, *J. Food Sci.*, 71, C298–C302.

Huang, D., Ou, R., and Prior, R. L., (2005), the chemistry behind antioxidant capacity assays, *J. Agric. Food Chem.*, 53, 1841–1844.

Ishikawa, M., et al., (1990), Effect of vapor pressure on the rate of softening of fish bone by super-heated steam cooking, *Nippon Suisan Gakkaishi*, 56, 1687–1691.Chakrabarti, R., (2002), Carotenoprotein from tropical brown shrimp shell waste by enzymatic process, *Food Biotechnol.*, 16, 81–85.

Jo, J. H., et al., (2004), Optimization of skate (*Raja avirostris*) cartilage hydrolysis for the preparation of chondroitin sulfate, *Food Sci. Biotechnol.*, 13, 622–627.

Jongjareonrak A., et al., (2006), Skin gelatin from bigeye snapper and brownstripe red snapper: chemical compositions and effect of microbial transglutaminase on gel properties, *Food Hydrocoll.*, 20, 1216–1219.

Jung, W.-K., Park, P. J., Byun, H. G., and Moon, S.-H., (2005), Preparation of hoki (*Johnius belengeri*), bone oligophosphopeptide with a high affinity to calcium by carnivorus intestine crude proteinase, *Food Chem.*, 91, 333–330.

Kajimoto, O., Suguro, S., and Takahashi, T., (2001), Clinical effects of glucosamine hydrochloride diet for dry skin, *J. Jap. Soc. Food Sci. Technol.*, 48, 335–339.

Kim, J. S., and Park, J. W., (2004), Characterization of acid-soluble collagen from Pacific whiting surimi processing byproducts, *J. Food Sci.*, 69, 637–641.

Kim, J. S., Choi, J. D., and Koo, J. G., (1998), Component characteristics of fish bone as a food source, *Agric. Chem. Biotechnol.*, 41, 67–71.

Kim, S.-K., Dominic, Y., Ravichandran, D., Khan, S. B., and Tae Kim, Y. T., (2008), Prospective of the cosmeceuticals derived from marine organisms, *Biotechnol. Bioprocess Eng.*, 813, 511–523.

Kim, S. K., and Mendis, E., (2006), Bioactive compounds from marine processing by-products—a review, *Food Res. Int.* 39, 383–393.

Ko, T. F., Weng, T., M., and Chiou, R. Y., (2002), Squalene content and antioxidant activity of *Terminalia catappa* leaves and seeds, *J. Agric. Food Chem.*, 50, 5343–5346.

Koch, C., Dolle, S., Metzger, M., Rasche, C., Jungclas, H., Ruhi, R., Renz, H., and Worm, M., (2008), Docosahexaenoic acid (DHA) supplementation in atopic eczema: a randomized, double-blind, controlled trial, *Br. J. Dermatol.*, 158, 786–792.

Kris-Etherton, P. M., Harris, H., and Appel, L. J., (2002), Fish consumption, fish oil, ω-3 fatty acids and cardiovascular disease, *Circulation*, 106, 2747–2751.

Kwon-Kim, S., et al., (2001), Isolation and characterization of antioxidative peptides from gelatin hydrolyzate of Alaska pollock skin, *J. Agric. Food Chem.*, 49, 1984–1987.

Lamy, F., Mas Chamberlin, C., Mondon, P., Peschard, O., and Lintner, K, (2006), Strategies of antiaging actives in sunscreen products, *Cosmetics Toiletries*, 121, 6, 45–56.

Lands, W. E. M., (2005), *Fish, ω-3 and human health*, 2nd ed., CRC Press, Boca Raton, FL, 235.

Lee, C. H., Singla, Y. and Lee, Y., (2001), Biomedical applications of collagen, *Int. J. Pharm.* 221, 1–22.

Li, G. Y., Fukunaga, S., Takenouchi, K., and Nakamura, F., (2005), Comparative study of the physiological properties of collagen, gelatin and collagen hydrolysate as cosmetic materials, *Int. J. Cosmetic Sci.*, 27, 101–106.

Lichtenstein, A. H., and Russell, R. M., (2005), Essential nutrients: food or supplements? Where should the emphasis be? *JAMA*, 294, 351–355.

Lintner, K., (2007), Peptides, amino acids and proteins in skin care? *Cosmetics Toiletries*, 122, 10, 26–34.

Lourith, N., and Kanlayavattanakul, M., (2009), Natural surfactants used in cosmetics: glycolipids, *Int. J. Cosmetic Sci.*, 31, 255–272.

Lu, X., Chapman, K. W., and Regenstein, J. M., (1997), Characterization of several fish gelatins, in *Seafood safety, processing and biotechnology*, ed. F. Shahidi, Y. Jones, and D. D. Kitts, Technomic, Lancaster, PA, 18.

Lupo, M. P., and Cole, A. L., (2007), Cosmeceutical peptides, *Dermatol. Ther.*, 20, 343–349.

Madhu, B., et al., (2008), Effect on carotenoid pigments during enzymatic extraction and freeze drying of carotenoprotein from shrimp head waste, *Lebensm. Wiss. Technol.*, 41, 227–231.

Manjusha, K. P., (2011), Isolation and characterization of glucosaminoglycans and a study of its bioactive potential in two commercially important species of cephalopods PhD thesis, Cochin University of Science and Technology, Kerala, India.

Mathews, I., (2010), Marine based ingredients, http://www.specialchem4cosmetics.com/.

McAlindon, T. E., et al., (2000), Glucosamine and chondroitin for treatment of osteoarthritis: a systematic quality assessment and meta-analysis, *JAMA*, 283, 1469–1471.

Minihane, A. M., and Lovegrove, J. A., (2006), *Health benefits of polyunsaturated fatty acids (PUFAs) in improving the fat content of foods*, ed. C. Williams, Woodhead, Cambridge, 560.

Moneto, P., Gómez-Guillén, M. C., and Borderías, A. J., (1999), Functional characterisation of muscle and skin collagenous material from hake (*Merluccius merluccius* L.), *Food Chem.*, 65, 55–59.

Morganti, P., et al., (2010), Cosmetic science in skin aging: achieving the efficacy by the chitin nano-structured crystallites, *SOFW-J.*, 136, 14–24.

Morimura, S., et al., (1999), Development of an effective process and evaluation for utilization of collagen contained in livestock and fish waste, *Proc. Biochem.*, 37, 1403.

Nagai, T., Izumi, M., and Ishi, M., (2004), Fish scale collagen. Preparation and partial characterization, *Int. J. Food Sci. Technol.*, 39, 239–242.

Nan, I. V., (2004) The osteoporosis prevention diet, *Food Science Central*, November 29, 2010, http://www.foodsciencecentral.com/

No, H. K., and Meyers, S. P., (1999), Preparation and characterization of chitin and chitosan—A review, *J. Aquatic Food Prod. Technol.*, 4, 27–41.

Ogawa, M., Portier, R. J., Moody, M. W., Bell, J., Schexnayder, M. A., Jack, N., and Losso, J. N., (2004), Biochemical properties of bone and scale collagens isolated from the subtropical fish black drum (*Pogonia cromis*) and sheepshead seabream (*Archosargus probatocephalus*), *Food Chem.*, 88, 495–501.

Peschard, O., and Lintner, K., (2000), Biologically active peptides: from a laboratory bench curiosity to a functional skin care product, *Int. J. Cosmetic Sci.*, 22, 207–218.

Philanto, A., and Korhonen, H., (2003), Bioactive peptides and proteins, *Adv. Food Nutr. Res.*, 47, 175–182.

Proffitt, F., (2007), A surprising use of fish waste, *Infofish Int.*, 6, 44.

Sachindra, N. M., Bhaskar, N., and Mahendrakar, N. S., (2005), Carotenoids in different body components of Indian shrimps, *J. Sci. Food Agric.*, 85, 67–71.

Santhanam, A., (2010), *Skin deep*, Harpercollins, Noida India.

Satterlee, C. R., (2004), Farmed salmon: caught in a number game, *Food Technol.*, 58(2), 108.

Shahidi, F., and Abuzaytoon, R., (2005), Chitin, chitosan and co-products: chemistry, production, applications and health effects, *Adv. Food Nutr. Res.*, 49, 94–112.

Shahidi, F., Metusalach, P., and Brown, J. A., (1998), Carotenoid pigments in seafoods and aquaculture, *Crit. Rev. Food Sci. Nutr.*, 38, 1–35.

Sharma, S., (2010) Cosmetics and toiletries industry in India, November 25, 2010. *Daily News and Analysis (DNA)*.

Subasinghe, S., (1999), Chitin from shell waste—health benefits overshadowing industrial benefits, *Infofish Int.*, 3, 58–62.

Sultanbawa, Y., and Aksnes, A., (2003), Tuna process waste—an unexploited resource, *Infofish Int.*, 3, 37–41.

Synowiecki, J., and Al-Khateeb, N. A., (2003), Production, properties, and some new applications of chitin and its derivatives, *Crit. Rev. Food Sci. Nutr.*, 43, 145–150.

Thankappan, T. K., (2002), Isolation of squalene from shark liver oil, in *Seafood safety*, ed. P. K. Surendran et al., Society of Fisheries Technologists, Cochin, India, 173.

Tharanathan, R. N., and Kuttur, F. S., (2003), Chitin—the undisputed biomolecule of great potential, *Crit. Rev. Food Sci. Nutr.*, 43, 61–70.

Venugopal, V., (2006), Influence of processing on nutrition, in *Seafood processing: adding value through quick freezing, retortable packaging and cook-chilling*, CRC Press, Boca Raton, FL, chap. 15.

Venugopal, V., (2008), Marine sources of vitamins and minerals, in *Marine Products for Healthcare*, chap. 8, p. 527, CRC Press, Boca Raton, FL.

Wang, T.-W., et al., (2006), The effect of gelatin-chondroitin sulfate-hyaluronic acid skin substitute on wound healing in SCID mice, *Biomaterials*, 27, 5689–5602.

Ward, K. B., Wang, Z., and Xu, S., (2001), Gelatin: a valuable protein for food and pharmaceutical industries, *Crit. Rev. Food Sci. Nutr.*, 41, 481–484.

Volkman, J. K., (1999), Australasian research on marine natural products: chemistry, bioactivity and ecology, *Marine Freshwater Res.*, 50, 761–771.

Willcox, J. K., Ash, S. L., and Catignani, G. L., (2004), Antioxidants and prevention of chronic disease, *Crit. Rev. Food Sci. Nutr.*, 44, 275–290.

Zhou, P., Mulvaney, S. J., and Regenstein, J. M., (2006), Properties of Alaska pollock skin gelatin: a comparison with tilapia and pork skin gelatins, *J. Food Sci.*, 71, C313, 2006.

18 Unlocking the Power of Marine Cosmeceuticals for Wrinkle-Free Skin

Dinanath B. Fulse, Spandana R. Kopalli, and Sushruta Koppula

CONTENTS

18.1 INTRODUCTION

The oceans encompass about 71% of the surface of our planet, and approximately 90% of the total global species live in the marine habitat (Bob, 2003). Marine ecosystems are complex and dynamic. Marine cosmeceuticals, currently a buzzword in the industry, describe beauty care products containing active pharmaceutical ingredients derived from marine sources that show physiological actions in the treatment of aging skin. Chronological skin aging, often called natural aging, is a universal and inevitable process (Ravichandran et al., 2005). Both men and women want to erase the ravages of time and restore their skin to its younger days. This is because of the never-ending quest for the perfect, wrinkle-free complexion and prolonged interest in eternal youth with minimum signs of aging in this beauty-conscious world. Thousands of antiwrinkle creams have flooded the market due to the increased desire of people to maintain the youthful appearance of their skin. The common ingredients used in antiwrinkle creams are retinol, hydroxyl acids, coenzyme Q10, copper peptides, kinetin, and tea extracts. Although many cosmeceutical products have been in existence

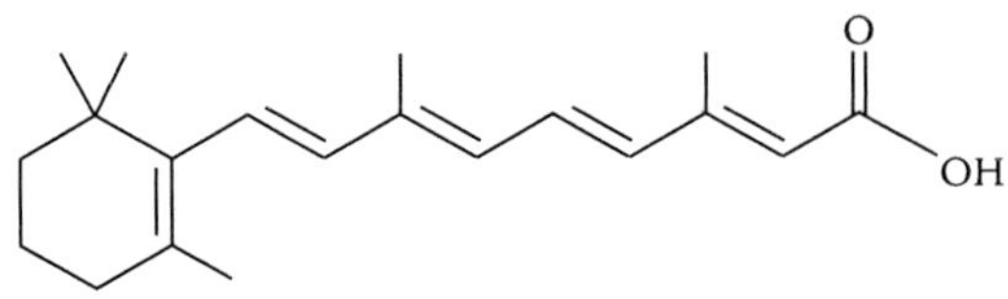

FIGURE 18.1 Chemical structure of tretinoin.

for centuries, the focus has always been given to products derived from marine sources claiming antiwrinkle effects. Bioactive substances derived from marine sources have diverse functional roles as a secondary metabolite, and these properties can be applied to the development of novel antiwrinkle cosmeceutical products. In this chapter, a thorough review of the existing and emerging agents used in the cosmeceutical industry for their role in producing wrinkle-free skin is discussed.

18.2 COMMON INGREDIENTS USED IN ANTIWRINKLE CREAMS

18.2.1 Retinol

Retinol is a component of vitamin A, which is widely used as a potent antioxidant and as an over-the-counter antiwrinkle cream (Reza et al., 2007). Free radicals generated by the unstable oxygen molecules that tend to break down the skin cells, further causing wrinkles, can be neutralized by using antioxidants. Tretinoin, a vitamin A derivative, was reported to be more potent than retinol (Kligman and Draelos, 2004) and was used as a topical application that was approved by the Food and Drug Administration (FDA) for treating wrinkles (Figure 18.1). Due to its higher potential effect than retinol, and to avoid possible side effects, tretinoin was made available only by prescription. Tretinoin works by removing dead skin cells from the surface and stimulating collagen production to form new skin. Application twice a week should produce positive results within 2 to 6 months.

Although retinol seems beneficial as an antiwrinkle ingredient, care should be taken to avoid its use during pregnancy, because it may increase the risk of birth defects in offspring.

18.2.2 Hydroxy Acids

Alpha hydroxy acids, beta hydroxy acids, and polyhydroxy acids are synthetic versions of acids derived from sugar-containing fruits. These acids are exfoliants—substances that remove the upper layer of old, dead skin and stimulate the growth of smooth, evenly pigmented new skin (Yaar and Gilchrest, 2001).

Alpha hydroxy acids may even stimulate the production of collagen and elastin. Alpha hydroxy acids are reported to improve wrinkling and roughness of skin after months of daily application. Because hydroxy acids increase susceptibility to sun damage, one should always wear sunscreen during use and for at least 1 week afterward (Lockman and Lockman, 2002).

18.2.3 Coenzyme Q10

Coenzyme Q10 is a nutrient that helps regulate energy production in cells. Some studies have shown reduction in fine wrinkles around the eyes with no side effects. Other studies show that application before sun exposure protects against sun damage (Prahl et al., 2008).

18.2.4 Copper Peptides

Copper is a trace element found in every cell. In products applied to the skin, which are combined with small protein fragments called peptides, copper peptides enhance wound healing. They

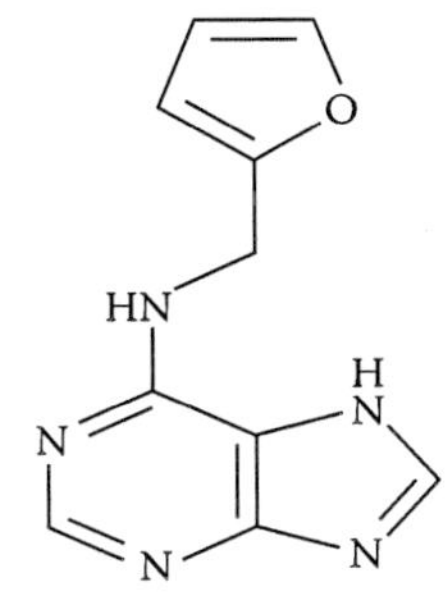

FIGURE 18.2 Chemical structure of kinetin.

also stimulate production of collagen and may enhance the action of antioxidants (Gorouhi and Maibach, 2009).

18.2.5 Kinetin

As a plant growth factor, kinetin may improve the appearance of wrinkles and uneven pigmentation with minimal irritation (Figure 18.2). It is unclear how it works, but it may help reduce wrinkles by helping skin retain moisture and by stimulating the production of collagen. It may also be a potent antioxidant (Rattan, 2002).

18.2.6 Tea Extracts

Green, black, and oolong teas contain compounds with antioxidant and anti-inflammatory properties. Green tea extracts are the ones most commonly found in wrinkle creams (Katiyar and Elmets, 2001).

18.3 MARINE-DERIVED AGENTS USED IN WRINKLE-FREE SKIN PRODUCTS

As the ocean is home to thousands of seaweed and algae species, one can expect a high tide of marine-based beauty products. Some of the important marine ingredients used for wrinkle-free skin are discussed here.

18.3.1 Astaxanthin

Astaxanthin is a carotenoid and belongs to a larger class of phytochemicals known as terpenes. It is classified as a xanthophyll ("yellow leaves"), which is a colorful, fat/oil-soluble pigment. Astaxanthin can be found in microalgae, yeast, salmon, trout, krill, shrimp, crayfish, crustaceans, and the feathers of some birds. Currently, the primary natural source for astaxanthin is *Haematococcus pluvialis* (microalgae).

UV light, smoke, pollution, and even oxygen radicals or free radicals damage the by-products, which make the skin appear older than it actually is (Suganuma et al., 2010). Closer examination reveals that these factors cause reactive oxygen species (ROS) to build up in skin and inflict damage at a cellular level. Astaxanthin (Figure 18.3) is the world's most powerful antioxidant that acts effectively against these environmental stressors. Astaxanthin works beneath the skin's surface and has a clinically documented antiwrinkle effect. This antioxidant scavenges free radicals and binds to them before they can do skin damage and makes skin wrinkle free (Dean et al., 2004).

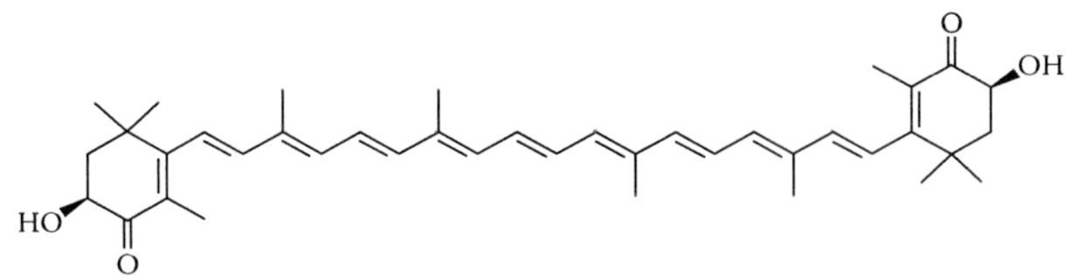

FIGURE 18.3 Chemical structure of astaxanthin.

18.3.2 L-Hydroxyproline

This amino acid is one of the important compounds used in cosmeceuticals, mostly collected from marine sources like fish and nonmarine sources. L-hydroxyproline also protects skin from free radicals caused by UV light radiation. The collagen builders include an exciting new nutrient called L-hydroxyproline (Figure 18.4), along with other nutrients, which may help reduce the appearance of fine lines and wrinkles on skin. The effects are attributed not only to its superior moisture retention, but also to the physiological activities of L-hydroxyproline, including stimulation of collagen synthesis by fibroblast cells, stabilization of collagen molecules, and normalization of epidermal cells by improving the health of the basal layer. Combining these ingredients with vitamins A, B_5, C, and E, and other antioxidants, in conjunction with topical use of UV filters, can create synergy and further increase the efficacy of cosmetics and skin care products. When an effective topical skin care regimen is supported by dietary supplementation, even greater improvements in skin moisture content and a decrease in fine lines and wrinkles can be achieved (Lik, 2007).

18.3.3 Antarcticine Glycoprotein

A glycoprotein is a macromolecule composed of a protein and a carbohydrate. Antarcticine is a combination of proteins (86%) and sugars (14%) produced by a recently discovered organism in a glacier in Antarctica, which is vital for the regular function of cells. It has been demonstrated to have many unique benefits when used in skin care formulas. With superb protection against dryness, antarcticine helps hydrate skin and provides a noticeably smoother appearance while shielding skin from the damaging effects of free radicals. It further helps to protect skin from extremely cold temperatures, facilitating skin's retention of natural water reserves. The organic substance thrives in the most extreme environmental conditions as a result of the cryoprotective (protection against freezing) properties of the molecule. The greatest advantage of antarcticine is its ability to cohere to the upper skin layers.

Antarcticine has great antiwrinkle effect, promotes the formation of both type I and type IV collagen, increases elastin formation, and promotes the formation of these proteins, showing visible antiwrinkle activities. This compound is mostly used in Kiehl's Ultra Facial Cream and Biotoxyl eye cream (Cosmetics.Pro, 2011).

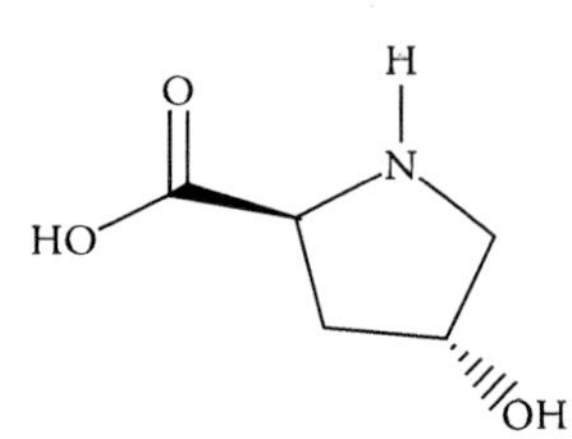

FIGURE 18.4 Chemical structure of L-hydroxyproline.

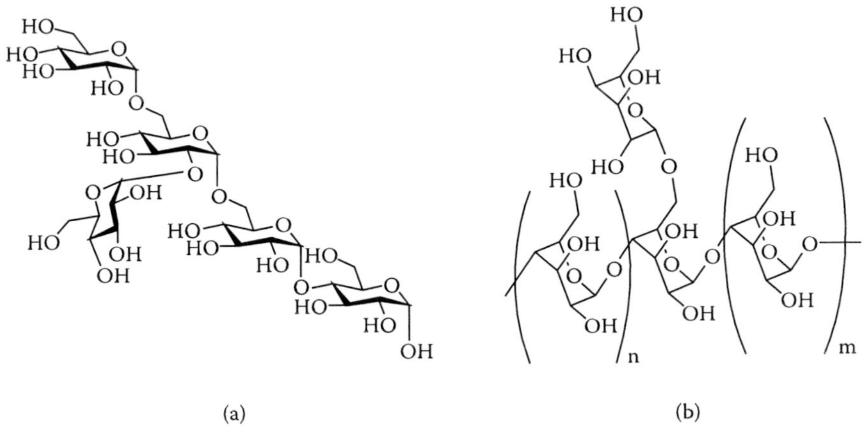

FIGURE 18.5 Chemical structures of (a) pentasaccharide and (b) galactomannan.

18.3.4 Homeostatine

Homeostatine is a novel hydrocolloidal three-dimensional matrix of galactomannans of Andean origin, derived from the seeds of *Caesalpinia spinosa* (Mol.) O. Kuntze, which sequentially releases a marine pentasaccharide derived from the algae *Enteromorpha compressa* (L.) Nees. This product helps recover and maintain homeostasis in the extracellular matrix, which results in the reduction of wrinkles.

Enteromorpha compressa is a member of the class Chlorophyceae, also known as green algae, which is the largest algae group. The chemical composition of *Enteromorpha* includes carbohydrates (48%), proteins (10–18%), and lipids (0.5–1.7%). Homeostatine activity is due to the combined action of its components: marine pentasaccharide (MPS) from *E. compressa* and galactomannan from *C. spinosa* (Figure 18.5). The action of homeostatine is aimed at preventing skin wrinkles.

Homeostatine increases the production of dermal collagen and other ECM components in fibroblasts. It inhibits the synthesis of metalloproteinases (MMPs) and pro-inflammatory mediators. Homeostatine appears to be a powerful inhibitor (35–40%) of IL-1α-induced stimulation. This action reduces the production of metalloproteinases by fibroblasts, which in turn could reduce the degradation of skin collagen and other extracellular matrix components. In this way, the formation of wrinkles may be prevented, and furthermore, reducing the production of cytokines impairs the amplification of inflammatory responses (Provital Group, 2006).

18.3.5 Sporopollenin

Sporopollenin is a component of special types of algae, which is also found in small amounts in the spores and pollen of plants such as ferns, club moss, and flowers. The production of sporopollenin in the form of microcapsules is described in German Preliminary Published Application 199 02 724 (Maack, 2007). Sporopollenin can be used in the form of dispersion but not as a solution in the mechanical support phase for the encapsulation of substances and cells. Moreover, sporopollenin, in its untreated condition, is tightly built into the cell walls of the alga and is not available as a separate biocomponent. However, by employing a special extraction process, it is possible to isolate, enrich, and preserve. Sporopollenin is an acetolysis-resistant inert biopolymer possessing a complicated structure with aliphatic (mainly isoprenoid) and aromatic components variably present.

This compound is used as an antimicrobial pharmaceutical composition for treating a microbial disease and heavy metals poisoning. It is a cosmetic composition for treating age-related or sunlight-induced wrinkle formation, and also a nutritional supplement for vertebrates and a supplemented

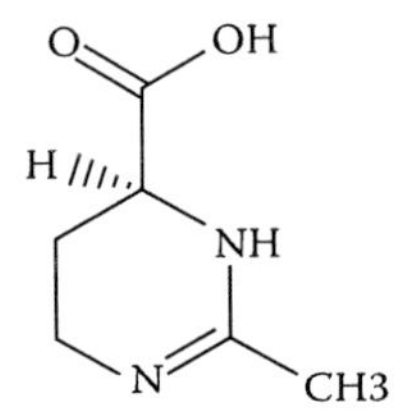

FIGURE 18.6 Chemical structure of ectoine.

beverage. Sporopollenin is able to bind various substances such as heavy metals, toxins, pesticides etc. Sporopollenin, manufactured by Kenrico in Japan, is the only natural ingredient in the world that can significantly reduce heavy metals, skin wrinkles, pain, stress, and depression (Dionisio-Sese, 2010).

Other marine ingredients used in antiwrinkle products are from brown seaweed, rich in vitamin A, and may aid in cell renewal. Chlorella, nutrient-packed green microalgae, restores the skin's pH balance and also firms skin by protecting collagen and elastin fibers against the enzymes, collagenase, and elastase that degrade the skin. Ectoine (Figure 18.6) is becoming a new trend in antiwrinkle products found in some marine extremophiles like bacteria.

Ectoine is emerging as an active ingredient in skin care and sun protection products. It is supposed to stabilize proteins and other cellular structures and protects the skin from stresses like UV irradiation and dryness. Research is being conducted into the antiaging effects of ectoine. Because of its outstanding water-binding activity, ectoine might be especially useful in preventing water loss in dry atopic skin and in recovering skin viability, thereby preventing skin aging via the elimination of wrinkles (Graf et al., 2008).

18.4 CONCLUSION

Thousands of antiwrinkle creams have flooded the market with the people's increased desire to maintain the youthful appearance of their skin over time. The ocean being a treasure trove of mineral-rich botanicals, it should be further explored to promote youthful and wrinkle-free skin. Green seaweed has collagen-encouraging properties that bring smoothening effects to the skin and reduce dark under-eye circles. Red algae have potential anti-inflammatory principles and fight free radicals responsible for sagging and mottled skin. Further, because of concerns over the side effects of various synthetic ingredients, research shows that cosmeceuticals derived from marine sources have increased greatly in popularity for their beneficial effects on skin. Despite the fact that many new species of marine life have still to be discovered, there is an optimistic view on the future potential for marine ingredients within the personal care and beauty markets. The potential for new marine-derived compounds and extracts is immense, and the cosmeceutical industry is poised for accelerated development in the near future. It is envisioned that continued exploration and research will lead to many organisms with unusual structures and compounds with great promise in the cosmetic world.

REFERENCES

Dean, A.F., Erik, W.K., Juris, L.E., Sean, O.M., Samuel, F.L. 2004. The efficient synthesis of disodium disuccinate astaxanthin (Cardax). *Org. Proc. Res. Dev.* 8: 796–801.

Dionisio-Sese, M.L. 2010. Aquatic microalgae as potential sources of UV-screening compounds. *Philippine J. Sci.* 1: 5–16.

Drogin, B. (October 24, 2003). Census of marine life maps an ocean of species. *Los Angeles Times*. http://www.latimes.com/new/la-na-fish2-2009aug02,0,5785256.story?page=1&track=ntothtml (retrieved August 18, 2009).

Gorouhi, F., Maibach, H.I. 2009. Role of topical peptides in preventing or treating aged skin. *Int. J. Cosmet. Sci.* 31: 327–345.

Graf, R., Anzali, S., Buenger, J., Pfluecker, F., Driller, H. 2008. The multifunctional role of ectoine as a natural cell protectant. *Clin. Dermatol.* 26: 326–333.

Katiyar, S.K., Elmets, C.A. 2001. Green tea polyphenolic antioxidants and skin photoprotection. *Int. J. Oncol.* 18: 1307–1313.

Kligman, D.E., Draelos, Z.D. 2004. High-strength tretinoin for rapid retinization of photoaged facial skin. *Dermatol. Surg.* 30: 864–866.

Lik, T.T. 2007. Bioactive natural products from marine cyanobacteria for drug discovery. *Phytochemistry* 68: 954–979.

Lockman, A.R., Lockman, D.W. 2002. Skin changes in the maturing woman. *Clin. Fam. Pract.* 4: 113–114.

Maack, A. 2007. Soluble composition containing sporopollenin and the use thereof. US Patent WO03/094942.

Prahl, S., Kueper, T., Biernoth, T., Wöhrmann, Y., Münster, A. 2008. Aging skin is functionally anaerobic: Importance of coenzyme Q10 for anti aging skin care. *Biofactors* 32: 245–255.

Provital Group. 2006. http://www.lotioncrafter.com/pdf/tech_lit_homeostatine.pdf.

Rattan, S.I.S. 2002. N^6-Furfuryladenine (kinetin) as a potential anti-aging molecule. *J. Anti-Aging Med.* 5: 113–116.

Ravichandran, G., Shivaram, B.V., Kolhapure, S.A. 2005. Evaluation of the efficacy and safety of "anti-wrinkle cream" in the treatment of facial skin wrinkles: A prospective, open, phase III clinical trial. *Antiseptic* 102: 65–70.

Reza, K., Kwak, H.S.R., Schumacher, W.E., Cho, S., Hanft, V.N., Hamilton, T.A., King, A.L., Neal, J.D., Varani, J., Fisher, G.J., Voorhees, J.J., Kang, S. 2007. Improvement of naturally aged skin with vitamin A (retinol). *Arch. Dermatol.* 143:606–612.

Skin Artica. Cosmetics.Pro. 2011. http://cosmetics.pro/skinarctica.html

Suganuma, K., Nakajima, H., Ohtsuki, M., Imokawa, G. 2010. Astaxanthin attenuates the UVA-induced up-regulation of matrix-metalloproteinase-1 and skin fibroblast elastase in human dermal fibroblasts. *J. Dermatol. Sci.* 58: 136–142.

Yaar, M., Gilchrest, B.A. 2001. Skin aging: Postulated mechanisms and consequent changes in structure and function. *Clin. Geriat. Med.* 17: 617–630.

19 Sea Water and Sea Mud
Cosmeceutical Applications

Se-Kwon Kim, Jayachandran Venkatesan, and P.N. Sudha

CONTENTS

19.1 INTRODUCTION TO SEA WATER AND SEA MUD

From ancient days, sea water and sea mud have been widely used for various applications, including fishery, algae aquaculture, medicine, food, drinks, cosmetics, and other economic value-added products (Liu et al. 2008). Cosmeceutical products often consist of water, which comes from different sources such as sea, river, flowers, and plants. Water from sea sources is promising and most widely used for cosmeceutical applications due to its rich mineral substances and trace metals. Recent research studies reveal that sea water and sea mud are useful for treating various skin diseases, such as psoriasis, acne, wrinkles, and photoaging effects.

Sea water used as a cosmetic agent is 100% natural with no side effects, fragrance-free, oil-free, alcohol-free, and preservative-free (Soto 2008). The minerals of sea water play a significant role in treating the various skin diseases. Sea water is highly effective on inflammatory skin diseases due to its constituent minerals, such as sodium, potassium, magnesium, strontium, and selenium (Lee and Maibach 2004, 2005).

19.2 COSMECEUTICALS FROM SEA WATER

The deep-sea water industry is characterized by rich mineral substances and trace metals suitable for multiapplication development. Deep ocean water is commonly referred to as the water below the photic zone and is typically cold, clean, stable, and rich in nutrients. Japan and the United States have been exploiting deep ocean water for utilization in pharmaceutical, cosmetic, and fishery industries for many years (Liu et al. 2008).

19.3 CONSTITUENTS OF SEA WATER MINERALS AND THEIR FUNCTIONS

Sea water is mainly composed of chloride, sodium, magnesium, calcium, potassium, and sulfate, which are present in the ionic state. The important constituents of sea water are shown in Table 19.1.

Minerals cannot be produced within the body; they need to be consumed through food, drink, or other means. Dead Sea water consists of naturally stable, purified, and enriched essential minerals, including sodium, magnesium, calcium, potassium, lithium, and rubidium. These elements play a necessary role in human body skin metabolism (Ma'Or et al. 1997). The Dead Sea is composed of organic and inorganic portions; the overall salt content of the Dead Sea is 320 g/L, and the average mineral salt content is carbonate, 23.20%; calcium, 20.40%; magnesium, 4.90%; chloride, 7.80%; sulfate, 7.80%; sodium, 1.70%; and potassium, 1.30%. Another mineral-rich constituent of the Dead

TABLE 19.1
Composition of Sea Water at 3.5% Salinity

S. No.	Element	Symbol	Atomic Weight	ppm[a]
1	Oxygen	O	15.9994	883,000
2	Hydrogen	H	1.00797	110,000
3	Chloride	Cl	35.453	19,400
4	Sodium	Na	22.9898	10,800
5	Magnesium	Mg	24.312	1,290
6	Sulfur	S	32.064	904
7	Calcium	Ca	40.08	411
8	Potassium	K	39.102	392
9	Bromine	Br	79.909	67.3
10	Carbon	C	12.011	28
11	Nitrogen	N	14.007	15.5
12	Fluorine	F	18.998	13
13	Strontium	Sr	87.62	8.1
14	Boron	Br	10.811	4.45
15	Silicon	Si	28.086	2.9
16	Argon	Ar	39.948	0.45
17	Lithium	Li	6.939	0.17
18	Rubidium	Rb	85.47	0.12

Source: From Turekian, K.K., *Oceans*, Prentice-Hall, Englewood Cliffs, NJ, 1968.

[a] Other elements are <0.1 ppm (mg/L).

TABLE 19.2
General Functions of Various Sea Water Minerals

S. No.	Element	General Function in Skin
1	Chloride	Vital for proper cell metabolism
2	Sodium	Helps fight free radicals formed in the skin; mitigates skin aging
3	Magnesium	Increases the production of antioxidants; insufficient levels of manganese may lead to dermatitis
4	Sulfur	Important for collagen synthesis and for healthy skin keratinization
5	Calcium	Strengthens cell membranes; aids skin cell metabolism; essential for healing wounds and preventing infection; helps cleanse pores
6	Potassium	Regulates the transfer of nutrients through cell membranes; helps prevent the formation of free radicals; helps prevent acne
7	Bromine	Stimulates the skin's natural repair process; acts as a natural antibiotic; helps relieve the discomfort associated with skin disorders
8	Boron	Plays a crucial role in maintaining transmembrane functionality
9	Fluorine	Promotes healthy skin and nails
10	Strontium	Plays a role in connective tissue formation, reduces sensory irritation such as stinging, burning, itching, and inflammation
11	Silicon	Important for maintaining proper skin thickness and strength, as well as for the production of collagen

Source: Obey Your Body, http://www.obeyyourbody.com/Catalog_47.html.

Sea is black mud, which is generally rich in organic portions and is also known as bituminous tar (Matz et al. 2003).

The skin-smoothing treatment can be very efficiently performed with 1% Dead Sea mineral solution, without mineral additives, and control gel without antiwrinkle agents (Ma'Or et al. 1997). Treatment of natural mineral-containing amorphous silica, aluminum oxide, ferric oxide, titanium oxide, calcium oxide, magnesium oxide, potassium oxide, sodium, and phosphorus oxide on human skin cells with keratinocytes after UV irradiation significantly decreased MMP-1, MMP-2, and IL-2 secretion and increased cell growth and laminin expression (Jung et al. 2009). The common minerals and their general role in skin treatment are explained in Table 19.2

19.3.1 Sodium

Sodium is present in the form of sodium chloride (NaCl) in a major portion of sea water, and present as 10,800 ppm level of water in the fourth position. The combination of sodium and chloride is a major constituent of water next to oxygen and hydrogen. The effect of sea water may be attributed to skin barrier preservation by sodium chloride and potassium chloride, and an emollient effect by only sodium chloride (Yoshizawa et al. 2001). Mineral water with 250 mM sodium chloride and 50 mM potassium chloride was the most effective for preventing disruption of the skin barrier and stratum corneum water content after cumulating irritation by sodium lauryl sulfate (Yoshizawa et al. 2003). Moreover, acute irritant contact dermatitis is frequently treated with cool water or saline compresses, as a result of which significant reduction has been found in transepidermal water loss, with no statistically significant difference between the efficacies of both waters (Levin and Maibach 2001). In mice challenged with 1-chloro-2,4-dinitrobenzene (DNCB) in combination with sodium chloride (14%) there was no statistically significant difference in the degree of ear swelling (Greiner and Diezel 1990).

19.3.2 Magnesium

Current research studies suggest that magnesium-containing sea water solution has a great beneficial effect on various skin diseases such as psoriasis, wrinkles, and irritant contact dermatitis. Magnesium bromide and magnesium chloride inhibit the well-known excessive proliferation of psoriatic keratinocytes (Levi-Schaffer et al. 1996). Bathing in sea water containing magnesium chloride has significantly increased skin hydration and reduced skin roughness, redness, and inflammation of the skin compared to tap water control (Proksch et al. 2005). *In vitro* and *in vivo* studies reveal that magnesium ions significantly inhibit the antigen-presenting function of human epidermal langerhans cells (Schempp et al. 2000).

The influence of magnesium ions on inflammation in allergic contact dermatitis was reduced in DNCB in BALB/c-induced mice. Animals challenged with 0.125% DNCB in the presence of magnesium chloride demonstrated significantly less pronounced contact dermatitis (ear swelling) than did animals challenged with DNCB alone (28 and 14%) (Greiner and Diezel 1990). However, a low concentration of magnesium salt compresses showed no significant effect on barrier disruption or dryness in the irritant (Yoshizawa et al. 2001).

19.3.3 Calcium

Calcium is the most abundant element in the human body when compared to other elements. It's normally present in bone, tissue, and blood. The majority is present in bone. The function of calcium for skin treatment is obscure. Sea water contains 411 ppm calcium.

19.3.4 Potassium

The role of potassium ions for the skin treatment from sea water remains obscure. However, a few experiments explain that a minimum concentration of potassium salt is useful for skin treatment. For example, 10 mM potassium chloride may be attributed to the prevention of skin barrier (Yoshizawa et al. 2003). The effect of sea water may be attributed to skin barrier preservation by potassium chloride (Yoshizawa et al. 2001). Potassium bromide solution in diluted Dead Sea brine reversibly inhibits cell proliferation in *in vitro* cell culture systems. Potassium bromide is more powerful as an inhibitor than sodium and magnesium chlorides (Shani et al. 1987).

19.3.5 Strontium

Strontium is present in sea water to the extent of 8.1 ppm. Strontium-related compounds have been proven for anti-inflammatory properties (Jellum et al. 2004, 2011). Strontium salts to intact skin produced potent suppression of sting, burn, and itch caused by many irritant chemicals (Hahn 1999). Strontium nitrate may act as a topical antipruritic agent in reducing histamine-mediated itch (Zhai et al. 2000). Strontium contributes to stabilization of the calcium phosphate structure in the bone and pointed out that osteoporosis is the lack of not only calcium in the bone but also phosphorus, silicon, magnesium, fluoride, and strontium, the other constituents of bone.

19.4 COSMECEUTICALS FROM SEA MUD

Sea mud contains various nutrients and minerals, and even antibacterial components. Therefore sea mud has been used for making skin care and cosmeceutical products. Sea mud has been added to various cosmetic products due to its beneficial effects. Dead Sea mud is especially known for its therapeutic properties for psoriasis and other skin-related disorders. Thus several beauty products are based on this mud. Mud therapy has a beneficial effect for skin due to the following reasons: it

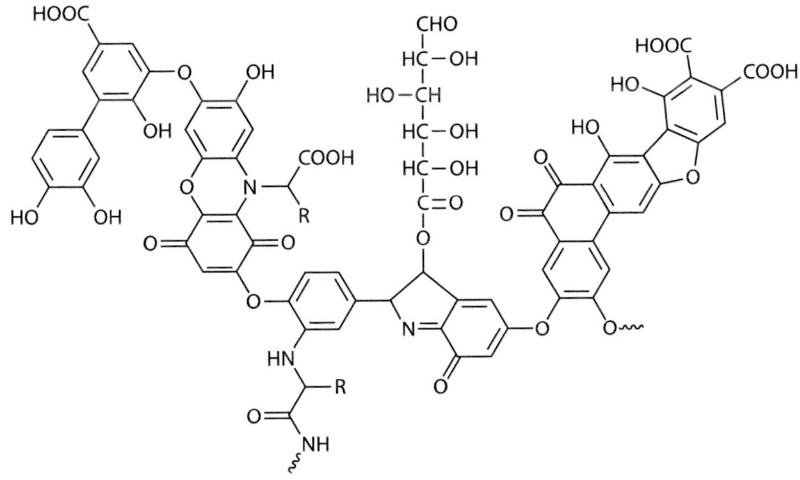

FIGURE 19.1 Structure of humic acid.

helps to retain water for a longer time, restores skin pH, and helps in blood circulation, acne repair and prevention, and antiaging effect.

Dead Sea mud and waters rich in minerals can be applied to either the entire body or just parts. Dead Sea mud masks stimulate, tighten, clean, and invigorate the skin with a refreshing glow. Dead Sea mud can also be used for the relief of various aches and pains, and because of these beneficial effects sea mud can be used in promoting health effects and hygiene. The mud from the Dead Sea in Jordan is the most widely talked about sea mud. Dead Sea mud is black in color (called Dead Sea black mud) (Momani et al. 2009). Dead Sea black mud is a homogeneous mixture of Dead Sea minerals and organic materials that may help in decreasing some skin diseases and rheumatism pains and rejuvenate skin cells.

The use of Dead Sea bath salts at home is an effective way to relax and absorb the minerals of the Dead Sea. Research has proven the efficiency of Dead Sea bath salts in alleviating skin diseases such as psoriasis, by treatment that can be administered at home by using about 2.2 lb for each bath. It is also highly beneficial for relieving muscle tension or for simply relaxing and enjoying a rejuvenating experience. The possible role of Dead Sea minerals on skin smoothing was studied. The surprising results were that Dead Sea minerals were found to be superior in comparison with 45 leading cosmetic brands, in smoothing the skin's surface and reducing wrinkle depth and skin roughness (Ma'Or et al. 1996, 1997).

19.5 CONSTITUENTS OF SEA MUD AND THEIR FUNCTIONS

Sea mud usually consists of organic and inorganic portions. The organic portion is mainly composed of humic acid (Figure 19.1) and the inorganic portion is rich in more than 19 different mineral components, such as Na^+, Mg^{2+}, Zn^{2+}, and so on. The potential use of the organic and inorganic substances from sea mud in various skin products is a safe biological substances for skin protective purposes (Kim et al. 2010).

Dead Sea mud is believed to be derived from older sediments or the red-brown soils that are swept into the sea during the wintertime. Sediment precipitations in the Dead Sea were investigated for the first time by Neev and Emery (1967). The chemical properties of Dead Sea mud samples are shown in Table 19.3. The pH values are between 5.7 and 6.1 (weak acid), whereas the electrical conductivity values are between 59 and 67 μS/cm. The above pH and electrical conductivity values classify the analyzed samples of the Dead Sea mud as weak electrolytes (Neev and Emery 1967).

TABLE 19.3
Chemical Properties of Different Sea Mud Samples

	Samples		
Property	**North**	**Middle**	**South**
Total surface area (m^2/g)	830	774	788
Cation exchange capacity (meq/100g)	45.75	29.31	35.57
pH at 28.5°C	6.1	6.01	5.7
EC μS/Cm	59	65	67
Different minerals (%)			
1. Clay	8.76	4.15	6.65
2. Dolomite	5.05	5.21	6.17
3. Gypsum	3.85	1.45	2.1
4. Kaolinite	12.64	9.89	12.07
5. Pyrite	8.93	7.21	3.32
6. Quartz	18.01	33.11	20.21
7. Carbonates	40.21	37.75	47.75
8. Total feldspar	2.55	1.23	1.73

19.5.1 Thermal Mud Therapy

Turkey is one of the top countries in the world in terms of thermal source richness, with almost 1,000 thermal springs throughout Anatolia. The temperature of these hot springs varies between 20 and 110°C. The pH of the thermal waters ranges from 6.60 to 8.70 and temperatures vary from 30 to 84°C (Çelik Karakaya et al. 2010). Thermal mud is a hyperthermal or hyperthermalized mud produced by the primary or secondary mixing of a solid component (usually a clayey geomaterial) with mineral-thermal water. After suitable preparation (maturation), the mixing leads to the production of the therapeutic mud (peloid) that is used as a poultice in medical practice. Nowadays, the clay/water ratio, degree of maturation, and workability of a peloid mud are matters of empirism, hitherto left to the judgment of the practitioner and not repeatable for different spas (Veniale et al. 2004).

There are three main items that determine the final characters of peloids: (1) the chemical and mineralogical features of the clay materials, (2) the chemistry of the waters used to mature and prepare the mud, and (3) maturation procedures that strongly influence microbiological activities and the availability of organic components (Veniale et al. 2007; Tateo et al. 2009). Although there is a general agreement on those premises, the mechanisms of action of muds are still unknown (Fioravanti et al. 2003), likely because several stimuli work on patients: thermic activity, salt composition and concentrations, osmotic pressure, and electric conductivity.

The therapeutic effects of peloids have been extensively studied, particularly in the fields of rheumatology and dermatology (Bellometti et al. 2000). Suitable properties and active factors of a mature thermal mud are water retention, consistency, bioadhesiveness, ease of handling, pleasant sensation while applied to the skin, cooling kinetics (thermoinsulation and heat retention), and exchange capacity through mud-skin interface (mobile-soluble chemical elements and organic substances produced by the metabolic activity of growing microorganisms).

19.5.2 Clays in Thermal Mud Therapy

From ancient times different types of clays have been under use for multiple therapeutic purposes (Viseras et al. 2007; Aguzzi et al. 2007; Veniale et al. 2007). Soil ingestion is a practice seen in people of all countries. Clay pastes and powders are used as solar protectors, in toothpastes, creams, powders and emulsions, bathroom salts, and deodorants, as well as in pelotherapy, for treating rheumatism, arthritis, and bone-muscle traumatic damages (Veniale et al. 2007; Carretero and Pozo 2010).

Bentonite, illite, montmorillonite, zeolite, clinoptolite, and kaolin are some of the major medical clays that are currently in use in human and veterinary medicine. The therapeutic activity of the clays mainly depends upon their physical and physicochemical properties as well as the chemical composition. The main properties that make them useful in medical applications are the structure, specific surface area, and grain size distribution, which determine absorption/adsorption and cation exchange capacity.

19.5.3 Mud Pack Therapy

Mud pack therapy is frequently used in daily rheumatologic practice. Mud pack application alone or in combination with balneotherapy has been found to be effective on painful arthritic processes. It has been shown to induce anti-inflammatory activity and influence the antioxidant system in patients with osteoarthritis (OA) (Bellometti et al. 1997). Mud pack application is another treatment choice for OA patients. Studies have shown that mud therapy affects several biochemical markers and has antirheumatic actions. It has a protective effect on cartilage homeostasis (Bellometti et al. 1997, 2000). It has a role in the production of pro-inflammatory cytokines and in modulating inflammatory reactions. Mud pack has a therapeutic activity due to both an anti-inflammatory component and thermal effect (Bellometti et al. 1997).

Pelotherapy is the local or generalized application of thermal muds (hot cataplasms called peloids) for treating rheumatism, arthritis, and bone-muscle traumatic damages. The recent development of cutaneous chemotherapy (treatments of seborrhoeic skin, antimicrobial and antifungal agents) has requested clay minerals with peculiar properties, aiming to realize new functional and durable medicaments (Veniale et al. 2004).

19.5.4 Therapeutic Effect of Sea Mud

Heat (47–50°C) plays a common fundamental role in the beneficial effects of thermal mud therapy. This kind of exogenous thermic therapy reduces muscle contractions; in turn, muscle relaxation is associated with relief from pain. Heat stimulates opioids and hypothalamic-pituitary-adrenal hormone secretions (Tateo et al. 2009), which modify abnormal muscle conditions. Heat also increases joint collagen extension, favoring stretching and reducing ankylosis and fibrotic retraction. Hence it acts as a good analgesic, rheumatologic arthropathy, and anti-inflammatory agent (Bellometti and Galzigna 1998).

With decreasing side effects better therapeutic results may be obtained by coupling pharmacological anti-inflammatory therapy and mud therapy, which reduces the amount of drug administration. The usefulness of this combined approach has been reported by a few studies on patients affected by OA (Hercogova et al. 2002; Evcik et al. 2007) or fibromyalgia (Fioravanti et al. 2007). Particularly, subjects with knee osteoarthritis (one of the most limiting diseases for deambulation in senescence) have shown pain relief and improvement of articular function, walking, and daily activities (Evcik et al. 2007).

From a clinical point of view, thermal therapy is considered very useful in the treatment of illnesses of the locomotor apparatus, such as primary and secondary osteoarthritis, extra-articular rheumatologic diseases (i.e., tendinitis, bursitis, neuritis, fibroneuritis), and also posttraumatic syndromes (Fioravanti et al. 2003). Elective thermal treatment for ostheoarthritis patients is represented by mud bath applications. Clinical observations describe the effects of mud therapy as antalgic (analgesic and muscle relaxant) and anti-inflammatory. Furthermore, experimental studies have shown regenerative actions of mud therapy on articular collagen and cartilage, which counteract ankylosis and fibrotic retraction and reduce joint stiffness (Tateo et al. 2009). Mud bath therapy also influences the condrocyte metabolism, controlling cartilage homeostasis, and significantly modulates the production of cytokines that damage cartilage.

Mud pack therapy has shown a significant decrease in the circulating levels of TNF-α and IL-1, correlated with joint pain relief (Bellometti et al. 1997), and increments in the serum levels of protective factors against the development and progression of the arthrosic degenerative processes, such as insulin-like growth factor I (IGF-I). Furthermore, recent studies (Mazzetti et al. 2001; Caraglia et al. 2005; Bellometti et al. 1996, 2000) demonstrated the effects of mud therapy against the activity of free radicals, such as nitric oxide and oxygen reactive metabolites, which are known to play a role in the pathogenesis of osteoarthritis.

19.6 COSMECEUTICALS FROM SEA SALT

Sea salt is not as refined as table salt, and it still contains traces of various minerals. Sea salt is perfect for creating bath salts, body scrubs, or custom-formulated therapeutic blends and used for food and cosmetic products. Sea salts have a low moisture content, which helps them to absorb fragrances and essential oils. Sea salt could be used in cosmetics for skin care (Blanco-Dávila 2000), skin lightening, deodorizing, hair conditioning, and cleansing (Gupta 2006).

19.6.1 Sea Water and Sea Mud for Skin Disease Treatments

In pharmaceutical formulations, thermal water is used as gastrointestinal protectors, oral laxatives, antidiarrheals, dermatological protectors, and cosmetics. The main factors that determine the nature of a peloid and its suitability for pelotherapeutic applications are low cooling rate, high absorption capacity, high cation-exchange capacity, good adhesiveness, ease of handling, and a pleasant feeling when applied to the skin. The composition and granulometry of the initial form of the clay and the composition of the mineral water used for the mixture are also important for getting better results in applying peloids (Veniale et al. 2007).

19.6.2 Psoriasis

Psoriasis has been cured by sea water (Shani et al. 1993; Montgomery 1979; Novotny 1977). Psoriasis is a chronic skin disease characterized by dry red patches covered with scales; it occurs particularly on the scalp, ears, genitalia, and the skin over bony prominences. The treatment of psoriasis has been achieved in many ways, including corticosteroids (Lezdey and Wachter 1993), oral administration of 8-methoxypsoralen followed by high-intensity long-wave ultraviolet light (Melski et al. 1977), and methotrexate (Saporito and Menter 2004).

Even though several options are available for the treatment of psoriasis, they each have disadvantages. Hence naturally derived products are highly recommended for psoriasis treatment, such as sea water, sea mud, and deep-sea minerals. Those products are with a consistency of solutions, emulsions, gels, suspensions, and salts, to be used depending on the specific purpose, and can be added to other cosmetic or pharmaceutical products. Balneotherapy with Dead Sea water has been reported as a successful treatment modality for psoriasis, atopic eczema, and vitiligo, but the precise mode of action has escaped definition so far. The saturating salt concentration (346 g/L) together with the unique UV spectrum has been suggested to trigger the release of pro-inflammatory and chemotactic mediators. The antioxidant property of Dead Sea water was observed by decomposition of hydrogen peroxide on pseudocatalse activity using Fourier transform Raman spectroscopy (Schallreuter et al. 2002).

The combination of salt water bath and irradiation yielded a significant decrease of the minimal erythema dose when compared to UVB alone (median 90 mJ/cm^2 vs. 130 mJ/cm^2, $p < 0.01$). Analysis of variance showed a significant influence of salt water bath on erythema ($p < 0.05$) but not on skin pigmentation. Within the minimal erythema dose test area the erythema index of the salt water exposed forearms was elevated significantly ($p < 0.05$), while skin pigmentation was not affected. Thus, bathing the skin in salt water leads to a decreased threshold level for the elicitation of UVB-induced erythema and a selective increase of the erythemal response. This sensitization to the effects of short-wave UVB radiation may increase immunosuppressive effects of UVB radiation and may lead to an increased efficacy of UVB phototherapy. However, there is also an increased sunburn risk when salt water baths are followed by exposure to UV radiation (Schempp et al. 1999; Nissen et al. 1998).

Sodium chloride (24%) solution is not useful for the treatment of psoriasis; by experiment no difference has been found between salt water and tap water for the treatment of psoriasis (Gambichler et al. 2001). However, the combination of salt water brine baths in different concentrations and subsequent UVA/B irradiation is used increasingly for the treatment of psoriasis or atopic dermatitis (Simon et al. 2000). The effectiveness of salt water baths and subsequent selective ultraviolet phototherapy was investigated in a prospective study of 40 patients with psoriasis vulgaris and atopic dermatitis. There were two groups with 20 patients each. The first group was treated with salt water (15%) that contained synthetic Dead Sea salt called Psorisal; the patients in the second group had a bath in a 3% NaCl solution. After 4 weeks of daily treatment, 80% of the patients in the group treated with Psorisal had significantly better results than the second group (Zimmermann and Utermann 1994).

Treatment of psoriasis vulgaris with tar- and oil-containing sea water baths has been reported and compared to that with sulfur baths. The superiority of the tar–sea water treatment compared to sulfur baths was confirmed. For the first time good experiences with a new oil bath are reported (Kasten 1978). Dermud is a commercial product, made with deep-sea water mud. Dermud has protective, antioxidant, and anti-inflammatory properties that can antagonize biological effects of UVB irradiation in skin and reduce skin photodamage and photoaging, and more generally reduce oxidative stress and inflammation in skin pathologies (Portugal Cohen and Soroka 2009).

Artificial balneophototherapy is an imitation of natural balneophototherapy using highly concentrated salt solutions (25%) and UVB radiation and is useful for psoriasis treatment (Roos et al. 2010)

19.6.3 Atopic Dermatitis

Atopic dermatitis is a chronic, eczematous skin disease that follows patients from early childhood. Sea water drinking may be highly useful to reduce atoptic dermatitis treatment (Hataguchi et al. 2005). The usefulness of the mineral water with 250 mM NaCl and 50 mM KCl as therapy was found more efficient for atopic dermatitis than for other chronic dermatitis (Yoshizawa et al. 2003). Short-term application of deep-sea water as found effective for treatment on atopic dermatitis (Adachi et al. 1998).

19.6.4 Antiagaing and Bleaching Agent

Cleopatra and the Queen of Sheba both reportedly bathed in the Dead Sea to keep their skin looking youthful. An Israeli company, Ahava, now has the rights to make Dead Sea mud products. The salty mud from the lowest sea on earth contains high concentrations of minerals such as magnesium, sodium, phosphate, and bromides, which work together to create a moisturizer that slows the aging process.

Photoaging is the term dermatologists use to describe the type of aging caused by exposure to the sun's rays. The amount of photoaging that develops depends on: (1) a person's skin color and (2) his or her history of long-term or intense sun exposure. People with fair skin who have a history of sun exposure develop more signs of photoaging than those with dark skin. In the darkest skin, the signs of photoaging are usually limited to fine wrinkles and a mottled complexion. Sea mud has a high cleansing ability; it annexes fats and dirt, discards dead skin cells, and exposes a fresh skin layer. It reduces skin laxity and even performs a certain stretch, soothes the skin, and gives it moisture, thus removing the problem of photoaging. In combination, these effects make the skin smoother, more flexible, and give it a clean, fresh, and silky appearance. The observations and significant conclusions of psoriasis treatment with sea water and sea mud with radiation therapy are explained in Table 19.4.

TABLE 19.4
Sea Water and Sea Mud with Radiation Therapy for Several Skin-Related Diseases

Source	No. of Patients	Type of Dermatitis	Therapy	Treatment Group	Control Group	Duration	Observation	Significance Conclusion
Sukenik et al. 1994	166	Psoriasis and psoriatic arthritis	Balneotherapy (mud packs and sulfur baths)	146	20	3 weeks	Morning stiffness, grip strength, effluent joints, PASI, cervical, thoracic, and lumbar spine pain	Significant results have been found between treated (Dead Sea) and untreated groups. Dead Sea treatment groups had reduction of spinal pain and range of movement in the lumbar spine, and significant improvement ($p < 0.001$ and $p = 0.022$, respectively) was observed in the study group only.
Elkayam et al. 2000	42	Psoriatic arthritis	Ultraviolet ray, Dead Sea bath, mud therapy, and sulfur bath	23	19	4 weeks	Stiffness, tender joints, swollen joints, right grip, left grip, PASI	Significance difference has been found between treated group and control group, such as PASI, swollen joints, and inflammatory neck and back pain. Addition of mud packs and sulfur bath to sun ultraviolet exposure and Dead Sea baths seem to prolong beneficial effects and improve inflammatory back pain.
Codish et al. 2005	28	Ankylosing spondylitis	Balneotherapy, climatic therapy at the Dead Sea area	14	14	2 weeks	—	Climatotherapy at the Dead Sea area can improve the condition of patients suffering from long-standing ankylosing spondylitis.
Shani et al. 1985	—	Psoriasis	Dead Sea bath	—	—	4 weeks	Serum Br, Rb, Ca, and Zn	The improvement of the psoriatic condition after bathing in the Dead Sea may be partly attributed to the minerals' effect on the psoriatic skin.
Portugal Cohen and Soroka 2009	*In vitro*	Psoriasis and inflammatory	Dead Sea mud	—	—	—	MTT, caspase 3, total antioxidant, oxygen radical absorbance	Dermud has protective, antioxidant, and anti-inflammatory properties that can antagonize biological effects of UVB irradiation in skin.
Katkhanova and Sakhnov 2009	556	Psoriasis and inflamamatory	Sun, air, and sea water baths	278	278	—	Psoriasis area and severity index	The study showed a 2.39-fold decrease of the psoriasis area and severity index (PASI) in the children of group 1 compared with its 1.38-fold decrease in group 2.

Dawe et al. 2005	60	Psoriasis	Dead Sea salt solution + ultraviolet B phototherapy	—	—	8 weeks	Erythema data	In this population the addition of pretreatment Dead Sea salt soaks to NB-UVB did not result in a clinically important improvement in clearance of psoriasis.
Abels et al. 1995	1,448	Psoriasis	Dead Sea psoriasis clinic	—	—	—	—	The overall response in a large cohort of psoriasis patients treated at the Dead Sea was excellent.
Halevy et al. 1997	30	Psoriasis	Dead Sea water and common salt	13	12	3 weeks	PASI score	The present study implies a beneficial effect to bathing with either Dead Sea bath salt or common salt as a sole therapy for psoriasis vulgaris. However, we observed an enhanced beneficial effect of Dead Sea bath salt compared to common salt.
Even-Paz et al. 1996	81	Psoriasis	Dead Sea water + ultraviolet radiation	—	—	—	—	In subjects who only bathed in Dead Sea water the improvement was 28%, in those who only sunbathed it was 73%, and in those who did both it was 83%.

19.6.5 Antioxidative Influence of Thermal Mud Therapy

There is also favorable influence of balneotherapy found on antioxidative processes. In a study conducted on patients with type II diabetes beneficial effect of balneotherapy on the glutathione metabolism of platelets was observed (Ohtsuka et al. 1996). Grabski et al. (2004) reported the reduction of superoxide dismutase activity in patients with rheumatoid arthritis undergoing treatment with sulfuric water. Bellometti et al. (1996, 2000) had reported in their papers on the antioxidant property of balneotherapy and medicinal mud, showing the reduction of nitrogen monoxide activity and myeloperoxidase and enhancement of glutathione peroxidase activity. Thus it can be inferred that a therapeutic effect of mud therapy in treating osteoarthritis is mediated by different mechanisms.

Sea mud is nontoxic to skin cells even though it consists of minerals. The water and methanol extract of sea mud is used as an anti-inflammatory agent with UVB-irradiated human keratinocytes by measuring PGE_2. The water extract showed higher inhibition of PGE_2 production from HaCaT cells (30% inhibition) than the methanol extract at 200 ppm (μg/g). The water extract mainly composed of minerals in the mud inhibited PGE_2 production by 83% at 200 ppm, which is more than a standard anti-inflammatory drug, such as (1 μM) indomethacin (Kim et al. 2010).

19.7 CONCLUSION

Traditionally, sea water and sea mud are applied for skin-related diseases such as psoriasis, acne, eczema, cellulite, and joint pain due to their high rich content of minerals. However, conflicts are also present in the usage of sea water and sea mud for skin-related treatment that need to be cleared in future research.

ACKNOWLEDGMENTS

This work was supported by a grant from the Marine Bioprocess Research Centre of the Marine Bio 21 Center funded by the Ministry of Land, Transport and Maritime, Republic of Korea. P.N. Sudha thanks the authorities of DKM College, Thiruvalluvar University, India, and her family for their support.

REFERENCES

Abels, D.J., T. Rose, and J.E. Bearman. 1995. Treatment of psoriasis at a Dead Sea dermatology clinic. *International Journal of Dermatology* 34 (2):134–137.

Adachi, J., H. Sumitsuzi, K. Endo, T. Fukuzumi, and T. Aoki. 1998. Evaluation of the effect of short-term application of deep sea water on atopic dermatitis. *Japanese Journal of Allergology* 47 (1):57–60.

Aguzzi, C., P. Cerezo, C. Viseras, and C. Caramella. 2007. Use of clays as drug delivery systems: possibilities and limitations. *Applied Clay Science* 36 (1–3):22–36.

Bellometti, S., and L. Galzigna. 1998. Serum levels of a prostaglandin and a leukotriene after thermal mud pack therapy. *Journal of Investigative Medicine* 46 (4):140–145.

Bellometti, S., M. Cecchettin, and L. Galzigna. 1997. Mud pack therapy in osteoarthrosis: changes in serum levels of chondrocyte markers. *Clinica Chimica Acta* 268 (1–2):101–106.

Bellometti, S., M. Cecchettin, A. Lalli, and L. Galzigna. 1996. Mud pack treatment increases serum antioxidant defenses in osteoarthrosic patients. *Biomedicine and Pharmacotherapy* 50 (1).

Bellometti, S., M. Poletto, C. Gregotti, P. Richelmi, and F. Berte. 2000. Mud bath therapy influences nitric oxide, myeloperoxidase and glutathione peroxidase serum levels in arthritic patients. *International Journal of Clinical Pharmacology Research* 20 (3–4):69–80.

Blanco-Dávila, F. 2000. Beauty and the body: the origins of cosmetics. *Plastic and Reconstructive Surgery* 105 (3):1196.

Caraglia, M., S. Beninati, G. Giuberti, et al. 2005. Alternative therapy of earth elements increases the chondroprotective effects of chondroitin sulfate in mice. *Experimental and Molecular Medicine* 37 (5):476–481.

Carretero, M.I., and M. Pozo. 2010. Clay and non-clay minerals in the pharmaceutical and cosmetic industries. Part II. Active ingredients. *Applied Clay Science* 47 (3–4):171–181.

Çelik Karakaya, M., N. Karakaya, S. Sarloglan, and M. Koral. 2010. Some properties of thermal muds of some spas in Turkey. *Applied Clay Science* 48 (3):531–537.

Codish, S., S. Dobrovinsky, M.A. Shakra, D. Flusser, and S. Sukenik. 2005. Spa therapy for ankylosing spondylitis at the Dead Sea. *Israel Medical Association Journal* 7 (7):443.

Dawe, R.S., S. Yule, H. Cameron, H. Moseley, S.H. Ibbotson, and J. Ferguson. 2005. A randomized controlled comparison of the efficacy of Dead Sea salt balneophototherapy vs. narrowband ultraviolet B monotherapy for chronic plaque psoriasis. *British Journal of Dermatology* 153 (3):613–619.

Elkayam, O., J. Ophir, S. Brener, et al. 2000. Immediate and delayed effects of treatment at the Dead Sea in patients with psoriatic arthritis. *Rheumatology International* 19 (3):77–82.

Evcik, D., V. Kavuncu, A. Yeter, and I. Yigit. 2007. The efficacy of balneotherapy and mud-pack therapy in patients with knee osteoarthritis. *Joint Bone Spine* 74 (1):60–65.

Even-Paz, Z., R. Gumon, V. Kipnis, D.J. Abels, and D. Efron. 1996. Dead Sea sun versus Dead Sea water in the treatment of psoriasis. *Journal of Dermatological Treatment* 7 (2):83–86.

Fioravanti, A., G. Perpignano, G. Tirri, et al. 2007. Effects of mud-bath treatment on fibromyalgia patients: a randomized clinical trial. *Rheumatology International* 27 (12):1157–1161.

Fioravanti, A., M. Valenti, E. Altobelli, et al. 2003. Clinical efficacy and cost-effectiveness evidence of spa therapy in osteoarthritis. The results of "Naiade" Italian Project. *Panminerva Medica* 45 (3):211.

Gambichler, T., S. Rapp, E. Senger, P. Altmeyer, and K. Hoffmann. 2001. Balneophototherapy of psoriasis: highly concentrated salt water versus tap water—a randomized, one blind, right/left comparative study. *Photodermatology, Photoimmunology and Photomedicine* 17 (1):22–25.

Grabski, M. 2004. Hydrogen sulfide water balneum effect on erythrocyte superoxide dismutase activity in patients with rheumatoid arthritis—in vitro study. *Przegl Lek* 61 (12):1405.

Greiner, J., and W. Diezel. 1990. Inflammation-inhibiting effect of magnesium ions in contact eczema reactions. *Der Hautarzt; Zeitschrift für Dermatologie, Venerologie, und verwandte Gebiete* 41 (11):602.

Gupta, SK. 2005. Cosmetic and pharmaceutical masks for skin improvement. US Patent 20060198805, September 7, 2006.

Hahn, G.S. 1999. Strontium is a potent and selective inhibitor of sensory irritation. *Dermatologic Surgery* 25 (9):689–694.

Halevy, S., H. Giryes, M. Friger, and S. Sukenik. 1997. Dead Sea bath salt for the treatment of psoriasis vulgaris: a double blind controlled study. *Journal of the European Academy of Dermatology and Venereology* 9 (3):237–242.

Hataguchi, Y., H. Tai, H. Nakajima, and H. Kimata. 2005. Drinking deep-sea water restores mineral imbalance in atopic eczema/dermatitis syndrome. *European Journal of Clinical Nutrition* 59 (9):1093–1096.

Hercogova, J., E. Stanghellini, E. Tsoureli Nikita, and G. Menchini. 2002. Inhibitory effects of Leopoldine spa water on inflammation caused by sodium lauryl sulphate. *Journal of the European Academy of Dermatology and Venereology* 16 (3):263–266.

Jellum, E., B. Fagerlund, C. Kjolberg, J. Klaveness, and T. Ege. 2004. Anti-inflammatory treatment based on strontium compounds. WO Patent WO/2004/084,920.

Jellum, E., B.J. Fagerlund, C.M. Kjolberg, and J. Klaveness. 2007. Strontium compound for treatment of subdermal soft tissue pain. US Patent 20110038953, February 17, 2007.

Jung, S.H., Y.K. Seo, M.Y. Youn, C.S. Park, K.Y. Song, and J.K. Park. 2009. Anti-aging and anti-inflammation effects of natural mineral extract on skin keratinocytes. *Biotechnology and Bioprocess Engineering* 14 (6):861–868.

Kasten, F. 1978. Treatment of psoriasis vulgaris with tar- and oil-containing seawater baths. *Hautarzt* 29 (3):158.

Katkhanova, O.A., and S.N Sakhnov. 2009. Natural physical factors of the Kuban Black Sea coastal area in the rehabilitative treatment of children with psoriasis and concomitant ophthalmoherpes. *Voprosy kurortologii, fizioterapii, i lechebnoÄ fizicheskoÄ kultury* (2):36.

Kim, J.H., J. Lee, H.B. Lee, J.H. Shin, and E.K. Kim. 2010. Water-retentive and anti-inflammatory properties of organic and inorganic substances from Korean sea mud. *Natural Product Communications* 5 (3):395–398.

Lee, I., and H. Maibach. 2004. Sea water salts: effect on inflammatory skin disease—an overview. *Dermatologie in Beruf und Umwelt* 52 (2):62–66.

Lee, I., and H.I. Maibach. 2005. Sea water salts: effect on inflammatory skin disease. An overview. *Textbook of Cosmetic Dermatology* 24:467.

Levin, C.Y., and H.I. Maibach. 2001. Do cool water or physiologic saline compresses enhance resolution of experimentally induced irritant contact dermatitis? *Contact Dermatitis* 45 (3):146–150.

Levi-Schaffer, F., J. Shani, Y. Politi, E. Rubinchik, and S. Brenner. 1996. Inhibition of proliferation of psoriatic and healthy fibroblasts in cell culture by selected Dead-Sea salts. *Pharmacology* 52 (5):321–328.

Lezdey, J., and A.J. Wachter. 1993. Treatment of psoriasis. US Patent 5190197, March 2, 1993.

Liu, T.K., H.H. Hwung, J.L. Yu, and R.C. Kao. 2008. Managing deep ocean water development in Taiwan: experiences and future challenges. *Ocean and Coastal Management* 51 (2):126–140.

Ma'or, Z., S. Magdassi, D. Efron, and S. Yehuda. 1996. Dead Sea mineral-based cosmetics: facts and illusions. *Israel Journal of Medical Sciences* 32 (JUL):S28–S35.

Ma'or, Z., S. Yehuda, and W. Voss. 1997. Skin smoothing effects of Dead Sea minerals: comparative profilometric evaluation of skin surface. *International Journal of Cosmetic Science* 19 (3):105–110.

Matz, H., E. Orion, and R. Wolf. 2003. Balneotherapy in dermatology. *Dermatologic Therapy* 16 (2):132–140.

Mazzetti, I., B. Grigolo, L. Pulsatelli, et al. 2001. Differential roles of nitric oxide and oxygen radicals in chondrocytes affected by osteoarthritis and rheumatoid arthritis. *Clinical Science* 101 (6):593–599.

Melski, J.W., L. Tanenbaum, J.A. Parrish, T.B. Fitzpatrick, and H.L. Bleich. 1977. Oral methoxsalen photochemotherapy for the treatment of psoriasis: a cooperative clinical trial. *Journal of Investigative Dermatology* 68 (6):328–335.

Momani, K., T. El-Hasan, S. Auaydeh, and K. Al-Nawayseh. 2009. Heavy metals distribution in the Dead Sea black mud, Jordan. *Jordan Journal of Earth Environmental Science* 2 (2):81–88.

Montgomery, B.J. 1979. Bathing for psoriasis in the Dead Sea. *Journal of the American Medical Association* 241 (3):227.

Neev, D., and K.O. Emery. 1967. The Dead Sea—depositional processes and environments of evaporites: Israel geology. *Survey Bulletin* 41:147.

Nissen, J.B., W.W. Avrach, E.S.Hansen, K.P.Stengaard, and K.Kragballe. 1998. Increased levels of enkephalin following natural sunlight (combined with salt water bathing at the Dead Sea) and ultraviolet A irradiation. *British Journal of Dermatology* 139:1012–1019.

Novotny, F. 1977. Experience with therapeutic baths in psoriasis. *Cesko-Slovenska Dermatologie* 52 (4):261–263.

Obey Your Body. http://www.obeyyourbody.com/Catalog_47.html.

Ohtsuka, Y., N. Yabunaka, I. Watanabe, H. Noro, and Y. Agishi. 1996. Balneotherapy and platelet glutathione metabolism in type II diabetic patients. *International Journal of Biometeorology* 39 (3):156–159.

Portugal Cohen, M., and Y. Soroka. 2009. Protective effects of a cream containing Dead Sea minerals against UVB induced stress in human skin. *Experimental Dermatology* 18 (9):781–788.

Proksch, E., H.P. Nissen, M. Bremgartner, and C. Urquhart. 2005. Bathing in a magnesium rich Dead Sea salt solution improves skin barrier function, enhances skin hydration, and reduces inflammation in atopic dry skin. *International Journal of Dermatology* 44 (2):151–157.

Roos, S., S. Hammes, and H.M. Ockenfels. 2010. Psoriasis: natural versus artificial balneophototherapy. *Der Hautarzt; Zeitschrift für Dermatologie, Venerologie, und verwandte Gebiete* 61 (8):683–690.

Saporito, F.C., and M.A. Menter. 2004. Methotrexate and psoriasis in the era of new biologic agents. *Journal of the American Academy of Dermatology* 50 (2):301–309.

Schallreuter, K.U., J. Moore, S. Behrens Williams, et al. 2002. *In vitro* and *in vivo* identification of 'pseudocatalase' activity in Dead Sea water using Fourier transform Raman spectroscopy. *Journal of Raman Spectroscopy* 33 (7):586–592.

Schempp, C.M., H.C. Dittmar, D. Hummler, et al. 2000. Magnesium ions inhibit the antigen-presenting function of human epidermal Langerhans cells *in vivo* and *in vitro*. Involvement of ATPase, HLA-DR, B7 molecules, and cytokines. *Journal of Investigative Dermatology* 115 (4):680–686.

Schempp, C.M., K. Müller, J. Schulte Mönting, E. Schöpf, and J.C. Simon. 1999. Salt water bathing prior to UVB irradiation leads to a decrease of the minimal erythema dose and an increased erythema index without affecting skin pigmentation. *Photochemistry and Photobiology* 69 (3):341–344.

Shani, J., S. Barak, D. Levi, et al. 1985. Skin penetration of minerals in psoriatics and guinea pigs bathing in hypertonic salt solutions. *Pharmacological Research Communications* 17 (6):501–512.

Shani, J., R. Sharon, R. Koren, and Z. Even-Paz. 1987. Effect of Dead-Sea brine and its main salts on cell growth in culture. *Pharmacology* 35 (6):339–347.

Shani, J., E. Tur, E. Wald, et al. 1993. Computerized morphometry of psoriatic keratinocytes after bathing in the Dead Sea and in Dead Sea bath salt solutions. *Journal of Dermatological Treatment* 4 (4):195–198.

Simon, J.C., D. Pfieger, and E. Schopf. 2000. Recent advances in phototherapy. *European Journal of Dermatology* 10 (8):642–645.

Soto, J.A. 2008. Purified ocean water as a cosmetic and a therapeutic solution. *SÖFW-Journal* 134 (4).

Sukenik, S., H. Giryes, S. Halevy, L. Neumann, D. Flusser, and D. Buskila. 1994. Treatment of psoriatic arthritis at the Dead Sea. *Journal of Rheumatology* 21 (7):1305–1309.

Tateo, F., A. Ravaglioli, C. Andreoli, et al. 2009. The *in-vitro* percutaneous migration of chemical elements from a thermal mud for healing use. *Applied Clay Science* 44 (1–2):83–94.

Turkian, K. K. 1968. Oceans. Prentice-Hall, Englewood Cliffs, NJ.

Veniale, F., E. Barberis, G. Carcangiu, et al. 2004. Formulation of muds for pelotherapy: effects of. *Applied Clay Science* 25 (3–4):135–148.

Veniale, F., A. Bettero, P.G. Jobstraibizer, and M. Setti. 2007. Thermal muds: perspectives of innovations. *Applied Clay Science* 36 (1–3):141–147.

Viseras, C., C. Aguzzi, P. Cerezo, and A. Lopez-Galindo. 2007. Uses of clay minerals in semisolid health care and therapeutic products. *Applied Clay Science* 36 (1–3):37–50.

Yoshizawa, Y., K. Kitamura, S. Kawana, and H.I. Maibach. 2003. Water, salts and skin barrier of normal skin. *Skin Research and Technology* 9 (1):31–33.

Yoshizawa, Y., H. Tanojo, S.J. Kim, and H.I. Maibach. 2001. Sea water or its components alter experimental irritant dermatitis in man. *Skin Research and Technology* 7 (1):36–39.

Zhai, H., W. Hannon, G.S. Hahn, R.A. Harper, A. Pelosi, and H.I. Maibach. 2000. Strontium nitrate decreased histamine-induced itch magnitude and duration in man. *Dermatology* 200 (3):244–246.

Zimmermann, J., and S. Utermann. 1994. Photo-brine therapy in patients with psoriasis and neurodermatitis atopica. *Der Hautarzt; Zeitschrift fÞr Dermatologie, Venerologie, und verwandte Gebiete* 45 (12):849.

20 Potential Cosmeceutical Applications of Phlorotannins and Fucoidans from Marine Algae in the Treatment of Atopic Dermatitis

Noel Vinay Thomas and Se-Kwon Kim

CONTENTS

20.1 INTRODUCTION

Atopic dermatitis (AD) is a pruritic inflammatory skin disease associated with a personal or family history of allergy. AD can occur at any age; most often it affects infants and young children. In some instances, it may persist into adulthood or actually first show up later in life. The prevalence of AD is on the rise and estimated at ~17% in the United States. The fundamental lesion in AD is a defective barrier that results in dry, itchy skin, and is aggravated by mechanical injury inflicted by scratching. This allows entry of antigens via the skin and creates a milieu that shapes the immune response to these antigens (Oyoshi et al., 2009). Clinical observations suggest that AD is the cutaneous manifestation of a systemic disorder that gives rise to asthma, food allergy, and allergic rhinitis (Novak and Bieber, 2003). This inflammatory disorder is exclusively characterized by elevated serum IgE levels and peripheral eosinophilia. At the molecular level, AD is categorized into two types: extrinsic, which involves IgE-dependent sensitization and makes up to 70–80% of the patients, and intrinsic, which involves IgE-mediated sensitization and makes up 20–30% of the patients worldwide (Novak and Bieber, 2003). Both forms of AD have associated eosinophilia. In extrinsic AD, memory T cells expressing the skin homing receptor, cutaneous lymphocyte-associated antigen (CLA), produce increased levels of Th2 cytokines. These include IL-4 and IL-13, which are known to induce isotype switching to IgE synthesis, as well as IL-5, which plays an important role in eosinophil development and survival. These CLA^+ T cells also produce abnormally low levels of IFN-γ, a Th1 cytokine known to inhibit Th2 cell function. Intrinsic AD is associated with less IL-4 and IL-13 production than extrinsic AD (Leung et al., 2004). On the whole, AD is considered one of the common, often chronic (long-lasting) skin diseases that affects a large percentage of the world's population and involves susceptibility genes, the environment, a defective skin barrier function, and immunological responses. From the biopsy analysis of the clinically unaffected skin of AD patients, compared with nonatopic skin, an increased number of Th2 cells expressing IL-4 and IL-13, but not IFN-γ, mRNA has been reported (Hamid et al., 1994). When the antigen-specific IgE binds to Fcε receptors, release of inflammatory mediators, such as histamine, leukotrienes, and prostaglandins, is induced in response to the binding of multivalent antigens, which subsequently cross-link the Fcε receptors. These factors cause the immediate hypersensitivity reaction, and the late phase inflammatory reaction involving T cells and eosinophils (Werfel et al., 1995). Recent studies reveal that coupling of the IgE onto the Fcε receptors without the help of antigens could elicit such inflammatory responses (Kitaura et al., 2003). Thus it is evident that inhibition of IgE production or reduction in the concentration of IgE would be the best therapeutic approach in treating atopic diseases like AD, which is one of the most challenging skin inflammatory diseases. However, recently a recombinant humanized anti-IgE monoclonal antibody (mAb), omalizumab, which is used to treat severe asthma, has shown a promising effect in the treatment of a few cases of AD (Sheinkopf et al., 2008). However, the challenging aspect with this treatment is that an enormous amount of (i.e., very expensive) mAb is required to remove IgE from patients with AD, since the level of serum IgE in many patients with AD is tens to hundred times higher than that in patients with asthma or allergic rhinitis. These challenges have prompted researchers to focus on the isolation of bioactive substances from natural resources that could potentially inhibit the production of IgE and become promising leads in the cosmeceutical industry. A schematic representation for the best therapeutic approaches for the effective management of AD is shown in Figure 20.1.

Until now several research groups have reported many natural compounds from terrestrial and marine organisms that exhibit anti-inflammatory and antiallergic properties. However, marine organisms have not been studied as extensively as terrestrial ones. Many reports suggest that marine organisms are richer sources for structurally diverse bioactive compounds with valuable nutraceutical, pharmaceutical, and cosmeceutical potential. In the present chapter, we have made an effort to discuss the possible cosmeceutical applications of phlorotannins and fucoidans that are found in marine algae.

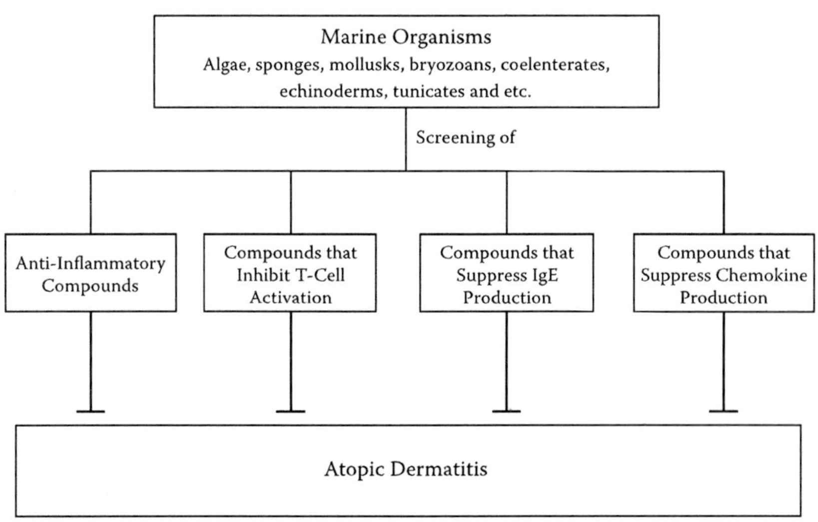

FIGURE 20.1 Screening of bioactive compounds from marine organisms that have anti-inflammatory activities, can regulate T cell activation, and suppress high concentrations of IgE and excess production of *chemokines* could prove to be a beneficial strategy for the management of AD in cosmeceutical pursuits.

20.2 PHLOROGLUCINOL DERIVATIVES AND THEIR POSSIBLE ROLE IN THE TREATMENT OF AD

Seaweeds possess enormous biologically important and beneficial ingredients that aid in the betterment of human health. Recently, the isolation and characterization of biologically active components from seaweeds have gained a lot of attention from various research groups across the world. Also, epidemiological and clinical studies have specified that consumption of plant-derived foods and drinks, for instance, tea, red wine, and soya bean products, could reduce the risk of oxidative damage-related diseases such as aging and other lifestyle diseases (Shibata et al., 2008). The marine algal phloroglucinol derivatives otherwise known as phlorotannins (Figure 20.2) are exclusively confined to marine brown algae. These phlorotannins are suggested to be formed by the polymerization of phloroglucinol (1,3,5-tryhydroxybenzene) monomer units and biosynthesized through

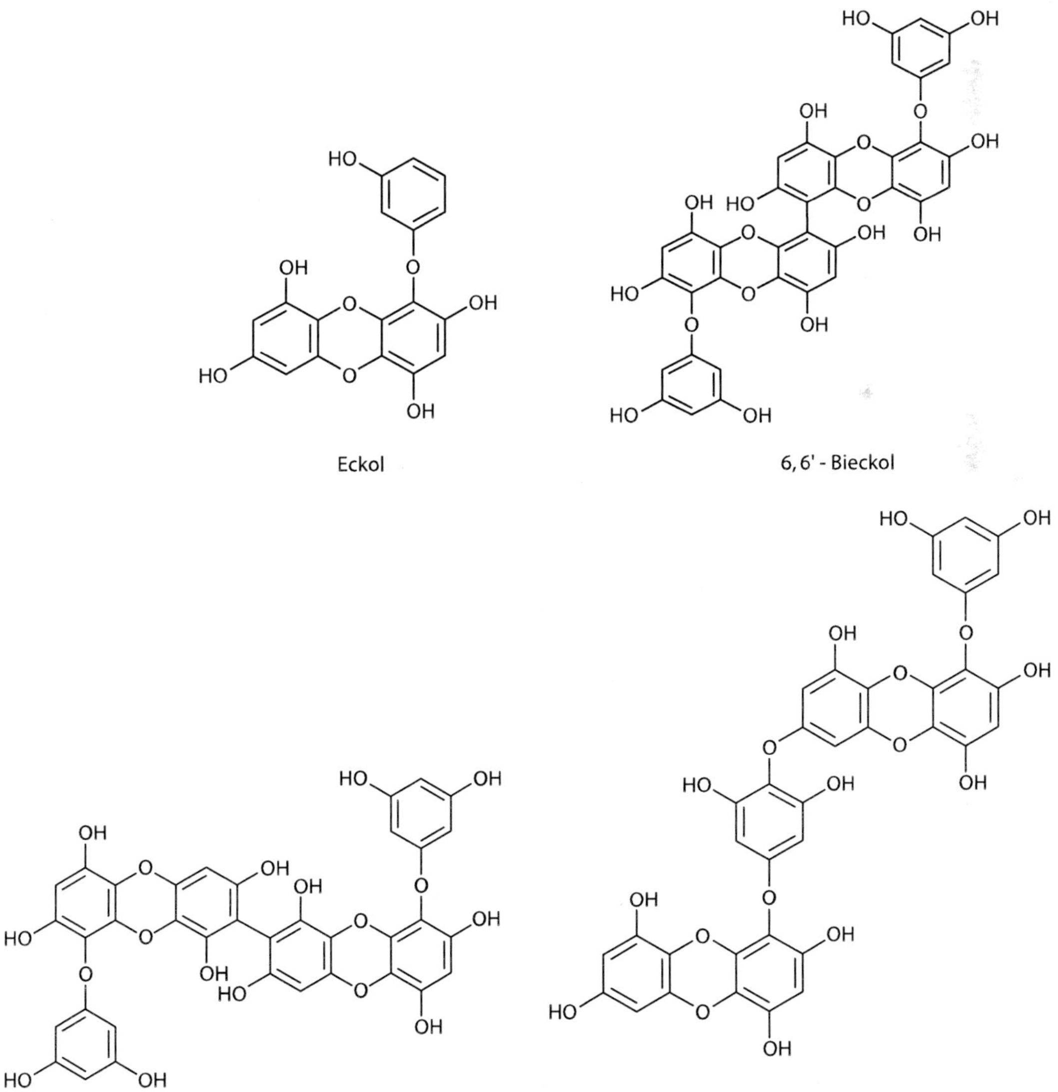

FIGURE 20.2 Chemical structures of *phlorotannins* from marine algae that have anti-inflammatory activities.

the acetate-malonate pathway, also known as the polyketide pathway. Phlorotannins are highly hydrophilic components with a wide range of molecular sizes ranging between 126 and 650 kDa (Ragan and Glombitza, 1986). Phlorotannins are tannin derivatives composed of several phloroglucinol units linked to each other in different ways and mostly isolated from brown algae (Singh and Bharate, 2006). Marine brown algae-derived phlorotannins have been investigated for their human beneficial aspects that include anti-inflammatory, hyaluronidase inhibitory activities. *In vitro* studies with the methanol extracts from marine brown algae *Eisenia arborea* have shown 74.3 ± 33.8% inhibition of histamine release from rat basophile leukemia cells (RBL-2H3) sensitized with anti-dinitrophenyl (DNP) IgE and stimulated with DNP-BSA. These observations suggest that methanol extract, which is rich in phlorotannins, exhibits the potential to treat histamine-related inflammatory diseases that include AD (Sugiura et al., 2006b). Hyaluronidase (HAase, EC.3.2.1.35), an enzyme that depolymerizes the polysaccharide hyaluronic acid (HA) in the extracellular matrix of connective tissue, is known to be involved in allergic effects. Six phlorotannins—phloroglucinol, an unknown tetramer, eckol (a trimer), phlorofucofuroeckol A (a pentamer), dieckol, and 8,8′-bieckol (hexamers)—obtained from brown algae *Eisenia bicyclis* and *Ecklonia kurome*, were tested *in vitro* for their ability to inhibit hyaluronidase activity. It was reported that these crude phlorotannins had a stronger inhibitory effect on hyaluronidase than well-known inhibitors such as catechins and sodium cromoglycate. According to these findings, 8,8′-bieckol has shown stronger hyaluronidase inhibatory effect with an IC_{50} value of 40 μM, which was was about seven times stronger than that of DSCG (a major and active component of antiallergic drugs) (Shibata et al., 2002), thus suggesting the importance of polyphenolic derivatives from marine algae as potential anti-inflammatory substances that could be useful leads in cosmeceuticals to treat AD.

Due to the abundant content of phloroglucinol derivatives in *Ecklonia cava*, it is used as a food ingredient and folk medicine against allergic diseases in Asian countries, especially Korea. Crude extract from *E. cava* was investigated for its antiallergic activity by Le et al. (2009). Their chemical investigation led to the isolation of the two main bioactive phlorotannin derivatives 6,6′-bieckol and 1-(3′,5′-dihydroxyphenoxy)-7-(2″,4″,6-trihydroxyphenoxy)-2,4,9-trihydroxydibenzo-1,4-dioxin for the first time from this genus, together with phloroglucinol and dieckol. These derivatives were assessed by histamine release assay on human basophilic leukemia (KU812) and rat basophilic leukemia (RBL-2H3) cultured cell lines, respectively. Furthermore, flow cytometric analysis indicated that the potential antiallergic mechanism is due to the suppression of binding activity between IgE and fragment crystallizable receptor I (FcεRI). Their investigation suggests that phloroglucinol derivatives are potential candidates for the functional foods, cosmetic, and pharmaceutical industries (Le et al., 2009).

In order to understand the cellular and molecular histopathological mechanisms and the expression pattern of chemokines in AD, *Dermatophagoides farinae* extract (DfE)-induced NC/Nga AD models are developed in mice and are considered indispensable for deriving a proper medicinal approach to treat this skin inflammation. Recently, the effects of methanolic extracts of *Ecklonia cava* (MEECs) on *Dermatophagoides farinae* antigen-induced NC/Nga mouse model have been evaluated by a research group at Biotoxtech, South Korea. The MEECs are considered to be rich in polyphenolic components, especially phlorotannins. These phlorotannins from marine algae are reported to exhibit anti-inflammatory activities by several research groups. The *Dermatophagoides farinae* antigen-induced NC/Nga mice were administered with MEECs (subcutaneous injection) at concentrations of 3.3, 10, and 30 mg/ml. These mice have shown a remarkable recovery after 21 days of administration of MEECs, and the results (as shown in Figure 20.3) are more promising than those of the commercial drug Betamethasone (BTT Study B09517, Report 2010). Moreover, there was a considerable decrease in the concentration of IgE in the experimental animal models. These findings confirm the effectiveness of polyphenolic compounds as potential cosmeceutical leads for the formulations of lotions and creams to cure AD.

A new phlorotannin, phlorofucofuroeckol-B is isolated from *Eisenia arborea*, an edible brown alga that is occasionally used as a folk medicine in gynecopathy in Japan. The *in vitro* studies on rat

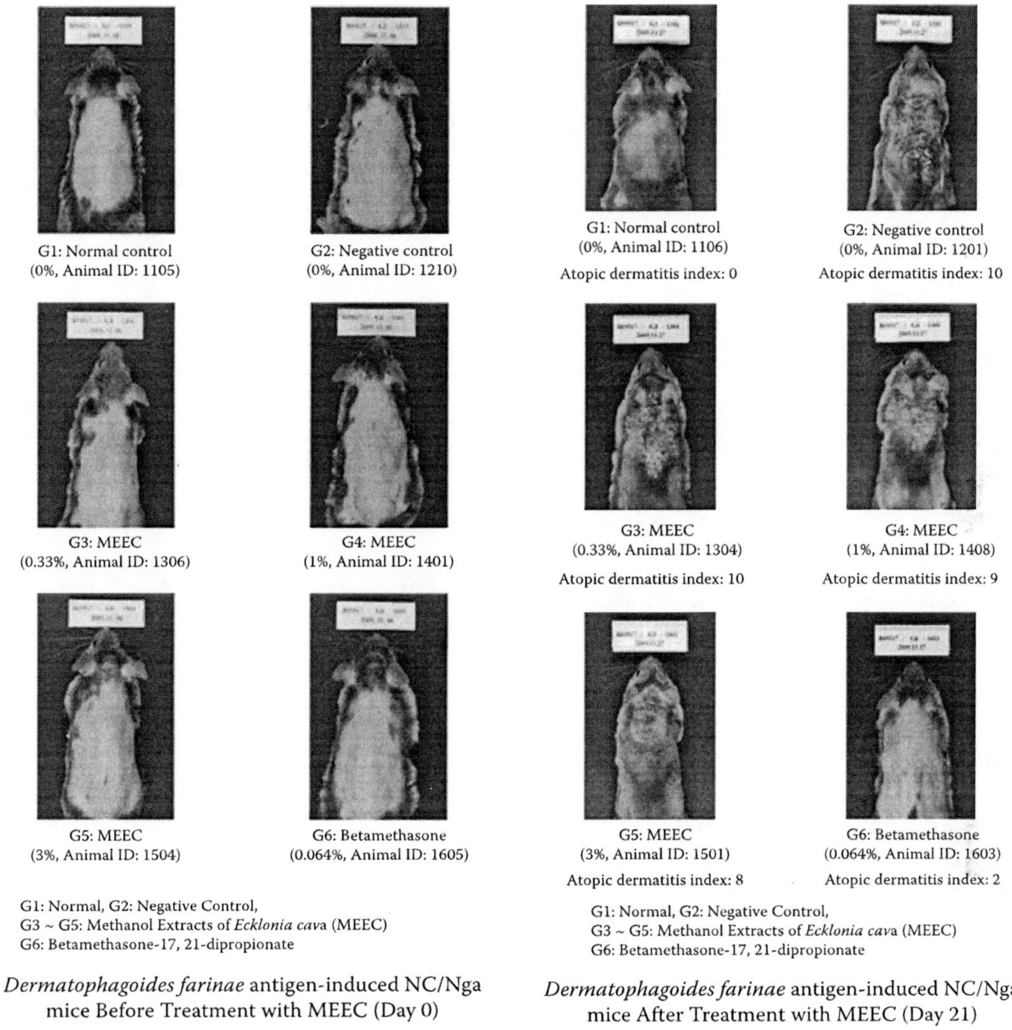

FIGURE 20.3 The effects of MEEC on *Dermatophagoides farinae* antigen-induced NC/Nga mice model. In the picture toward the left, the condition of AD mice at day 0 is shown. In the picture toward the right, the recovered condition of AD mice after 21 days of cutaneous administration with MEECs is shown. The MEEC at 3 mg/ml has shown maximum potential in curing the AD in these mice models, compared to the one treated with *Betamethasone* (commercial drug).

basophile leukemia (RBL)-2H3 cells confirmed that this phlorotannin is capable of inhibiting histamine release, ensuring the antiallergic property (Sugiura et al., 2006a). This inhibitory property of phlorotannins against histamine could make them promising cosmeceutical agents, especially in the treatment of skin allergies like AD.

Eckol, 6,6′-bieckol, 6,8′-bieckol, 8,8′-bieckol, phlorofucofuroeckol-A, and phlorofucofuroeckol-B obtained from *Eisenia arborea* have been reported to exhibit activities similar to or greater than those of the typical inhibitor for allergies, epigallocatechin gallate. Phlorofucofuroeckol-B showed the greatest activity among the tested phlorotannins, at 2.8 times greater than that of epigallocatechin gallate (Sugiura et al., 2007), hence suggesting advanced scientific investigations and more animal model studies to unravel the molecular mechanism of phlorotannins as anti-inflammatory substances.

FcεRI, a high-affinity receptor for IgE, is expressed by basophils and mast cells on the cell surface, and they act as effector cells in allergic-related reactions. Shim et al. (2009) have investigated the inhibitory effect of *Ecklonia cava* (EC) methanolic extract on the expression of FcεRI in human basophilic KU812F cells. Their research work revealed that EC extract caused a concentration-dependent reduction in the cell surface expression of FcεRI. The extract was also capable of reducing the binding between IgE or serum IgE and cell surface FcεRI. RT-PCR analysis revealed that EC extract reduced the mRNA expression of total cellular FcεRI α-chain. Fluorescence spectrophotometry studies showed that the extract inhibited the FcεRI-mediated release of histamine in a concentration-dependent manner. Therefore these results suggest that EC extract might exert its antiallergic activity through negative regulation of FcεRI expression and a decrease in histamine release (Shim et al., 2009).

On the other hand, the effects of 80% methanol extracts from frozen samples of 41 macroalgae and one sea grass collected in the Ise-Shima region of Japan were investigated on histamine release from rat basophile leukemia cells (RBL-2H3) sensitized with antidinitrophenyl (DNP) IgE and stimulated with DNP-BSA. Of the 21 brown algae, 5 green algae, 15 red algae, and 1 sea grass tested, extracts only from 7 brown algae have shown histamine release inhibitory activity from RBL cells. The cytotoxic effects of the seven brown algal extracts were investigated by Trypan blue staining; only *Eisenia arborea* and *Sargassum thunbergii* did not show cytotoxic effects, hence suggesting *E. arborea* and *S. thunbergii* may contain compounds that have antiallergic effects without inducing cell death (Sugiura et al., 2006b). According to these investigations, it is understood that phloroglucinol derivatives possess the ability to reduce the production of IgE and other inflammatory responses. Thus it is suggested that more advanced molecular studies, animal model studies, are highly recommended to establish phlorotannins as potential cosmeceutical leads for the treatment of AD. Moreover, there are several reports confirming the skin disease treatment abilities of phlorotannins. For example, phlorotannins eckol and dieckol isolated from *E. stolonifera* have attenuated the expression of matrix metalloproteinase-1 (MMP-1) in human dermal fibroblasts. These findings reveal that the inhibition of MMP-1 (which is an interstitial collagenase, mainly responsible for the degradation of dermal collagen in the human skin aging process) expression by *E. stolonifera*-derived phlorotannins was in correlation with the inhibition of both NF-κB and activator protein-1 (AP-1) reporter activity (Joe et al., 2006). This unique feature of phlorotannins in repairing skin damage from various allergens could be exploited for the better treatment of ever-challenging AD.

20.3 FUCOIDANS AND THEIR POSSIBLE ROLE IN THE TREATMENT OF AD

Fucoidan is an ingredient of seaweed that is composed of a polymer of α1→3-linked l-fucose, with sulfate groups on some of the fucose residues at the four positions (Patankar et al., 1993). It is reported that fucoidan exhibits a variety of biological effects in mammals, including the prevention of leukocyte rolling by binding to L-selectin (Preobrazhenskaya et al., 1997) and the inhibition of low-density lipoprotein uptake by macrophage through the prevention of macrophage binding to the scavenger receptor (Nishikawa et al., 1990). The role of fucoidans in reducing the concentration of IL-4 and IL-13 in bronchoalveolar lavage fluid, and also the inhibition of IgE in ovalbumin-induced (OVA-induced) mouse airway hypersensitivity, was reported by Maruyama et al. (2005). They have extracted mekabu fucoidan from the marine algae *Undaria pinnatifida*, and demonstrated its role in the augmentation of type 1 T-helper (Th1) cell response in normal BALB/c mice and the suppression of production of Th2 cytokines that have relieved pulmonary inflammation in the experimental animals, thus suggesting marine-derived fucoidan's possible candidacy as a potential lead in the treatment of inflammatory skin disorders like AD (Maruyama et al., 2005). A generalized pathway of how fucoidan could be helpful in reducing the risk of AD is shown in Figure 20.4. Many recent research articles strongly suggest that inhibition of IgE production could prove a major therapeutic breakthrough for treating AD. One research group has reported that fucoidan inhibited the production of IgE and Cε germline transcription in murine B cells induced by IL-4 (100 ng/ml) and anti-CD40 antibodies (10 μg/ml). Also, it had a significant effect on the production of IgE. These results suggest that fucoidan inhibited IgE production by preventing the NFκB p52-mediated pathways

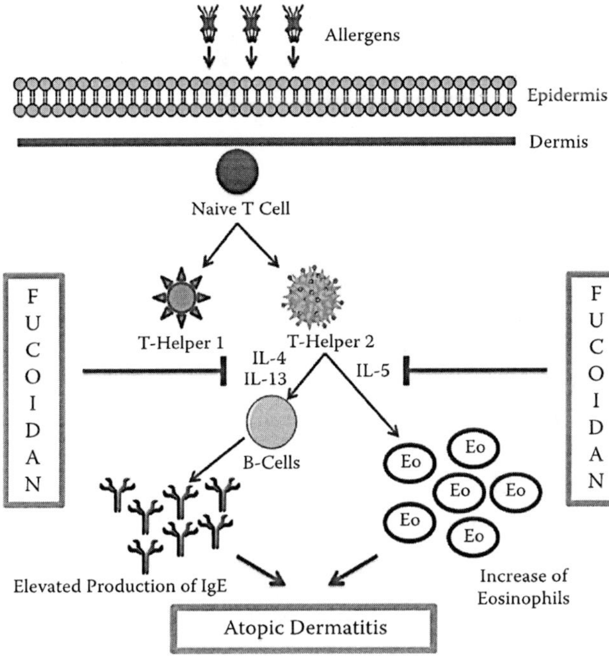

FIGURE 20.4 Fucoidan's ability in suppressing the activation of Th-2 cells that prevents the elevated production of IgE and eosinophils, which in turn helps in the control of AD.

activated by CD40 (Oomizu et al., 2006). Another research group suggested that the OVA-induced increase of plasma IgE was significantly suppressed when fucoidan was intraperitoneally administered in OVA-sensitized mice. The production of IL-4 and IFN-γ in response to OVA in spleen cells isolated from OVA-sensitized mice treated with fucoidan *in vivo* was lower than that from mice treated without fucoidan. Moreover, the flow cytometric analysis and ELISpot assay data revealed that the administration of fucoidan suppressed a number of IgE-expressing and IgE-secreting B cells, respectively (Yanase et al., 2009). These results indicate that fucoidan inhibits the increase of plasma IgE through the suppression of the IgE-producing B cell population.

In the search for natural anti-inflammatory compounds there has been a tremendous effort made on terrestrial resources, and many terrestrial plant sources have been used in antiallergic compounds that could cure skin-related inflammations, for instance, konjac glucomannan, which is a highly viscous, water-soluble, and high molecular weight polysaccharide consisting of a single chain of d-glucose and d-mannose joined by β-1,4-linkage with some branches (Kato et al., 1998). This polysaccharide is rich in konjac, and is obtained from the tubers of the potato-like plant *Amorphophallus konjac*. Interestingly, oral administration of pulverized konjac glucomannan (PKGM) that is reduced to a size of less than 105 μm in average diameter by means of a pulverizer prevented the elevation of plasma IgE, as well as the development of eczematous lesions and scratching behavior in NC/Nga mice, a model for AD (Onishi et al., 2005). However, when compared to the terrestrial atmosphere, the marine atmosphere is considered to be most challenging in terms of turbidity, salinity, and several other factors, and importantly, seaweeds do dwell in such harsh environments. This makes them an ideal choice for the screening of biologically important components, and more importantly, research done on these organisms is very sparse.

Fucoidan is reported to have the ability to cure the skin burns caused by ultraviolet radiation. Pretreatment of human skin fibroblast (HS68) cells with fucoidan inhibited UVB-induced MMP-1 expression *in vitro*. Extracellular signal-regulated kinase (ERK) activation was markedly inhibited by treatment with fucoidan. Fucoidan has significantly reduced MMP-1 mRNA expression, thus suggesting it as a potential agent for the prevention and treatment of skin photoaging (Moon et al., 2008). These scientific findings ensure the role of fucoidan's ability to repair skin damage, and hence it might be a potential agent in the cosmeceutical pursuit of an enriched treatment of AD.

20.4 CONCLUSIONS AND FUTURE PROSPECTS

AD is one of the most challenging skin diseases that is still in need of effective and efficient lead compounds for a complete cure. Several topical formulations are already available for the management of this skin disease; however, a permanent cure for this disease still has not been completely achieved. The marine environment is enriched with different kinds of organisms that harbor a wide range of biologically important compounds that are useful for cosmeceutical benefits for humans. Until now, only a few organisms have been exploited for the screening of cosmeceutical compounds from marine species. Marine algae have recently become a favorite study material for many researchers across the globe because of their unparalleled biologically important functions that could truly enhance human health systems. In this chapter, we have brought to the forefront the possible cosmeceutical applications of phlorotannins and fucoidans from marine algae as potential leads that could be used for the treatment of AD. However, more studies need to focus on the screening of novel compounds from marine algae that could find themselves a prominent place in the treatment of not only AD, but also various other skin inflammations. In addition, as the marine environment includes sponges, mollusks, bryozoans, coelenterates, echinoderms, tunicates, and other marine microorganisms, and reports suggest that most of them do possess anti-inflammatory substances within themselves, they could definitely serve researchers with a wide choice of cosmeceutical compounds that could cure AD.

REFERENCES

BTT Study B09517, Report 2010. *Effects of methanolic extracts from Ecklonia cava on Dermatophagoides farinae antigen induced NC/Nga mice*. Biotoxtech, South Korea.

Hamid, Q., M. Boguniewicz, and D.Y. Leung. 1994. Differential *in situ* cytokine gene expression in acute versus chronic atopic dermatitis. *Journal of Clinical Investigation* 94 (2):870–876.

Joe, M.J., S.N. Kim, H.Y. Choi, et al. 2006. The inhibitory effects of eckol and dieckol from *Ecklonia stolonifera* on the expression of matrix metalloproteinase-1 in human dermal fibroblasts. *Biological and Pharmaceutical Bulletin* 29 (8):1735–1739.

Kato, J., M. Nagaya, N. Niimi, and S. Tanaka. 1998. Venovenous extracorporeal membrane oxygenation in newborn infants using the umbilical vein as a reinfusion route. *Journal of Pediatric Surgery* 33 (9):1446–1448.

Kitaura, J., J. Song, M. Tsai, et al. 2003. Evidence that IgE molecules mediate a spectrum of effects on mast cell survival and activation via aggregation of the Fc RI. *Proceedings of the National Academy of Sciences of the United States of America* 100 (22):12911.

Le, Q.T., Y. Li, Z.J. Qian, M.M. Kim, and S.K. Kim. 2009. Inhibitory effects of polyphenols isolated from marine alga *Ecklonia cava* on histamine release. *Process Biochemistry* 44 (2):168–176.

Leung, D.Y.M., M. Boguniewicz, M.D. Howell, I. Nomura, and Q.A. Hamid. 2004. New insights into atopic dermatitis. *Journal of Clinical Investigation* 113 (5):651–657.

Maruyama, H., H. Tamauchi, M. Hashimoto, and T. Nakano. 2005. Suppression of Th2 immune responses by mekabu fucoidan from *Undaria pinnatifida* sporophylls. *International Archives of Allergy and Immunology* 137 (4):289–294.

Moon, H.J., S.R. Lee, S.N. Shim, et al. 2008. Fucoidan inhibits UVB-induced MMP-1 expression in human skin fibroblasts. *Biological and Pharmaceutical Bulletin* 31 (2):284–289.

Nishikawa, K., H. Arai, and K. Inoue. 1990. Scavenger receptor-mediated uptake and metabolism of lipid vesicles containing acidic phospholipids by mouse peritoneal macrophages. *Journal of Biological Chemistry* 265 (9):5226.

Novak, N., and T. Bieber. 2003. Allergic and nonallergic forms of atopic diseases 1. *Journal of Allergy and Clinical Immunology* 112 (2):252–262.

Onishi, N., S. Kawamoto, M. Nishimura, et al. 2005. A new immunomodulatory function of low-viscous konjac glucomannan with a small particle size: its oral intake suppresses spontaneously occurring dermatitis in NC/Nga mice. *International Archives of Allergy and Immunology* 136 (3):258–265.

Oomizu, S., Y. Yanase, H. Suzuki, Y. Kameyoshi, and M. Hide. 2006. Fucoidan prevents C [epsilon] germline transcription and NF [kappa] B p52 translocation for IgE production in B cells. *Biochemical and Biophysical Research Communications* 350 (3):501–507.

Oyoshi, M.K., R. He, L. Kumar, J. Yoon, and R.S. Geha. 2009. Cellular and molecular mechanisms in atopic dermatitis. *Advances in Immunology* 102:135–226.

Patankar, M.S., S. Oehninger, T. Barnett, R.L. Williams, and G.F. Clark. 1993. A revised structure for fucoidan may explain some of its biological activities. *Journal of Biological Chemistry* 268 (29):21770–21776.

Preobrazhenskaya, M.E., A.E. Berman, V.I. Mikhailov, et al. 1997. Fucoidan inhibits leukocyte recruitment in a model peritonial inflammation in rat and blocks interaction of P selectin with its carbohydrate ligand. *IUBMB Life* 43 (2):443–451.

Ragan, M.A., and K.W. Glombitza. 1986. Phlorotannins, brown algal polyphenols. *Progress in Phycological Research* 4:129–241.

Sheinkopf, L.E., A.W. Rafi, L.A.T. Do, R.M. Katz, and W.B. Klaustermeyer. 2008. Efficacy of omalizumab in the treatment of atopic dermatitis: a pilot study. *Allergy and Asthma Proceedings* 29 (5):530–537.

Shibata, T., K. Fujimoto, K. Nagayama, K. Yamaguchi, and T. Nakamura. 2002. Inhibitory activity of brown algal phlorotannins against hyaluronidase. *International Journal of Food Science and Technology* 37 (6):703–709.

Shibata, T., K. Ishimaru, S. Kawaguchi, H. Yoshikawa, and Y. Hama. 2008. Antioxidant activities of phlorotannins isolated from Japanese Laminariaceae. *Journal of Applied Phycology* 20 (5):705–711.

Shim, S.Y., L. Quang-To, S.H. Lee, and S.K. Kim. 2009. *Ecklonia cava* extract suppresses the high-affinity IgE receptor, Fc [epsilon] RI expression. *Food and Chemical Toxicology* 47 (3):555–560.

Singh, I.P., and S.B Bharate. 2006. Phloroglucinol compounds of natural origin. *Natural Product Reports* 23 (4):558–591.

Sugiura, Y., K. Matsuda, Y. Yamada, et al. 2006a. Isolation of a new anti-allergic phlorotannin, phlorofucofuroeckol-B, from an edible brown alga, *Eisenia arborea. Bioscience, Biotechnology, and Biochemistry* 70 (11):2807.

Sugiura, Y., K. Matsuda, Y. Yamada, et al. 2007. Anti-allergic phlorotannins from the edible brown alga, *Eisenia arborea. Food Science and Technology Research* 13 (1):54–60.

Sugiura, Y., Y. Takeuchi, M. Kakinuma, and H. Amano. 2006b. Inhibitory effects of seaweeds on histamine release from rat basophile leukemia cells (RBL-2H3). *Fisheries Science* 72 (6):1286–1291.

Werfel, S., W. Massey, L.M. Lichtenstein, and B.S. Bochner. 1995. Preferential recruitment of activated, memory T lymphocytes into skin chamber fluids during human cutaneous late-phase allergic reactions. *Journal of Allergy and Clinical Immunology* 96 (1):57–65.

Yanase, Y., T. Hiragun, K. Uchida, et al. 2009. Peritoneal injection of fucoidan suppresses the increase of plasma IgE induced by OVA-sensitization. *Biochemical and Biophysical Research Communications* 387 (3):435–439.

21 Application of Chitin Nanofibrils and Collagen of Marine Origin as Bioactive Ingredients

Pierfrancesco Morganti, Paola del Ciotto, Gianluca Morganti, and Véronique Fabien-Soulé

CONTENTS

21.1 INTRODUCTION

The use of polysaccharides in the pharmaceutical, biomedical, and cosmetic fields has ubiquitous appreciation of those ingredients, as they are derived from marine organisms, e.g., chitin and its derivatives. While considering many biomedically and cosmetically acceptable chitin/chitosan compounds, one finds their morphology and functionality span polymer technology and activity. They may act as rheology modifiers, suspending agents, hair and skin conditioners, as well as active compounds used for drug, cosmetic, food, and textile delivery systems, wound healing agents, scaffold for bone regeneration, derivatives for wastewater treatment, and many other known and unknown applications (Kim, S.K., 2011). Because of their massive and ready availability from biowastes of sea origin, researchers continue to struggle and explore the relationship between polysaccharide behavior and their performance. In addition, it is very important to understand how they may interact with other ingredients to create innovative bio- and eco-compatible formulations, based on the use of sustainable chemicals. According to the European Best Innovator Club, in fact, "sustainability will be one of the potential global challenges for business and political leaders in the next 50 years," as well as innovation having to address the AT-Kearney triple bottom line of economic success, social well-being, and environmental protection (Gerhardt and Hubbert, 2009). Similarly, production and use of chitosan nanofibrils (CNs) and CN complexes go in this direction. This chapter provides an overview of the structure and metabolism of the skin, the development of new CN-related complexes, and their use in the cosmetic and textile sectors (Morganti and Morganti, 2007; Morganti et al., 2007; Morganti, 2008, 2010a, 2010b).

21.2 SKIN STRUCTURE AND METABOLISM

21.2.1 Skin Barrier

As an interface of the body and the environment, skin is subjected to continuous abrasion by chemical and physical injury. This multicellular organ comprises three distinct major compartments: epidermis, dermis, and hypodermis.

The *epidermis*, containing three main viable cell types, the keratinocyte, the melanocyte, and the Langerhans cell, is covered by the outermost nonviable cell layer, the stratum corneum (SC).

At this level of the epidermis, the principal and protective keratinocytes produce keratins and traverse from the deeper *basal layer* of the skin to the upper one to reproduce themselves. Another important function of keratinocytes is to produce cytokines (peptide messengers) following stimulation, thus representing a key part of the body's early immunological responses. The melanocytes, by producing melanin, protect the skin from UV aggression, while Langerhans cells take part in the immune surveillance process.

The antigen-presenting Langerhans cells, in fact, can act as a *shuttle* for cytokines from the skin to the lymphatic system, or trigger the appearance of activated T-lymphocytes in the epidermis (Shroot, 1993). During the epidermis migration process, cells undergo a number of changes by synthesizing various proteins and lipids, involving the formation of the skin barrier (Figure 21.1). In normal skin, this cellular transit time from the basal layer to the SC is about 1 month, and in addition, the nature and distribution pattern of skin lipids changes markedly along with the cell dimension and organization. The intercellular lipid content is, however, pivotal for a competent

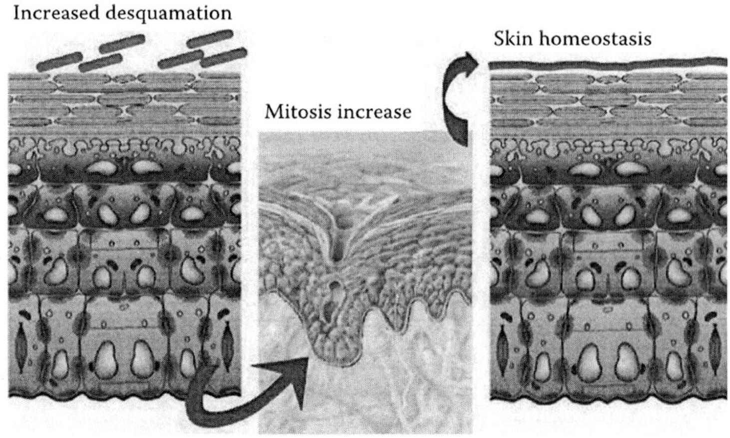

FIGURE 21.1 The skin turnover involving its barrier formation.

skin barrier and forms the only continuous domain in the SC (Nemaniak and Elias, 1980). It follows a tortuous path within the stratum corneum that may account in part for the barrier properties of the skin (Potts and Francoeur, 1991; Schaefer et al., 1999). The structure of the skin changes, in fact, from a close-packed aggregate of more or less cubital viable cells communicating with each other freely (and with many desmosomes), to flat dead cells organized as a bricks and mortar structure, the stratum corneum, where the lipids (lipid lamellae) are between the cells forming the mortar (Grayson and Elias, 1982). The cells in the SC become elongated, flattened, and filled with folded sheets of keratin, which gives each cell a significant rigidity and impermeability reinforced by the presence of the lipid lamellae. Differently from lipids of the other skin layers, composed prevalently of phospholipids, the SC lamellae are based on apolar ceramides, cholesterol, and free fatty acids.

Therefore the presence of keratins and the transformation of the lipids from polar to nonpolar compounds provide the outer layers of the skin with a hydrophobic waxy impermeable lipid barrier, based on components much more stable to enzymes, ultraviolet light, oxidation, and temperature. The horny layer is, in fact, responsible for preventing water loss, protecting the skin from the environmental aggressions, and also having mechanical and insulating properties reinforced by the other skin layers (Elias and Friend, 1975). Thus the mixture of keratins and lipid lamellae in the SC provides a superbly adapted cover for the underlying tissues, being chemically unreactive, hard and waterproof (from inside and outside), elastic, and resistant to abrasion and physical insult.

The *dermis*, positioned below the epidermis, supports the skin and adnexal structures, including hair, pilosebaceous units, and sweat glands. It is formed by cells, fibers, and ground substance (extracellular matrix) and, unlike the epidermis, is richly innervated and vascularized. The blood vessels lie in two main horizontal layers with vertical connections and terminate in fine capillary networks just beneath the epidermis. There are no blood vessels in the epidermis; however, nutrients required by the epidermal cells and any cellular waste products will pass through the capillary walls and the basement membrane (Figure 21.2). Fibrillar components of the dermis (collagen and elastic fibers), composed of specific amino acids, support the internal structure of the skin, while the ground substance is the *cementing* component of the dermis (macromolecules like glycosaminoglycans, hyaluronic acid, etc.), with the ability to bind numerous molecules of water.

The *hypodermis* (subcutaneous tissue) has a fundamental protective and storage function. To perform its basic function as a barrier, the skin needs *mechanical support*, provided by a framework of *extracellular matrix* (ECM), secreted by fibroblasts. In addition, it needs blood supply by vessels lined with *endothelial* cells to bring nutrients and oxygen, and to remove carbon dioxide and waste products. Moreover, it has to be supported by other cells, as *macrophage*, *Langerhans*, and *dendritic* cells, components of the immune system, necessary for its own defense against infection. Finally, the skin has an organization of nerves adapted to the proximity of the organ to the external environment, necessary for its own physiological needs.

21.2.2 The Cell Network Organization

The cell is organized as a close feedback loop living by many inputs, continuously coming from its own interior and from the exterior environment. Each cell has a membrane, containing the cellular fluid, organized as a molecular *soup* (cytosol) and composed of (1) nutritive chemicals, necessary to nourish all its own structures; (2) little factory centers, where the basic components of the cell structures are made; and (3) many other specialized organelles, necessary to store raw materials, recycle metabolites, or produce energy to support all the cellular necessities (Blobel, 1980). The *cell membrane* accepts the nutritive ingredients and eliminates the unwanted substances (Prakash and Matouschek, 2004).

Every cell has a well-distinguished *cytosol*, or cellular fluid, and the *nucleus* and *nucleolus*, containing the genetic information, where RNAs are produced and other RNA-protein complexes

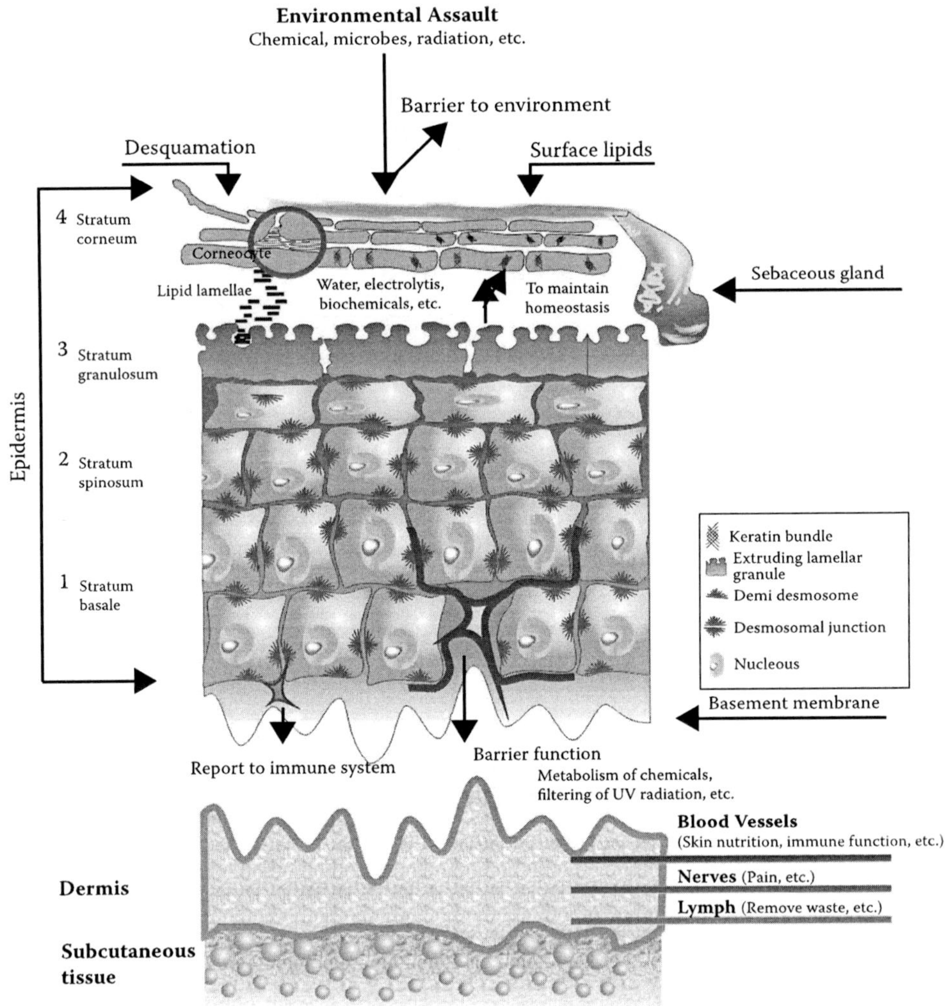

FIGURE 21.2 The skin: structure and function.

are assembled (Figure 21.3). The *endoplasmic reticulum* (ER) is a membrane organelle in which proteins and carbohydrates are processed, whereas the *Golgi apparatus* is a special storage bag for the generation and processing of sugar nucleotides and glycoproteins. The *mitochondrion*, which houses protein synthesis, generates adenosine triphosphate (ATP) energy via the oxidation of food molecules. The *peroxisomes* perform the oxidative functions that are not taken by mitochondria, which are the sites of oxygen utilization. The *lysosomes*, necessary to recycle many ingredients and control digestion of macromolecules, represent meeting places where several streams of intracellular traffic converge (Foerster and Zopf, 1962; Margulis, 1993).

The cellular global pattern is therefore a very complex network where all the cellular components are interconnected to each other by a feedback loop in a net involving thousands of biological processes.

Thus the common aim in skin biology is to understand the function of genes within their particular pathways (developmental and metabolic) that form skin, connective tissues, and organs, and also trying to know what makes cells different from one another.

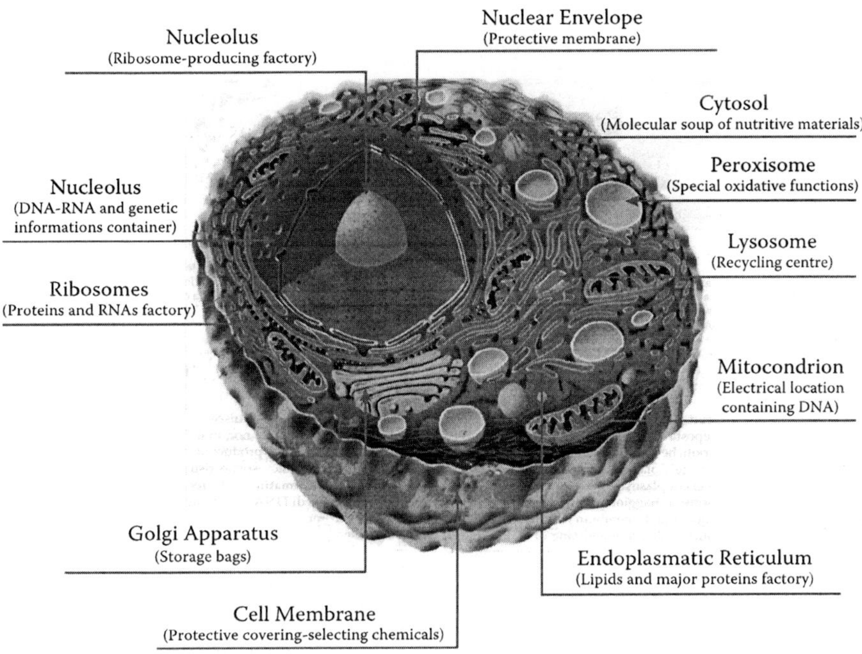

FIGURE 21.3 Cell organization.

21.2.3 The Cell as Self-Reproducing Unit

A human cell is the minimal self-reproducing unit of a living organism and is a vehicle for the transmission of genetic information by input signals. These signals are necessary to start up a succession of cellular events via the production of different proteins that activate many genes indispensable for cell development and turnover. A bidirectional communication mechanism between cells and genes exists by an overlapping of protein signals influenced from the interior and exterior environment of both cells and tissues (Figure 21.4). The presence of anomalous events and signals provoked by a lack of environment equilibrium interfering, for example, in the skin redox network of enzymatic and nonenzymatic antioxidant systems may lead to oxidative damage of many cell constituents (Thiele and Elsner, 2000).

For these reasons tocopherol, carotenoids, ascorbate, melatonin, and other natural or synthetic antioxidants used by cosmeceutical and nutraceutical formulations have been shown to exert an antioxidant effect in skin or skin cells, protecting cell lipids, proteins, or DNA (Morganti et al., 2008a).

However, every cell stores its genetic information in the same chemical form, as double-stranded DNA, and replicates itself by transcribing a portion of this information into molecules of the RNA. These in turn guide the synthesis of protein molecules by complex machinery (Braden and Touze, 1999; Bebenek et al., 2001; Albert et al., 2008). Thus the specific function of each protein depends on its amino acid sequence, coded by a specific gene (the functional unit of heredity) carried by the chromosomes. From a chemical point of view, proteins are by far the most structurally complex and sophisticated molecules known, so that the properties and functions of a cell are determined largely by the proteins it is able to make (Batzer and Deiminger, 2002; Strachan and Read, 2004).

But the genes also are controlled by the proteins. And the proteins in turn are coded by other genes, whose activation depends on other proteins coded by other genes, all involved together in complex *gene-protein nets*. This is because human evolution has organized the music of genes, the harmony of the cells, and the symphony of the different steps of life (Noble, 2006).

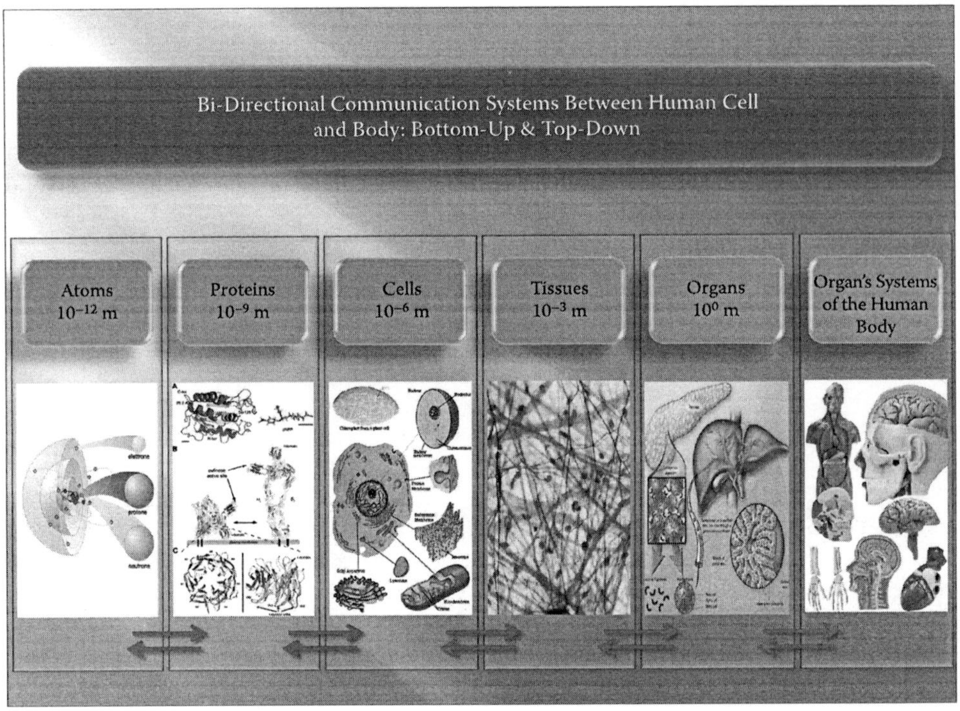

FIGURE 21.4 Continuous signal communication between human cells and tissues.

21.2.4 Signal Molecules and Cell Feedback Loops

Target cells use various intracellular mechanisms, including feedback loops, by which they respond to extracellular signals. And these signal molecules act in various combinations to regulate the behavior of the cell. Most of them act, in fact, as local mediators, which are secreted and then rapidly taken up, destroyed, or immobilized, so that they act only on neighboring cells. Others remain bound to the surface of the cell and mediate contact-dependent signaling. However, endocrine and synaptic signaling are the two distinct long-lasting and long-distance signaling systems, respectively. In endocrine signaling, hormones secreted by endocrine cells are carried in the blood to target cells throughout the body; in synaptic signaling, neurotransmitters, secreted and contacted by neural cell axons, act locally on postsynaptic cells.

The communication network in each cell is therefore essential to regulate the relative development of tissue and organs through bidirectional and different signals. Cytokines, for example, can be regarded as the alphabet of the biologic regulatory language (Corrk et al., 1993; Breer, 2003; Berridge, 2005). Thus for the cells to function properly, they must organize themselves in space and interact mechanically with their environment, rearranging their internal components as they grow, divide, and adapt to changing circumstances.

21.2.5 NICE Approach and Traditional Chinese Medicine for Mind-Body Skin Care

These cell signaling compounds, composed of short protein molecules (peptides), require not only extracellular signal molecules but also a complementary set of receptor proteins expressed by the target cells that specifically bind the same signal molecules (Breer, 2003; Berridge, 2005). Thus on one side small *hydrophobic* signal molecules, including steroid and thyroid hormones, may diffuse across the lipophilic plasma membrane of the target cell and activate intracellular receptor proteins, directly regulating the transcription of specific genes. On the other hand, most of the extracellular signal molecules cannot pass through the plasma membrane because of their hydrophilic characteristics. Therefore they activate the cell surface receptor proteins, which act as signal transducers, converting extracellular into intracellular signals, altering the behavior of the target cell. However, it is interesting to underline how tissues, cells, and organs may produce many different signal molecules, acting on and binding with specific protein receptors diffused all along the body. In fact, a continuous incoming of signals exists among the nervous, immune, cutaneous, and endocrine systems. These signals for a short distance act as direct *transmitters*, while for a long distance they require the production of mediating compounds such as catecholamine, oestrogenic, and androgenic compounds, acting through the bloodstream. In conclusion, they represent the biochemical expression of the emotions, have a fundamental role to coordinate all the activity of the immune and endocrine systems, and link each other to integrate all the mental and biological activities of both brain and skin. For all these reasons, at the level of cosmenutriceutical formulations, the nervous, immune, cutaneous, and endocrine (NICE) approach has evolved to demonstrate experimentally how the nervous, immune, cutaneous, and endocrine systems work together to give us global wellness and beauty (Morganti, 2009a, 2009b; Morganti et al., 2008a, 2010a, 2010b). This innovative approach, representing a bridge between the skin and our image recorded by our brain, gives the possibility to formulate, make, and control cosmetic products with a mind-body activity (Morganti, 2009a, 2009b, 2011b). In this way, the skin, alerted by the right signals and supplied with the active ingredients necessary to its cell life, may regularize its turnover, modulate the function of fibroblasts, and normalize the metabolic activity of adipocytes. Thus by the daily use of new-generation NICE-oriented cosmeceuticals, it is possible to obtain the regularization of the turnover of keratinocytes, protection of the Langerhans cells, and an increase in ATP production (Figures 21.5, 21.6). This leads to the possibility to use them on skin in the best way to retain its lipid and water content and rehydrate all cells at 360°, for a long period of time (Figures 21.7) (Morganti, 2009b; Morganti et al., 2010a). Moreover, a more regular production of collagen and elastin has been shown (Figure 21.8), resulting in a revitalized and more toned skin, and a reduction of face wrinkling to a

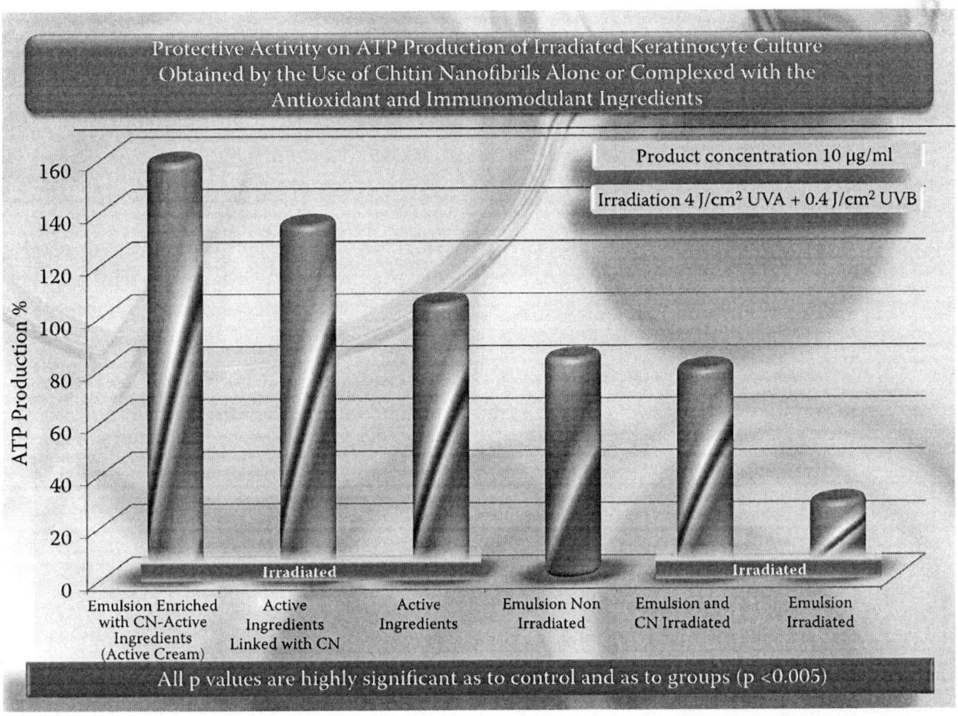

FIGURE 21.5 The *in-vitro* activity on ATP production of a cosmetic formulation (NICE approach).

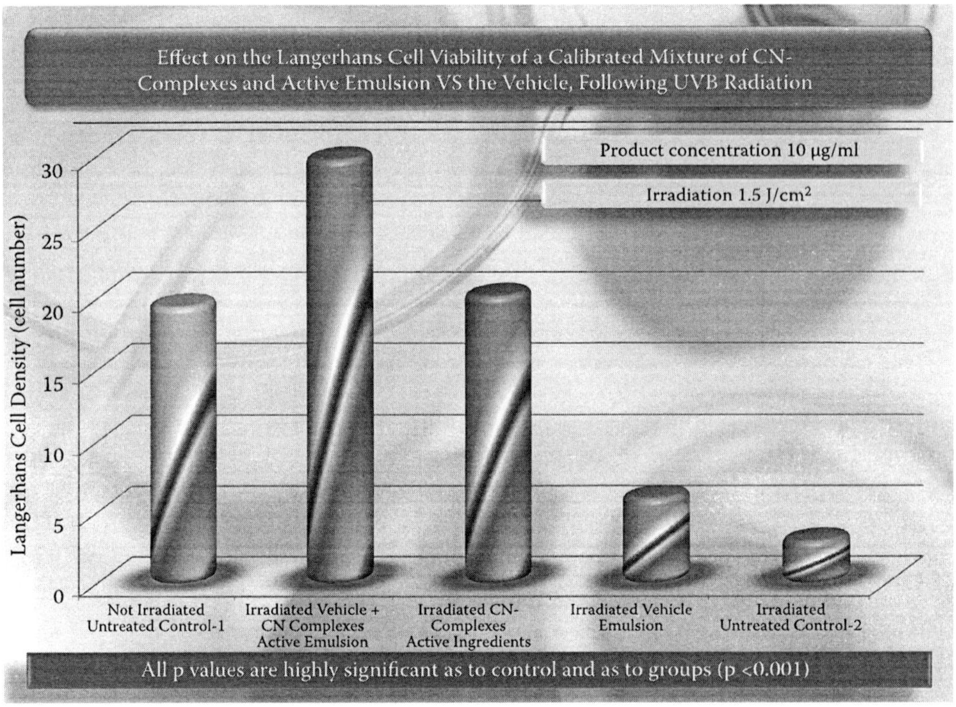

FIGURE 21.6 The *in-vitro* activity on Langerhans cells of a cosmetic formulation (NICE approach).

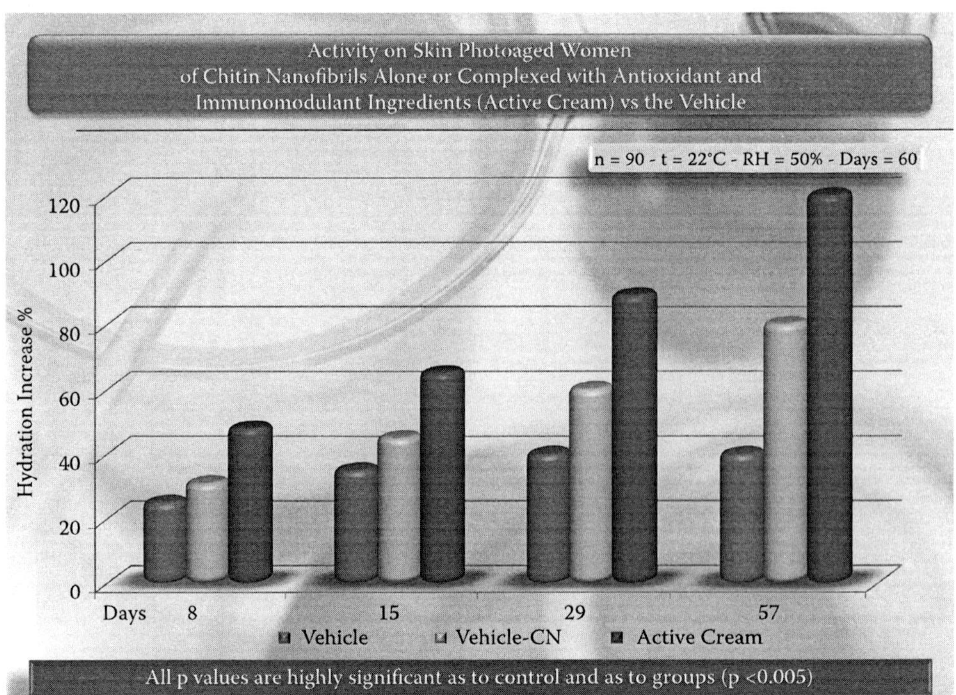

FIGURE 21.7 The *in-vivo* moisturizing activity of a cosmetic formulation (NICE approach).

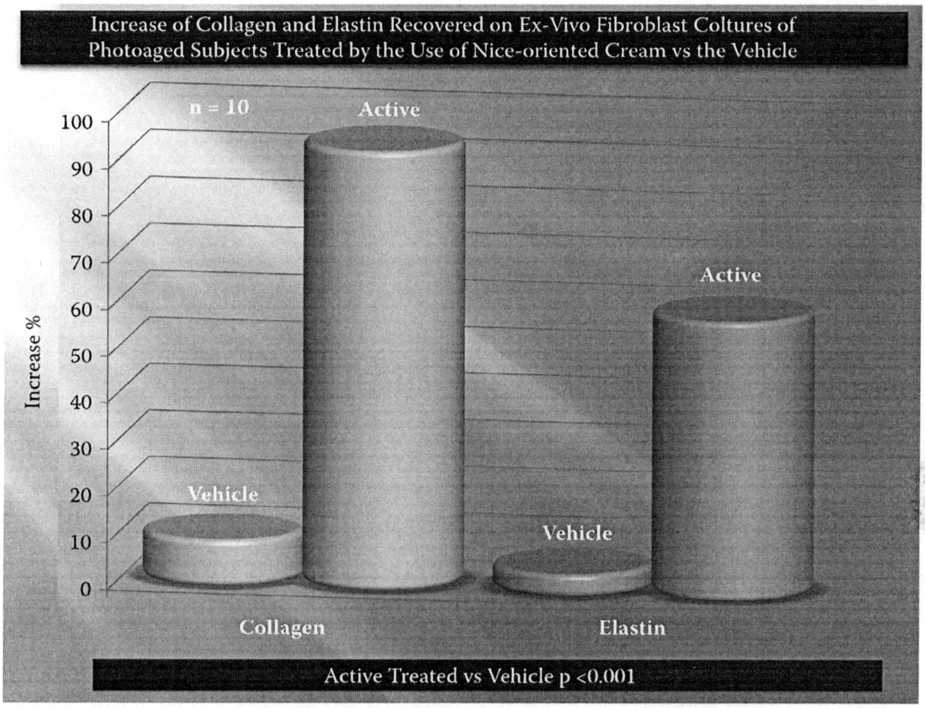

FIGURE 21.8 The *ex-vivo* anti-aging activity of a cosmetic formulation (NICE approach).

high extent (Figure 21.9). Therefore the NICE approach brings the possibility to elaborate cosmetic and dietetic formulations acting in perfect harmony with the daily life of the skin cells, according also to the Eastern concern of well-being of traditional Chinese medicine (TCM) (Morganti and Li, 2010).

21.3 VEHICLE COMPOSITION AND COSMETIC PENETRATION

21.3.1 The Vehicle as Delivery System

The ability to predict how the vehicle's composition will affect the absorption of the active ingredients into the skin depends on the knowledge of (1) the typology of the emulsion used, (2) the physicochemical characteristics of the ingredients used, (3) the facility with which the active can diffuse within the vehicle, and (4) the solubility of the active ingredients in the vehicle vs. the stratum corneum (SC). The primary purpose of the vehicle, as delivery system, is, in fact, to enable cosmetic active ingredients to be conveniently spread upon the skin and to have a good penetration throughout its layers. According to Wiechers (2008), the role of a delivery system is to ensure that the right concentration of the right chemical is reaching the right site in the body for a sufficiently long and correct period of time. And these are the so-called 4Rs of delivery. However, the delivery concept allows different considerations of the site of action. Therefore it has to be distinguished among *epidermal* delivery, where the site of action is the SC, or the viable epidermis; *transfollicular* delivery, where the site of action is represented by the hair follicle and the sebaceous glands; and *transdermal* delivery, where the target site is beyond the dermis and hypodermis. The state of hydration of the SC remains one of the most important factors in determining the rate of percutaneous absorption of a given ingredient, while the level of hydration is a function of the water concentration gradient between the dermis and the skin surface, as well as the ability of the SC to bind water. Thus the

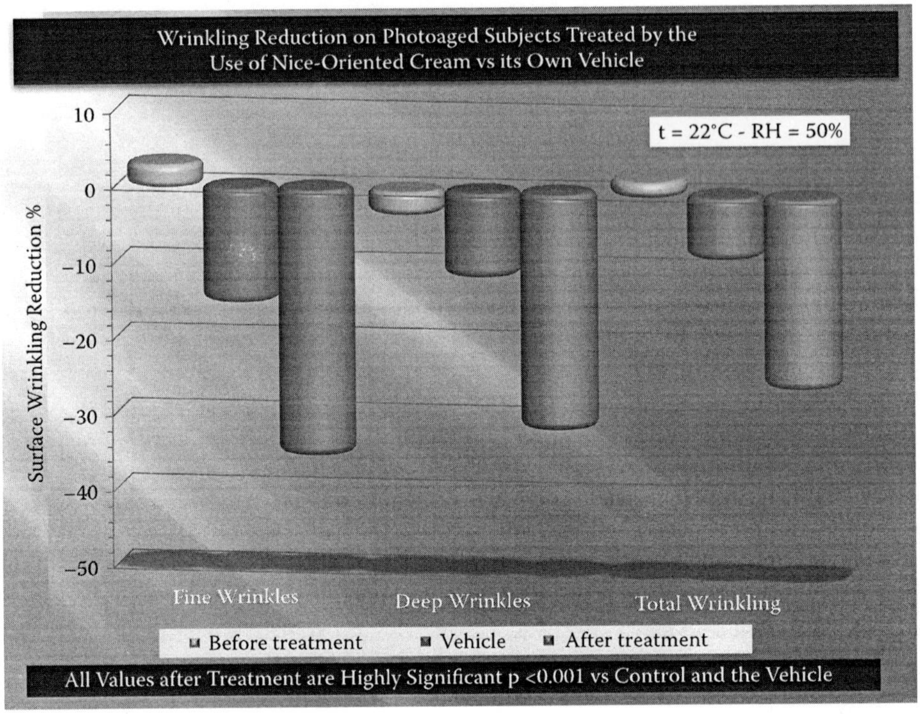

FIGURE 21.9 The *in-vivo* anti-aging activity of a cosmetic formulation (NICE approach).

hydration of the skin can be affected by both the ambient environmental conditions and the effects of applied agents on the properties of the SC. For example, topical vehicles may act as *occlusive dressings* if they contain fats and oils, reducing water loss to the atmosphere. This phenomenon increases the water content of the SC, promoting the penetration of the active ingredients. For this reason, greases or oils, particularly hydrophobic, inducing generally occlusion, increase skin hydration. Moreover, recent studies have suggested, for example, that glycerol, used as a humectant in cosmetic formulations, has the ability to maintain intercellular lipids in a lipid crystalline state, exhibiting a variety of biological skin effects. It increases the hydration of the SC, ameliorates skin turnover, enhancing desmosome degradation, and may also act as a penetration enhancer (Rawlings et al., 1994, 2004; Lodèn, 2001; Fluhr et al., 2005). However, the penetration rate of applied emulsions through normal or diseased skin is not well defined in terms of the combined effects of the vehicle and ingredients on the water content of the SC and skin penetrability, and also if the moisturizing activity of the ingredients used is considered of primary importance. The advantages of *moisturizing units* also include the substantivity of these ingredients to skin keratin (adherence to the skin), improved absorption of atmospheric water, and enhanced delivery of linked water through the skin (Martin, 1993).

21.3.2 Chitin Nanofibrils as Safe Carrier

Therefore many agents, including surfactants, phospholipids, pyrrolidones, lanolin derivatives, and different glycols, are used as vehicle components, because of their capacity to enhance the penetration of active ingredients through the SC. In conclusion, the use of the right delivery system is aimed at improving both the performance of presently used active ingredients and the *commercial* appeal of the product (Batt et al., 1998; Froebe et al., 1990). According to Magdassi and Touitou (1999), owing to the complexity of simultaneous biological activities required, it seems most likely that future cosmetic products will combine components of various delivery systems in one formulation.

To this purpose, it is interesting to underline how chitin nanofibrils (CNs) can be used as both drug and cosmetic carriers, thanks to their interesting ability to complex many active ingredients by ionic charges and facilitate their delivery across the skin (Morganti et al., 2006). In conclusion, CNs may be incorporated in a number of biological agents capable of different combined functions because of their high safety and the ability of their nanofibrils to easily travel through the intercellular space of the SC. This activity is probably due to their diffusion along the polar head group of the intercellular lipids (lipid lamellae). The lamellae pathway provides a primary barrier to the passive diffusion of water-soluble and lipid-soluble molecules across the SC, whose porous/polar organization may favor their permeation. The ample evidence of chitin biocompatibility, and in particular its favorable effect on human tissues and cells, supports the applicability of CNs in the pharmaceutical, food, cosmetic, and textile industries (Biagini et al., 2008). Finally, a cosmetic formulation, according to the NICE approach, has to be considered and studied not as a collection of single ingredients, but as a complex net of interrelations between the vehicle and all the other ingredients into a unified final product. The product and all its ingredients have, in fact, to enter in symbiosis with the harmonic composition of the human skin/body (Morganti, 2011b).

21.4 CHITIN NANOFIBRILS, COLLAGEN, AND RELATED COMPLEXES

21.4.1 Influence of Chitin Nanofibrils on Skin Cell Biology

The worldwide production of chitin has been estimated to be between 10^{10} and 10^{12} tons/year, making this natural biomarine raw material one of the most abundant by-products of industrial use (Kurita, 2006; Brück et al., 2011). In a period characterized by the necessity of producing goods by a sustainable chemistry and a green economy, to find the way for utilizing any kind available, natural biomass has became a must for all scientists and industrial makers. This is the reason why the industrial use of CNs represents an interesting and new opportunity for the pharmaceutical, cosmetic, and textile sectors. CN is one of the natural compounds and is industrially obtainable from the raw material chitin. As reported elsewhere, it is represented by nanocrystals of an average size of 240 × 7 × 5 nm capable of complexing different kinds of molecules by the formation of ionic bonds, even binding to a quantity of water (Morganti, 2010a, 2011a). However, for its polysaccharide character, CN can constitute a cell microenvironmental stimulus, influencing the correct trophism of both the skin and its appendages, and also controlling the molecular relationship of the mesenchymal epithelium and the hair follicle cycle (Frank et al., 2006; Biagini et al., 2008). Skin cells interact, in fact, with the extracellular environment (whether natural or artificial) via surface proteins such as integrins, which trigger various metabolic pathways of important roles in processing their shape, mobility, and proliferation (Rosso et al., 2004; Ng et al., 2005; Toh et al., 2006). Adequate extracellular inputs therefore prompt a local cellular response that may be accompanied by a diffuse response, while extracellular adhesion, cell proliferation, and migration can lead to their dynamic rearrangement (Tran et al., 2002; Ong et al., 2003). The influence of CNs on cell biology seems to be related to the osmotic stress induced by the particular hydrophilicity of some of its complexes obtained, for example, by the use of hyaluronic acid and collagen or other active ingredients, such as melatonin, alpha-lipoic acid, and ectoin. When skin cells are exposed, in fact, to a hypotonic extracellular fluid, they initially swell up with water, but subsequently shrink back to close their previous volume through the inflow of organic ions and osmolytes, which induce water outflow and skin hydration.

21.4.2 Chitin Nanofibril Complexes: A Two-Step Method of Preparation

In light of these considerations, we have realized different kinds of chemical complexes to be used as antiaging agents to ameliorate the turnover and life cycle of skin cells, and therefore to better the general appearance of the face and body. These complexes have been obtained by simply pouring the different kinds of ingredients into the suspension of CNs at ambient temperature, and stirring for at least 2 hours. Concentration, pH, shape, and electrical charges of the ingredients are important characteristic features for the self-assembly of the designed complex. However, different kinds of complexes are obtainable in relation to the prevalent electrical charges of the ingredient used. Because CNs are prevalently and positively charged, it is possible to obtain different kinds of polymeric microspheres or microfibrils of chitin hyaluronan by pouring hyaluronic acid (negatively charged) into the water suspension of CNs (Morganti and Del Ciotto, 2011). This complex may be spray-dried or cryo-spray-dried and controlled by x-ray diffraction. What is interesting to underline is the possibility, for example, to form a complex with CN (CN-lutein complex) by the use of oil-soluble lutein, and soon after to pour hyaluronan (HY) into the obtained suspension, for obtaining CN-HY polymeric microspheres or microfibrils. Other complexes may be easily obtained by the use of hydrosoluble molecules. However, the high solution viscosity generated by hyaluronic acid even at low concentration is the primary result of chain entanglement of its high molecular weight. Furthermore, strong intra- and intermolecular hydrogen bonding interactions allow CN and CN-HY to bind weights several times higher than their own in free water. Finally, it has been shown that combinations of HY with cationically charged polymers like CN can improve its binding to keratinized skin surfaces (Pavlichko, 1990; Morganti and Del Ciotto, 2011). Thus there is a possibility to obtain different block copolymers encapsulating both oil-soluble and water-soluble ingredients by using a two-step method. The obtained CN block polymers of excellent biocompatibility and composition resemble those of the skin structure, and have the ability to act both as transepidermal carriers and active compounds capable of rejuvenating and revitalizing different skin cells (Morganti et al., 2008a). The CN water cationic suspension, obtainable by our industrial patented method (Morganti, 2005), may be mixed with one or a series of anionic polysaccharide solutions producing precipitated block polymers and washed by water, and should be directly poured into different cosmetic emulsions. These block polymers may be prepared in a two-step way, encapsulating and thus protecting from oxidation, liposoluble, and sensible molecules, such as, for example, PUFA or vitamins C and E, facilitating also their skin delivery (Morganti et al., 2008a, 2008b).

21.4.3 Antiaging Activity of the Chitin Nanofibril Complexes

CN complexes were obtained respectively by the use of lutein, melatonin, glycine-gelatine, and hydrolyzed marine collagen (antioxidant/hydrating compounds) and ectoin (immunomodulating compound). The different active ingredients poured into gel/micro- or nanoemulsions, at 40°C by speed stirring were studied both *in vitro* and *in vivo* to evaluate their antiaging activity (Morganti et al., 2008a). From the obtained results (Figures 21.10–21.12) it is possible to underline how the contemporary use of these different complexes has an interesting antiaging effect, in comparison with the vehicle alone, also increasing skin hydration and elasticity, and ameliorating the appearance of roughness and wrinkles. As we have shown with previous studies, in fact, CN can not only complex both hydro- and liposoluble active ingredients, but also has the capacity to favor their activity, increasing their own efficacy and skin penetration (Cardillo and Morganti, 1994; Morganti et al., 2008a, 2008b, 2009). These examples underline the activity CN block polymers seem to have, as new active compounds and innovative delivery systems, allowing particular ionic bonds and connections among water, active molecules, and the cells of both the stratum corneum and the viable skin, positively influencing the feedback of the cell's signals.

21.4.4 Probable Metabolic Pathways of Chitin Nanofibril Complexes

The inter- and intramolecular bindings of CNs seem to be mediated by a group of noncovalent attractions (ionic and hydrogen bonds) that are individually weak, but whose energies can sum each other to create an effective force between two or more separate molecules or natural polymers and other cell components, according to cell necessities. The polysaccharide CN, and these block

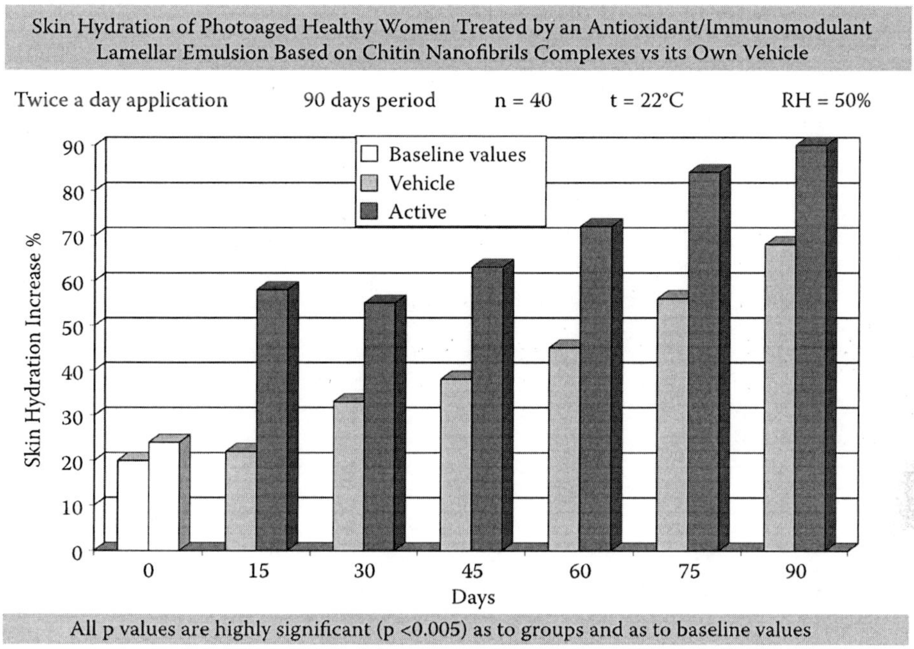

FIGURE 21.10 The *in-vivo* activity on skin hydration of chitin nanofibrils complexes versus the vehicle.

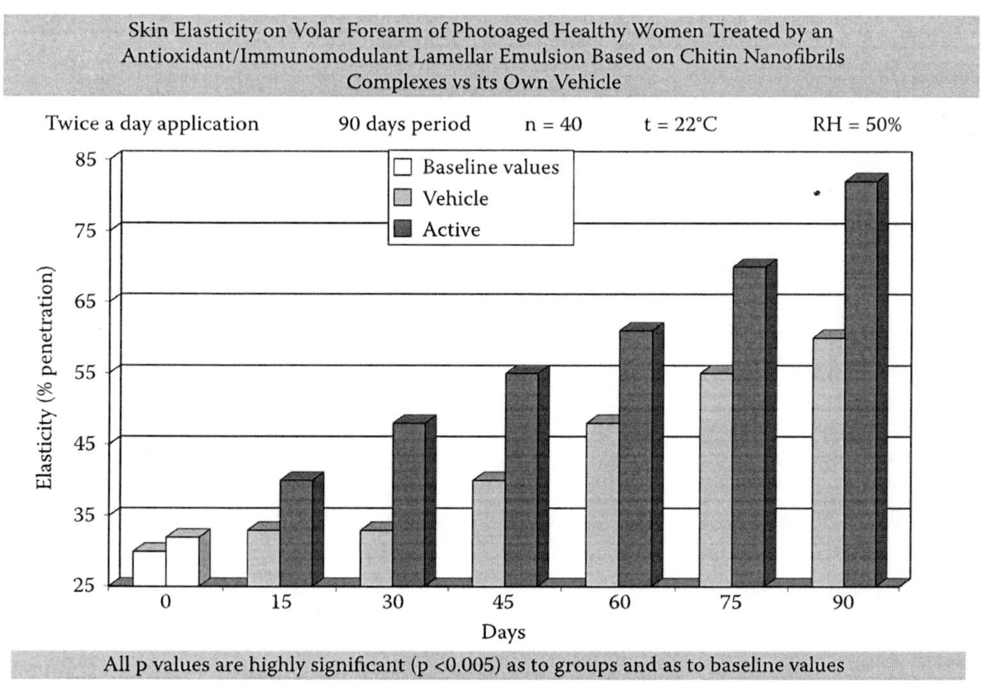

FIGURE 21.11 The *in-vivo* activity on skin elasticity of chitin nanofibrils versus the vehicle.

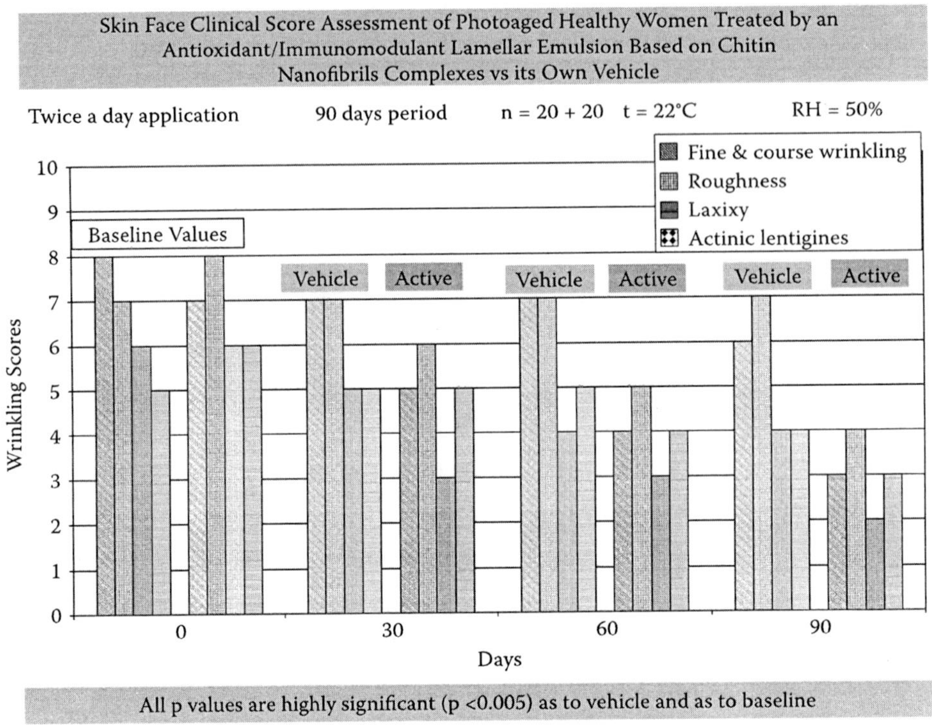

FIGURE 21.12 The anti-aging activity of chitin nanofibrils versus the vehicle.

polymers, recognized from the cell components as natural and physiological compounds, may have different fates (Morganti, 2010a, 2001b, 2011b; Morganti et al., 2011; Morganti and Del Ciotto, 2011). One part, totally hydrolyzed by the skin enzymes, may be used as a monomer subunit to construct the giant polymeric macromolecules of the cell, such as large polysaccharides. Others may act as an energy source or key component of messenger signals, and therefore are broken down and transformed into small molecules, in a maze of intracellular metabolic pathways. It is, in fact, important to remember that both glucosamine and N-acetyl-glucosamine, components of CN, are fundamental molecules of our body, while CN-bond carotenoids, collagen, melatonin, vitamin C, and alpha-lipoic acid are among the principal components of the protective and antioxidant network of the cell. In conclusion, CN, as a sugar-like polymer, may be considered a safe and biocompatible molecule for its particular character to easily complex and bind water together with many other active ingredients, and for its facility to penetrate different skin layers. For these reasons, it may be useful not only as a cosmetic and drug carrier, but also as an active component to formulate innovative cosmetic, food, and pharmaceutical products (Morganti et al., 2010a). There seem to be many other possibilities for using CN alone or bonded with other active ingredients, for formulating innovative nutraceuticals, or making new medical devices with wound healing activity and innovative fibers for biofunctional textiles (Morganti et al., 2006, 2007, 2008a, 2008b, 2010a, 2011; Muzzarelli et al., 2007; Morganti and Morganti, 2007; Morganti, 2009a, 2009b; Biagini et al., 2008; Mezzana, 2008; Morganti, 2008). For better understanding of this fascinating and rediscovered biopolymer, it is necessary to make further multidisciplinary studies, introducing new technologies and innovative biocomplexes. These new biocomplexes could have the capacity to easily interchange messenger signals with human cells. In order to understand their feedback loops, it is important to slow down their aging processes for ameliorating the quality of life and general well-being. The skin appears, in fact, as a network of interrelationships between its own different cell layers and the other cells of the body, all connected in a unique whole of nets. Unfortunately, we cannot precisely know when a particular cellular event should happen at the level of the skin, or what kind of cell energy in that moment necessary in this unique system. Therefore, we may only calculate the probability we have to obtain success in formulating and controlling, at best, the activity and efficacy of our products (Morganti et al., 2009; Morganti and Del Ciotto, 2010; Morganti, 2011b).

21.5 USE OF CHITIN NANOFIBRILS AND MARINE COLLAGEN IN INNOVATIVE COSMECEUTICALS

Natural and biodegradable polymers have attracted a great deal of interest in both therapeutic and cosmetic delivery because of their easy enzymatic degradation under normal physiological conditions. Among them, collagen and chitin nanofibrils (CNs) are amply used in the cosmetic field because of their unique properties, and they provide flexibility in the development of innovative delivery systems or in the protection of labile and easily oxidizable substances. Collagen, in fact, is the principal constituent of connective tissues like skin, tendon, and bone, representing 70% of the dermis weight. However, in order to make collagens suitable proteins for use in water-based cosmetic products and in a wide range of other applications, we need to convert them into a soluble form, which is usually obtained by enzymatic hydrolysis (Teglia and Secchi, 1999).

For our studies we used Rousselot® Peptan™ MR, marine collagen peptides of fish origin, mainly derived from type 1 collagen and with a molecular weight of 5,000 Da after hydrolysis. This commercially available collagen peptide was selected because of its particularly high amino acid content and is free of additives and preservatives. Moreover, Peptan marine is characterized by a specific amino acid content and is high in glycine, proline, hydroxyproline, and glutamic acid, which represent about 50% of the total amino acid content and give to the collagen peptides their unique properties. According to the *in vitro* studies of Bonnegarde et al. (2009), Peptan MR used in a fibroblast culture (10 days, 0.01 mg/ml) has been shown to have an effect on the differentiation of fibroblasts (Figure 21.13). Moreover, the concentration of neosynthesized collagen measured by the colorimetric method (Sirius red dye) increases 7.5% in 2 days when the fibroblast culture contains Peptan MR marine compared with a control culture containing bovine serum albumin

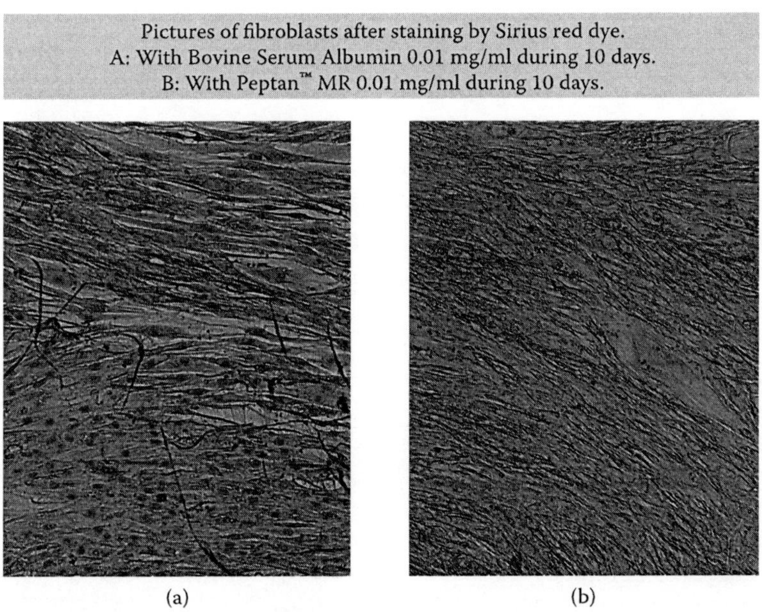

FIGURE 21.13 Differentiation of fibroblasts by the use of Peptan™ MR (b) compared with a control culture (a).

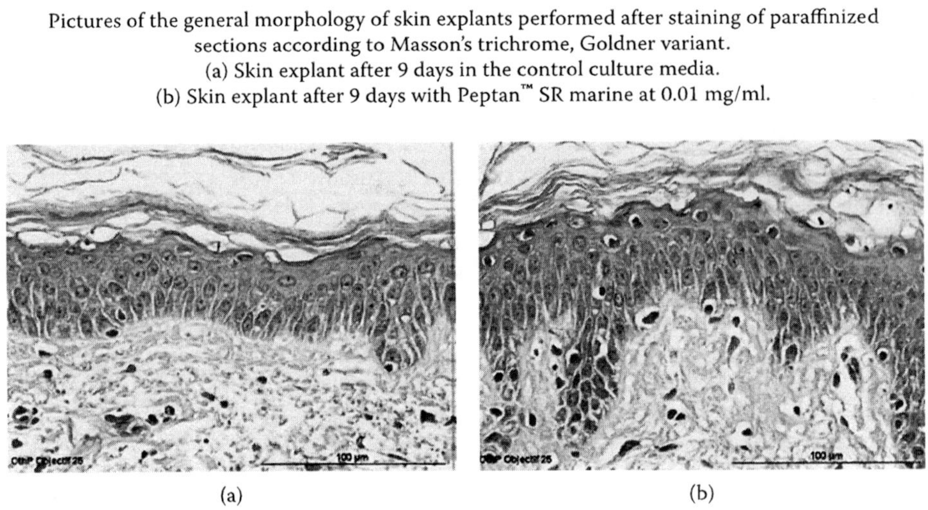

FIGURE 21.14 Skin explant treated by Peptan™ SR (b) compared with a control culture (a).

in the same conditions. When another type of Peptan from fish is used, Peptan SR marine with a molecular weight of 2,000 Da, the increase of neosynthesized collagen is 22.5% compared to the same control. More recently, *in vitro* studies carried out by Rousselot with Peptan SR marine on human skin explants indicate that fish-hydrolyzed collagen increases the density of collagen located in the papillary dermis of the skin (Figure 21.14) (unpublished results). It is shown by the increase of the collagen coloration by Masson's trichrome, Goldner variant, when Peptan SR marine is added to the culture medium at 0.01 mg/ml for 9 days. These different results indicate that fish collagen peptides, when present in the culture medium, increase collagen production in the skin by stimulating fibroblasts. The hypothesis of the *in vitro* effect of Peptan marine is that those collagen peptides are recognized by the fibroblasts as markers of endogenous collagen destruction. Hence a cascade of mechanisms in reaction to this destruction stimulates the production of collagen by the fibroblasts themselves (Ohara et al., 2010). By our recent studies *ex vivo* we have seen the possibility Peptan has to be complexed by CN (Morganti et al., 2011), notably increasing its efficacy (Figures 21.15 and 21.16).

21.5.1 Bioadhesiveness, Biodegradability, and Biocompatibility of Collagen–Chitin Nanofibril Complexes

The increased efficacy of collagen (CN) is probably due to the fact that both keratinocytes and fibroblasts should utilize more easily some of the amino acids of Peptan MR marine positively interfering with their cell networks. The resultant CN collagen oligomers have a low viscosity and a high solubility in water due probably to the short chain length and reduced molecular weight. Thus the efficacy of a complex with hydrolyzed collagen with a lower molecular weight like Peptan SR marine should be even higher. Moreover, because of CN having a structure similar to that of glycosaminoglycans, the obtained CN-Peptan complexes should be more easily recognized by the cell components to be utilized as a nutritive substratum. Due to the more frequent demand for recyclable and ecological material to save the environment, blends of chitosan, CN, and complexes obtained, for example, with polyvinyl alcohol (PVA) also may be used for producing polymeric films for food, cosmetic masques, or other specific purposes (Srinivasa et al., 2003; Morganti and Del Ciotto, 2011). Furthermore, because of their demonstrated bioadhesiveness, biodegradability, biocompatibility, and nonallergenicity, both CNs and these innovative complexes have the possibility to be used both as carrier and active compounds, offering a

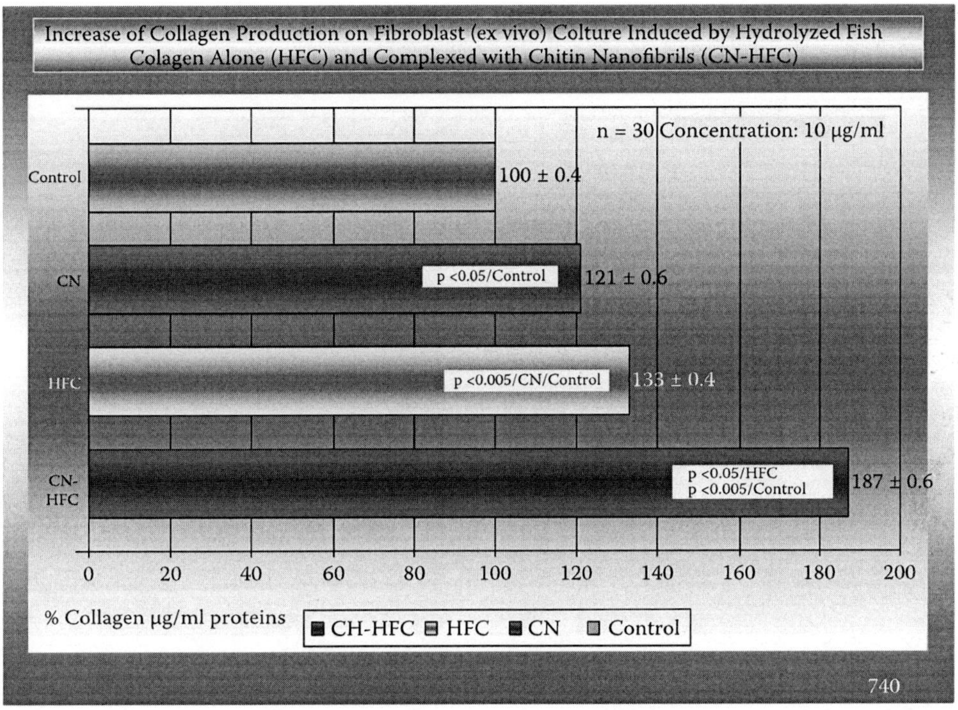

FIGURE 21.15 Increase of collagen production induced by Peptan™ SR versus Peptan™ SR complexed with chitin nanofibrils.

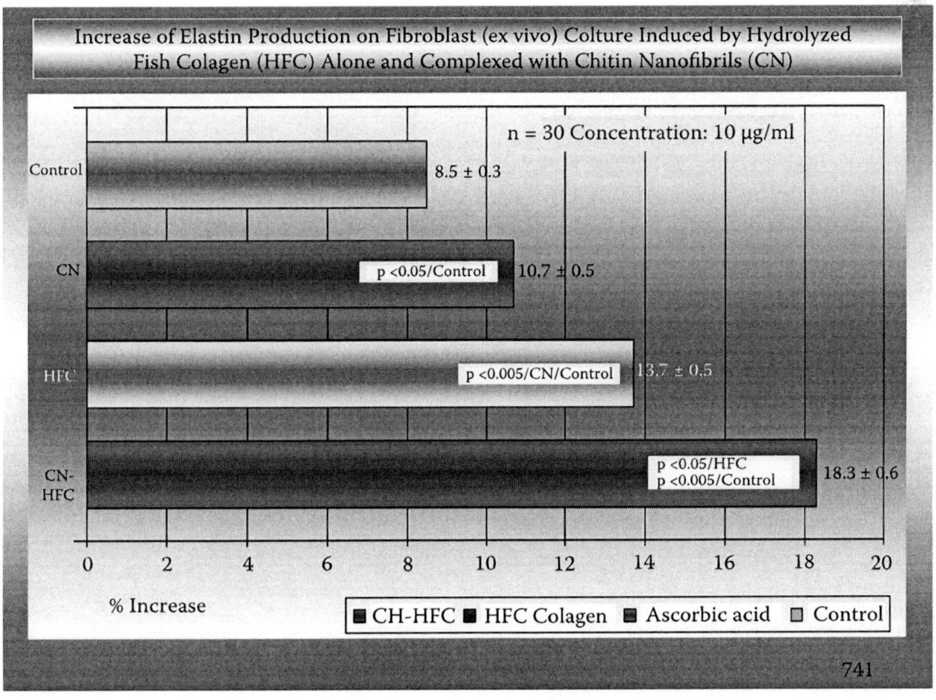

FIGURE 21.16 Increase of elastin production induced by Peptan™ SR versus Peptan™ SR complexed with chitin nanofibrils.

foundation for later research (Morganti et al., 2006, 2007, 2008b, 2009; Morganti and Morganti, 2007). Thus it should be useful to achieve a better knowledge of their molecular organization approach via a quantitative structure-activity relationship (QSAR). In this way we may rationalize the modality and process to make the complexes, optimizing their biological use (Lai and Lin, 2009). However, according to Kim and our own experiences (Kim, W.J., 2011; Morganti et al., 2010a, 2011a,b), the effective formation of these polyelectrolyte complexes takes place at a certain pH range that is within the pK_a values of the two polymers used. Generally the charge ratio of anionic to cationic block polymer, the degree of neutralization, the ionic strength, as well as the valence of single ions present in the electrolyte solution/suspension, play important roles in the formation and regulation of physicochemical properties of polymer complexes (Kekkonen et al., 2001; Gamsazade and Nasibov, 2002).

21.6 CHITIN NANOFIBRILS AND RELATED COMPLEXES FOR BIOFUNCTIONAL TEXTILES

Biofunctional and smart textiles are special fabrics that, capable of sensing changes in environmental conditions or during body functions, positively respond to these changes (Hipler and Elsner, 2006). Thus such textiles are able to adsorb substances or release therapeutic or cosmetic compounds to the skin, interacting with this protective and dynamic human organ. Because of these attributes, biofunctional textiles may guard against bacterial attack, enhance the protective function of the skin's stratum corneum, reduce skin irritation, and improve the skin's barrier properties by providing breathable, antibacterial, antiaging, and anti-UV characteristics.

According to the recent report on nanotechnology and textiles (Mantovani et al., 2009), the textile sector may be distinguished in these main sectors: fashion, interior textiles, and technical textiles, with the strongest growth in the last sector, ranging from 65 to 85 billion euros in the last 10 years. In terms of tonnage, Asia consumes 8.5 billion tons of technical textiles, while the United States and Europe consume 5.8 and 4.8 billion tons, respectively. However, both outdoors/sports and the medical textiles businesses, using nanotechnology, are rapidly developing, introducing to the market textiles with new functions, such as the reduction of unpleasant smells, absorption of moisture, UV blocking, and antiaging activity. Moreover, the application of nanotechnology has given rise to integrate textiles and electronics, for obtaining fabrics with sensing and communicating functionalities. Currently in the medical/cosmetic textiles business our research group is developing special antibacterial and cicatrizing gels and textile fibers capable of increasing wound dressing or blocking UVA rays, thus also having an antiaging approach. The special textiles obtained, by use of CNs and CN complexes, have shown the capacity to modulate the skin release of different chemicals used. At this purpose, by in-progress studies, we are introducing CNs, CN-Ag+, CN-lutein, and CN-hyaluronic acid (HY) into textiles, following two different approaches: (1) directly introducing the nanocomplexes into the cellulose raw material for the fiber production, and (2) using new electrospinning methodologies to obtain nonwoven fabrics. On the obtained tissues we controlled the stability in time of CNs alone, CN-lutein, CN-HY, and CN-Ag and the quantity of the active ingredients linked into the fibers. CNs or their complexes are, in fact, added as water suspensions at the beginning of the process step, preceding the spinning of the cellulose solution. From the first results we have seen that both lutein and Ag+, firmly bound to CN, remain anchored and distributed across the fiber matrix of the cellulose (Figure 21.17), not being affected by conventional cleansing methods (Morganti et al., 2010c). Thus we fastened our attention on CN-Ag+ and CN-lutein to have the possibility of manufacturing, respectively, antibacterial/antifungal textiles and antiaging fabrics (Biagini et al., 2008; Muzzarelli et al., 2007). It is well known that *silver's* (Ag+) antimicrobial efficacy is proportional to the availability of its bioactive released Ag+ ion, capable of interacting with bacterial or fungal cell membranes. Therefore we have prebonded to CN a nanostructured Ag+, which has demonstrated the highest efficacy with the lowest toxicity in comparison with the normal

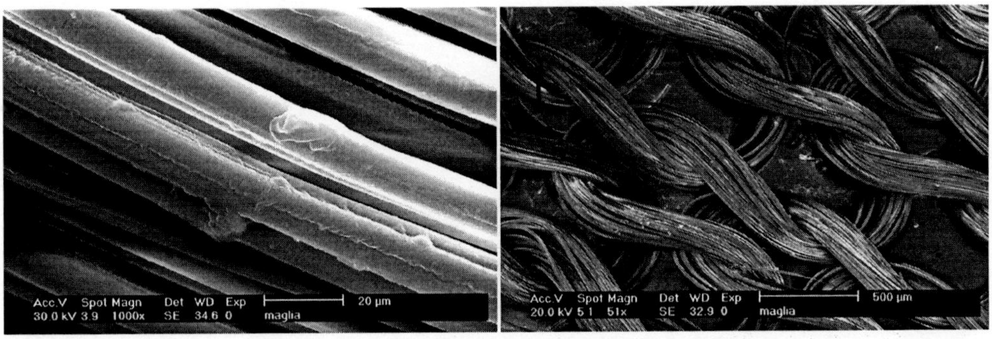

FIGURE 21.17 Cellulose-CN fibers and textile lutein bonds.

Ag^+ compounds used. This is because its nanodimension gives the molecule the possibility to act at very low concentrations (10 ppm/L), thus exhibiting minimal human risk (Landsdown, 2006; Morganti and Carezzi, 2010b).

Lutein has been selected because of its protective activity vs. both eyes and skin. Particularly, it has been recently demonstrated how lutein seems able to increase skin hydration, surface skin lipids, and skin elasticity, demonstrating an interesting photoprotective activity (Palombo et al., 2007; Morganti et al., 2008a; Morganti, 2009). What is interesting to underline is the high efficacy the complex lutein-CN seems to have, as antioxidant, photoprotective, and antiaging compound (Morganti and Morganti, 2007; Morganti et al., 2008a, 2008b). At the moment dermatologists worldwide are in fact looking for clothing that has the ability to protect the skin from incident solar radiation, reflecting, absorbing, and scattering solar wavelength. Fabrics, in fact, differ in the ability to attenuate light in this way because they differ in fiber composition and moisture content, as well as in type and concentration of dye, optical whiteners, or UV-scattering molecules absorbed and linked to the fibers. Today, there are not in the market fibers and textiles capable of neutralizing especially UVA rays (UVA_1 and UVA_2). It is known that UVA_1 rays are specially cancer promoting and UVA_2 age increasing. Thus cellulose fibers processed with chitin nanofibrils (CNs) linked with lutein represent an innovative way to protect the skin from the deleterious activity of UVA rays and blue light, when used to produce special clothing (Morganti and Morganti, 2007; Morganti et al., 2007). Thus we are presently testing the stability and regular distribution of the CN-lutein complex over the produced textile surface together with the protective and anti-UVA *in vitro* and *in vivo* efficacy of both the fiber and the finished clothing (Morganti and Berardesca, 2010). On the other side, the obtained cellulose-CN-Ag textiles, controlled *in vitro* by the agar diffusion test (Infante et al., 1996; Morganti et al., 2010a, 2010b), have demonstrated an interesting antimicrobial and antifungal activity (Morganti et al., 2010b), not affected by five standard washing cycles at 40°C.

By the second electrospinning method we have produced a nonwoven tissue comprising CNs, chitosan polymers, and hyaluronic acid (HY) (Figure 21.18). In fact, electrospinning has recently gained much interest in tissue engineering applications because of the possibility to obtain nanofibers capable of mimicking the native state of the human extracellular matrix (ECM). It is important to remember that when the diameters of polymer fiber material are shrunk from micrometers to nanometers, there appear several amazing characteristics, such as a very large surface area to volume ratio, flexibility in surface functionalities, and superior mechanical activities, which can lead to innovative nonwoven fabrics with high performances for advanced applications. Therefore the CN-HY's nanofibers are very interesting because of their geometrical and topological architecture, which seems to mimic the native state of the ECM in living tissues (Morganti and Berardesca, 2010).

Thus the introduction of nanotechnologies into the textile world has opened the path to a large number of other applications, introducing new features, such as, for example, the combination

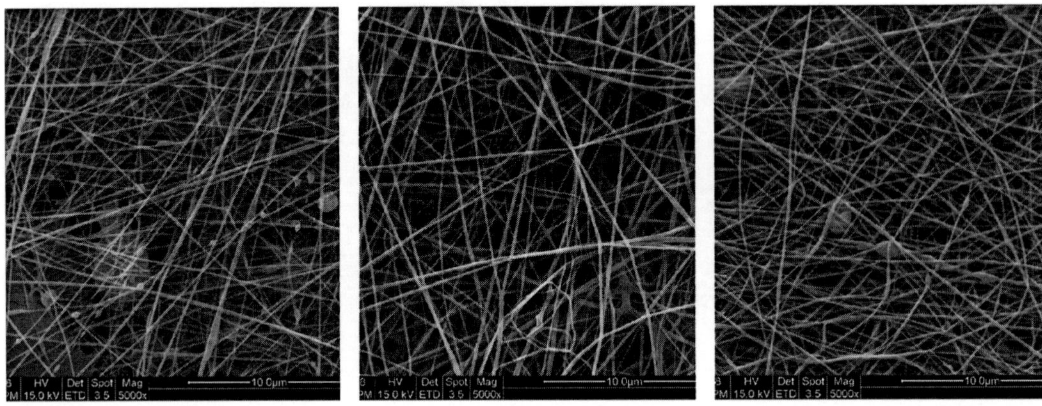

FIGURE 21.18 Nonwoven tissue comprising CN, chitosan, and HY.

of sport textiles with sensor devices to continuously monitor an athlete's physiological condition (Williams, 2002; Zhou, 2008).

CN, as a natural and biodegradable polysaccharide capable of making complexes with many active molecules, presents exciting opportunities to develop not only innovative cosmetics, food, and therapeutics, but also new electrospun nanofibers, with unexpected functional performances, to be used as vehicles for drug and cosmetic delivery and wound healing activity (Basit, 2005; Lukas et al., 2008; Morganti, 2008, 2010a, 2010b; Morganti et al., 2009).

21.7 CONCLUSION AND FUTURE PERSPECTIVES

As stated, the SC layer is considered to be crucial for maintaining the skin's barrier function. Due to this *impermeable* barrier, great is the difficulty of delivering active substances through it. Colloidal natural systems have been used for improving the cosmetic/therapeutic performance by targeting the active ingredients to the site of action, and also increasing the patent life of the product. Among them, CN and its complexes as block polymers seem to open new frontiers for improving the general delivery and efficacy of cosmetics, food, fibers, and therapeutics, when applied on the skin as emulsions or as biofunctional textiles.

These colloidal polymers, when nanostructured, are considered *smart* biomaterials. They seem to change their properties in response to selective biological recognition events, instigating biochemical signaling events capable of controlling directly some cellular behavior, according to the NICE approach. Modification of the CN and CN complex properties, amount of water, cross linking established with other chemicals, and biological skin components, density, dimension, and morphological forms of the polymers may influence the diffusion of the active ingredient in the polymeric matrix and in the cell network. Naturally, all these physicochemical phenomena have to be well known to formulate skin-friendly products, according to the NICE concept. However, the effectiveness of CN or CN complexes at the level of the cell depends not only on hand picking the ingredients and the right emulsion, but also on verifying their emollient activity, the specific emulsifier, and the methodology used in making the emulsion, as well as the organization and relative size of the micelles, and lamellae obtained. All these parameters strongly influence the release from the emulsion/textile and the transcutaneous transfollicular or transdermal penetration through the skin layers (Morganti et al., 2008a, 2008b, 2010, 2011).

In conclusion, all chitins and their derivatives have to be well characterized from the chemical and biochemical standpoint, being fully biocompatible with human cells and the environment. There are many more interesting possibilities for industrial utilization of this raw material for producing at large scale not only the different kinds of chitosan and glucosamine compounds, but also

CN and its complexes, useful for the therapeutic, food, cosmetic, and textile sectors. Recently our group has verified the possibility of using CN as a filter to purify water, removing not only organic matter but also any kind of microorganisms (Sudha, 2011; Morganti et al., 2010c). Thus nanofiber-based membranes should represent a novel series of filters of last generation to be used in place or together with present ultra-filtration membranes. In addition to its small pore size and high permeability, which make CN appropriate for a wide range of filter applications, this new membrane offers a high specific surface area, good interconnectivity of pores, and the potential to incorporate active chemicals on a nanoscale and at low cost. A more extensive use of these innovative CN compounds in the therapeutic, cosmetic, and food sectors is desirable to combine different positive results. On one hand, it should be possible to eliminate from the environment the large presence of chitin by-products. On the other hand, the obtained bio- and eco-compatible new ingredients, used as innovative compounds in the therapeutic, cosmetic, food, and textile sectors, and as biopolymers adsorbent for water and air decontamination, should ameliorate our future quality of life (Morganti and Berardesca, 2009; Morganti et al., 2010c). The final challenge will be to industrially produce these new CN filters for sanitizing the water in different areas of the world, such as many African countries, with it being possible to produce these filters at low cost and make them easy to use for inexperienced staff.

REFERENCES

Albert, B., Johnson, A., Lewis, J., Raff, M., Roberts, K., and Walter, P. 2008. *Molecular biology of the cell*. 5th ed. Garland Science, New York.

Basit, A.W. 2005. Advances in colonic drug delivery. *Drugs* 65(14): 1991–2007.

Batt, M.D., Davis, W.B., Fairhust, E., Gerrard, W.A., and Ridge, B.D. 1988. Changes in physical properties of the stratum corneum following treatment with glycerol. *J. Soc. Cosmet. Chem.* 39: 367–381.

Batzer, M.A., and Deiminger, P.L. 2002. ALU repeats and human genomic diversity. *Nature Rev. Genet.* 3: 370–379.

Bebenek, K., Matsuda, T., Masutani, C., Hanoaka, K., and Hunkel, T.A. 2001. Proofreading of DNA polymerase eta-dependent replication errors. *J. Biol. Chem.* 276: 2317–2320.

Berridge, M.J. 2005. Unlocking the secrets of cell signalling. *Ann. Rev. Physiol.* 67: 1–21.

Biagini, G., Zizzi, A., Giantomassi, F., Orlando, F., Lucarini, G., Mattioli Belmonte, M., and Tucci, M.G. 2008. Cutaneous absorption of nanostructured chitin associated with natural synergistic molecules (lutein). *J. Appl. Cosmetol.* 26: 69–80.

Blobel, G. 1980. Intracellular protein topogenesis. *Proc. Natl. Acad. Sci. USA* 77: 1496–1500.

Bonnegarde, A., Fabien-Soulé, V., Tomé, D., and Blais, A. 2009. A sensitive screening system for human type 1 collagen synthesis by fibroblasts, poster. Nutrevent, Lille, France.

Braden, C., and Tooze, J. 1999. *Introduction to protein structure*. 2nd ed. New York, Garland Science.

Breer, H. 2003. Sense of smell: recognition and transduction of olfactory signals. *Biochem. Soc. Trans.* 31: 113–116.

Brück, W.M., Slater, J.W., and Carney, B.F. 2011. Chitin and chitosan from marine organisms. In S.-K. Kim (ed.), *Chitin, chitosan oligosaccharides and their derivates*. New York, CRC Press, 11–23.

Cardillo, A., and Morganti, P. 1994. A fast non-invasive method for skin hydration control. *J. Appl. Cosmetol.* 12: 11–13.

Corrk, M.J., Mee, J.B., and Duff, G.W. 1993. Cytokines. In Priestley, G.S. (ed.), *Molecular aspect of dermatology*. New York, Johan Wiley & Son, 129–146.

Elias, P.M., and Friend, D.S. 1975. The permeability barrier in mammalian epidermis. *J. Cell. Biol.* 65 (1): 180–192.

Fluhr, J.W., Bornkessel, A., and Berardesca, E. 2005. Glycerol—just moisturizer? Biological and biophysical effects. In M. Loden and H.I. Maibach (eds.), *Dry skin moisturizer*. London, CRC Press, 227–244,

Foerster von, H., and Zopf, G.F. 1962. *Principles of self-organization*. Pergamon Press, New York.

Frank, L., Lebreton-Decoster, C., Godeau, G., Coulomb, B., and Jozeconvicz, J. 2006. Dextran derivates modulate collagen matrix in dermal equivalent. *J. Biomater. Sci. Polym.* 17: 499–517.

Froebe, C.L., Simion, F.A., Ohlneyer, H., Rein, L.D., Mattai, J., Cagan, R.H., and Friberg, S.E. 1990. Prevention of stratum corneum lipid phase transitions *in-vitro* by glycerol—an alternative mechanism for skin moisturization. *J. Soc. Cosmet. Chem.* 41:51–65.

Gamsazade, A.I., and Nasibov, S. 2002. Formation and properties of polyelectrolyte complexes of chitosan hydrochloride and sodium dextra sulphate. *Carbohydr. Polym.* 50: 339–343.

Gerhardt, C., and Hubbert, J. 2009. Innovation, substainability and complexity. A.T. Kearney meeting, Berlin, June 18–19. http://www.atkearney.de/content/veranstaltungen/bestinnovator/club.php.

Grayson, S., and Elias, P.M. 1982. Isolation and lipid biochemical characterization of stratum corneum membrane complexes: implication for the cutaneous permeability barrier. *J. Invest. Dermatol.* 78: 128–135.

Hipler, U.C., and Elsner, P. 2006. *Biofunctional textiles and the skin*. Basel, Karger.

Infante, M.R., Diz, M., Mauresa, A., Pinazo, A., and Erra, P. 1996. Note: microbial resistance of wool fabric treated with bis-Quats compounds. *J. Appl. Bateriol.* 81: 212–216.

Kekkonen, J., Lattu, H., and Stenius, P. 2001. Adsorption kinetics of complexes formed by appositely charged polyelectrolytes. *J. Colloid Interface Sci.* 234: 384–388.

Kim, S.-K. 2011. *Chitin, chitosan, oligosaccharides and their derivatives*. Boca Raton, FL, CRC Press.

Kim, W.J. 2011. Elecrostatic properties of chitosan. In S.-K. Kim (ed.), *Chitin, chitosan, oligosaccharides and their derivatives*. New York, CRC Press, 117–125.

Kurita, K. 2006. Chitin and chitosan: functional biopolymers from marine crustaceans. *Mar. Biotechnol. Chem.* 8: 203–226.

Lai, W.F., and Lin, M.C.M. 2009. Nucleic acid delivery with chitosan and its derivatives. *J. Controlled Release* 134: 158–168.

Landsdown, A.B.C. 2006. Silver in health care: antimicrobial effects and safety in use. In H.C. Hipler and P. Elsner (eds.), *Biofunctional textiles and the skin*. Basel, Karger.

Loden, M. 2001. Hydrating substances. In A.O. Barel, M. Paye, and H.I. Maibach (eds.), *Handbook of cosmetic science and technology*. New York, Marcel Dekker, 399–415.

Lukas, D., Sarkar, A., and Pokorny, P. 2008. Self organization of jets in electrospinning from free liquid surface. A generalized approach. *J. Appl. Physics* 103 (8): 1–7.

Magdassi, S., and Touitou, E. 1999. *Novel cosmetic delivery systems*. New York, Marcel Dekker.

Mantovani, E., Zappelli, P., Conde, J., Sitja, R., and Periales, F. 2009. ObservatoryNANO. Report on nanotechnology and textiles. AIRI/NANOTEC IT and Bax & Willems Consulting. Textiles report, May 2009, final proof. www.observatorynano.eu/project/filesystem/file/TS10.

Margulis, L. 1993. *Symbiosis in cell evolution*. 2nd ed. San Francisco, Freeman.

Martin, G.P. 1993. Phospholipids as skin penetration enhancers. In K.A. Walters and J. Hadgraft (eds.), New York, Marcel Dekker, 57–93.

Mezzana, P. 2008. Clinical efficacy of a new nanofibrils-based gel in wound healing. *Acta Chirurgiae Plasticae* 50: 81–84.

Morganti, P. 2005. Int patent PCT/IB2005/053576.

Morganti, P. 2008. Leather and textile chemicals: chitin nanofibrils in textiles. *Specialty Chem. Mag.* 29 (9): 26.

Morganti, P. 2009a. Dal controllo delle emozioni la bellezza globale. *Farmacia News*, November.

Morganti, P. 2009b. The photoprotective activity of nutraceuticals. *Clin. Dermatol.* 27: 166–174.

Morganti, P. 2010a. Use and potential of nanotechnology in cosmetic dermatology. *Clin. Cosmet. Investig. Dermatol.* 3: 5–13.

Morganti, P. 2010b. Chitin-nanofibrils in skin treatment. *J. Appl. Cosmetol.* 27: 251–270.

Morganti, P. 2011a. Chitin-nanofibrils and their derivatives as cosmeceuticals. In S.-K. Kim (ed.), *Chitin, chitosan oligosaccharides and their derivates*. New York, CRC Press, 531–541.

Morganti, P. 2011. NICE melody for innovative mind-body skin care. *Cosmet. Sci. Technol.* 14 (2): 27–33.

Morganti, P., and Berardesca, E. 2010. In-progress study.

Morganti, P., Carezzi, F., and Del Ciotto, P. 2010b. Unpublished data.

Morganti, P., Hong-Duo, C., Gao, X.H., and Li, Y. 2009. Nanoscience challenging cosmetics, healthy food and biotextiles. *Söfw J.* 135 (4): 2–7

Morganti, P., Chianese, A., and Stoller, M. 2010c. Unpublished data. Studies in progress.

Morganti, P., and Del Ciotto, P. 2011. Chemical physical properties and biocompatibility of chitin nanofibrils complexes. *J. Appl. Cosmetol* (in press).

Morganti, P., Del Ciotto, P., Fabrizi, G., Palombo, M., Palombo, P., and Morganti, G. 2011. Anti-ageing activity of polymeric films produced by the use of collagen-chitin nanofibrils complexes and PVA. *J. Appl. Cosmetol.*, in press.

Morganti, P., Fabrizi, G., Palombo, P., Palombo, M., Ruocco, E., Cardillo, A., and Morganti, G. 2008b. Chitin nanofibrils: a new active cosmetic carrier. *J. Appl. Cosmetol.* 26: 113–128.

Morganti, P., and Li, Y.H. 2010. Wellness and beauty outside in: China West—working together. *Söfw J.* 136(n11): 1–5.

Morganti, P., and Morganti, G. 2007. Fabricating a good treatment. *SPC* 80(4): 85–86.

Morganti, P., Morganti, G., Fabrizi, G., and Cardillo, A. 2008a. A new sun to rejuvenate the skin. *J. Appl. Cosmetol.* 26: 159–168.

Morganti, P., Morganti, G., and Li, Y. 2007. Nanostructured products: technology and future. *J. Appl. Cosmetol.* 25: 161–178.

Morganti, P., Muzzarelli, R.A.A., and Muzzarelli, C. 2006. Multifunctional use of innovative chtitin nanofibrils for skin care. *J. Appl. Cosmetol.* 24: 105–114.

Morganti, P., Palombo, M., Palombo, P., Cardillo, A., Carezzi, F., Morganti, G., Ruocco, E., and Dziergowski, S. 2010a. Cosmetic science in skin aging: achieving the efficacy by the chitin nano-structured crystallites. *Söfw J.* 136(n3): 14–24.

Muzzarelli, R.A.A., Morganti, P., Morganti, G., Palombo, P., Palombo, M., Biagini, G., Mattioli Belmonte, M., Giantomassi, F., Orlandi, F., and Muzzarelli, C. 2007b. Chitin nanofibrils/chitosan glycolate composites as wound medicaments. *Carbohydr. Polym.* 70(3): 274–284.

Nemaniak, M.K., and Elias, P.M. 1980. *In situ* precipitation: a novel cytochemical technique for visualization of permeability pathways in mammalian stratum corneum. *J. Histol. Cytochem.* 28: 573–578.

Ng, S., Wu, Y.N., Zhou, Y., Toh, Y.E., Ho, Z.Z., Chia, S.M., Zhu, J.H., Mao, H.Q., and Yu, H. 2005. Optimization of 3-D hepatocyte culture by controlling the physical and chemical properties of extra-cellular matrices. *Biomaterials*, 16: 3153–3163.

Noble, D. 2006. *The music of life. Biology beyond the genome.* Oxford University Press, Oxford.

Ohara, H., Ichikawa, S., Matsumoto, H., Akiyama, M., Fujimoto, N., Kobayashi, T., and Tajima, S. 2010. Collagen-derived dipeptide, proline-hydroxyproline, stimulates cell proliferation and hyaluronic acid synthesis in cultured human dermal fibroblasts. *J. Dermatol.* 37: 330–338.

Ong, L.L., Er, C.P.N., Ho, A., Aung, M.T., and Yu, H. 2003. Kinectin anchors the translation elongation factor-1 delta to the endoplasmic reticulum. *J. Biol. Chem.* 278: 32115–123.

Palombo, P., Fabrizi, G., Ruocco, M., Ruocco, V., Flühr, J., Roberts, R., and Morganti, P. 2007. Beneficial long-term effects of combined oral/topical antioxidant treatment with carotenoids lutein and zeaxanthin on human skin: a double-blinded, placebo-controlled study in humans. *Skin Pharmacol. Physiol.* 20: 199–210.

Pavlichko, J.P. 1990. Polymer interactions to enhance the function of hyaluronic acid. *Drug. Cosmet. Int.* 147: 21–24.

Potts, R.O., and Francoeur, M.L. 1991. The influence of stratum corneum morphology on water permeability. *J. Invest. Dermatol.* 96: 495–499.

Prakash, S., and Matouschek, A. 2004. Protein unfolding in the cell. *Trends Biochem. Sci.* 29: 593–600.

Rawlings, A.V., Canestrari, D.A., and Doblowski, B. 2004. Moisturizing technology versus clinical performance. *Dermatol. Ther.* 17: 49–56.

Rawlings, A.V., Scott, I.R., and Harding, C.R. 1994. Stratum corneum moisturization at the molecular level. *J. Invest. Dermatol.* 103: 731–740.

Rosso, F., Giordano, A., Barbarisi, M., and Barbarisi, A. 2004. From cell-ECM interactions to tissue engineering. *J. Cell Physiol.* 199: 174–180.

Schaefer, H., Redelmeier, T.E., and Benech-Kieffer, F. 1999. The skin and its permeability. In S. Madgassi and E. Touitou (eds.), *Novel cosmetic delivery systems.* New York, Marcel Dekker, 9–49.

Shroot, B. 1993. Medical chemistry in dermatology. In G.C. Priestley (ed.), *Molecular aspects of dermatology.* New York, John Wiley & Son, 189–209.

Srinivasa, P.C., Ramesh, M.N., Kumar, K.R., and Tharanathon, R.N. 2003. Properties and sorption studies of chitosan-polyvinyl alcohol blend films. *Carbohydr. Polym.* 53: 431–438.

Strachan, T., and Read, A.P. 2004. *Human molecular genetics.* New York, Garland Science.

Sudha, P.N. 2011. Chitin/chitosan and derivatives for waste water treatment. In S.-K. Kim (ed.), *Chitin, chitosan, oligosaccharides and their derivatives.* New York, CRC Press, 561–585.

Teglia, A., and Secchi, D. 1999. Proteins in cosmetics. In E.D. Goddard and J.V. Gruber (eds.), *Principles of polymer science and technology in cosmetics and personal care.* New York, Marcel Dekker, 391–464.

Thiele, J., and Elsner, P. 2000. *Oxidants and antioxidant in cutaneous biology.* Basel, Switzerland, Karger.

Toh, Y.C., Ng, S., Khong, Y.M., Zhang, X., Zhu, Y., Lin, P.C., Te, C.M., Sun, W., and Yu, H. 2006. Cellular response to nanofibrous environment. *Nano Today* 1: 34–43.

Tran, H., Pankov, R., Tran, S.D., Hampton, B., Burgess, W.H., and Yamada, K.M. 2002. Integrin clustering induces kinectin accumulation. *J. Cell. Sci.* 115: 2031–40.

Wiechers, J.W. 2008. *Science and application of skin delivery systems.* Carol Stream, IL, Allured Publ. Co.

Williams, D. 2002. The enabling technological advantages of chitin. January/February 8–9. www.medicaldeviceserviceonline.com.

Zhou, X. 2008. Project report ECG53. zhouxinan0129@gmail.com.

22 Screening Strategies for the Discovery of Marine Microbial Cosmeceuticals

Se-Kwon Kim and Ira Bhatnagar

CONTENTS

22.1 INTRODUCTION

Finding novel and efficient compounds of natural origin has been a major point of concern for research in the pharmaceutical and cosmeceutical sciences. This is mainly because of their high affinity to the target, little loss of entropy when they bind to a protein, and their bioavailability. Moreover, natural compounds are quite flexible in conformational acquisition in aqueous and lipophilic environments (Bhatnagar and Kim, 2010). Nature is a storehouse of a vast variety of flora and fauna, including microbial inhabitants that contribute chemically diverse compounds with great therapeutic and cosmeceutic potential. These compounds further have a giant impact on the drug discovery regime of the present day (Da Rocha et al., 2001).

The marine environment harbors a variety of ecosystems with inter- and intraspecies interactions leading to the production of novel compounds in response to mechanisms of predation and self-defense. Such natural compounds cover a wide range of bioactivity, such as photoprotective (Pallela et al., 2010), antihelmintic, antibacterial, anticoagulant, antifungal, anti-inflammatory, antimalarial, antiprotozoal, antituberculosis, antiviral, and other miscellaneous mechanisms of action (Mayer et al., 2009). For some decades, marine pharmaceutics has been a developing field in the cosmeceutic production and development arena. The various inhabitants of the marine environment, including bacteria, fish, algae, corals, crustaceans, and even sea mud, are known to have tremendous potential to be utilized as cosmeceutical agents (Kim et al., 2008). Marine antibacterial and antifungal compounds (Table 22.1) can serve as excellent sources of cosmeceutic agents viz. skin creams, ointments, and

TABLE 22.1
List of Selective Antimicrobials Produced by Marine Microorganisms for Probable Use in the Cosmeceutical Industry

Compound Name	Chemical Structure	Class of Microorganism	Activity	Reference
BE-43472B	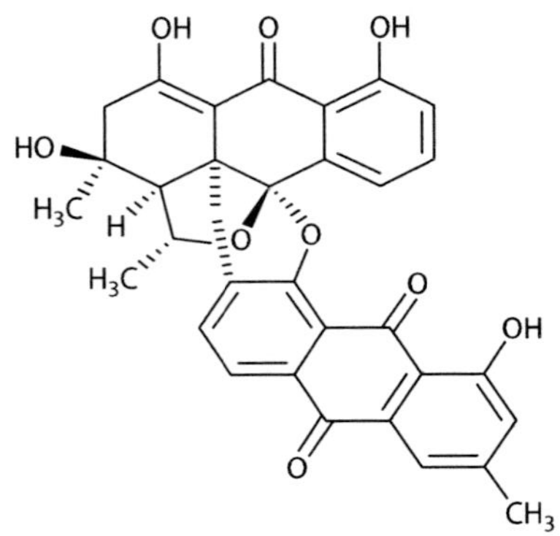	Symbiotic microorganism	Antibacterial (anti-MRSA and anti-VRE)	Nicolaou et al., 2009
DAPG	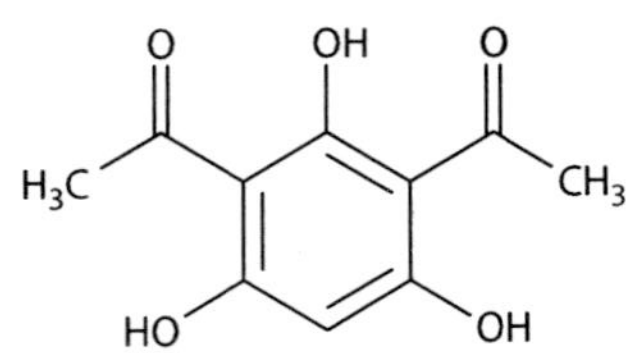	Symbiotic microorganism	Antibacterial (anti-MRSA, anti-VRSA, and anti-VRE)	Kamei and Isnansetyo, 2003
Macrolactin V	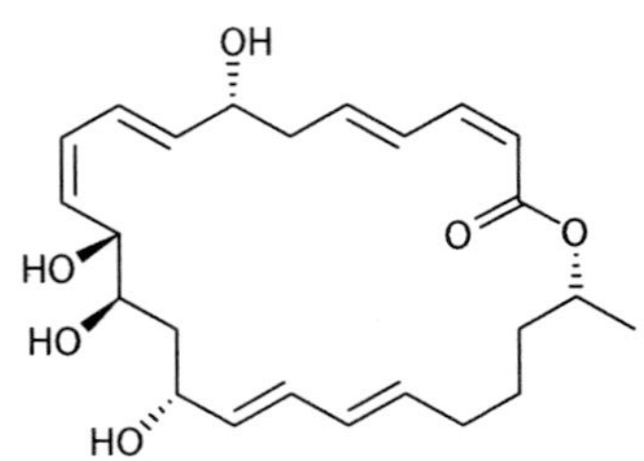	Symbiotic microorganism	Antibacterial and antilarval	Gao et al., 2010

Alternaramide	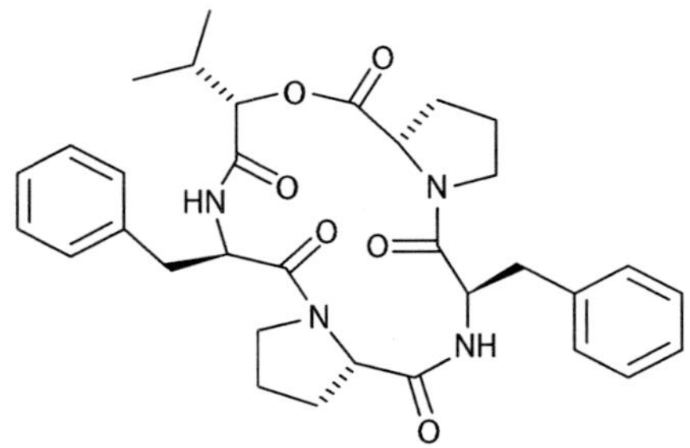	Fungi	Antibacterial	Kim et al., 2009
Pyrone	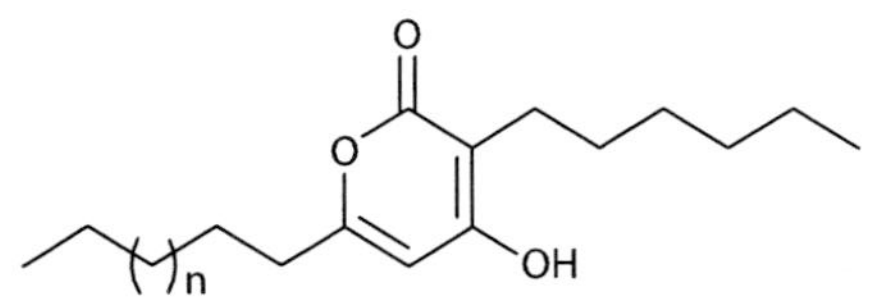	Bacteria	Antibacterial	Maya et al., 2003
MC21-B	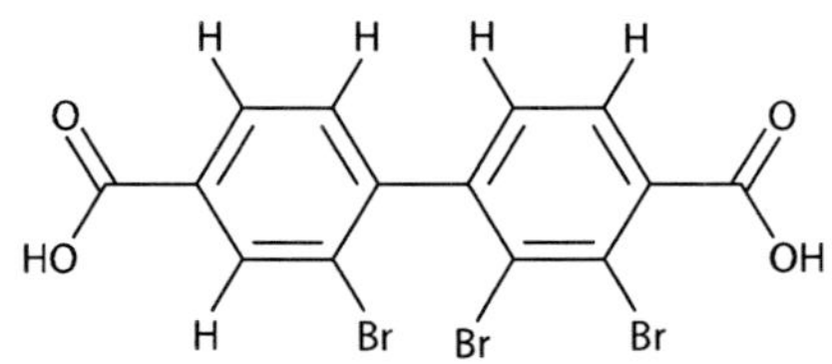	Bacteria	Antibacterial	Isnansetyo and Kamei, 2009
Macrolactin S	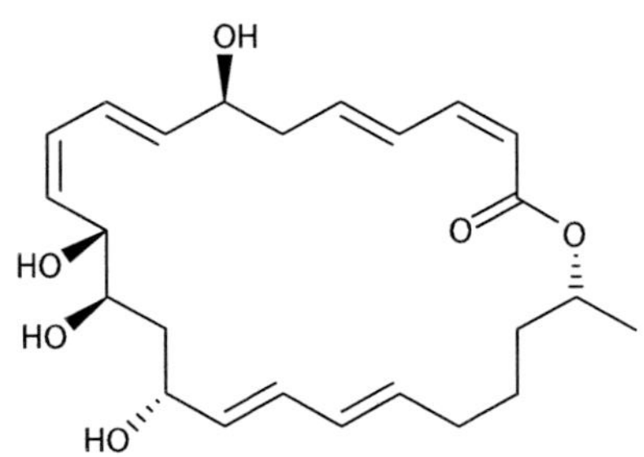	Bacteria	Antibacterial	Lu et al., 2010.

(continued)

TABLE 22.1 (Continued)
List of Selective Antimicrobials Produced by Marine Microorganisms for Probable Use in the Cosmeceutical Industry

Compound Name	Chemical Structure	Class of Microorganism	Activity	Reference
Marinomycin A	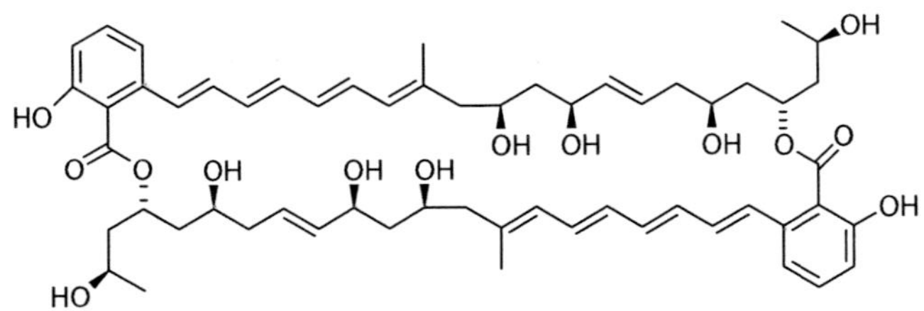	Actinomycetes	Antibacterial	Kwon et al., 2006
Resistoflavine	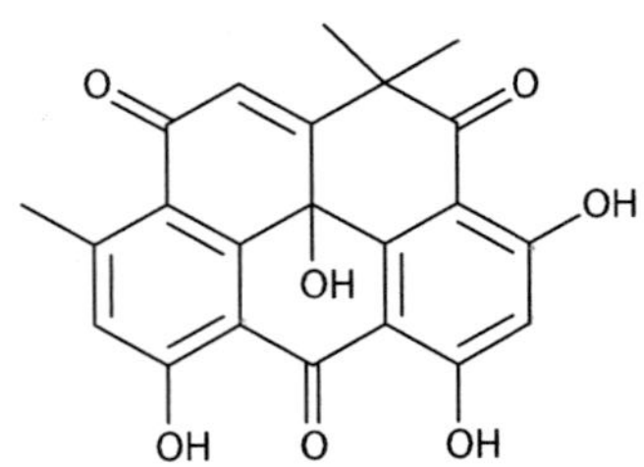	Actinomycetes	Antibacterial	Gorajana et al., 2007
Toyocamycin	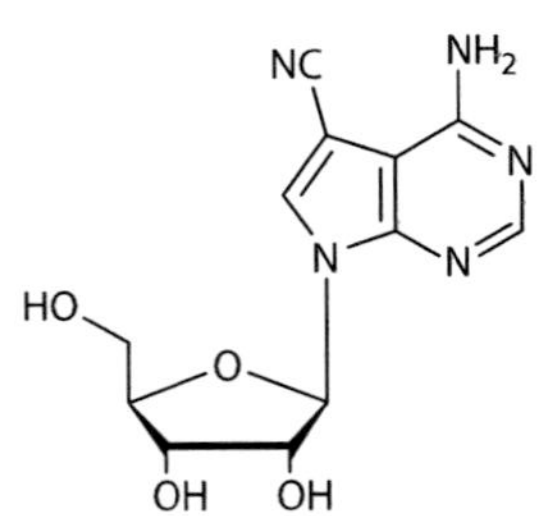	Cyanobacteria	Antifungal	Burja et al., 2001

oral supplements against microbial agents such as bacteria and fungi. Despite a great deal of research in the field of biotechnology, marine exploration for novel and more efficient cosmeceutic agents has been a drought area. Various aspects have to be looked into in order to ensure a continuous and consistent supply of marine cosmeceutics (Figure 22.1). Some of these aspects are discussed in this chapter.

22.2 POTENTIAL OF MARINE MICROBES AS COSMECEUTIC PRODUCERS

A unique source in the world's oceans, the deep ocean and geothermal vents, is now becoming the focus of considerable microbiological interest (CORHH, 1999). Marine microbiota not only leads in quantitative measures, but also has tremendous potential to represent a biomedical resource of unknown magnitude and great promise (Figure 22.2). All that is needed is the correct approach to increase the production of cosmeceuticals of marine microbial origin.

Planococcus maitriensis Anita I is known to produce exopolymers, and they can be used for bioremediation, enhanced oil recovery, and as cosmetic additives (Kumar et al., 2007). Yet another exopolysaccharide (EPS) produced by marine bacterium isolated from the marine annelid *Alvinella pompejana* is a major constituent of the cosmetic formulation market: DEEPSANE™. The composition of this EPS includes glucose, galactose, rhamnose, fucose, and mannose as neutral sugars, along with glucuronic and galacturonic acids and a diacidic hexose identified as a 3-O-(1-carboxyethyl)-D-glucuronic acid (Cambon-Bonavita et al., 2002). Ectoines are a class of compounds that can be used for the protection of human skin against damaging environmental influences. Certain halophilic bacteria are known to produce such ectoines that provide multiple cosmetic benefits such as immune protection, cell protection, UV protection, and membrane protection (Khmelenina et al., 2000). Apart from these, UV protectants find extensive application in cosmetic sciences. Biopterin glucose, a UV-A protective pigment produced from a marine planktonic cyanobacterium, is now being used in sunscreen formulations (Matsunaga et al., 1993). Marine bacteria belonging to the genus *Microbulbifer* produce 4-hydroxybenzoate. The alkyl derivatives of these hydroxybenzoates, parabens, are extensively used as preservatives in a wide variety of food, pharmaceutical, and cosmeceutical industries, owing to their low toxicity to human cells and antimicrobila activity (Kim et al., 2008). Adding to the range of photodamage protectants, scytonemin is a carotenoid produced by marine cyanobacteria and can be used as a UV sunscreen cosmetic material (Takamatsu et al., 2003). Some marine bacteria produce hyaluronic acid, which finds its use in the cosmetic industry. Certain enzymes have been isolated from the extremophile *Thermus thermophilus* and are used for developing commercial products for skin protection by Sederma, Inc. (France) and California Tan, Inc. (United States), which are effective in the treatment of damaged skin exposed to UV radiation and heat (Lintner et al., 2002). Algae (micro and macro) are also known to be excellent sources of UV protective compounds and enzymes.

22.3 ISOLATION OF MARINE MICROORGANISMS

As has been listed in various research articles and reviews, marine microbes are an excellent source of secondary metabolites of various degrees of bioactivity. It is not easier to select an efficient candidate for cosmeceutic production, just on the basis of literature survey and expectations. The path to a good cosmeceutical agent involves many pitfalls and has to cross many milestones, from the time of conception of the idea of cosmeceutic generation. The first step in this regard is the isolation of microorganisms. The enormity of the marine environment and its uncountable microbial population makes it even more difficult to choose the proper candidate. More than half of the marine microbial consortium remains undisclosed owing to certain restrictive factors. One such factor is lack of knowledge of isolation procedures for marine microorganisms. With growing advances in the field of microbiology and biotechnology, new and innovative techniques have come to light in recent years to increase the efficiency of the isolation of novel microorganisms from the marine biosphere.

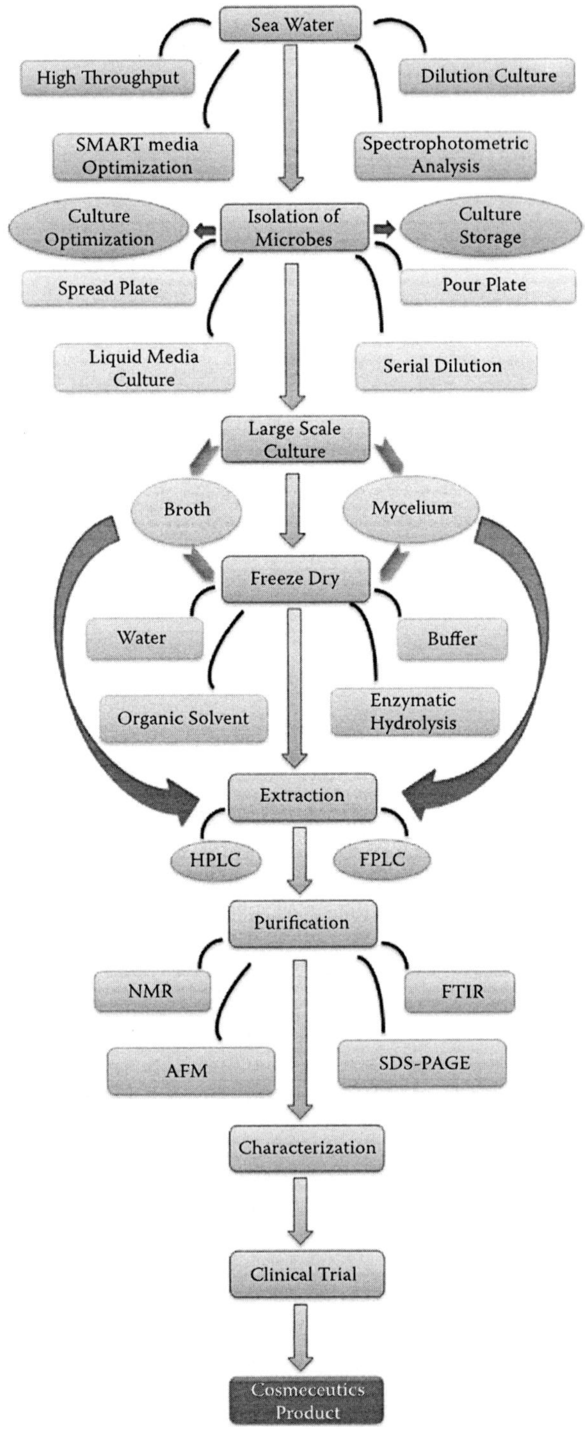

FIGURE 22.1 Schematic representation of steps involved in cosmeceutics production from the marine environment.

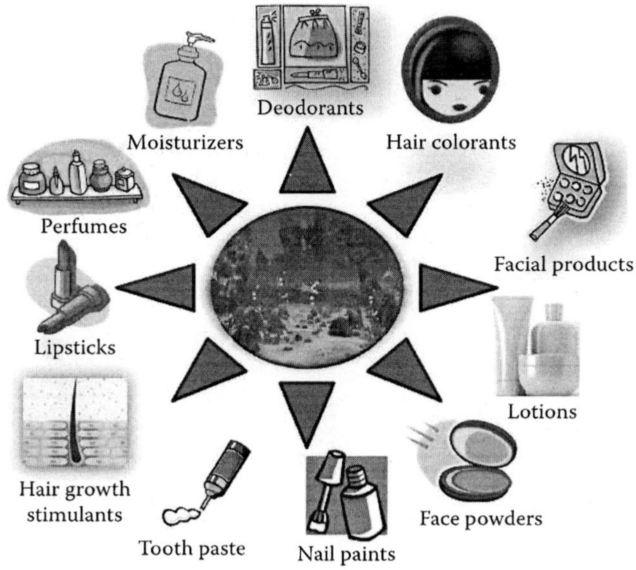

FIGURE 22.2 Range of cosmeceutical products from the marine environment.

These include growth media modifications, high-throughput screening, and endpoint dilution methods using microtiter dish plate formats (Connon and Giovannoni, 2002; Rappe et al., 2002; Giovannoni et al., 2005).

22.3.1 High-Throughput Screening

The most commonly used culture medium for isolation of marine bacteria has been marine agar, formulated by Claude ZoBell (1946). The limitation of this medium lies in the higher concentration of organic carbon, compared to natural environments and short incubation periods, thus giving a false indication of growth patterns under natural conditions, as cultures in this medium grow comparatively faster than their native counterparts. Mincer et al. (2002) have addressed the high carbon content issue and the relatively short incubation times and have significantly improved the techniques for the isolation of novel marine actinomycete strains from sediments. They have employed several concentrations of organic carbon from diverse sources, longer incubation times, and the pretreatment of the samples with heat shock to enrich spore-forming microorganisms (Mincer et al., 2002). As a result, novel lineages of marine actinomycetes have recently been isolated and cultured from diverse marine sediments (Toledo et al., 2006). Further, based on data generated by culture-independent studies on the *in situ* diversity as seen by ribosomal analysis of the actinomycetes, Maldonado et al. (2005) selected carbon sources for their targeted isolation. Yet another novel method has recently come into the picture, when Diversa Corporation developed a high-throughput cultivation (HTC) technology that employs agarose microcapsules to encapsulate single cells directly from environmental samples. The microcapsules are then transferred into mini-fermentation columns for growth and microcolony development. Culture medium containing nutrient concentrations in accordance with the natural environment is then added to the column. This method has been suggested to be suitable for massively parallel cultivation of microorganisms for natural product screening and drug discovery (Keller and Zengler, 2004) for effective use in the cosmeceutical industry.

22.3.2 Culture Media Optimization

The most important aspect of microbiology and microbial culture isolation is media optimization. It is easier to formulate media for culturing terrestrial microbes owing to their obvious nutrient requirements, but since the oceanic environment is so diverse and high in salt content, designing an efficient medium is a tough task. More than two decades ago, Kaneyama et al. (1987) used media supplemented with sardine and cuttlefish powder to culture marine mud-associated bacteria. In fact, the use of minimal media that mimic the natural marine environment, supplemented with complex micro- and macronutrients required by the marine microbial consortia, would be the best choice for efficient isolation of marine microorganisms and the cosmeceutical product thereafter (Gal and Muller-Feuga, 1998).

Way back in 1988, Bechet and Blondeau (1988) reported that submerged cultures of *Streptomyces* sp. produce tetracycline when grown in an iron-deficient medium. A decade later, Imada et al. (1998) reported that addition of selenomethionine to the marine bacterial cultures resulted in the enhanced production of antibiotics. Media containing sea water seem to be the best source of providing necessary nutrients in appropriate amounts for the isolation of producer microorganisms. The type of sugar present may be altered from glucose to tryptone or mannose to mannitol, depending upon the choice of cell component to be isolated. Addition of antibiotics to the medium may also be very effective for the specific isolation of fungi (cycloheximide sensitive) and bacteria (ampicillin, penicillin, and streptomycin sensitive). Variations in the type of peptone, such as SOY peptone or MEAT peptone, may also be a good aspect to look into. A pronounced increase in cultivation was achieved by Bruns et al. (2002) when they added cyclic AMP (cAMP), N-butyryl homoserine lactone, or N-oxohexanoyl-DL-homoserine lactone at a low concentration of 10 μM to the cultivation medium for heterotrophic bacteria from the central Baltic Sea. The most effective inducer was found to be cAMP, which led to cultivation efficiencies of up to 100% of total bacterial counts (Bruns et al., 2002). A recent work has reported the use of SMART as a new method for detection and isolation of bacteria. SMART stands for selective medium-design algorithm restricted by two constraints. The authors have used this novel strategy for designing highly selective media based on two selective agents, a carbon source and antimicrobials (Kawanishi et al., 2011).

22.3.3 Microtiter Plate Method

Another approach to culture the uncultured phylotypes from oligotrophic marine ecosystems is a high-throughput culturing procedure that utilizes the concept of extinction culturing to isolate cultures in small volumes of low-nutrient media. This technique was used by Connon and Giovannoni (2002) to isolate marine bacteria that were cultivated at *in situ* substrate concentrations—typically three orders of magnitude less than common laboratory media. As per their report, microtiter plates and a newly developed procedure for making cell arrays were employed to raise the throughput rate and lower detection sensitivity, permitting cell enumeration from 200 μl aliquots of cultures with densities as low as 10^3 cells/ml. This study led to the screening of approximately 2,500 extinction cultures from 11 separate samplings of marine bacterioplankton. This technique was helpful in isolating four unique cell lineages of marine Proteobacteria clades that were previously known to be unculturable (Connon and Giovannoni, 2002). The count of cells cultivated by this technique is far beyond those obtained by traditional microbiological culturing techniques.

22.4 MICROBIOLOGICAL AND BIOTECHNOLOGICAL STRATEGIES FOR COSMECEUTICAL PRODUCTION

Once the microbial population is isolated in a pure form, the next challenge remains to select the most suitable candidate for the production of the cosmeceutical of choice. The task of making a choice is quite tricky, as one organism may be easy to grow but not very effective in production of some bioactive compounds, whereas other producer organisms may be difficult to grow and yet produce bioactive substances in bulk. Certain approaches are now on the market for selection of appropriate candidates for cosmeceutics production, some of which are discussed here.

22.4.1 Spectrophotometric Method for Identifying Microbial Populations

The idea of developing newer approaches for enhancing marine microbial cosmeceutical production is to bring out a novel approach that is simple, rapid, accurate, and inexpensive. The spectrophotometric approach can be used with both crude extracts and pure compounds. This is an ideal method for implementation in a chemistry laboratory by nonmicrobiologists, can be run under nonaseptic conditions, requires only small amounts of materials, and gives results in 2 hours.

This technique works on the principle that nonspecific esterases occur widely among microbes (Lundgren, 1981) and have been shown to hydrolyze colorless fluorescein diacetate (FDA) to fluorescein, a yellowish green compound that absorbs strongly in the visible range (Grasselli and Ritchey, 1975). FDA hydrolysis/staining has found application in the determination of microbial populations from various sources (Swisher and Carroll, 1980; Sugar et al., 1983; Ferderle et al., 1990; McCarthy, 1987), including sea water (Gilbert et al., 1992).

Swisher and Carroll (1980) demonstrated, by spectrophotometric means, that the amount of fluorescein produced by hydrolysis of FDA provided a good approximation of the cumulative biomass of fungi, bacteria, and algae growing on Douglas fir (*Pseudotsuga minziesii*) foliage. The fluorescein absorbance was directly proportional to the microbial population, and a standardized method was developed to determine microbial biomass. Gilbert et al. (1992) showed that fluorescein is released quantitatively as a function of the number of cells of the marine microalga *Tetraselmis suecica*.

22.4.2 Calorimetric Broth Microdilution Method as Growth Indicator

Cultures produced by dilution of the original population to near extinction of the ability to grow are called dilution cultures. These appear to have advantages for studying oligobacteria because culture selection then favors the most abundant, rather than the most nutrient-tolerant organisms. Dilution culture, a method for growing the typical small bacteria from natural aquatic assemblages, was developed in 1992 when Button and co-workers attempted to isolate and culture aquatic bacteria (Button et al., 1993). Microbial populations were measured, diluted to a small and known number of cells, inoculated into unamended sterilized sea water, and examined three times for the presence of 104 or more cells per ml over a 9-week interval.

The Alamar Blue (resazurin) assay incorporates a colorimetric and fluorometric growth indicator based on detection of the metabolic activity of cells. The Alamar colorimetric system is easy to use and the results are easy to read. It appears to be an acceptable method for antimicrobial susceptibility testing of *Staphylococci* and *Enterococci*, which in turn helps in their use in combinatorial therapy with other cosmetic agents and drugs of a cosmeceutical nature (Baker and Tenover, 1996). The colorimetric oxidation-reduction-based broth microdilution test was indeed found to be reproducible when six laboratories determined the minimum inhibitory concentrations (MICs) of fluconazole and amphotericin B independently for 100 coded yeast isolates using this method, and the results were 99.4% similar with fluconazole and 98.6% similar with amphotericin B (Pfaller et al., 1994). Tiballi et al. (1995) examined the reliability and accuracy of the colorimetric assay using Alamar Blue reagent in the performance of susceptibility tests for *Candida albicans*. They compared the broth macrodilution method recommended by the National Committee for Clinical Laboratory Standards (NCCLS) with a macrodilution method modified with the Alamar reagent and a microdilution method modified with the Alamar reagent. Their results suggested that the methods modified with the Alamar reagent yielded clear endpoints and excellent correlation for the broth macrodilution and microdilution methods (Tiballi et al., 1995).

22.4.3 Metagenomic Approaches

Natural compounds, especially those derived from marine microbial populations, are an excellent source of cosmeceutic agents. However, the development of these compounds into drugs has been held back by difficulties in achieving a sustainable supply of these often complex molecules for preclinical and clinical development. An approach that has been successfully employed to identify the biosynthetic source of secondary metabolite natural products has been the use of metagenomics. Metagenomics offers a biotechnological solution to this supply problem. This method involves the genomic analysis of unculturable microorganisms by direct extraction and cloning of DNA from an assemblage of microorganisms (Handelsman, 2004). Heterologous expression approaches have also led to the isolation of secondary metabolism gene clusters from uncultured microbial symbionts of marine invertebrates. Combining a metagenomic approach with heterologous expression holds much promise for the sustainable exploitation of the chemical diversity present in the sponge microbial community (Kennedy et al., 2007). In the late 1980s, direct analysis of rRNA gene sequences had shown that the vast majority of microorganisms present in the environment had not been captured by culture-dependent methods (Handelsman, 2004). This approach has been successfully employed with marine sponge-associated microflora by generating a sponge metagenomic library and subsequently screening this library with gene fragments or gene clusters from biosynthetic pathways involved in the production of secondary metabolites. Given the fact that polyketides are an important class of bioactive bacterial secondary metabolites, efforts have been focused on isolating polyketide synthase (PKS) gene clusters from sponge-associated bacterial metagenomic libraries (Schirmer et al., 2005; Kim and Fuerst, 2006).

22.5 CONCLUDING REMARKS

The saga of marine microbial bioactive metabolites is continuing, with new compounds being added day by day. Our knowledge, however, is still a miniscule of what exists deep in the oceans. The use of improved cultivation approaches for the discovery of marine natural products from marine microbes is of paramount importance for the development of new pharmaceuticals. Marine microbial bioactive substances have indeed been a rich source of lead molecules in cosmeceutics and drug discovery, but because of underlying difficulties in obtaining high-quality natural products or in applying modern screening assays, their use in the cosmeceutical industry has been discouraged. Fortunately, new technologies in mass spectrometry (electrospray ionization ion cyclotron resonance fourier transform mass spectrometer [ESI-ICR FTMS], fluorescence-activated cell sorting coupled with mass spectrometer [FACS-MS), nuclear magnetic resonance [NMR], structure-activity relationships by NMR [SAR by NMR], saturation transfer difference NMR [STD-NMR]) and other spectroscopic techniques can greatly facilitate the efficient creation of high-quality natural product libraries, bimolecular target or cell-based screening, and early-hit characterization. The use of mass spectrometry and high performance liquid chromatography-mass spectromoetry (HPLC-MS) together with spectral databases is a powerful tool in the chemometric profiling of biosources for natural product production. To improve and optimize cosmeceutics development from marine microbial substances we need to ensure proper longtime storage of cultures that may be facilitated by cryopreservation so that the most promising agents do not get lost. In order to avoid extra work and repetitive isolation errors, DNA profiling of producer organisms should be performed and maintained for identification and dereplication purposes. Newer and effective extraction procedures should be followed, and better yield should be ensured by maintaining appropriate growth conditions. As we move forward with these technologies and develop them even further, we can expect a higher number of efficient natural marine microbial compounds in the cosmeceutical industry and drug development arena.

ACKNOWLEDGMENTS

This research was supported by the Marine Bioprocess Research Center of the Marine Bio 21 Center funded by the Ministry of Land, Transport and Maritime, Republic of Korea. The authors are thankful to Mr. Jayachandran Venkatesan for providing the chemical structures in Table 22.1.

REFERENCES

Baker, C. N., and Tenover, F. C. 1996. Evaluation of Alamar colorimetric broth microdilution susceptibility testing method for *Staphylococci* and *Enterococci*. *Journal of Clinical Microbiology* 34:2654–59.

Bechet, M., and Blondeau, R. 1988. Iron deficiency-induced tetracycline production in submerged cultures by *Streptomyces aureofaciens*. *Journal of Applied Microbiology* 84:889.

Bhatnagar, I., and Kim, S.-K. 2010. Marine antitumor drugs: status, shortfalls and strategies. *Marine Drugs* 8:2702–20.

Bruns, A., Cypionka, H., and Overmann, J. 2002. Cyclic AMP and acyl homoserine lactones increase the cultivation efficiency of heterotrophic bacteria from the central Baltic Sea. *Applied Environmental Microbiology* 68:3978–87.

Burja, A. M., Banaigs, B., Abou-Mansour, E., Burgess, J. G., and Wright, P. C. 2001. Marine cyanobacteria—a prolific source of natural products. *Tetrahedron* 57:9347–77.

Button, D. K., Shut, F., Quang, P., Martin, R., and Robertson, B. 1993. Viability and isolation of marine bacteria by dilution culture: theory, procedures, and initial results. *Applied and Environmental Microbiology* 59:881–91.

Cambon-Bonavita, M. A., Raguenes, G., Jean, J., Vincent, P., and Guezennec, J. 2002. A novel polymer produced by a bacterium isolated from a deep-sea hydrothermal vent polychaete annelid. *Journal of Applied Microbiology* 93:310–15.

Connon, S. A., and Giovannoni, S. J. 2002. High-throughput methods for culturing microorganisms in very-low-nutrient media yield diverse new marine isolates. *Applied Environmental Microbiology* 68:3878–85.

CORHH: Committee on the Ocean's Role in Human Health, National Research Council. 1999. *Monsoons to microbes: understanding the ocean's role in human health*, 78. Washington, DC: National Academies Press.

Da Rocha, A. B., Lopes, R. M., and Schwartsmann, G. 2001. Natural products in anticancer therapy. *Current Opinions in Pharmacology* 1:364–69.

Ferderle, T. W., Ventullo, R. M., and White, D. C. 1990. Spatial distribution of microbial biomass, activity, community structure and the biodeterioration of linear alkylbenzene sulfonate (LAS) and linear alcohol ethoxylate (LAE) in the subsurface. *Microbial Ecology* 20:297–13.

Gal, Y. L., and Muller-Feuga, A. 1998. Marine microorganisms for industry. In *Actes de Colloques*, ed. Y. L. Gal and A. Muller-Feuga, 21. Editions Quae Versailles, France.

Gao, C.-H., Tian, X.-P., Qi, S.-H., Luo, X.-M., Wang, P., and Zhang, S. 2010. Antibacterial and antilarval compounds from marine gorgonian-associated bacterium *Bacillus amyloliquefaciens* SCSIO 00856. *Journal of Antibiotics* 63:191–93.

Gilbert, F., Galgani, F., and Cadiou, Y. 1992. Rapid assessment of metabolic activity in marine microalgae: application in ecotoxicological tests and evaluation of water quality. *Marine Biology* 1:199–205.

Giovannoni, S. J., Tripp, H. J., Givan, S., Podar, M., Vergin, K. L., Baptista, D., Bibbs, L., Eads, J., Richardson, T. H., Noordewier, M., Rappe, M. S., Short, J. M., Carrington, J. C. and Mathur, E. J. 2005. Genome streamlining in a cosmopolitan oceanic bacterium. *Science* 19:1242–1245.

Gorajana, A., Venkatesan, M., Vinjamuri, S., Kurada, V. V. S. N. B., Peela, S., Jangam, P., et al. 2007. Resistoflavine, cytotoxic compound from a marine actinomycete, *Streptomyces chibaensis* AUBN1/7. *Microbiological Research* 162:322–27.

Grasselli, J. G., and Ritchey, W. M. 1975. *Atlas of spectral data and physical constants for organic compounds*. 2nd ed., vol. 3. Cleveland, OH: CRC Press.

Handelsman, J. 2004. Metagenomics: application of genomics to uncultured microorganisms. *Microbiology and Molecular Biology Reviews* 68:669–85.

Imada, C., Hotta, K., and Okami, Y. 1998. A novel marine *Bacillus* with multiple amino acid analog resistance and selenomethionine-dependent antibiotic productivity. *Journal of Marine Biotechnology* 6:189–92.

Isnansetyo, A., and Kamei, Y. 2009. Anti-methicillin-resistant *Staphylococcus aureus* (MRSA) activity of MC21-B, an antibacterial compound produced by the marine bacterium *Pseudoalteromonas phenolica* O-BC30T. *International Journal of Antimicrobial Agents* 34:131–35.

Kamei, Y., and Isnansetyo, A. 2003. Lysis of methicillin-resistant *Staphylococcus aureus* by 2,4-diacetylphloroglucinol produced by *Pseudomonas* sp. AMSN isolated from a marine alga. *International Journal of Antimicrobial Agents* 21:71–74.

Kameyama, T., Takahashi, A., Kurasawa, S., Ishizuka, M., Okami, Y., Takeuchi, T., and Umezawa, H. 1987. Bisucaberin, a new siderophore, sensitizing tumor cells to macrophage-mediated cytolysis. I. Taxonomy of the producing organism, isolation and biological properties. *Journal of antibiotics* 40:1664–1670.

Kawanishi, T., Shiraishi, T., Okano, Y., Sugawara, K., Hashimoto, M., Maejima, K., et al. 2011. New detection systems of bacteria using highly selective media designed by SMART: selective medium-design algorithm restricted by two constraints. *PLoS One* 6:e16512.

Keller, M., and Zengler, K. 2004. Tapping into microbial diversity. *Nature Reviews Microbiology* 2:141–50.

Kennedy, J., Marchesi, J. R., and Dobson, A. D. 2007. Metagenomic approaches to exploit the biotechnological potential of the microbial consortia of marine sponges. *Applied Microbiology and Biotechnology* 75:11–20.

Khmelenina, V. N., Sakharovskii, V. G., Reshetnikov, A. S., and Trotsenko, I. A. 2000. Synthesis of osmoprotectors by halophilic and alkalophilic methanotrophs. *Mikrobiologiia* 69:465–70.

Kim, M. Y., Sohn, J. H., Ahn, J. S., and Oh, H. 2009. Alternaramide, a cyclic depsipeptide from the marine-derived fungus *Alternaria* sp. SF-5016. *Journal of Natural Products* 72:2065–68.

Kim, S.-K., Ravichandran, D. Y., Khan, S. B., and Kim, Y. T. 2008. Prospective of the cosmeceuticals derived from marine organisms. *Biotechnology and Bioprocess Engineering* 13:511–23.

Kim, T. K., and Fuerst, J. A. 2006. Diversity of polyketide synthase genes from bacteria associated with the marine sponge *Pseudoceratina clavata*: culture-dependent and culture-independent approaches. *Environmental Microbiology* 8:1460–70.

Kumar, A. S., Mody, K., and Jha, B. 2007. Evaluation of biosurfactant/bioemulsifier production by a marine bacterium. *Bulletin of Environmental Contamination and Toxicology* 79:617–21.

Kwon, H. C., Kauffman, C. A., Jensen, P. R., and Fenical, W. 2006. Marinomycins A-D, antitumor-antibiotics of a new structure class from a marine actinomycete of the recently discovered genus "Marinispora." *Journal of American Chemical Society* 128:1622–32.

Lintner, K., Lamy, F., Mas-Chamberlin, Mondon, C. P., Scocci, S., Buche, P., and Girard, P. 2002. Heat-stable enzymes from deep sea bacteria: a key tool for skin protection against UV-A induced free radicals. *IFSCC Magazine* 5:195–200.

Lu, X. L., Xu, Q. Z., Shen, Y. H., Liu, X. Y., Jiao, B. H., Zhang, W. D., and Ni, K. Y. 2010. Macrolactin S, a novel macrolactin antibiotic from marine *Bacillus* sp. *Natural Product Research* 22:342–47.

Lundgren, B. 1981. Fluorescein diacetate as a stain of metabolically active bacteria in soil. *Oikis* 36:17–22.

Maldonado, L. A., Stach, J. E., Pathom-aree, W., Ward, A. C., Bull, A. T., and Goodfellow, M. 2005. Diversity of cultivable actinobacteria in geographically widespread marine sediments. *Antonie Van Leeuwenhoek* 87:11–18.

Matsunaga, T., Burgess, J. G., Yamada, N., Komatsu, K., Yoshida, S., and Wachi, Y. 1993. An ultraviolet (UVA) absorbing biopterin glucoside from the marine planktonic cyanobacterium *Oscillatoria* sp. *Applied Microbiology and Biotechnology* 39:250–53.

Maya, S. P., Kong, F., Jeffrey, J. E., Daniel, A. A., Paola, S. A., Valerie, B. S., Peter, P. J., Weiss, W. J., et al. 2003. Novel alpha-pyrones produced by a marine *Pseudomonas* sp. F92S91: taxonomy and biological activities. *Journal of Antibiotics* 56:1033–44.

Mayer, A. M., Rodríguez, A. D., Berlinck, R. G., and Hamann, M. T. 2009. Marine pharmacology in 2005–6: marine compounds with anthelmintic, antibacterial, anticoagulant, antifungal, anti-inflammatory, antimalarial, antiprotozoal, antituberculosis, and antiviral activities, affecting the cardiovascular, immune and nervous systems, and other miscellaneous mechanisms of action. *Biochimica Biophysica Acta* 1790:283–308.

McCarthy, B. J. 1987. Rapid methods for the detection of biodeterioration in textiles. *International Biodeterioration* 23:357–64.

Mincer, T. J., Jensen, P. R., Kauffman, C. A., and Fenical, W. 2002. Widespread and persistent populations of a major new marine actinomycete taxon in ocean sediments. *Applied and Environmental Microbiology* 68:5005–11.

Nicolaou, K. C., Lim, Y. H., and Becker, J. 2009. Total synthesis and absolute configuration of the bisanthraquinone antibiotic BE-43472B. *Angewandte Chemie International Edition* 48:3444–48.

Pallela, R., Yoon, N.-Y., and Kim, S.-K. 2010. Anti-photoaging and photoprotective compounds derived from marine organisms. *Marine Drugs* 8:1189–202.

Pfaller, M. A., Vu, Q., Lancaster, M., Espinel-Ingroff, A., Fothergill, A., Grant, C., et al. 1994. Multisite reproducibility of colorimetric broth microdilution method for antifungal susceptibility testing of yeast isolates. *Journal of Clinical Microbiology* 32:1625–28.

Rappe, M. S., Connon, S. A., Vergin, K. L., and Giovannoni, S. J. 2002. Cultivation of the ubiquitous SAR11 marine bacterioplankton clade. *Nature* 418:630–633.

Schirmer, A., Gadkari, R., Reeves, C. D., Ibrahim, F., DeLong, E. F., and Hutchinson, C. R. 2005. Metagenomic analysis reveals diverse polyketide synthase gene clusters in microorganisms associated with the marine sponge *Discodermia dissoluta*. *Applied Environmental Microbiology* 71:4840–49.

Sugar, A. M., Brummer, E., and Stevens, D. A. 1983. Comparisons and applications of methods for assessing the cellular viability of *Blastomyces dermatitidis*. *Journal of Microbiological Methods* 1:127–32.

Swisher, A. R., and Carroll, C. 1980. Fluorescein diacetate hydrolysis as an estimator of microbial biomass on coniferous needle surfaces. *Microbiology and Ecology* 6:217–26.

Takamatsu, S., Hodges, T. W., Rajbhandari, I., Gerwick, W. H., Hamann, M. T., and Nagle, D. G. 2003. Marine natural products as novel antioxidant prototypes. *Journal of Natural Products* 66:605–8.

Tiballi, R. N., He, X., Zarins, L. T., Revankar, S. G., and Kauffman, C. A. 1995. Use of a colorimetric system for yeast susceptibility testing. *Journal of Clinical Microbiology* 33:915–17.

Toledo, G., Green, W., Gonzalez, R. A., Christoffersen, L., Podar, M., Chang, H. W. et.al. 2006. High throughput cultivation for isolation of novel marine microorganisms. *Oceanography* 19:120–25.

ZoBell, C. E. 1946. *Marine microbiology, a monograph on hydrobacteriology*. Waltham: Chronica Botanica Company.

23 Cosmeceutical Properties of Brown Algae and Its Industrial Applications

You-Jin Jeon, W.A.J.P. Wijesinghe, and Se-Kwon Kim

CONTENTS

23.1 INTRODUCTION

Algae belong to a group of organisms that have enormous ecological importance and represent a significant proportion of the world's biodiversity. Seaweeds have always been of great interest in Asian culture as marine food sources (Rioux et al., 2009). In contrast, seaweeds represent an important economical resource mostly in the Asia and Pacific countries as well as in India, where they are not only largely harvested but also intensively and largely employed in human nutrition (Caliceti et al., 2002). They can be classified as green algae (Chlorophyta), red algae (Rhodophyta), and brown algae (Phaeophyta) depending on their nutrient and chemical composition (Dawczynski et al., 2007). Total world seaweed production is estimated to be 8.5 million metric tons. In addition, the seaweed industry uses 7.5–8 million metric tons of wet seaweeds annually, either from the wild or from cultivated crop (Dhargalkar and Verlecar, 2009). Seaweeds or marine macroalgae are a potential renewable resource in the marine environment and are known to be an extremely rich source of bioactive compounds (Chandini et al., 2008). Therefore algae can be a very interesting natural source of new compounds with biological activities that could be used as functional ingredients (Plaza et al., 2008). Biological activities are correlated to the presence of chemical compounds, particularly secondary metabolites. The presence of these compounds may assist in predicting some traditional uses of medicinal plants (Kamatou et al., 2008).

Phaeophyceae, or brown algae, are a large group of multicellular algae, including many types of seaweeds in colder Northern Hemisphere waters. They play an important role in marine environments, both as food and for the habitats they form. Most brown algae contain the pigment fucoxanthin and various pheophycean tannins, which are responsible for the distinctive greenish brown color that gives them their name. Worldwide there are about 1,500 species of brown algae (Davis et al., 2003). Brown algae produce vast numbers of secondary metabolites (Reddy and Urban, 2009). Some species are of sufficient commercial importance that they have become subjects of extensive research in their own right.

Cosmeceuticals are cosmetic products with biologically active ingredients purporting to have medical or drug-like benefits. Dermatological research suggests that the bioactive ingredients used in cosmeceuticals do indeed have benefits beyond the traditional cosmetics, such as creams, lotions, and ointments (Chen et al., 2005). Focusing on bioproducts, recent trends in cosmeceutical research from natural sources suggest that algae are a promising group for novel biochemically active substances (Heo et al., 2009). The purpose of this chapter is to review the current literature on cosmeceutical properties of brown seaweeds with their industrial applications. In addition, attempts have been made to update the information covered in recent articles on the subject. A large percent of the communication therefore aims toward the assessment of brown algal secondary metabolites, and special attention was also paid to highlight the potential brown seaweed species for commercial exploration for their unique use as cosmeceutical ingredients.

23.2 SKIN AND COSMETICS

The skin is the largest organ in the body, and it is the outer covering of the body. Thus, skin is the body's first line of defense against the environment. Unlike other organs, skin is in direct contact with the environment, and therefore undergoes aging as a consequence of environmental damage (Fisher et al., 2002). Nowadays skin care cosmetic products of various types are proliferating on the market. This is due to demands of consumers all over the world. Cellulite, wrinkles, acne, rashes, dermatitis, and other skin ailments are among the many skin problems we need to look out for. Skin care products are manufactured with the sole purpose of taking care of the skin.

Cosmeceuticals are a category of cosmetic products claimed to have biologically active ingredients with medicinal or drug-like benefits (Gao et al., 2008). They are topically applied products that are more than merely cosmetic, yet are not true drugs that have undergone rigorous placebo-controlled studies for safety and efficacy (Reszko et al., 2009). Furthermore, they satisfy the needs of beauty and health. At present, a wide range of cosmetic products with biologically active ingredients have been developed and marketed. Protection, whitening, tanning, antiwrinkling, deodorants, antiaging, and nail and hair care are the main intended activities of cosmeceuticals. Many of the biological effects of brown algae are illustrated in Table 23.1. Chemically synthesized synthetic substances or extracted substances from plants or animals can be used as functional ingredients in the production of functional cosmetics (Gao et al., 2008).

23.3 POTENTIAL COMPOUNDS FROM BROWN ALGAE AS COSMECEUTICAL INGREDIENTS

23.3.1 Phlorotannins (Brown Algal Polyphenols)

Polyphenolic compounds are a large and diverse group of naturally occurring compounds containing phenolic functionality and secondary metabolites that exist in both terrestrial and aquatic

TABLE 23.1
Summary of the Biological Activities of Brown Algae with Reference to Cosmeceutical Applications

Brown Seaweed	Active Component	Activity	Potential Cosmeceutical Application	Reference
E. cava	Eckol	Antioxidant	Antiaging	Ahn et al., 2007; Kang et al., 2005
E. cava	6,6′-Bieckol, dieckol, and fucodiphloroethol	Antioxidant	Antiaging	Li et al., 2009
E. cava	Triphlorethol-A	Antioxidant	Antiaging	Kang et al., 2007
E. cava	Triphlorethol-A	Antioxidant	Antiaging	Kang et al., 2006
S. saliquastrum	Fucoxanthin	Antioxidant	Cytoprotective	Heo et al., 2008
S. marginatum, *P. tetrastomatica*, and *T. conoides*	Polyphenols	Antioxidant	Antiaging	Chandini et al., 2008
E. cava	Phlorotannins	Anti-inflammatory		Ryu et al., 2008
E. cava	Phlorotannins	Antiallergic		Lee et al., 2009
S. saliquastrum	Fucoxanthin	Antioxidant	Protect effect against oxidative stress induced by UV-B	Heo and Jeon, 2009
E. cava	Dieckol	Antioxidant	Protective effect against photooxidative stress	Heo et al., 2009; Ko et al., 2010
I. okamurae	Diphlorethohydroxycarmalol	Antioxidant	Protective effect against UV light-induced skin damage	Heo et al., 2010
E. cava and *S. saliquastrum*		Reduction of melanin synthesis and tyrosinase activity	Whitening	Cha et al., 2011
E. cava	7-Phloroeckol	Inhibition of melanin formation	Whitening	Yoon et al., 2009
E. cava	Dieckol	Tyrosinase inhibitor	Whitening	Heo et al., 2009
I. okamurae	Diphlorethohydroxycarmalol	Tyrosinase inhibitor	Whitening	Heo et al., 2010
E. cava	Phlorotannins	MMP inhibitor	Antiwrinkling	Kim et al., 2006
E. cava	Ethanolic extracts	Anti-inflammatory		Jung et al., 2009

environments (Shibata et al., 2002; Susanto et al., 2009). Tannins are defined as naturally occurring plant polyphenolic compounds and are widespread among terrestrial and marine plants. In contrast to terrestrial tannins, phlorotannins have been found only in marine brown algae. They are structurally less complex than terrestrial tannins, and are produced entirely by polymerization of phloroglucinol (Glombitza and Hauperich, 1997; Wang et al., 2008). Generally, the polyphenols

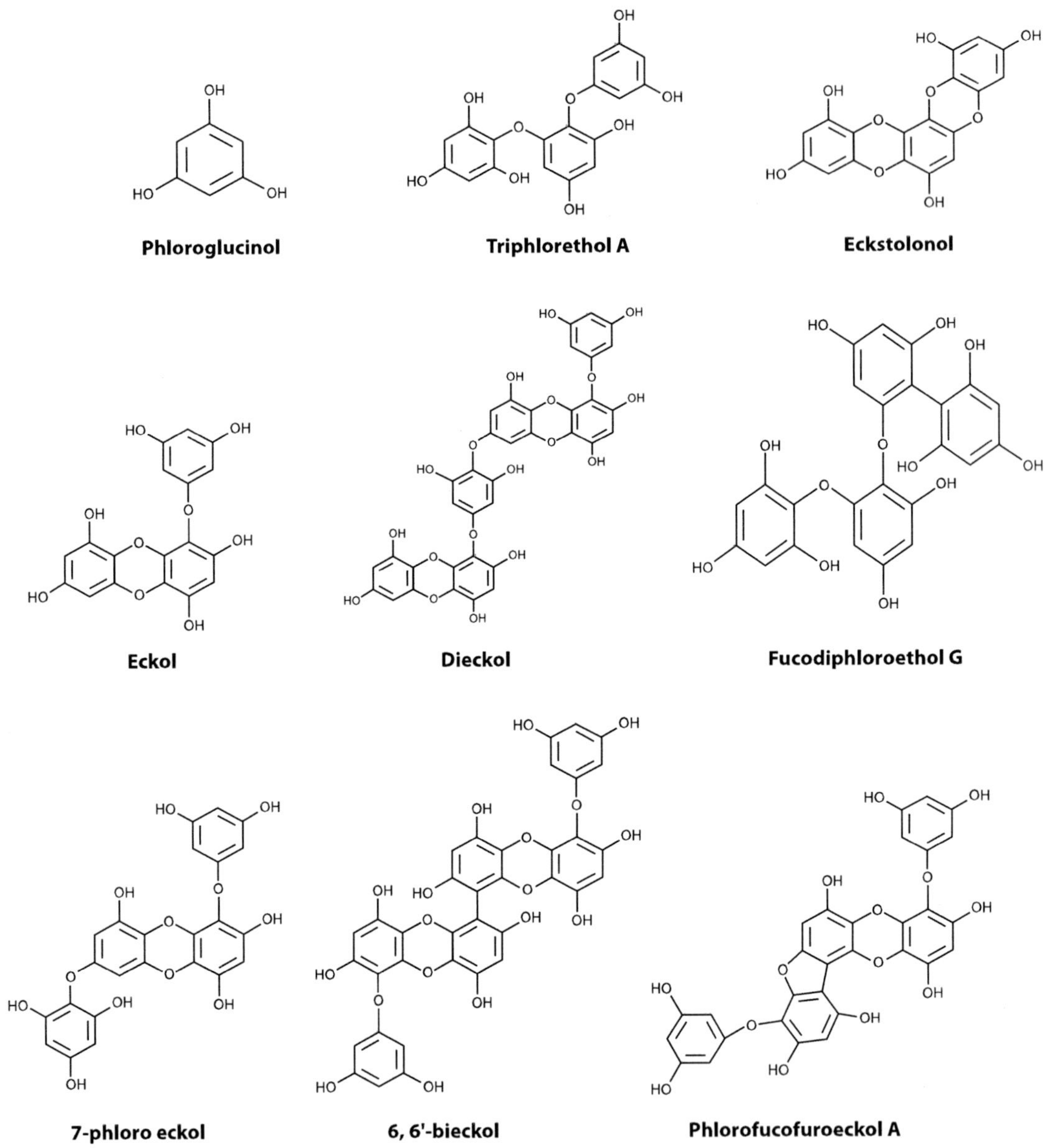

FIGURE 23.1 The chemical formulae of phlorotannin compounds isolated from marine brown algae.

of terrestrial plants, including tannins, consist of flavonoids or gallic acid and chlorogenic acid, but these polyphenols have not been found in marine algae (Shibata et al., 2002). The term *polyphenolic compounds* usually means compounds containing multiple phenolic functionalities (Handique and Baruah, 2002). Since these naturally occurring polyphenols are known to have numerous biological activities, they are found to be potential candidates for a number of industrial applications, such as functional foods, pharmaceuticals, nutraceuticals, and cosmeceuticals. Phlorotannins (brown algal polyphenols) are polyphenolic compounds found exclusively in brown algae. Phlorotannins, a subgroup of tannins, are produced entirely by polymerization of phloroglucinol units (Koivikko et al., 2005; Kang et al., 2007). During the last two decades, the roles and functions of phlorotannins have been the subject of many studies. Figure 23.1 shows the chemical formulae of some phlorotannins isolated from brown algae.

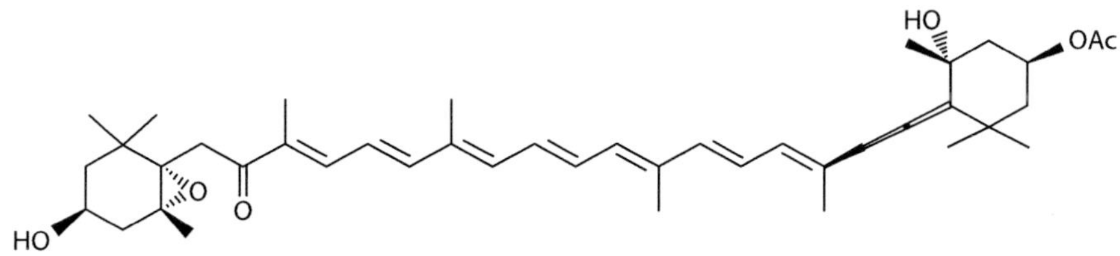

FIGURE 23.2 Chemical structure of fucoxanthin.

23.3.2 Fucoxanthin

Fucoxanthin is a pigment that falls under the broad category of carotenoids. It is found in edible brown algae such as *Undaria pinnatifida*, *Sargassum fulvellum*, *Laminaria japonica*, and *Hizikia fusiformis*, and along with h-carotene, which is one of the most abundant carotenoids found in nature (Hosokawa et al., 2004; Woo et al., 2010). Seaweeds containing fucoxanthin are quite easily found, as they often live in shallow waters, and can usually be found attached to underwater rock surfaces. The presence of fucoxanthins in all algae species can be readily externally determined by examining their color. Fucoxanthin is an accessory pigment in the chloroplasts of brown algae and most other heterokonts, giving them a brown or olive-green color, as the name suggests. Currently, much research is being conducted on how to extract fucoxanthin from certain species of brown algae, which would allow use of the substance in many functional applications. Some metabolic and nutritional studies carried out on rats and mice at Hokkaido University indicate that fucoxanthin promotes fat burning within fat cells in white adipose tissue by increasing the expression of thermogenin (Maeda et al., 2005). Figure 23.2 shows the chemical structure of fucoxanthin.

23.3.3 Alginic Acid

Alginic acid, also called algin or alginate, is a naturally occurring hydrophilic colloidal polysaccharide distributed widely in the cell walls and intracellular material of marine brown algae. It is the major structural polysaccharide of brown seaweeds (Leal et al., 2008). Acid polysaccharides found in brown algal cell walls are mainly composed of alginic acid and fucoidan. Alginic acid is a gelling polysaccharide containing 1,4-linked β-D-mannuronic and α-L-guluronic acid residues arranged in a nonregular, blockwise order along the chain (Andrade et al., 2004; Moore et al., 2004). Alginate produced by brown seaweed is widely used in various pharmaceutical preparations, and also it is a useful additive in the food industry for preparation of thickening drinks, jellies, and ice cream (Cardozo et al., 2007; Yamasaki et al., 2005). Alginate absorbs water quickly, which makes it useful as an additive in cosmetic formulations. Thus the cosmetic industry uses alginic acid as a thickening and water-binding agent. Figure 23.3 shows the chemical structure of alginic acid.

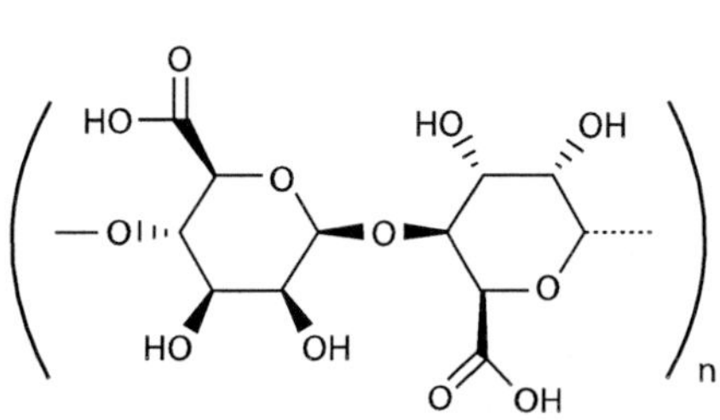

FIGURE 23.3 Chemical structure of alginic acid.

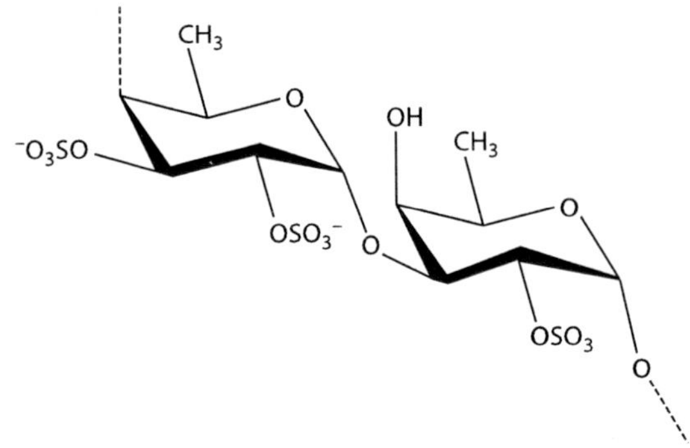

FIGURE 23.4 Chemical structure of fucoidan.

23.3.4 Fucoidan

Fucoidan refers to a type of polysaccharide that contains substantial percentages of L-fucose and sulfate ester groups, found mainly in various species of brown seaweeds (Matou et al., 2002; Li et al., 2008). Among the sulfated polysaccharides, fucoidans, found in marine brown algae, are well known to have potent biological activities such as anticoagulant, antivirus, antitumor, immunomodulatory, anti-inflammatory, and antioxidant (Jiang et al., 2010). Biological activities of fucoidans seem to be determined by their high degree of sulfation, although these activities depend on the fine structure and molecular weight (Zvyagintseva et al., 2003).

Fucoidan is capable of exhibiting skin cosmeticizing effects. Polygalactosides react with the protective outer surface of the skin, and the ion-ion interaction forms a protective moisturizing complex, while the fucose polymers are hygroscopic and act as hydrating agents. With this in mind, brown algal sulfated polysaccharide fucoidan would be helpful for the maintenance of the skin's elasticity by increasing hydration and thereby improving the skin's elasticity. Figure 23.4 shows the chemical structure of fucoidan.

23.4 BIOLOGICAL ACTIVITIES OF BROWN ALGAE WITH REFERENCE TO COSMECEUTICAL APPLICATION

23.4.1 Antioxidants

Antioxidants are compounds that protect cells against the damaging effects of reactive oxygen species, such as singlet oxygen, superoxide, peroxyl radicals, hydroxyl radicals, and peroxynitrite. An imbalance between antioxidants and reactive oxygen species results in oxidative stress, leading to cellular damage. Oxidative stress has been linked to cancer, aging, atherosclerosis, ischemic injury, inflammation, and neurodegenerative diseases. Flavonoids may help in providing protection against these diseases by contributing, along with antioxidant vitamins and enzymes, to the total antioxidant defense system of the human body. Epidemiological studies have shown that flavonoid intake is inversely related to mortality from coronary heart disease and to the incidence of heart attacks. Antioxidants have great popularity as primary ingredients in cosmetics, and they are commonly added to cosmetics as functional ingredients (Lupo, 2001). Therefore natural antioxidant extracts have attracted increasing interest in cosmeceutical applications (Sun and Ho, 2005).

Antioxidant activities of three phlorotannins—phloroglucinol, eckol, and dieckol—purified from brown seaweed *Ecklonia cava* were reported (Ahn et al., 2007). In this study, antioxidant activities were measured by the electron spin resonance (ESR) spectrometry technique for scavenging effects of free radicals such as 1,1-diphenyl-2-picrylhydrazyl (DPPH), alkyl, hydroxyl, and superoxide anion radical. Comet assay was used to determine the protecting effects against H_2O_2-mediated

DNA damage. Furthermore, eckol protects V79-4 cells against oxidative damage by enhancing the cellular antioxidant activity and modulating a cellular signal pathway (Kang et al., 2005). In addition, promising antioxidant properties of *E. cava* were reported with novel phlorotannin derivatives (Li et al., 2009). The antioxidant properties of all phlorotannins (phloroglucinol, eckol, fucodiphloroethol G, phlorofucofuroeckol A, 7-phloro eckol, dieckol, and 6,6′-bieckol) were assessed and 6,6′-bieckol, dieckol, and fucodiphloroethol G showed significantly higher activity than the others.

Triphlorethol-A phlorotannin found in *E. cava* protects V79-4 cells against oxidative stress-induced cell death through upregulation of HO-1 (Kang et al., 2007). Furthermore, triphlorethol-A protects cells against oxidative damage induced by radiation through reducing ROSs (Kang et al., 2006).

A cytoprotective effect of fucoxanthin isolated from the brown alga *Sargassum siliquastru* was reported (Heo et al., 2008). By contrast, a protective effect against oxidative stress-induced cell apoptosis was also demonstrated via nuclear staining with Hoechst dye. These results clearly indicate that fucoxanthin isolated from *S. siliquastrum* possesses prominent antioxidant activity.

Chandini et al. (2008) reported *in vitro* antioxidant activities of three selected Indian brown seaweeds, *Sargassum marginatum*, *Padina tetrastomatica*, and *Turbinaria conoides*. Among the seaweeds, the total methanolic extract of *T. conoides* had a significantly higher phenol content than the other two species. *In vitro* antioxidant activity of methanolic extracts from all three seaweeds showed an increase with increasing concentrations, indicating the dose dependency of these properties.

Free radical damage is one of the principal mechanisms of aging. Skin suffers additional free radical damage from sunlight and pollutants. When it comes to wrinkles, free radical damage causes collagen and other vital skin functions to break down. Topical antioxidants provide some protection against environmental damage to the skin and may be somewhat effective in slowing down skin aging. Cosmeceuticals are commercially available products that can not only improve the appearance of the skin but also benefit the health of the skin because of their antioxygenation (Guan et al., 2005). A great skin care product, whether it comes in a liquid, gel, serum, lotion, or cream, should contain a potent assortment of stable antioxidants to interrupt free radical damage and keep it from harming the skin. More and more people realize that one of the most effective ingredients of cosmeceuticals is the antioxidant, which can interrupt the radical chain processes and help the skin to repair systems.

23.4.2 Anti-Inflammatory and Antiallergic Potentials

Inflammation is the complex biological response of vascular tissues to harmful stimuli and a protective attempt by the organism to remove the injurious stimuli as well as the healing process for the tissue. Macrophages are key players in inflammation (Kazłowska et al., 2010). Oxidative stress-induced inflammation is mediated by the activation of NF-kB and AP-1. The undesired effects of oxidative stress have been found to be controlled by the antioxidant or anti-inflammatory effects of dietary polyphenols (Han et al., 2007). Antiallergic agents can be used to treat allergic reactions. Most of these agents act by preventing the release of inflammatory mediators or inhibiting the actions of released mediators on their target cells.

Ryu et al. (2008) successfully investigated the potent effect of compounds from *E. cava*, eckol, 2-dieckol, 6,6′-bieckol, and 1-(3′,5′-dihydroxyphenoxy)-7-(2″,4″,6″-trihydroxyphenoxy)-2,4,9-trihydroxydibenzo-1,4,-dioxin, on a host of commonly occurring diseases that possess an inflammatory component, including osteoarthritis, atherosclerosis, and cancer. These compounds not only inhibited the proinflammatory cytokine-induced expression of MMP-1, -3, and -13 in MG-63 human osteoblasts, but also exhibited converse regulatory activation of ERK and p38 MAP kinases in NO-induced dedifferentiation. In addition, ethanolic extracts of *E. cava* inhibited LPS-induced nitric oxide (NO) and prostaglandin E_2 (PGE_2) production in a concentration-dependent manner and inhibited inducible nitric oxide (iNOS) and cyclooxygenase

(COX)-2 in BV2 microglia (Jung et al., 2009). According to their results, they have further reported that *E. cava* extracts exert anti-inflammatory effects by suppressing proinflammatory cytokines.

Another recent study revealed that phlorotannins isolated from the brown alga *E. cava* exhibited anti-allergic activities (Le et al., 2009). Furthermore, flow cytometric analysis indicated that the potential anti-allergic mechanism is the suppression of binding activity between IgE and FceRI, and dieckol and 6,6′-bieckol showed the strongest inhibitory effects among the isolated compounds.

23.4.3 Prevention of the Ultraviolet (UV) Effect

Sunlight has a profound effect on the skin, causing premature skin aging, skin cancer, and a host of skin changes. UV-B radiation is responsible for a wide range of potentially damaging human and animal health effects, primarily related to the skin, eyes, and immune system. In addition, the UV component of sunlight is now recognized as the major environmental deleterious factor to human health. Whereas there is evidence that UV-B radiation interacts directly with cellular targets like DNA (de la Coba et al., 2009), human exposure to UV-B depends upon an individual's location, the duration and timing of outdoor activities, and precautionary behavior such as use of sunscreen, sunglasses, or protective clothing. An individual's skin color and age can influence the occurrence and severity of some of the health effects from exposure to UV-B. Topical application of sunscreens is widely advocated for prevention of UV-induced sunburn, photoaging, and skin cancer (Torres et al., 2006). Recent studies have focused on marine organisms as a source of natural bioactive molecules, and some UV-absorbing algal compounds are under investigation as candidates for new natural sunscreens (de la Coba et al., 2009).

According to Heo and Jeon (2009), fucoxanthin isolated from the marine brown alga *Sargassum siliquastrum* has the ability to protect against oxidative stress induced by UV-B radiation, which might be applied to antioxidant and cosmeceutical industries. In contrast, they investigated and confirmed the protective effect of fucoxanthin on UV-B-induced cell injury in human fibroblasts via 20,70-dichlorodihydrofluorescein diacetate (DCFH-DA), 3-(4,5-dimethylthiazol-2-yl)2,5-diphenyltetrazolium bromide (MTT), and comet assays. In addition, fucoxanthin significantly scavenged the ROSs generated by exposure to UV-B radiation.

Dieckol, a kind of phlorotannin isolated from a brown alga, *E. cava*, exhibited a prominent protective effect against photooxidative stress induced by UV-B radiation, which might be used in the cosmeceutical industry (Heo et al., 2009; Ko et al., 2010). Dieckol, phenolic moiety, is a kind of phlorotannin that consists of six phloroglucinol units (hexamer) and functional hydroxyl groups.

In another experiment the carmalol derivative, diphlorethohydroxycarmalol, isolated from the brown algae *Ishige okamurae*, showed a strong protective effect against intracellular reactive oxygen species (ROSs) induced by UV-B radiation (Heo et al., 2010).

Skin is the preferred target of oxidative stress as it is continuously exposed to UV radiation from sunlight and environmental oxidizing pollutants. It is well known that overexposure to UV radiation provokes acute sunburn reaction (Thiele et al., 1997). As mentioned in the previous reports with phlorotannins, they represent an interesting class of active polyphenolic compounds in the protection of UV light-induced skin damage.

23.4.4 Effect on Tyrosinase Activity and Melanin Synthesis

Melanin is a pigment that is primarily responsible for the color of skin, eyes, and hair. Melanocytes are melanin-producing cells located in the bottom layer of the skin's epidermis and the middle layer of the eye. Through a process called melanogenesis, these cells produce melanin. There are both basal and activated levels of melanogenesis; lighter-skinned people generally have low basal levels of melanogenesis. Exposure to UV-B radiation causes an increased melanogenesis as a response to DNA photodamage (Agar and Young, 2005). The difference in skin color between fair people and dark people is due not to the number of melanocytes in their skin, but to the melanocytes' level of activity, which is a process under hormonal control.

Melanogenesis takes place in the melanosome, a distinct cytoplasmic organelle of the melanocyte (Schallreuter et al., 1998). Tyrosinase is an enzyme that is required for melanocytes to produce melanin from the amino acid tyrosine. Thus in mammals, tyrosinase is responsible for pigmentation of the skin, eyes, and hair (Likhitwitayawuid, 2008). The study of tyrosinase inhibitory activity became of interest in recent years because of the significant industrial and economic impact of the inhibitors of this protein (Fails et al., 2009). In contrast, inhibition of tyrosinase activity might be useful in cosmeceutical applications for depigmentation.

Cha et al. (2010) reported that extracts of marine brown algae *E. cava* and *S. saliquastrum* reduced melanin synthesis and tyrosinase activity both *in vitro* and in the zebra fish animal model. Hence the results proved that both brown algal extracts could be used as an ingredient for whitening cosmetics.

Yoon et al. (2009) suggest that 7-phloroeckol, a phlorotannin isolated from *E. cava* might prove useful as a novel inhibitor of melanin formation in cosmetic applications. In their study, they assessed the feasibility of phlorotannins isolated from *E. cava* as inhibitors of melanin formation. They successfully demonstrated the inhibitory effect of isolated phlorotannins on mushroom tyrosinase activity and melanin formation in mouse B16F10 melanoma cells.

In another recent experiment it is suggested that dieckol, a phlorotannin isolated from *E. cava*, has the ability to inhibit tyrosinase activity in a relatively higher manner than the commercial tyrosinase inhibitor kojic acid (Heo et al., 2009). The results also proved the ability of the specific phlorotannin compound to be used in cosmeceutical applications.

The potential inhibitory effect of the marine brown algae *I. okamurae* on melanogenesis was assessed via the tyrosinase inhibitory effect (Heo et al., 2010). According to the results, the active compound diphlorethohydroxycarmalol exhibited higher levels of activity than a commercial whitening agent.

It now appears that important actions can be attributed to some secondary metabolites of brown algae, and evidence indicates that their potential whitening effect could be employed in cosmeceutical applications.

23.4.5 Matrix Metalloproteinase (MMP) Inhibitors

Matrix metalloproteinases (MMPs) are zinc-dependent endopeptidases. The MMPs belong to a larger family of proteases known as the metzincin superfamily. Collectively they are capable of degrading all kinds of extracellular matrix proteins, but also can process a number of bioactive molecules. They are known to be involved in the cleavage of cell surface receptors and the release of apoptotic ligands (Lint and Libert, 2007). MMPs are also thought to play a major role in cell behaviors such as cell proliferation, migration, differentiation, angiogenesis, apoptosis, and host defense.

The functional properties of skin depend on the integrity of collagen in the dermis. Moreover, collagen is tightly controlled by matrix metalloproteinases (Fujii et al., 2008). Cellular changes as well as qualitative and quantitative alterations of dermal extracellular matrix proteins are involved, resulting in loss of recoil capacity and tensile strength with wrinkle formation (Kochanek et al., 2000). The skin matrix is responsible for the skin's mechanical properties, including firmness, strength, suppleness, and elasticity. With increasing age, collagen synthesis becomes lower and MMP-1 (interstitial collagenase) levels become higher in naturally aged human skin, and these alterations cause changes such as skin wrinkling and loss of elasticity (Moon et al., 2004). Therefore control of collagen metabolism may be useful for cosmeceutical applications.

Inhibitory effects of *E. cava* extract on MMP activities have been reported (Kim et al., 2006). For the first time they reported a detailed study on the inhibitory effects of phlorotannins in the brown algae *E. cava* on MMP activities in cultured human cell lines. Further, *E. cava* extract did not exert any cytotoxic effect on the cell line at tested concentrations, anticipating its potential use as a safe MMP inhibitor, which can be utilized in cosmeceutical applications.

23.5 POTENTIAL BROWN ALGAE SPECIES FOR COMMERCIAL EXPLORATION

23.5.1 *Ecklonia cava*

E. cava is a kind of brown alga (Laminariaceae) found abundantly in the subtidal regions of Jeju Island, Korea, and Japan (Kim et al., 2006, 2008). It is plentifully produced on Jeju Island of Korea (30,000 tons per year) for commercial purposes. In addition, *E. cava* has long been utilized as a traditional food and also as a traditional folk herb (Li et al., 2009; Shim et al., 2009). Even this valuable seaweed is not available around the world; *E. cava* may expand its value in various industries due to its wide range of biological activities all over the world (Ahn et al., 2007). In addition, an increasing number of scientific papers have been published in the last few years highlighting the various biological activities and possible industrial applications of this brown seaweed. *E. cava* has a variety of compounds, including peptides, carotenoids, fucoidans, and phlorotannins, showing different biological activities (Heo et al., 2009). *E. cava* is a richer source of total polyphenolic compounds (phlorotannins) than other brown algae (Heo et al., 2003, 2005). Polyphenols are plant-derived antioxidants that have anti-inflammatory, photoprotective, and anticarcinogenic properties. Flavonoids are a subgroup of polyphenols that are popular ingredients in many cosmeceuticals (Choi and Berson, 2006). Therefore the secondary or primary metabolites produced by the brown seaweed *E. cava* might be bioactive compounds of interest in the cosmeceutical industry.

23.5.2 *Sargassum siliquastrum*

Sargassum is a brown to olive green macro alga of the brown algae family, and numerous species are distributed throughout the temperate and tropical oceans of the world, and are often the most obvious macrophyte in near-shore areas where *Sargassum* beds often occur near coral reefs. The plants grow subtidally, and they attach to coral, rocks, or shells in moderately exposed or sheltered rocky or pebble areas. In some cases there are floating populations (Nanba and Okuda, 1993). *Sargassum* is also cultivated and cleaned for use as an herbal remedy. It is reported that many Chinese herbalists prescribe powdered *Sargassum*. *S. siliquastrum* is also common in Pacific Indian Ocean water, and it is well known that this brown alga is a rich source of natural antioxidants (Lim et al., 2002).

23.5.3 *Ishige okamurae*

Ishige okamurae is a kind of edible brown macroalga with narrow fronds, a thick cortical layer, and acute apices; it belongs to the family Ishigeaceae (Chordariales, Phaeophyta). This seaweed is widespread on rocks in the upper and middle intertidal zone on rough open coasts and generally forms highly persistent populations in clear waters (Cho et al., 2005; Heo and Jeon, 2009; Zou et al., 2009).

23.6 CONCLUSION

It is well established that marine organisms are very important not only as resources of food, feed, and energy-rich sources, but also as rich sources of structurally novel and biologically active metabolites with valuable pharmaceutical, biomedical, and cosmeceutical potentials. In this chapter, an overview of cosmeceutical properties of brown algae and its industrial applications has been carried out. In contrast, brown seaweeds produce a variety of remarkable polyphenolic compounds (phlorotannins) collectively referred to as secondary metabolites. The relevant substances are diverse in their chemical structure and physiological function. The term *biological activities* generally refers to secondary metabolites that attract the attention of both scientists and industrialists. In conclusion, brown seaweeds have been identified as potential and easily accessible producers of a wide spectrum of natural substances of vital industrial needs. In contrast, since the brown seaweeds are a magnificent source of various biologically active compounds, they seem to be an endless source of new chemical constituents, which could be used as functional ingredients in cosmeceuticals. We believe that the information provided here can be useful to many research groups considering the huge interest in the search for new natural sources of functional ingredients.

REFERENCES

Agar, N., Young, A.R. (2005). Melanogenesis: a photoprotective response to DNA damage? *Mut Res* 571: 121–32.

Ahn, G.N., Kim, K.N., Cha, S.H., Son, C.B., Lee, J., Heo, M.S., Yeo, I.K., Lee, N.H., Jee, Y.H., Kim, J.S., Heu, M.S., Jeon, Y.J. (2007). Antioxidant activities of phlorotannins purified from *Ecklonia cava* on free radical scavenging using ESR and H_2O_2-mediated DNA damage. *Eur Food Res Technol* 226: 71–79.

Andrade, L.R., Salgado, L.T., Farina, M., Pereira, M.S., Mourão, P.A.S., Amado-Filho, G.M. (2004). Ultrastructure of acidic polysaccharides from the cell walls of brown algae. *J Struct Biol* 145: 216–25.

Caliceti, M., Argese, E., Sfriso, A., Pavoni, B. (2002). Heavy metal contamination in the seaweeds of the Venice lagoon. *Chemosphere* 47: 4436–54.

Cardozo, K.H.M., Guaratini, T., Barros, M.P., Falcao, V.R., Tonon, A.P., Lopes, N.P., Campos, S., Torres, M.A., Souza, A.O., Colepicolo, P., Pinto, E. (2007). Metabolites from algae with economical impact. *Comp Biochem Phys* C146:60–78.

Cha, S.H., Ko, S.C., Kim, D., Jeon, Y.J. (2011). Screening of marine algae for potential tyrosinase inhibitor: those inhibitors reduced tyrosinase activity and melanin synthesis in zebra fish. *J Dermatol* 38: 354–363.

Chandini, S.K., Ganesan, P., Bhaskar, N. (2008). *In vitro* antioxidant activities of three selected seaweeds of India. *Food Chem* 107: 707–13.

Chen, Z., Seo, J.Y., Kim, Y.K., Lee, S.R., Kim, K.H., Cho, K.H., Eun, H.C., Chung, J.H. (2005). Heat modulation of tropoelastin, fibrillin-1, and matrix metalloproteinase-12 in human skin *in vivo*. *J Invest Dermatol* 124: 70–78.

Cho, J.Y., Choi, J.S., Kang, S.E., Kim, J.K., Shin, H.W., Hong, Y.K. (2005). Isolation of antifouling active pyroglutamic acid, triethyl citrate and di-*n*-octylphthalate from brown seaweed *Ishige okamurae*. *J Appl Phycol* 17: 431–35.

Choi, C.M., Berson, D.S. (2006). Cosmeceuticals. *Semin Cutaneouse Med Surg* 25: 163–68.

Davis, T.A., Volesky, B., Mucci, A. (2003). A review of the biochemistry of heavy metal biosorption by brown algae. *Water Res* 37: 4311–30.

Dawczynski, C., Schubert, R., Jahreis, G. (2007). Amino acids, fatty acids, and dietary fibre in edible seaweed products. *Food Chem* 103: 891–99.

de la Coba, F., Aguilera, J., de Galves, M.V., Alvarez, M., Figueroa, F.L., Herrera, E. (2009). Prevention of ultraviolet effects on clinical and histopathological changes, as well as the heat shock protein-70 expression in mouse skin by topical application of algal UV-absorbing compounds. *J Dermatol Sci* 55: 161–69.

Dhargalkar, V.K., Verlecar, X.N. (2009). Southern ocean seaweeds: a resource for exploration in food and drugs. *Aquaculture* 287: 229–42.

Fails, A., Corda, M., Era, B., Fadda, M.B., Matos, M.J., Quezada, E., Santana, L., Picciau, C., Podda, G., Delogu, G. (2009). Tyrosinase inhibitor activity of coumarin-resveratrol hybrids. *Molecules* 14: 2514–20.

Fisher, G.J., Kang. S., Varani, J., Csorgo, Z.B., Wan, Y., Datta, S., Voorhees, J.J. (2002). Mechanisms of photoaging and chronological skin aging. *Arch Dermatol* 138: 1462–70.

Fujii, T., Wakaizumi, M., Ikami, T., Saito, M. (2008). Amla (*Emblica officinalis* Gaertn.) extracts procollagen production and inhibits matrix metalloproteinase-1 in human skin fibroblasts. *J Etnopharmacol* 119: 53–57.

Gao, X.H., Zhang, L., Wei, H., Chen, H.D. (2008). Efficiency and safety of innovative cosmecuticals. *Clin Dermatol* 26: 367–74.

Glombitza, K.W., Hauperich, S. (1997). Phlorotannins from the brown alga *Cystophora torulosa*. *Phytochemistry* 46: 735–40.

Guan, Y., Chu, Q., Fu, L., Ye, J. (2005). Determination of antioxidants in cosmetics by micellar ekectrokinetic capillary chromatography with electrochemical detection. *J Chromatogr A* 1074: 201–204.

Han, X., Shen, T., Lou, H. (2007). Dietary polyphenols and their biological significance. *Int J Mol Sci* 8: 950–988.

Handique, J.G., Baruah, J.B. (2002). Polyphenolic compounds: an overview. *Reactive Funct Polym* 52: 163–188.
Heo, S.J., Jeon, Y.J. (2009). Protective effect of fucoxzathin isolated from *Sargassum siliquastrum* on UV-B induced cell damage. *J Photochem Photobiol B Biol* 95: 101–107.
Heo, S.J., Ko, S.C., Cha, S.H., Kang, D.H., Park, H.S., Choi, Y.U., Kim, D., Jung, W.K., Jeon, Y.J. (2009). Effect of phlorotannins isolated from *Ecklonia cava* on melanogenesis and their protective effect against photo-oxidative stress induced by UV-B radiation. *Toxicol In Vitro* 23, 1123–1130.
Heo, S.J., Ko, S.C., Kang, S.M., Kang, H.S., Cha, S.H., Lee, S.H., Kang, D.H., Jung, W.K., Affan, A., Oh, C., Jeon, Y.J. (2010). Inhibitory effect against UV-B radiation-induced cell damage. *Food Chem Toxicol* 48: 1355–1361.
Heo, S.J., Ko, S.C., Kang, S.M., Kang, H.S., Kim, J.P., Kim, S.H., Lee, K.W., Kim, J.P., Kim, S.H., Lee, K.W., Cho, M.G., Jeon, Y.J. (2008). Cytoprotective effect of fucoxanthin isolated from brown algae *Sargassum siliquastrum* against H_2O_2-induced cell damage. *Eur Food Technol* 228Z: 145–151.
Heo, S.J., Lee, K.W., Song, C.B., Jeon, Y.J. (2003). Antioxidant activity of enzymatic extracts from brown seaweeds. *Algae* 18: 71–81.
Heo, S.J., Park, E.J., Lee, K.W., Jeon, Y.J., 2005. Antioxidant activities of enzymatic extracts from brown seaweeds. *Bioresour Technol* 96: 1613–1623.
Hosokawa, M., Kudo, M., Maeda, H., Konho, H., Tanaka, T., Miyashita, K. (2004). Fucoxanthin induces apoptosis and enhances the antiproliferative effect of the PPAR ligand, troglitazone, on colon cancer cells. *Biochim Acta* 1675: 113–119.
Jiang, Z., Okimura, T., Yokose, T., Yamasaki, Y., Yamaguchi, K., Oda, T. (2010). Effect of sulfated fucan, ascophyllan, from the brown alga *Ascophyllum nodosum* on various cell lines: a comparative study on ascophyllan and fucoidan. *J Biosci Bioeng* 110: 113–117.
Jung, W.K., Ahn, Y.W., Lee, S.H., Choi, Y.H., Kim, S.K., Yea, S.S., Choi, I., Park, S.G., Seo, S.K., Lee, S.W., Choi, I.W. (2009). *Ecklonia cava* ethanolic extracts inhibit lipopolysaccharide-induced cyclooxygenase-2 and inducible nitric oxide synthase expression in BV2 microglia via the MMP kinase and NF-κB pathways. *Food Chem Toxicol* 47: 410–417.
Kamatou, G.P.P., Zyl, R.L.V., Vuuren, S.F.V., Figueiredo, A.C., Barroso, J.G., Pedro, L.G., Viljoen, A.M. (2008). Seasonal variation in oil composition, oil toxicity and the biological activity of solvent extracts of three South African *Salvia* species. *South Afr J Botany* 74: 230–237.
Kang, K.A., Lee, K.H., Chae, S., Zhang, R., Jung M.S., Lee, Y., Kim, S.Y., Kim, H.S., Hong, G.J., Park, J.W., Ham, Y.M., Lee, N.H., Hyun, J.W. (2005). Eckol isolated from *Ecklonia cava* attenuates oxidative stress induced cell damage in lung fibroblast cells. *FEBS Lett* 579: 6295–6304.
Kang, K.A., Lee, K.H., Park, J.W., Lee, N.H., Na, H.K., Surh, Y.J., You, H.J., Chung, M.H., Hyun, J.W. (2007). Triphlorethol-A induces heme oxygenase-1 via activation of ERK and NF-E2n related factor 2 transcription factor. *FEBS Lett* 581: 2000–2008.
Kang, K.A., Zhang, R., Lee, K.H., Chae, S., Kim, B.J., Kwak, Y.S., Park, J.W., Lee, N.H., Hyun, J.W. (2006). Protective effect of triphloretheol-A from *Ecklonia cava* against ionizing radiation *in vitro*. *J Radiat Res* 47: 61–68.
Kazlowska, K., Hsu, T., Hou, C.C., Yang, W.C., Tsai, G.J. (2010). Anti-inflammatory properties of phenolic compounds and crude extract from *Porphyra dentata*. *J Ethanopharmacol* 128: 123–130.
Kim, M.M., Ta, Q.V., Mendis, E., Rajapakse, N., Jung, W.K., Byun, H.G., Jeon, Y. J, Kim, S.K. (2006). Phlorotannins in *Ecklonia cava* extract inhibit matrix metalloproteinase activity. *Life Sci* 79: 1436–1443.
Kim, S.K., Lee, Y., Jung, W.K., Kim, J.H., Choi, I., Park, S.G., Seo, S.K., Lee, S.W., Lee, C.M., Yea, S.S., Choi, Y.H., Choi, I.W. (2008). Effect of *Ecklonia cava* ethanolic extracts on airway hyperresponsiveness and inflammation in a murine asthma model: role of suppressor of cytokine signaling. *Biomed Pharmacother* 62: 289–296.
Ko, S.C., Cha, S.H., Heo, S.J., Lee, S.H., Kang, S.M., Jeon, Y.J. (2010). Protective effect of *Ecklonia cava* on UVB-induced oxidative stress: *in vitro* and *in vivo* zebrafish model. *J Appl Phycol*, DOI 10.1007/s10811-010-9565-z.
Kochanek, K.S., Brenneisen, P., Wenk, J., Herrmann, G., Ma, W., Kuhr, L., Meewes, C., Wlaschek, M. (2000). Photoaging of the skin from phenotype to mechanisms. *Exp Gerontol* 35: 307–316.
Koivikko, R., Loponen, J., Honkanen, T., Jormalainen, V. (2005). Contents of soluble, cell-wall-bound and exuded phlorotannins in the brown alga *Fucus vesiculosus*, with implications on their ecological aspects. *J Chem Ecol* 31: 195–212.
Leal, D., Rossi, M., Caruso, F. (2008). FT-IR spectra of alginic acid block fractions in three species of brown seaweeds. *Carbohydrate Res* 343: 308–316.

Le, Q.T., Li, Y., Qian, Z.J., Kim, M.M., Kim, S.K. (2009). Inhibitory effects of polyphenols isolated from marine algae *Ecklonia cava* on histamine release. *Process Biochem* 44: 168–176.

Lee, S.H., Young, L., Karadeniz, F., Kim, M.M., Kim, S.K. (2008). Alpha-glucosidase and alpha-amylase inhibitory activities of phlorotannin derivatives from *Ecklonia cava* (Abstract). *J Biotechnol* 136S: S588.

Li, B., Lu, F., Wei, X., Zhao, R. (2008). Fucoidan: structure and bioactivity. *Molecules* 13: 1671–1695.

Li, Y., Quian, Z.J., Ryu, B., Lee, S.H., Kim, M.M., Kim, S.K. (2009). Chemical components and its antioxidant properties *in vitro*: an edible marine brown alga, *Ecklonia cava*. *Bioorg Med Chem* 17: 1963–1973.

Likhitwitayawuid, K. (2008). Stilbenes with tyrosinase inhibitory activity. *Curr Sci* 94: 44–52.

Lim, S.N., Cheung, P.C.K., Ooi, V.E.C., Ang, P.O. (2002). Evaluation of antioxidative activity of extracts from a brown seaweed, *Sargassum siliquastrum*. *J Agric Food Chem* 50: 3862–3866.

Lint, V.P., Libert, C. (2007). Chemokine and cytokine processing by matrix metalloproteinases and its effect on leukocyte migration and inflammation. *J. Leukoc. Biol.* 82: 1375–1381.

Lupo, M.P. (2001). Antioxidants and vitamins in cosmetics. *Clin Dermatol* 19:467–473.

Maeda, H., Hosokawa, M., Sashima, T., Funayama, K., Miyashita, K. (2005). Fucoxanthin from edible seaweed, *Undaria pinnatifida*, shows antiobesity effect through UCP1 expression in white adipose tissues. *Biochem Biophys Res Commun* 332: 392–397.

Matou, S., Helley, D., Chabut, D., Bros, A., Fisher, A.M. (2002). Effect of fucoidan on fibroblast growth factor-2-induced angiogenesis *in vitro*. *Thombosis Res* 106: 213–221.

Moon, H., Chung, J.H., Lee, J.K., Zee, O.P. (2004). Triterpenoid saponin from *Viola hondoensis* W. Becker et H Boss and their effect on MMP-1 and type 1 procollagen expression. *Arch Pharm Res* 27: 730–733.

Moore, D.E., Miao, W.G., Benikos, C. (2004). Quantitative determination of alginic acid in pharmaceutical formulations using capillary electrophoresis. *J Pharm Biomed Anal* 34: 233–238.

Nanba, N., Okuda, T. (1993). Growth of main branches of *Sargassum siliquastrum* at Tsuyazaki. *Nippon Suisan Gakkaishi* 59: 1675–1678.

Plaza, M., Cifuentes, A., Ibanez, E. (2008). In the search of new functional food ingredients from algae. *Trends Food Sci Technol* 19: 31–39.

Reddy, P., Urban, S. (2009). Meroditerpenoids from the southern Australian marine brown alga *Sargassum fallax*. *Phytochemistry* 70: 250–255.

Reszko, A.E., Berson, D., Lupo M.P. (2009). Cosmeceuticals: practical applications. *Dermatol Clin* 27: 401–416.

Rioux, L.E., Turgeon, S.L., Beaulieu, M. (2009). Effect of season on the composition of bioactive polysaccharides from the brown seaweed *Saccharina longicruris*. *Phytochemistry* 70: 1069–1075.

Ryu, B. Li, Y., Qian, Z.J., Kim, M.M., Kim, S.K. (2008). Exhibitory effects of compounds from brown alga *Ecklonia cava* on the human osteoblasts (Abstract). *J Biotechnol* 136S: S588.

Schallreuter, K., Slominski, A., Pawelek, J.M., Jimbow, K., Gilchrest, B.A. (1998). What controls melanogenesis? *Exp Dermatol* 7: 143–150.

Shibata, T., Yamaguchi, K., Nagayama, K., Kawagushi, S., Nakamura, T. (2002). Inhibitory activity of brown algal phlorotannins against glycosidases from the viscera of the turban shell *Turbo cornutus*. *Eur J Phycol* 37, 4: 493–500.

Shim, S.Y., To, L.Q., Lee, S.H., Kim, S.K. (2009). *Ecklonia cava* extract suppresses the high-affinity IgE receptor, FcRIε expression. *Food Chem Toxicol* 47: 555–560.

Sun, T., Ho, C.T. (2005). Antioxidant activities of buckwheat extracts. *Food Chem* 90: 743–749.

Susanto, S., Feng, Y., Ulbricht, M. (2009). Fouling behavior of aqueous solutions of polyphenolic compounds during ultrafiltration. *J Food Eng* 91: 333–340.

Thiele, J.J., Podda, M., Packer, L. (1997). Tropospheric ozone an emerging environmental stress to skin. *Biol Chem* 378: 1299–1305.

Torres, A., Enk, C.D., Hochberg, M., Srebnic, M. (2006). Porphyra-334, a potential natural source for UVA protective screens. *Photochem Photobiol Sci* 5: 432–435.

Wang, Y., Xu, Z., Bach, S.J., McAllister, T.A. (2008). Effects of phlorotannins from *Ascophyluum nodosum* (brown seaweed) on *in vitro* ruminal digestion of mixed forage or barley rain. *Anim Feed Sci Technol* 145: 375–395.

Woo, M.N., Jeon, S.M., Kim, H.J., Lee, M.K., Shin, S.K., Shin, Y.C., Park, Y.B., Choi, M.S. (2010). Fucoxanthin supplementation improves plasma and hepatic metabolism and blood glucose concentration in high-fat fed C57BL/6N mice. *Chemicobiol Interact* 186: 316–322.

Yamasaki, M., Ogura, K., Hashimoto, W., Mikami, B., Murata, K. (2005). A structural basis for depolymerization of alginate by polysaccharide lyase family-7. *J Mol Biol* 352: 11–21.

Yoon, N.Y., Eom, T.K., Kim, M.M., Kim, S.K. (2009). Inhibitory effect of phlorotannins isolated from *Ecklonia cava* on mushroom tyrosinase activity and melanin formation in mouse B16F10 melanoma cells. *J Agric Food Chem* 57: 4124–4129.

Zou, Y., Li, Y., Kim, M.M., Lee, S.H., Kim, S.K. (2009). Ishigoside, a new glyceroglycolipid isolated from the brown alga *Ishige okamurae*. *Biotechnol Bioprocess Eng* 14: 20–26.

Zvyagintseva, T.N., Shevchenko, N.M., Chizhov, A.O., Krupnova, T.N., Sundukova, E.V., Isakov, V.V. (2003). Water-soluble polysaccharides of some far-eastern brown seaweeds. Distribution, structure and their dependence on the developmental conditions. *J Exp Marine Biol Ecol* 294: 1–13.

24 Potential Applications of Chitosan as a Marine Cosmeceutical

Elena Lecumberri Lima, Laura Calderón Muñoz, Ruth Expósito Harris, and Ángeles María Heras Caballero

CONTENTS

24.1 INTRODUCTION

24.1.1 Cosmeceuticals: Potential Role of Chitosan as a Cosmeceutical

24.1.1.1 Definition and Regulatory Aspects

The term *cosmeceutical* was introduced by Albert Kligman more than 20 years ago to define those topically applied personal care products that are not merely cosmetics yet are not true drugs, i.e., moisturizers with skin protection factor (SPF) (Kligman, 2000). In fact, the term results from the combination of cosme(tics) + (pharma)ceuticals, and thus is considered to be designed for the enhancement of health and beauty, influencing the biological function of the skin (Amer and Maged, 2009).

It is well known that every country applies extensive and rigid regulations over cosmetics and drugs, imposed by correspondent government agencies. In contrast, despite the strong controversy caused since this new concept was proposed, there is still no official recognition or legal regulation. A legal framework is mandatory because cosmeceuticals are the fastest growing segment of the natural personal care industry (Draelos, 2008).

24.1.1.2 Chitosan as a Potential Marine Origin Cosmeceutical

Vitamins, herbs, oils, and botanicals from different origins may be used in cosmeceuticals. A new interesting field of knowledge emerges in relation to the research of potential marine sources for cosmeceutical purposes. Chitosan, mainly obtained from crustacean exoeskeletons, stands out as a marine origin potential cosmeceutical not only because of its skin moisturizing and regeneration properties but also because it possesses remarkable versatility and optimal properties to vehiculize active agents. The use of this natural biopolymer as an active agent carrier in a cosmeceutical formulation may provide beneficial properties of both chitosan and a vehiculized agent in a symbiotic way, optimizing the results of the skin treatment. Different aspects of the role of chitosan as a potential marine origin cosmeceutical are discussed in this chapter.

24.1.2 The Skin: An Overview of Critical Points Related to the Use of Cosmeceuticals

24.1.2.1 Skin Structure as a Barrier

The skin is the largest organ in the body and is considered an eligible route for drug delivery and cosmetic or cosmeceutical activities.

The effectiveness of a cosmeceutical is determined not only by its formulation and active agents but also by the concentration of cosmeceutical agents that reach the target in an active form, which is highly limited by the skin barrier (Sheffer et al., 2008). This barrier effect is the result of three main layers: the outer to inner *epidermis*, with the sublayers stratum corneum (horny layer), stratum granulosum (granular layer), stratum spinosum (prickly cell layer), malpighion layer (pigment layer), and stratum germinativum or stratum basale (growing layer); the *dermis*, which serves as a location for appendages of skin; and the *hypodermis* or subcutaneous layer.

In the epidermis, cells from lower layers travel upward during their life cycle and become flat, keratinized, water-resistant dead cells of the corneum. In healthy skin, basal cells in the stratum germinativum have a high replicating rate, constantly maintaining the loss of dead horny cells from the skin surface. The malpighion layer, also named pigment layer, is responsible for protection against radiation by means of melanin produced in melanocytes. In the stratum spinosum, the intercellular fine prickles form bridge desmosomes that connect altered basal cells. These links maintain the integrity of the epidermis. The stratum corneum, at the final stage of differentiation, is the most superficial layer of the epidermis. Cells in this layer slough off (this process in known as desquamation), ensuring skin renewal about every 2 weeks. This keratinized layer is responsible for retaining water and provides protection against pathogens and harmful chemicals.

The dermis is responsible for the elasticity of the skin and for the sense of touch and heat. This region, lying between the epidermis and the subcutaneous layer, consists mainly in a dense network of structural protein fiber (i.e., collagen and elastin) included in a mucopolysaccharidic matrix. In this region, there are located mechanoreceptors as well as hair follicles (with associated sebaceous glands), sweat glands, apocrine glands, lymphatic vessels, and blood vessels.

Below the dermis lies the hypodermis or subcutaneous tissue, which consists of a fat-rich tissue (hypodermis contains 50% of body fat) and attaches the skin to underlying bone and muscle (Proksch et al., 2008; Goswami et al., 2008; Wiechers, 2008a, 2008b).

24.1.2.2 The Skin as a Storing and Aesthetic Organ

Every single component and process within the skin layers is responsible for the complexity and optimal functionality of the skin. Thus the improvement of healthy and normal functionality of every cell and process within the skin results in an improved status and appearance of this organ, which is the characteristic aim of cosmeceuticals.

The functions of skin include such important events as (Proksch et al., 2008; Goswami et al., 2008):

- Protection against external agents, including chemicals, pathogens, and radiation
- Containment of body fluids, preventing excessive water loss
- Temperature regulation
- Biochemical synthesis (i.e., vitamin D)
- Reception of stimuli (pressure, heat, cold, pain)
- Interindividual/interspecies identification/sexual attraction
- Storage for lipids and water

Besides the protective and storing purposes, the skin plays an important aesthetic function equally related to the status of the skin components and metabolism, as well as adequate water content (Yu and Grekin, 2007; Verdier-Sévrain and Bonté, 2007; Boissieux et al., 2000; McCullough and Kelly, 2006). The enhancement of this function constitutes the main objective of cosmeceuticals.

Topical cosmeceuticals must contribute to maintain balanced water content and adequate nutritional state of the skin by improving the aesthetic appearance.

The formulations prepared for topical application can be divided into three groups, depending on the site where the effect occurs:

1. Epidermic formulations: With no penetration or absorption.
2. Endodermic formulations: With penetration but not absorption. The local action occurs at the corneal layer, skin annexes, germinative layer, or dermis level.
3. Diadermic formulations: Penetration and absorption take place; therefore their action is systemic (Trommer and Neubert, 2006).

Major cosmeceutical formulations may be included in the first two groups. When designing endodermic formulations, more suitable absorption routes to penetrate through the skin barrier must be taken into account to achieve optimal effects.

24.1.2.3 Main Absorption Routes

To optimize the effect of endodermic cosmeceutical products, active agents should penetrate the skin barrier to reach their target. There are three main absorption routes: penetration across the stratum corneum (transcellular or intercellular), shunt route through sweat ducts, and shunt route along the hair follicle with its associated sebaceous gland.

The sinuous route through the lipoidal intercellular matrix appears to be the main penetration way to cross the stratum corneum for lipophilic and amphiphilic substances. This pathway seems to be the main absorption route. But a transcellular route may be taken into account, especially for little polar hydrophilic molecules. Regardless of whether a molecule is penetrating via the transcellular or intercellular route, it always will have to cross the lipid bilayer that fills the intercellular spaces between keratinocytes of the stratum corneum (Benson, 2005; Wiechers, 2008b).

Penetration through shunt routes appears to be minimal when compared to transepidermal pathways. However, these pathways become critical if skin appendages are the target of active agents of the topical formula (i.e., antiacne products).

24.2 CHITOSAN AND ITS DERIVATIVES AS VALUABLE COSMECEUTICALS

Chitin is the second most important natural polymer in the world following cellulose. The main sources are two marine crustaceans, shrimp and crabs. Chitosan is a natural, biocompatible, biodegradable, and nontoxic polymer that is easily obtained by deacetylation of chitin (Galed et al., 2005). Chitosan is the most important derivative of chitin and is a promising material for many applications, including food, biomedical, pharmaceutical, cosmetics, and as discussed in this chapter, cosmeceutical. Moreover, chitosan is the only pseudonatural cationic polymer. Being a semicrystalline polymer in the solid state, it becomes a soluble polyelectrolyte in acidic aqueous solutions. These unique properties along with the possibility to generate a number of derivatives with specific physicochemical properties (i.e., improved solubility for specific pH values) supply high potentiality to this polymer. Furthermore, chitosan shows a wide versatility with respect to the shapes that can be designed with this polymer, including gels, solutions, films, fibers, sponges, or micro- and nanoparticles. As an example, scanning electron microscopy (SEM) images of chitosan-based delivery systems are collected in Figure 24.1. All these manipulatable properties and multiple shaping possibilities provide chitosan a high added value for the design of devices with specific purposes and targets under specific conditions (Rinaudo, 2006; Mourya and Inamdar, 2008; Paños et al., 2008, Satpathy, 2008). Detailed information about the chemical properties of chitosan has been presented in Chapter 3.

The relevance of chitosan in the development of cosmeceuticals is that two different but complementary roles appear. Chitosan exerts properties that define this polymer as a cosmeceutical, as it is capable of improving the health and beauty of skin through modifications or interactions with biological subjacent processes. On the other hand, chitosan stands out as an eligible vehicle for active cosmeceutical agents in relation to its unique properties and versatility, as explained above (Caiqin

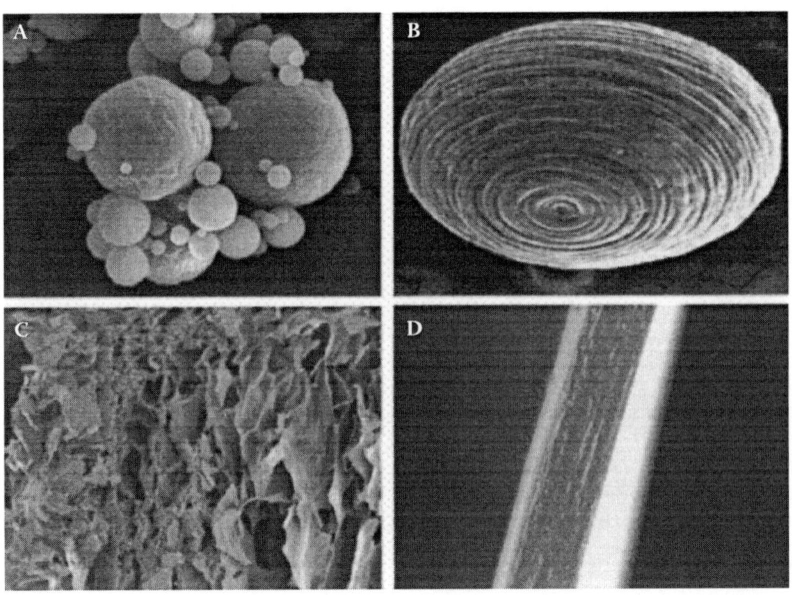

FIGURE 24.1 Scanning electron microscopy (SEM) images of different chitosan-based delivery systems. **(A)** Chitosan microspheres. **(B)** Chitosan bead. **(C)** Cross section of dehydrated chitosan hydrogel. **(D)** Cross section of chitosan film.

et al., 2006; He et al., 2008; Felt et al., 1998; Emir and Raphael, 2006). The properties of chitosan and its derivatives in relation to their remarkable role as cosmeceutical agents and as valuable cosmeceutical vehicles are discussed below.

24.3 SPECIFIC PROPERTIES OF CHITOSAN AND ITS DERIVATIVES AS ELIGIBLE COSMECEUTICAL AGENTS AND CARRIERS

24.3.1 Hydrating Properties

Some authors have demonstrated the durable moisturizing effect of chitosan and its derivatives on the skin. Increased water-binding capacity and skin moisture have been reported in studies. Film/gel-forming hydrophilic nature implies a high capacity of binding water. As reported by Chen et al. (2002), 6-carboxymethyl groups in the molecular structure of chitosan seem to be the main active site responsible for moisture retention ability, which has also been related to molecular weight (the higher the Mw, the higher the moisture retention ability). The superiority of the moisturizing effect of chitosan and its derivatives over some compounds makes them preferable to include in cosmetic formulations. Moreover, these studies also show an increased skin elasticity related to the use of chitosan-based compounds (Jayakumar et al., 2010).

Another advantage of chitosan as a hydration agent in comparison with other compounds is its nonocclusive nature (Khan and Peh, 2003; Wittaya-areekul and Prahsarn, 2006). Thus chitosan allows for normal skin function, including water and gas flux, while contributing to the maintenance of adequate hydration grade of the treated skin.

Moisture retention is a well-known process related to the enhancement of health and physiology of skin. When miniaturization takes place, the skin appearance is improved by the mitigation of some age-related signs. The beneficial effects of moisture retention are also evidenced by a reduced dry look in hair. Based on their physicochemical properties, chitosan and its derivatives appear to be valuable moisture retention agents, to be included in cosmeceuticals focused on treatment or prevention of skin aging or hair regeneration (Hashizume, 2004; Rawlings and Harding, 2004; Chen et al., 2002, 2003).

Besides the potential role of chitosan as a hydrating cosmeceutical, the use of this polymer may improve the effect of other active cosmeceuticals when applied together. There can be found a considerable number of studies supporting the outstanding role of skin hydration in the effectiveness of the delivery of topically applied active agents. High hydration status has been related to the enhancement of skin penetration. This is a key point in the optimization of cosmeceutical formulations because it increases the total amount of active agent that crosses the stratum corneum, finally achieving the target site (Behl et al., 1980; Williams and Barry, 2004). Thus the moisture retention ability of chitosan and derivatives supports the valuable relevance of this marine origin polymer as an active cosmeceutical delivery vehicle.

24.3.2 Film-Forming Ability and Bioadhesiveness

Film-forming capacity, one of the most outstanding properties of chitosan, is a highly valuable characteristic feature of cosmeceuticals. Besides film-forming ability, the strong positive charge of chitosan allows it to bind to negatively charged surfaces such as hair and skin, providing bioadhesiveness, and thereby it becomes a very useful hair and skin care ingredient. A remarkable advantage provided by filmogenic cosmeceutical agents is their contribution to a smooth aspect of the treated skin by mitigating the microimperfections, contributing to evoke a more youthful appearance on the application site. Besides the pleasant feeling of smoothness, the formed film also contributes to protect the skin from environmental conditions (i.e., UV radiation). Additionally, the film-forming ability is a valuable property for including this polymer in hair cosmeceutical formulations as antifrizzing and hair softener agents (Pandya et al., 2007; Wittaya-areekul and Prahsarn, 2006; Jayakumar et al., 2010).

In chitosan and its derivatives the film-forming ability takes place along with a strong bioadhesiveness. This contributes to the relevance of these compounds as an advantageous choice in the design of cosmeceutical formulations because bioadhesive delivery systems may prolong the contact between the active agent and the stratum corneum. This reservoir effect enhances the penetration of active agents in a long-term controlled manner and at the same time reduces the washing of active agent from the skin surface. Moreover, the undesirable ghosting effect is avoided and there is absence of a residue feeling on the skin (Henriksen et al., 1996; Cattaneo, 2005).

24.3.3 Antibacterial Activity

Antibacterial properties of chitosan and its derivatives are well documented, and the effect of molecular modification of native chitosan in this regard has been widely studied (Mengibar et al., 2010; Jayakumar et al., 2010; Gerasimenko et al., 2004; Liu et al., 2001a). As a cosmeceutical agent carrier, the preservative role protecting cosmeceutical formulation stability is the main relevance that has been attributed to the antibacterial nature of this polymer. However, its role as a cosmeceutical active agent should be taken into account in the design of cosmeceutical formulations for antiacne purposes. Antibacterial activity together with anti-inflammatory activity of chitosan supports the valuable potential of this polymer as an active agent in this type of cosmeceutical formulation. Furthermore, as bacterial resistances to antibiotics are increasing, alternative treatments have gained in importance, and chitosan should be taken into account in this regard, especially in long-term acne treatments (Landeck et al., 2010; Lee et al., 2006; Huang et al., 2007). In this case, the shunt route would be the main effective pathway for penetrating the skin barrier.

The antibacterial activity of chitosan and its derivatives also plays an important role in its application for the promotion of optimal scarring and skin regeneration, especially in disrupted skin, as explained below.

24.3.4 Antifungal Properties

Antifungal properties of chitosan and its derivatives have been recently reported by a considerable number of researchers (Landeck et al., 2010; Martínez et al., 2010; Sebti et al., 2005; Roller and Covill, 1999). Besides the application of chitosan as a protective agent against fungi and degradation of cosmeceutical components in a formulation, it could be included as a basic ingredient with physicochemical properties in antifungal nail lacquers. In this kind of product, chitosan would act not only as a simple vehicle but also as a cosmeceutical active agent improving optimum nail condition and nail regeneration in a nonocclusive manner and enhancing the permeation of active agents (Monti et al., 2005). Moreover, by means of its antifungal activity, chitosan and its derivatives could act as active agents in the prevention or treatment of nail fungi infections alone or with other antifungal compounds, with the improved penetration role as mentioned above.

24.3.5 Skin Regeneration

Recently, the use of natural products as cosmeceuticals for wound healing and skin regeneration has gained more importance. Some examples of these natural substances are β-glucans, honey, aloe, cocoa, oak bark extracts, etc. (Davis and Pérez, 2009).

The wound healing process consists of three different phases. The initial inflammatory phase is characterized by platelet activation and release of growth factors and cytokines attracting epithelial cells, endothelial cells, and fibroblasts to the injured area. The next phase is the proliferative one, where growth factors are secreted and cell proliferation is enhanced. Finally, in the remodeling phase, collagen production and organization take place, leading to the mature scar. To complete each one of these phases, achieving complete wound healing and skin reconstruction, a number of events must take place in an effective and coordinated manner. Chitosan has shown promoting activities for many of these healing events, such as chemotaxis, phagocytosis, or collagen remodeling enhancing functions of macrophages and fibroblasts and promoting reepithelization of the wounded area (Chung et al., 1994; Bladrick, 2010; Huang and Fu, 2003).

Even in primary stages, chitosan may play a crucial role mediated by its markedly hemostatic nature, and promotes rapid blood clotting, thereby preventing blood loss, which triggers the healing process. This property has led this polymer to gain approval to be used in bandages and other hemostatic devices. Moreover, chitosan bandages are successfully used on battlefields (Okamoto et al., 2003; Kozen et al., 2007; Wedmore et al., 2006).

A large number of *in vitro* and *in vivo* studies support the remarkable role of this natural biopolymer in the promotion of wound healing. Improved proliferation of normal human skin and keratinocytes has been demonstrated by numerous authors (Chatelet et al., 2001; Kweon et al., 2003; Yun et al., 2010; Cho et al., 1999).

Chitosan-based wound dressings have proved to promote the regeneration of normal skin structures, leading to complete skin functionality, including skin appendages, with minimal scar tissue. The avoidance of the formation of hypertrophic scars or colored tissue in the healed area was improved by the chitosan, which promotes it as a skin regenerating cosmeceutical. Moreover, chitosan has even proved its effectiveness as a scar remover (Kweon et al., 2003; Stone et al., 2000; Delorino and Cresidio, 2009).

Based on the valuable implications of chitosan and its derivatives in the reconstruction of skin explained above, the potential use of this natural marine origin polymer as a cosmeceutical for skin regeneration can be valuable for wound healing, including burn treatment, proper scar formation, and wrinkle prevention. In this aim, chitosan-mediated inhibition of activation and expression of matrix metalloproteinase-2, strongly related to wrinkle formation, must be taken into account to support an antiwrinkle effect of chitosan derivatives, especially chitooligosaccharides (Kim and Kim, 2006).

Apart from the properties directly related to skin regeneration improvement, others support the use of chitosan and its derivatives in cosmeceuticals for skin healing and reconstruction. Antibacterial, antifungal, film-forming ability, and bioadhesiveness, previously discussed, contribute to their potential role as eligible wound healing and skin regenerative agents (Ong et al., 2008; Muzzarelli et al., 1990; Muzzarelli, 1989).

Furthermore, nonocclusive retention capacity and moisturization, discussed previously in this chapter, also contribute to the role of chitosan and its derivatives in wound healing and skin regeneration. It has been demonstrated that the use of semiocclusive devices leads to increases in reepithelization rates, maintaining local moisture in the wound environment, together with normal water flux and the elimination of excess exudate production (Turner, 1979; McColl et al., 2007). Due to their physicochemical properties, chitosan-based films and hydrogels are optimal devices for this aim, as shown in numerous studies (Ge et al., 2000; Wittaya-areekul and Prahsarn, 2006; Loke et al., 2000).

Moreover, biocompatibility, nontoxicity, and degradability by enzymes such as lysozyme or glucosamidases into harmless glucosamine and N-acetylglucosamine support its use even in disrupted skin without any risk of undesirable toxic consequences (Kean and Thanou, 2009; Huang et al., 2010).

Chitosan can be used alone or with specific drugs or agents for skin regeneration purposes. Some approaches have been carried out *in vitro* and *in vivo* with specific drugs, growth factors, or even human fibroblasts and keratinocytes with successful results (Lim et al., 2010; Sung et al., 2010; Wittaya-areekul and Prahsarn, 2006; Alemdaroglu et al., 2006).

The use of chitosan and its derivatives in a cosmeceutical formulation, together with other drugs or active agents with regenerative properties, would lead not only to a controlled drug/active agent release and dose optimization, but also to an enhanced penetration with an improved regenerating effect over the skin. To improve the control in the vehiculized active agent release from chitosan-based devices, the use of natural cross-linking agents as such genipin has gained interest to achieve adequate dose/time release for each particular case (Harris et al., 2010; Mi, 2005; Mi et al., 2003; Muzzarelli, 2009).

24.3.6 Antioxidant Activity

Chitosan and its derivatives, especially chitooligosaccharides, show antioxidant activity that offers protection against free radicals. Therefore the use of chitosan in cosmeceutical formulations would play a significant role in protecting the skin from oxidative damage that involves several processes and implications in skin aging (Harish et al., 2007; Anraku et al., 2009).

Thus not only the above-discussed skin regenerating properties but also the antioxidant activity of chitosan support the use of this natural marine origin polymer for the design of antiaging cosmeceutical formulations.

The antioxidant activity of chitosan cannot be limited to antiaging effects but extended to anticarcinogenic and UVB-induced photodamage (Pinnel, 2003; Pelle et al., 1999; Kogan et al., 2004).

Moreover, in the role of chitosan as a vehicle for active agents, the antioxidant capacity of the polymer leads to protection of active compounds vs. oxidation, thus contributing to the stability of the cosmeceutical formulation (Jeffries, 2008).

24.3.7 Enhancement of Skin Penetration

24.3.7.1 Factors Affecting Skin Penetration

As mentioned earlier, the success of a cosmeceutical treatment is highly dependent on the total amount of active agent that crosses the skin barrier. The outer layer, the stratum corneum of the epidermis, seems to be mainly responsible for the resistance to penetration.

Factors affecting skin penetration can be classified into two main groups: those related to the physicochemical properties of the cosmeceutical active agent and those depending on the skin barrier properties.

Partition coefficient, pH condition, solubility, concentration, particle size and shape, and molecular weight are the main properties of active agents affecting skin penetration. The encapsulation in chitosan may play an important role in enhancing the penetration of cosmeceutical agents. This point is discussed in the following sections.

The characteristics of the skin related more to resistance to transepidermal penetration are moisture content, thickness of the stratum corneum (variations can be registered in different body parts and between races), and general skin health, as well as temperature. It can be highlighted that stratum corneum integrity is highly dependent on factors like age, skin care, and environment, thus affecting penetration across the epidermis (Hashizume, 2004; Rawlings and Harding, 2004).

24.3.7.2 Methods for Skin Penetration Enhancement: Role of Chitosan

Main absorption routes, physicochemical properties of the active agent, and skin barrier are critical to reach maximum penetration of a cosmeceutical. Therefore it is mandatory to consider these interrelated factors to reach the optimum effect for each cosmeceutical formulation.

Some strategies have been developed to bypass the stratum corneum, including physical techniques, chemical methods, and strategies for the vehiculization of active agents.

24.3.7.2.1 *Physical Techniques*

The most remarkable physical enhancement techniques can be divided into structure based (microneedles), electrically based (iontophoresis, electroporation, ultrasound, and photochemical wave), and velocity based (jet propulsion). All these physical enhancement technologies are not useful for most of the topical formulations, especially those for household use, because of the high costs and necessity of sophisticated equipment and qualified staff to use them (Cross and Roberts, 2004; Bharkatiya and Nema, 2009).

24.3.7.2.2 *Chemical Methods: Implications of Chitosan*

Enhancing methods for chemical penetration increase skin permeability by altering the physicochemical nature of the stratum corneum to reduce its diffusion resistance. The more commonly used methods enhance skin penetration mainly by increasing the hydration or by solvent or denaturalization action over the intercellular lipids and lipoproteins in the stratum corneum (Sharma, 2008).

Enhancers of skin penetration provoke structural changes in the stratum corneum to increase skin permeation. Solvents and surfactants traditionally used as permeation enhancers have been associated with undesired reactions such as irritation, allergy, or inflammation. Moreover, skin conditions have changed in the last 30–40 years by raising the incidence of skin disorders such as eczema or atopic dermatitis (Wiechers, 2008a, 2008b; Pons-Guiraud, 2004; Welss et al., 2004; Farage et al., 2006). This change increases the necessity to carry out an exhaustive and careful selection of the excipients included in topical formulations to prevent harmful side effects.

Since the most commonly used chemicals in topical formulations as penetration enhancers induce irritation, it is mandatory to replace them with nonirritating compounds. Many studies have shown that chitosan-mediated applications decrease electrical resistance, with an associated increase in skin permeability (Kumar et al., 2004).

The role of chitosan as a permeation enhancer may be related to the chitosan-induced reversible tight junction disruption, thus increasing the permeation of active agents through the paracellular route (Vllasaliu et al., 2010; Smith et al., 2004). Nonirritant permeation enhancement of chitosan has been evidenced by numerous authors (Wattanarak et al., 2008; Amorim et al., 2010). This supports the valuable role of the polymer as an effective and harmless replacement of commonly used chemical penetration enhancers. Moreover, some studies have demonstrated that chitosan-based topical devices are more effective penetration enhancers than commonly used delivery systems (Kumar et al., 2004).

Hydration is an efficient and safe method to improve skin penetration. Increased water content within the stratum corneum could modify partitioning from the topical formulation into the membrane, promoting absorption (Bharkatiya and Nema, 2009; Benson, 2005). As explained above, chitosan is a moisturizing agent, and this property may be useful for applying this natural biopolymer as a permeation enhancer, especially in the form of a hydrogel.

Super saturation is another permeation enhancer method without alteration of the stratum corneum structure (Bharkatiya and Nema, 2009). Film-forming ability, discussed above, supports the use of chitosan and derivatives as a permeation enhancer, as it involves a reservoir effect that forces the solute active agent into the stratum corneum while prolonging the residence time and activity period of active ingredients (Henriksen et al., 1996; Cattaneo, 2005).

24.3.7.2.3 *Vehicle-Based Strategies of Chitosan*

Multiple shaping possibilities allow chitosan-based cosmeceuticals to be designed in a wide variety of delivery devices, including gels, solutions, films, fibers, sponges, or micro- and nanospheres (Lorenzo-Lamosa et al., 1998; Liu et al., 2001b; Krajewska, 2004; Aranaz et al., 2009). This versatility provides chitosan a highly added value, as it facilitates the design of optimal devices for specific cosmeceutical formulations. Chitosan vehiculization may also provide increased skin permeation, as it improves penetration of active agents that have limited absorption rates when applied alone.

Chitosan cosmeceutical carrier devices are in the form of films, gels, solutions, sponges, and hydrogels that provide a moisturizing and reservoir effect, nonocclusive, skin regeneration, antifungal, antibacterial, and antioxidant properties, together with increased skin penetration through opening tight junctions, as explained above. All these properties support the use of the chitosan-based devices as cosmeceutical vehicles, as they can be included in the definition of a well-designed vehicle. As defined by Epstein (2009), an optimal vehicle protects the active agent from environmental damages, promotes entry of topically applied substances into the stratum corneum, hydrates the skin, and is aesthetically elegant and nonirritant (Epstein, 2009).

Besides these properties, the encapsulation of cosmeceutical active agents into chitosan matrices to form capsules, microcapsules, beads, and micro- or nanoparticles provides added properties that may be useful for cosmeceutical purposes. The possibility of separating the visual perception of the "active agents" from the matrix may improve eligibility for consumers. On the other hand, encapsulation at the level of nanometers may be useful for making a formulation where a homogeneous appearance is needed. Highly controlled loading and delivery capacities, enhanced protection of labile active agents, improvement of selectivity in the release, separation of noncompatible ingredients that will release at specific time or location, and the improvement of skin penetration, even of components that alone would not cross the stratum corneum, are some of the advantages associated with the use of chitosan-based encapsulated formulations (Poncelet, 2008).

Moreover, the right selection of the specific chitosan in terms of Mw and deacetylation grade is very important to design an encapsulated delivery system that implies control in the shape, size, size distribution, and grade of hydrophobicity, among other factors, controlling the degree of diffusivity of the entrapped active agent. The technique employed for the active agent encapsulation is also critical for the delivery process (Aranaz et al., 2009; Paños et al., 2008).

The control in the release of cosmeceutical active agents from chitosan-based devices is one of the most advantageous characteristics supporting the valuable role of this polymer to be used in delivery systems, as shown in a considerable number of *in vitro* and *in vivo* studies (Felt et al., 1998; Paños et al., 2008; Kofuji et al., 2000; Harris et al., 2010a).

Besides the physicochemical properties related to chitosan's high control for the release of active agents, cross-linking appears to be a useful tool to improve a controlled delivery of chitosan-based vehicles. Like other factors related to controlled release, Mw and deacetylation grade of specific chitosan are critical for the degree of cross-linking. In the last decades, cross-linked chitosan-based delivery systems have gained much attention as the optimal choice, with a wide range of potential applications, including cosmetics and cosmeceuticals. Cross-linking is considered to be an easily controllable and effective tool to improve control in release from chitosan-based carriers (Mi et al., 2000; Jameela et al., 1998).

Several cross-linkers, including tripolyphosphate, ethylene glycol, and glutaraldehyde, have been studied, demonstrating that the type and concentration of the cross-linking agent retard in a different grade the release from the chitosan matrix. Concerns about the toxicity of these compounds prompted the investigation of natural nontoxic cross-linkers (Yuan et al., 2007; Mi et al., 2002). Genipin, isolated from the plant *Gardenia jasminoides* Ellis, appears as a promising natural origin nontoxic cross-linker agent, as it effectively improves the control in the release from chitosan-based devices, as shown in Figure 24.2 (Butler et al., 2003; Mi et al., 2003; Harris et al., 2010b).

The degree of cross-linking of genipin-chitosan seems to be easily controllable independent of pH, temperature, genipin concentration, or reaction time (Yuan et al., 2007), which facilitates an adequate cross-linking for each cosmeceutical formulation.

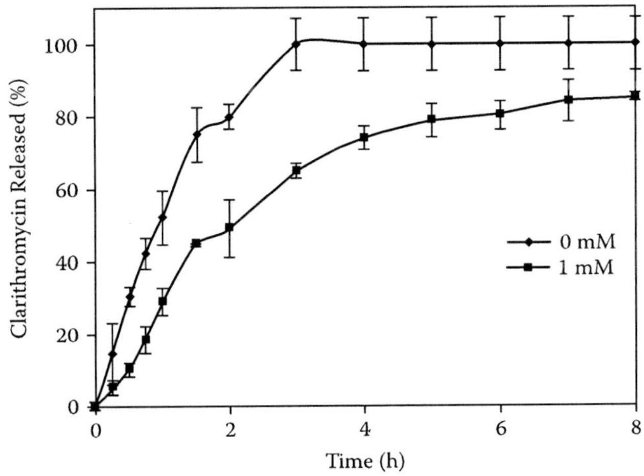

FIGURE 24.2 Release profiles of clarithromycin in simulated gastric fluid from chitosan hydrochloride (5 mg/ml) microspheres loaded with clarithromycin (50% w/w) without genipin (0 mM) and cross-linked with 1 mM genipin at 50°C during 5 h. (From Harris, R., et al., *Mar. Drugs*, 8, 1750–1762, 2010. With permission.)

24.4 CONCLUSIONS

The valuable role of chitosan and its derivatives in the cosmeceutical field is related not only to the moisturizing ability and enhancement of skin regeneration but also to the remarkable versatility and optimal properties to vehiculize the active agents.

As discussed in this chapter, chitosan-based cosmeceutical formulations would exert their beneficial effects by two complementary ways: a potent cosmeceutical role, enhancing the beauty of skin, hair, or nails, and as a high-value eligible cosmeceutical carrier. Chitosan-based delivery devices improve the effectiveness of vehiculized activities, promoting the penetration across the skin barrier and optimizing their beneficial effects via the highly controlled loading and release in the target. It can be concluded that chitosan and its derivatives stand out as valuable marine-origin cosmeceuticals.

REFERENCES

Alemdaroglu, C., Degim, Z., Celebi, N., Zor, F., Ozturk, S., and Erdogan, D. 2006. An investigation on burn wound healing in rats with chitosan gel formulation containing epidermal growth factor. *Burns* 32:319–327.

Amer, M., and Maged, M. 2009. Cosmeceuticals versus pharmaceuticals. *Pharm. Clin. Dermatol.* 27:428–430.

Amorim, C.D., Couto, A.G., Netz, D.J., de Freitas, R.A., and Bresolin, T.M. 2010. Antioxidant idebenone-loaded nanoparticles based on chitosan and N-carboxymethylchitosan. *Nanomedicine: Nanotechnology, Biology and Medicine* 6:745–752.

Anraku, M., Fujii, T., Furutani, N., Kadowaki, D., Maruyama, T., Otagiri, M., Gebicki, J.M., and Tomida, H. 2009. Antioxidant effects of a dietary supplement: reduction of indices of oxidative stress in normal subjects by water-soluble chitosan. *Food Chem. Toxicol.* 47:104–109.

Aranaz, I., Mengibar, M., Harris, R., Paños, I., Miralles, B., Acosta, N., Galed, G., and Heras, A. 2009. Functional characterization of chitin and chitosan. *Curr. Chem. Biol.* 3(2):203–230.

Behl, C.R., Flynn, G.L., Kurihara, T., Harper, N., Smith, W., Higuchi, W.I., Ho, N.F.H., and Pierson, C.L. 1980. Hydration and percutaneous absorption. 1. Influence of hydration on alkanol permeation through hairless mouse skin. *J. Invest. Dermatol.* 75:346–352.

Benson, H.A.E. 2005. Transdermal drug delivery: penetration enhancement techniques. *Curr. Drug Del.* 2:23–33.

Bharkatiya, M., and Nema, R.K. 2009. Skin penetration enhancement techniques. *Pharmaceutics* 1:110–115.

Bladrick, P. 2010. The safety of chitosan as a pharmaceutical excipient. *Regul. Toxicol. Pharm.* 56:290–299.

Boissieux, L., Kiss, G., Magnenat-Thalmann, N., and Kalra, P. 2000. Simulation of skin aging and wrinkles with cosmetics insight. Paper presented at Proceedings of Eurographics Workshop on Animation and Simulation (EGCAS 2000), Interlaken, August 21–22.

Butler, M.F., Ng, Y.-F., and Pudney, P.D.A. 2003. Mechanism and kinetics of the crosslinking reaction between biopolymers containing primary amine groups and genipin. *J. Polym. Sci. A* 41:3941–3953.

Caiqin, Q., Huirong, L., Qi, X., Yi, L., Juncheng, Z., and Yumin, D. 2006. Water-solubility of chitosan and its antimicrobial activity. *Carbohydr. Polym.* 63:367–374.

Cattaneo, M.V. 2005. Encapsulation. In M. R. Rosen (Ed.), *Delivery system handbook for personal care and cosmetic products. Technology, applications, and formulations*, 273–282. William Andrew, Norwich, NY.

Chatelet, C., Damour, O., and Domard, A. 2001. Influence of the degree of acetylation on some biological properties of chitosan films. *Biomaterials* 19:261–268.

Chen, L., Du, Y., Wu, H., and Xiao, L. 2002. Relationship between molecular structure and moisture-retention ability of carboxymethyl chitin and chitosan. *J. Appl. Polym. Sci.* 83:1233–1241.

Chen, L., Du, Y., and Zeng, X. 2003. Relationships between the molecular structure and moisture-absorption and moisture-retention abilities of carboxymethyl chitosan. II. Effect of degree of deacetylation and carboxymethylation. *Carbohydr. Res.* 338(4):333–340.

Cho, Y.W., Cho, Y.N., Chung, S.H., Yoo, G., and Ko, S.W. 1999. Water-soluble chitin as a wound healing accelerator. *Biomaterials* 20:2139–2145.

Chung, L.Y., Schmidt, R.J., Hamlyn, P.F., Sagar, B.F., Andrews, A.M., and Turner, T.D. 1994. Biocompatibility of potential wound management products: fungal mycelia as a source of chitin/chitosan and their effect on the proliferation of human F1000 fibroblasts in culture. *J. Biomed. Mater. Res.* 28:463–469.

Cross, S.E., and Roberts, M.S. 2004. Physical enhancement of transdermal drug application: is delivery technology keeping up with pharmaceutical development? *Curr. Drug Del.* 1:81–92.

Delorino, M.A., and Cresidio, S.P. 2009. Investigation of chitosan from squid pen as scar remover. *World Appl. Sci. J.* 5:98–103.

Draelos, Z.D. 2008. The cosmeceutical realm. *Clin. Dermatol.* 26:627–632.

Davis, S., and Pérez, R. 2009. Cosmeceuticals and natural products: wound healing. *Clin. Dermatol.* 27:502–506.

Emir, B.D., and Raphael, M.O. 2006. Perspectives on chitosan drug delivery systems based on their geometries. *J. Bioact. Compat. Polym.* 21:351–368.

Epstein, H. 2009. Cosmeceuticals vehicles. *Clin. Dermatol.* 27:453–460.

Farage, M.A., Katsarou, A., and Maibach, H.I. 2006. Sensory, clinical and physiological factors in sensitive skin: a review. *Contact Dermatitis* 55:1–14.

Felt, O., Buri, P., and Gurny. 1998. Chitosan: a unique polysaccharide for drug delivery. *Drug. Dev. Ind. Pharm.* 24:979–993.

Galed, G., Miralles, B., Paños, I., Santiago, A., and Heras, A. 2005. N-Deacetylation and depolimerization reactions of chitin/chitosan: influence of the source of the chitin. *Carbohyd. Polym.* 62:316–320.

Ge, J., Ciu, Y., Yan, Y., and Jiang, W. 2000. The effect of structure on pervapouration of chitosan membrane. *J. Membr. Sci.* 165:75–81.

Gerasimenko, D.V., Avdienko, I.D., Bannikova, G.E., Zueva, O.Y., and Varmalov, V.P. 2004. Antibacterial effects of water-soluble low-molecular weight chitosans on different microorganisms. *Appl. Biochem. Microbiol.* 40:253–257.

Goswami, J.P., Sharma, S., Gupta, G.D., Mustafa, A., and Chaudhary, D. 2008. Topical drug delivery systems: a review. *Pharmaceut. Rev.* 6(5).

Harish Prashanth, K.V., and Tharanathan, R.N. 2007. Chitin/chitosan: modifications and their unlimited application potential overview. *Trends Food. Sci. Technol.* 18, 117–131.

Harris, R., Lecumberri, E., and Heras, A. 2010b. Chitosan-genipin microspheres for the controlled release of drugs: clarithromycin, tramadol and heparin. *Mar. Drugs* 8:1750–1762.

Harris, R., Lecumberri, E., Mateos-Aparicio, I., Mengíbar, M., and Heras, A. 2010a. Chitosan nanoparticles and microspheres for the encapsulation of natural antioxidants extracted from *Ilex paraguariensis*. *Carbohydr. Polym.*, doi:10.1016/j.carbpol.2010.07.003a.

Hashizume, H. 2004. Skin aging and dry skin. *J. Dermatol.* 31: 603–609.

He, W., Guo, X., and Zhang, M. 2008. Transdermal permeation enhancement of N-trimethyl chitosan for testosterone. *Int. J. Pharm.* 356:82–87.

Henriksen, I., Green, K.L., Smart, J.D., Smista, G., and Karlsen J. 1996. Bioadhesion of hydrated chitosans: an *in vitro* and *in vivo* study. *Int. J. Pharm.* 145:231–240.

Huang, S., and Fu, X. 2010. Naturally derived materials-based cell and drug delivery systems in skin regeneration. *J. Control Release* 142:149–159.

Huang, Y.C., Vieira, A., Yeh, M.K., and Chiang, C.H. 2007. Pulmonary anti-inflammatory effects of chitosan microparticles containing betamethasone. *J. Bioact. Biocompat. Polym.* 22:30–41.

Jameela, S.R., Kumary, T.V., Lal, A.V., and Jayakrishnan, A. 1998. Progesterone-loaded chitosan microspheres: a long acting biodegradable controlled delivery system. *J. Control Release* 52:17–24.

Jayakumar, R., Prabaharan, M., Nair, S.V., Tokura, S., Tamura, H., and Selvamurugan, N. 2010. Novel carboxymethyl derivatives of chitin and chitosan materials and their biomedical applications. *Prog. Mater. Sci.* 55:675–709.

Jeffries, N. 2008. Delivering on antiaging potential. In J.W. Wiechers (Ed.), *Skin barrier: chemistry of skin delivery systems*, 37. Allured Publishing Corporation, Carol Stream, IL.

Kean, T., and Thanou, M. 2009. Biodegradation, biodistribution and toxicity of chitosan. *Adv. Drug Deliv. Rev.* 62:3–11.

Khan, T.A., and Peh, K.K. 2003. A preliminary investigation of chitosan film as dressing for punch biopsy wounds in rats. *J. Pharm. Pharmaceut. Sci.* 6:20–26.

Kim, M.-M., and Kim, S.-K. 2006. Chitooligosaccharides inhibit activation and expression of matrix metalloproteinase-2 in human dermal fibroblasts. *FEBS Lett.* 580:2661–2666.

Kligman, A. 2000. Cosmeceuticals: do we need a new category? In P. Elsner and H. Maibach (Eds.), *Cosmeceuticals*, 1. Marcel Dekker, New York.

Kofuji, K., Ito, T., Murata, Y., and Kawashima, S. 2000. The controlled release of a drug from biodegradable chitosan gel beads. *Chem. Pharm. Bull.* 48:579–581.

Kogan, G., Skorik, Y.A., Zitnanova, I., Krizkova, I., Durackova, Z., Gomes, C.A.R., Yatluk, Y.G., and Krajcovic, J. 2004. Antioxidant and antimutagenic activity of N-(2-carboxyethyl)chitosan. *Toxicol. Appl. Pharmacol.* 201:303–310.

Kozen, B., Kircher, S., Henao, J., Godinez, F., and Johnson., A. 2007. An alternative field hemostatic agent? Comparison of a new granule dressing to existing chitosan wafer, zeolite and standard dressinngs in a lethal hemorrhagic groin injury. *Annals of Emergency Medicine* 50:S60–S6.

Krajewska, B. 2004. Application of chitin and chitosan based materials for enzyme immobilizations: a review. *Enzyme Microb. Technol.* 35:126–139.

Kumar, M.N.V.R., Muzzarelli, R.A.A., Muzzarelli, C., Sashiwa, H., and Domb, A.J. 2004. Chitosan chemistry and pharmaceutical perspectives. *Chem. Rev.* 104:6017–6084.

Kweon, D.-K., Song, S.-B., and Park, Y.-Y. 2003. Preparation of water-soluble chitosan/heparin complex and its application as wound healing accelerator. *Biomaterials* 24:1595–1601.

Landeck, L., Sabath, S., John, S.M., Gediga, G., and Schurer, N.Y. 2010. Antibacterial effects of a chitosan-containing spray: results of a pilot study. *J. Drugs Dermatol.* 9:1237–1240.

Lee, D.-W., Shirley, S.A., Lockey, R.F., and Mohapatra, S.S. 2006. Thiolated chitosan nanoparticles enhance anti-inflammatory effects of intranasally delivered theophylline. *Respir. Res.* 7:112.

Lim, C.K., Yaacob, N.S., Ismail, Z., and Halim, A.S. 2010. *In vitro* biocompatibility of chitosan porous skin regenerating templates (PSRTs) using primary human skin keratinocytes. *Toxicol. In Vitro* 24:721–727.

Liu, X., Yun, L., Dong, Z., Zhi, L., and Kang, D. 2001a. Antibacterial action of chitosan and carboxymethyl chitosan. *J. Appl. Polym. Sci.* 79:1324–1335.

Liu, X.D., Nishi, N., Tokura, S., and Saikiri, N. 2001b. Chitosan coated cottonfiber: preparation and physical properties. *Carbohyd. Polym.* 44: 233–238.

Loke, W.K., Lau, S.K., Yong, L.L., and Sum, C.K. 2000. Wound dressing with sustained anti-microbial capability. *J. Biomed. Mater. Res.* 53:8–17.

Lorenzo-Lamosa, M.L., Remuñàn, C., Vila-Jato, J.L., and Alonso, M.J. 1998. Design of microencapsulated chitosan microspheres for colonic drug delivery. *J. Contr. Release* 52:109–118.

Martínez, L.R., Mihu, M.R., Tar, M., Cordero, R.J., Han, G., Friedman, A.J., Friedman, J.M., and Nosanchuk, J.D. 2010. Demonstration of antibiofilm and antifungal efficacy of chitosan against candidal biofilms, using an *in vivo* central venous catheter model. *J. Infect. Dis.* 201:1436–1440.

McColl, D., Cartlidge, B., and Connolly, P. 2007. Real-time monitoring of moisture levels in wound dressings *in vitro*: an experimental study. *Int. J. Surg.* 5:316–322.

McCullough, J.L., and Kelly, K.M. 2006. Prevention and treatment of skin aging. *Ann. N.Y. Acad. Sci.* 1067:323–331.

Mengíbar, M., Ganan, M., Miralles, B., Carrascosa, A.V., Martínez-Rodríguez, A.J., Peter, M.G., and Heras, A. 2010. Antimicrobial activity of chitosan oligosaccharides (COS) against *Campylobacter jejuni*. *Carbohydr. Polym.*, doi:10.1016/j.carbpol.2010.04.042.

Mi, F.-L. 2005. Synthesis and characterization of a novel chitosan-gelatin bioconjugate with fluorescence emission. *Biomacromolecules* 6:975–987.

Mi, F.-L., Sung, H.-W., and Shyu, S.-S. 2000. Synthesis and characterization of a novel chitosan-based network prepared using naturally occurring crosslinker. *J. Polym. Sci. A* 38:2804–2814.

Mi, F.-L., Sung, H.-W., Shyu, S.-S., Su, C.-C., and Peng, C.-K. 2003. Synthesis and characterization of biodegradable TPP/genipin co-crosslinked chitosan gel beads. *Polymers* 44:6521–6530.

Mi, F.-L., Tan, Y.-C., Liang, H.-F., and Sung, H.-W. 2002. *In vivo* biocompatibility and degradability of a novel injectable-chitosan-based implant. *Biomaterials* 23:181–191.

Monti, D., Saccomani, L., Chetoni, P., Burgalassi, S., Saettone, M.F., and Mailland, F. 2005. *In vitro* transungual permeation of ciclopirox from a hydroxypropyl chitosan-based, water-soluble nail lacquer. *Drug Dev. Ind. Pharm.* 31:11–17.

Mourya, V.K., and Inamdar, N.N. 2008. Chitosan—modifications and applications: opportunities galore. *Reactive Funct. Polym.* 68:1013–1051.

Muzzarelli, R.A.A. 1989. In G. Sjak-Break, T. Anthonsen, and P. Sandford (Eds.), *Chitin and chitosan*, 79. Elsevier, New York.

Muzzarelli, R.A.A. 2009. Genipin-crosslinked chitosan hydrogels as biomedical and pharmaceutical aids. *Carbohydr. Polym.* 77:1–9.

Muzzarelli, R.A.A., Tarsi, R., Filippini, O., Giovanetti, E., Biagini, G., and Varaldo, P.E. 1990. Antimicrobial properties of N-carboxybutyl chitosan. *Anti. Agents Chem.* 34:2019–2025.

Okamoto, Y., Yano, R., Miyatake, K., Tomohiro, I., Shigemasa, Y., and Minami, S. 2003. Effects of chitin and chitosan on blood coagulation. *Carbohydr. Polym.* 53:337–342.

Ong, S.-Y., Wu, J., Moochhala, S.M., Tan, M-H., and Lu, J. 2008. Development of a chitosan-based wound dressing with improved hemostatic and antimicrobial properties. *Biomaterials* 29:4323–4332.

Pandya, S., Harinarayana, D., Jain, D., Bidkar, S.J., Kulthe, S., and Kadam, N.M. 2007. An attractive biocompatible polymer for pharmaceutical application in various dosage forms—chitosan. *Pharm. Rev.* 5(3).

Paños, I., Acosta, N., and Heras, A. 2008. New drug delivery systems based on chitosan. *Curr. Drug Discov. Technol.* 5:333–341.

Pelle, E., Muizzuddin, N., Mammone, T., Marenus, K., and Maes, D. 1999. Protection against endogenous and UVB-induced oxidative damage in stratum corneum lipids by an antioxidant containing cosmetic formulation. *Photodermatol. Photoimmunol. Photomed.* 15(3–4):115–119.

Pinnel, S.H. 2003. Cutaneous photodamage, oxidative stress, and topical antioxidant protection. *J. Am. Acad. Dermatol.* 48:1–19.

Poncelet, D. 2008. Microencapsulation and cosmetics. In J.W. Wiechers (Ed.), *Science and applications of skin delivery systems*, 251–267. Allured Publishing Corporation, Carol Stream, IL.

Pons-Guiraud, A. 2004. Sensitive skin: a complex and multifactorial syndrome. *J. Cosmet. Dermatol.* 3:145–148.

Proksch, E., Brandner, J.M., and Jensen, JM. 2008. The skin: an indispensable barrier. *Exp. Dermatol.* 17:1063–1072.

Rawlings, A.V., and Harding, C.R. 2004. Moisturization and skin barrier function. *Dermatol. Ther.* 17:43–48.

Rinaudo, M. 2006. Chitin and chitosan: properties and applications. *Prog. Polym. Sci.* 31:603–632.

Roller, S., and Covill, N. 1999. The antifungal properties of chitosan in laboratory media and apple juice. *Int. J. Food Microbiol.* 47:67–77.

Satpathy, T.K. 2008. Chitosan used in pharmaceutical formulations: a review. *Pharm. Rev.* 6(3).

Sebti, I., Martial-Gros, A., Carnet-Pantiez, A., Grelier, S., and Coma, V. 2005. Chitosan polymer as bioactive coating and film against *Aspergillus niger* contamination. *J. Food Sci.* 70:M100–M104.

Sharma, S. 2008. Topical preparations are used for the localized effects at the site of their application by virtue of drug penetration into the underlying layers of skin or mucous membranes. *Latest Rev.* 6(5).

Sheffer, A., Charlene, N.G., and Sheffer, S. 2008. Nanotechnology enhances bio-adhesion and release after rinse-off. In J.W. Wiechers (Ed.), *Skin barrier: chemistry of skin delivery systems*, 419–426. Allured, Carol Stream, IL.

Smith, J., Wood, E., and Dornish, M. 2004. Effect of chitosan on epithelial cell tight junctions. *Pharm. Res.* 21:43–49.

Stone, C.A., Wright, H., Devaraj, V.S., Clarke, T., and Powell, R. 2000. Healing at skin graft donor sites dressed with chitosan. *Br. J. Plast. Surg.* 53:601–606.

Sung, J.H., Hwang, M.-R., Kim, J.O., Lee, J.H., Kim, Y.I., Kim, J.H., Chang, S.W., Jin, S.G., Kim, J.A., Lyoo, W.S., Han, S.S., Ku, S.K., Yong, C.S., and Choi, H-G. 2010. Gel characterisation and *in vivo* evaluation of minocycline-loaded wound dressing with enhanced wound healing using polyvinyl alcohol and chitosan. *Int. J. Pharm.* 392:232–240.

Trommer, H., and Neubert, R.H.H. 2006. Overcoming the stratum corneum: the modulation of skin penetration. A review. *Skin Pharmacol. Physiol.* 19:106–121.

Turner, T.D. 1979. Hospital usage of absorbent dressings. *Pharm. J.* 222:421–424.

Verdier-Sévrain, S., and Bonté, F. 2007. Skin hydration: a review on its molecular mechanisms. *J. Cosmet. Dermatol.* 6:75–82.

Vllasaliu, D., Exposito-Harris, R., Heras, A., Casettaria, L., Garnetta, M., Illum, L., and Stolnik, S. 2010. Tight junction modulation by chitosan nanoparticles: comparison with chitosan solution. *Int. J. Pharm.* 400:183–193.

Wattanarak, P.B., Ingkaninan, K., Khorana, N., and Viyoch, J. 2008. Development of curcuminoids hydrogel patch using chitosan from various sources as controlled-release matrix. *Int. J. Cosmet. Sci.* 30:205–218.

Wedmore, I., McManus, J.G., Pusateri, A.E., and Holcomb, J.B. 2006. A special report on the chitosan-based hemostatic dressing: experience in current combat operations. *Trauma* 60:655–658.

Welss, T., Basketter, D.A., and Schröder, K.R. 2004. *In vitro* skin irritation: facts and future. State of the art review of mechanisms and models. *Toxicol. In Vitro* 18:231–243.

Wiechers, J.W. 2008a. Skin delivery: what it is and why we need it. In J.W. Wiechers (Ed.), *Skin barrier: chemistry of skin delivery systems*, 1–21. Allured, Carol Stream, IL.

Wiechers, J.W. 2008b. Avoiding transdermal cosmetic delivery. In J.W. Wiechers (Ed.), *Skin delivery systems*, 12–22. Allured Publishing Corporation, Carol Stream, IL.

Williams, A.C., and Barry, B.W. 2004. Penetration enhancers. *Adv. Drug Deliv. Rev.* 56(5):603–618.

Wittaya-areekul, S., and Prahsarn, C. 2006. Development and *in vitro* evaluation of chitosan-polysaccharides composite wound dressings. *Int. J. Pharm.* 313:123–128.

Yu, S.S., and Grekin, R.C. 2007. Aesthetic analysis of Asian skin. *Facial Plastic Surg. Clin. N. Am.* 15:361–365.

Yuan, Y., Chesnutt, B.M., Utturkar, G., Haggard, W.O., Yang, Y., Ong, J.L., and Bumgardner, J.D. 2007. The effect of crosslinking of chitosan microspheres with genipin on protein release. *Carbohydr. Polym.* 68:561–567.

Yun, Y.-R., Won, J.E., Jeon, E., Lee, S., Kang, W., Jo, H., Jang, J.-H., Shin, U.S., and Kim, H.-W. 2010. Fibroblast growth factors: biology, function, and application for tissue regeneration. *J. Tissue. Eng.*, Volume 2010, Article ID 218142, 18 pp.

25 Industrial Applications of Marine Cosmeceuticals

Janak K. Vidanarachchi and Maheshika S. Kurukulasuriya

CONTENTS

25.1 INTRODUCTION

The term *cosmeceuticals* was first introduced by Reed (1962), with the introduction of science to the development of the cosmetic industry. This concept was further popularized with the discovery of active substances that cannot be considered cosmetics or drugs by Albert Klingman in the late 1970s (Newburger, 2009; Manela-Azulay and Bagatin, 2009). The *cosmeceutical*, which is the word used to describe these active substances, is derived from the words *cosmetic* and *pharmaceutical*. Including the meaning of both words, the word *cosmeceutical* has been defined as cosmetic products having drug-like benefits that enhance or protect the appearance or order of the human body (Kim et al., 2008).

These human health benefits are achieved from the active ingredients present in the cosmeceuticals, such as peptides, vitamins, polysaccharides, growth factors, phytochemicals, antioxidants, enzymes, and essential oils that have been chemically synthesized, naturally extracted, or obtained by certain biotechnological techniques.

However, the usage of certain cosmeceuticals has exhibited undesirable problems, such as irritability of the skin, photosensitivity, hyperpigmentation, hypopigmentation, infectivity, and carcinogenicity (Gao et al., 2008). Therefore the safety of cosmeceutical products is an important consideration in cosmeceutical development. Since natural sources are recognized as safe and economical, active ingredients from natural sources have attracted the attention of the cosmeceutical industry during the recent past, especially ingredients extracted from plants (Draelos, 2009). Similar to the contribution from herbal extracts, the active substances derived from marine organisms have greater potential to be applied as cosmeceuticals, as these compounds are naturally found inside marine organisms, protecting them from the harmful external environment. These marine organisms include algae, microalgae, bacteria, cyanobacteria, fungi, certain fish species, and crustaceans. The different secondary metabolites isolated from this heterogenous group of organisms have shown various biological activities, such as antimicrobial (Demirel et al., 2009), anticancerous (Hwang et al., 2006; Althunibat et al., 2009), anti-inflammatory (Shibata et al., 2003; Hwang et al., 2006), antioxidative (Anggadiredja et al., 1997; Shanab, 2007; Matanjun et al., 2008), and antiphotoaging (Moisan and Mitchell, 2001; Reimer et al., 2007), which made it possible to use them in many applications, including the cosmeceutical industry. These special properties of marine-derived substances have been applied to cosmeceuticals for the enhancement of the health of skin by providing nutrients, preventing possible damage to the skin, delivering agents of other important active ingredients, and preserving or coloring cosmeceuticals. Furthermore, the research and development that have been focused on marine-derived substances during the recent past have discovered novel bioactive substances as well as novel biological properties from them that can be applied to the development of novel cosmeceuticals. Therefore this chapter was written with the objective of reviewing the different bioactive substances derived from the diverse group of marine organisms and their potential applications in the cosmeceutical industry.

25.2 DIVERSITY OF MARINE ORGANISMS

The massive area of the earth's surface (71%) is covered with oceans (Baharum et al., 2010), providing habitats for many marine organisms. This diverse marine environment is a biological diversity-rich source where many species and their potentials have yet to be discovered. However, certain marine organisms have been identified as resources in modern industries. Especially some species of algae, microalgae, bacteria, crustaceans, and other organisms have been widely used as resources in the cosmeceutical industry.

25.2.1 Diversity of Marine Algae and Microalgae

Out of the marine organisms, algae, including microalgae, are the main group of organisms that have a great contribution in the production of cosmeceuticals. These industrially important marine algae are mainly constituted into three groups: green algae (Chlorophyta), brown algae (Phaeophyta), and red algae (Rhodophyta). According to the recent review of Kim et al. (2008), among these three groups, brown and red algae are widely used in current cosmetic products. Despite the groups, both macroalgae and microalgae have great economical potential due to their fast growth rate, availability of high amounts of minerals, and ability to produce pigments, enzymes, fatty acids, vitamins, and other bioactive substances using natural resources that are abundantly available in the marine environment. Therefore some of the marine algal species such as *Haematococcus*, *Dunaliella salina*, and *Dunaliella bradawil* have been cultured commercially for the mass production of industrially important active ingredients. However, among the algae that are grown commercially, the different species of *Dunaliella* have received special attention in the production of these high-value compounds. *Dunaliella* is a halotolerant, unicellular green alga that has wide geographical distribution (Oren, 1998). This organism is able to survive in extreme environmental conditions such as high salinity, low pH, high irradiance, and subzero temperatures (Kim et al., 2008). Many studies have been conducted on the optimum growth conditions of the different species of *Dunaliella*, as this information is very important in the cultivation of this organism commercially. Similarly, research has been carried out and active substances such as various pigments, peptides, and polysaccharides have been isolated from many other algal and microalgal species thriving in the marine environment for the development of future cosmeceuticals (Table 25.1).

25.2.2 Diversity of Marine Microorganisms

The progress in industrial microbiology has extended the range of microorganisms that have the potential to be used as cosmeceuticals from microalgae to other marine microorganisms, which belong to bacteria, cyanobacteria, and fungi. Approximately 10^6 of bacterial taxa are found in the oceans (Pedros-Alio, 2006), which have been adapted to stressful marine environments such as abyssal depths of oceans. Since the discovery of the first hydrothermal vent in 1977 on the Galapagos Ridge, many Gram-negative rods including bacterium strain HYD 1545 (Vincent et al., 1994), *Alteromonas macleodii* subsp. *fijiensis* (Raguenes et al., 1996), *Vibrio diabolicus* (Raguenes et al., 1997a), *Alteromonas infernus* (Raguenes et al., 1997b), bacterium strain 657, which is called *Alteromonas macleodii* subsp. *fijiensis* biovar deepsane, and the thermophilic bacterium *Thermus thermophilus* (Lintner et al., 2009) have been isolated from this type of special marine ecosystem. In contrast to the high-temperature environment found in hydrothermal vents, the other marine extreme environment, sea ice, also provides a habitat for some halotolerant, psychrotolerant bacteria belonging to the genus *Pseudoalteromonas* (Nicholas et al., 2005a). The adaptive mechanisms of these bacteria to stressful conditions have resulted in the production of bioactive expolysaccharides and certain enzymes that have many applications in the cosmetic and medicinal fields.

TABLE 25.1
Diversity of Marine Algae and Microalgae and the Distribution of Cosmeceutically Important Secondary Metabolites

Algal Group	Species of Algae	Secondary Metabolite	Reference
Green algae (Chlorophycota)	*Dunaliella salina*	β-Carotene	Cifuentes et al., 1992; Marin et al., 1998
	Dunaliella bardawil	β-Carotene	Ben-amotz et al., 1989
	Codium dimorphum	Linoleic acid	Goecke et al., 2010
	C. fernandezianum	Linoleic acid	Goecke et al., 2010
	C. fragile	Linoleic acid	Goecke et al., 2010
Brown algae (Phaeophycota)	*Sargassum polycystum*	Carotenoids/terpenoids	Anggadiredja et al., 1997
	Sargassum pakkidum	Phlorotannin + vitamin E + vitamin C	Ye et al., 2009
	S. ringgoldianum	Phlorotannin	Nakai et al., 2006
	Fucus vesiculosus	Phlorotannin	Koivikko et al., 2008
	Ecklonia cava	Phlorotannin	Shibata et al., 2004; Senevirathne et al., 2006; Ahn et al., 2007; Shibata et al., 2008
	E. kurome	Phlorotannin	Shibata et al., 2004, 2008
	E. bicyclis	Phlorotannin	Shibata et al., 2004, 2008
	Fucus vesiculosus	Fucoidan	Patankar, 1993
	Laminaria japonica	Fucoidan	Wang et al., 2008
Red algae (Rhodophycota)	*Laurencia obtuse*	Carotenoids/terpenoiuds	Anggadiredja et al., 1997
	Haematococcus	Astaxanthin	Fabregas et al., 2001
	Porphyra umbilicalis	MAA—porphyra-334, shinorine	Schmid et al., 2004
	Porphyra yezoensis	MAA—porphyra-334	Zhang et al., 2005a, 2005b
	Gracilaria cornea	MAA—porphyra-334, shinorine	Sinha et al. 2000

Similar to the bacteria, cyanobacteria that is a phylogenetically primitive group of Gram-negative prokaryotes such as *Oscillatoria* sp. (Matsunga et al., 1993) and *Anacystis nidulans* (Eker et al., 1990) have exhibited production of pigments or enzymes as adaptations to survive in extreme climates, including osmotic and UV stress. These photoprotective active compounds produced in marine microorganisms may have a wide role in the development of novel photoprotective cosmeceuticals. However, the isolation of these microorganisms from their natural habitats, which belong to extreme marine environments, is a challenge for the usage of them in the cosmeceutical industry.

25.2.3 The Other Marine Sources

Even though the availability of active substances from marine microorganisms is a limiting factor, extracts from certain fish species and crustaceans that contain various polysaccharides, proteins, peptides, pigments, and vitamins are abundant for the production of cosmetics, as these materials are results of the fish and shrimp processing industries. The global demand for shrimp products is increasing, and this has resulted in accumulation of shrimp waste consisting of a large amount of exoskeletons. These exoskeletons of crustaceans contain high amounts of polysaccharides called chitin, which has many applications in the cosmetic and pharmaceutical fields. Similarly, fish processing by-products contain a considerable amount of fish skin, which is a protein-rich source that can be used in cosmeceutical skin preparations. Apart from fish skin, shark bile is also considered an important ingredient in products that are present in the current cosmetic market (Sankaran and Mouly, 2007). Since sharks are a threat in the wild, the large amount of shark bile that is needed for the mass production of cosmetics is problematic, and novel ingredients with similar properties need to be discovered.

25.3 MARINE-DERIVED SUBSTANCES USED IN THE COSMECEUTICAL INDUSTRY

Despite the source of origin, bioactive substances derived from the diverse group of marine organisms have been widely applied and have the potential to be applied as active ingredients in cosmeceutical products. These bioactive substances can be mainly categorized as carotenoids, polysaccharides, proteins or peptides, polyphenols, vitamins, fatty acids, and enzymes.

25.3.1 Carotenoids

Carotenoids are the main marine-derived pigments that have received the attention of the cosmeceutical industry. These are lipid-soluble, natural pigments with a 40-carbon structure that is derived from the enzymatic polymerization of five carbon isoprene units (Figure 25.1).

25.3.1.1 β-Carotene

Of the different carotenoids synthesized, β-carotene is the first cosmeceutically important carotenoid produced in the carotenogenesis pathway. The structure of this molecule consists of eight isoprene units that have undergone cyclation at each end of the structural chain (Figure 25.2). This complex structure is produced as a photoprotective agent in many marine microalgae, especially in the different species of *Dunaliella* that are exposed to extreme environmental conditions. Though the production of β-carotene in a favorable environment is minimal, *Dunaliella* accumulates more than 10% of the dry cell weight of β-carotene in oil globules located in the interthylakoid space of the chloroplast under high salinity (Borowitzka et al., 1990; Hadi et al., 2008), nutrient starvation such as low nitrogen availability (Jimenez and Niell, 1991; Pisal and Lele, 2005), and high irradiance (Salguero et al., 2005). However, the effect of available nitrogen for the production of β-carotene is in debate, having both positive and negative points. According to Jimenez and Niell (1991), a low level of nitrogen such as 5 mM NO_3^- increases the carotenogenesis in *Dunaliella*, whereas Marin et al. (1998) have reported that there was no effect of using low nitrogen levels to induce the high rate of carotenogenesis in *Dunaliella*. This observed difference of required nitrogen content for carotenogenesis may have contributed to the differences in the other environmental conditions or differences in the species of *Dunaliella* selected for their studies. Similarly, the optimum salinity level for high production of β-carotene varies with the species of *Dunaliella*, although all species need a high salinity. Furthermore, within the species of *Dunaliella salina*, differences in carotenogenesis have been observed by Cifuentes et al. (1992) under the same environmental conditions, and the authors suggested that this difference may be due to the variations in the strains of *Dunaliella salina* that have been selected for their study. This within-species variation of β-carotene production is further confirmed by the recent study of Kleinegris et al. (2010), as the strain 19/18 of *Dunaliella* produces a high amount of β-carotene, whereas no carotenoids were extracted from *Dunaliella salina* strain 19/25 under the same environmental conditions. Therefore the selection of a suitable species as well as a suitable strain of *Dunaliella* that produces a high level of β-carotene is essential prior to establishment of a commercial culturing system of them.

Currently the commercial preparation of β-carotene is available using *D. salina* and *D. bardawil*. Even though *D. salina* and *D. bardawil* produce β-carotene efficiently at high salinities such as 25% (>4 M) sodium chloride, their growth at this salinity is minimum. According to Hadi et al. (2008),

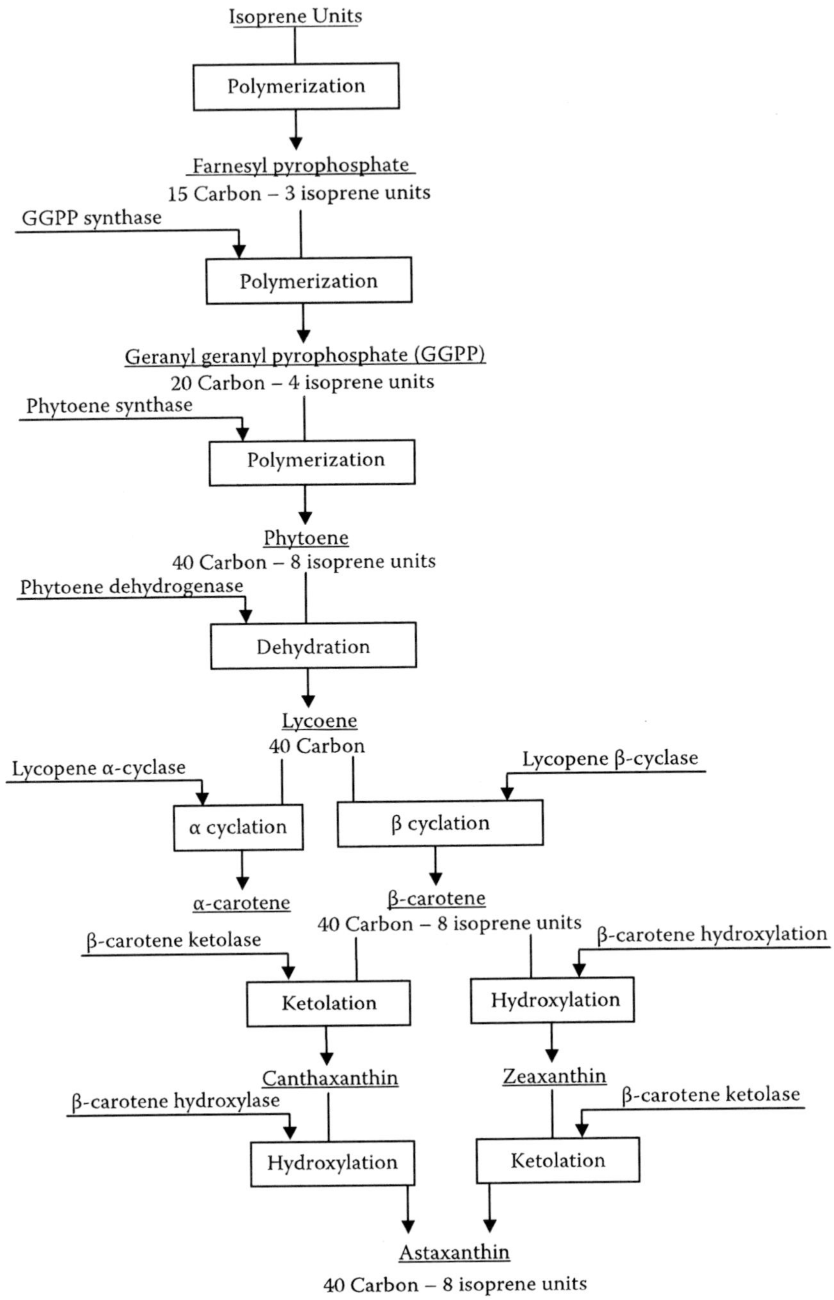

FIGURE 25.1 Schematic diagram of the carotenoid biosynthesis pathway. (Adapted from Lee, J. H., et al., *Biotechnology and Bioprocessing Engineering*, 12, 312–317, 2007.)

the optimum growth of *D. salina* is achieved at 2 M (11.6%) sodium chloride, while *D. bardawil* show their optimum growth at 1 M (5.8%) sodium chloride level (Gomez et al., 2003). Therefore prior to induction of carotenogenesis in *Dunaliella*, the number of cells available for the production of β-carotene should be increased by providing optimum growth conditions for them. The salinity of the culturing ponds needs to be maintained around a 2 M (11.6%) sodium chloride level to obtain the optimum density of *Dunaliella* followed by the provision of a stressful environment to achieve

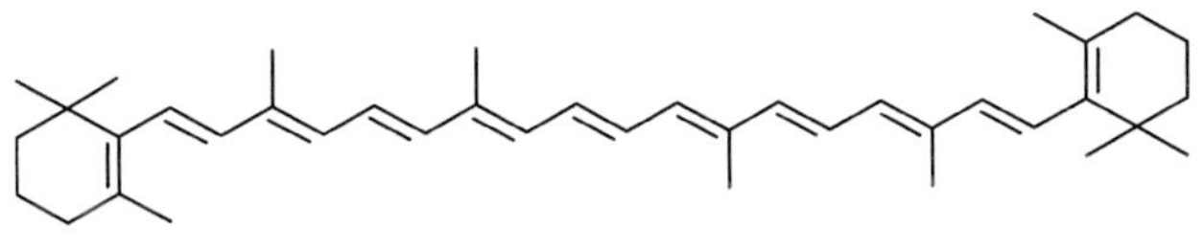

(a)

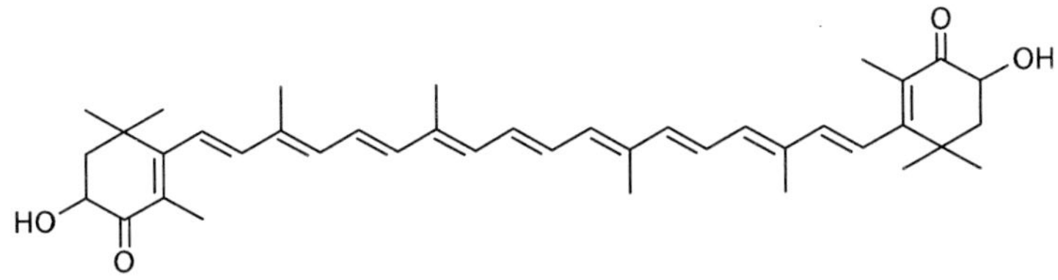

(b)

FIGURE 25.2 Structure of carotenoids: (a) β-carotene and (b) astaxanthin.

the maximum β-carotene production from them (Table 25.2). This type of two-stage production system of *Dunaliella salina* has been carried out by Ben-amotz (1995), and 300 mg of β-carotene m^{-2} day^{-1} production has been achieved from 0.8 million *Dunaliella* cells, which were obtained from the first stage of this system. Furthermore, the simultaneous production and extraction of β-carotene in a two-phase bioreactor have been reported from *D. salina* (Hejazi et al., 2002) and *D. bardawil* (Kleinegris et al., 2010). Since the current natural β-carotene production is only sufficient to fulfill 10% of the total demand, this type of product-maximizing system can be applied to mass production of β-carotene, as the demand for natural β-carotene is increasing. This commercially produced β-carotene is mainly used as pigments in the food, pharmaceutical, and cosmetic industries and to enhance the pigmentation of fish, poultry, and eggs (Jin and Melis, 2003). Moreover, as this pigment is synthesized by the microalgae as a photoprotective agent, the role of β-carotene as a pigment has been extended to an antioxidative agent in cosmeceuticals due to the human health benefits that have been observed in this molecule.

25.3.1.2 Astaxanthin

Similar to β-carotene, astaxanthin is also applied in the cosmeceutical industry as a pigment and as an antioxidant agent. Astaxanthin, which has the chemical structure 3′3-dihidroxy-4′4′diketo-β-carotene, is synthesized from β-carotene in the carotenogenesis pathway. β-Carotene is converted to astaxanthin enzymatically, through either canthazanthin or zeaxanthin (Figure 25.1). The final

TABLE 25.2
Optimum Growth and Carotenogenesis Conditions Required for Microalgae *Dunaliella*

	Optimum Growth Conditions			Optimum Carotenogenesis Conditions			
Species	Salinity (NaCl %)	Temp.	N. Content	Salinity (NaCl %)	Temp.	N. Content	Reference
D. salina	NA	NA	NA	>4 M (25%)	NA	NA	Borowitzka et al., 1990
D. salina	2 M (11.8%)	NA	0.75 mM	3 M (17.4%)	NA	0.00[a]	Pisal and Lele, 2005
D. salina	<2 M (10%)	22°C	NA	NA	38°C	NA	Garcia et al., 2007
D. bardawil	1 M (5.8%)	NA	0.65 mM	2 M (11.8%)	NA	0.00[b]	Gomez et al., 2003

Note: Temp., temperature; N. content, nitrogen content; NA, not available.
[a] No additional N is provided after achieving the maximum growth by 0.75 mM nitrogen content.
[b] No additional N is provided after achieving the maximum growth by 0.65 mM nitrogen content.

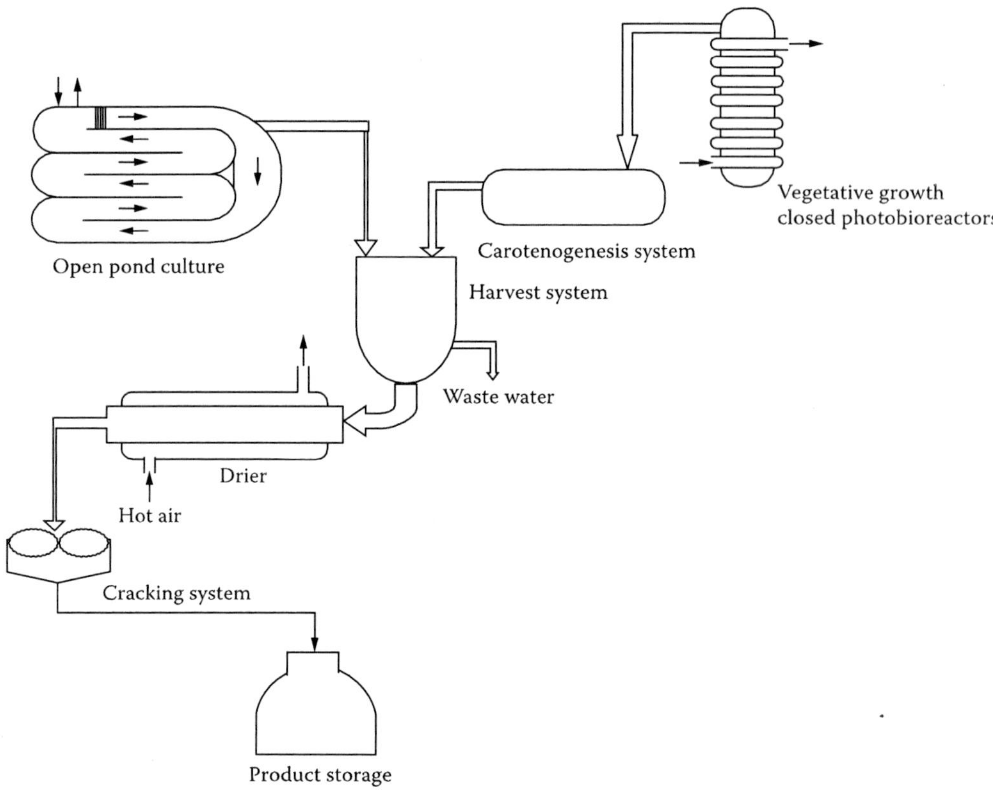

FIGURE 25.3 Schematic diagram of the process of commercial production of natural astaxanthin from *Haematococcus* through an open-pond culture system and photobioreactors. (Adapted from Lorenz, R. T., and Cysewski, G. R., *Trends in Biotechnology*, 18, 160–167, 2000.)

structure of this carotenoid, astaxanthin possesses an additional hydroxyl and ketone group on each β-ionone ring, which increases the biochemical properties of the astaxanthin molecule, including its antioxidant activity (Figure 25.2) (Capelli and Cysoewski, 2006). Since this molecule has important properties to survive in the marine environment, many marine fish, algae, and crustaceans have accumulated this lipid-soluble red pigment abundantly. Though these organisms have accumulated astaxanthin through the food chains, this pigment is primarily produced in bacteria and certain algae found in the marine environment.

Out of the different astaxanthin producers, *Haematococcus*, which accumulates a large amount of astaxanthin in the haematocysts under stressful environment conditions, is used for the commercial production of astaxanthin. Recently, completely closed photobioreactors and large-scale open culture ponds have been successfully established for the production of astaxanthin from *Haematococcus* species (Figure 25.3). Since this microalgae also produces astaxanthin efficiently under stressful environmental conditions such as high temperature, high light intensity, and nutrient starvation, a two-step process where *Haematococcus* is grown under favorable conditions and transferred to a stressful environment for the production of astaxanthin is used in a large-scale open-pond culture system (Lorenz and Cysewski, 2000; Fabregas et al., 2001).

Apart from the microalgae species, different bacteria have been isolated from the marine environment which have the potential for the production of natural astaxanthin. Furthermore, the genes of these bacteria that are responsible for the production of astaxanthin have been isolated and the ability of producing astaxanthin from these genes has been studied by Lee et al. (2007), exploring the potential of producing astaxanthin biotechnologically.

25.3.2 Polysaccharides

Polysaccharides are a major constituent in plant and microbial cell walls that have been used in many industries. Especially chitin, which is found in the exoskeletons of crustaceans, and exopolysaccharides, which are secreted from marine bacteria, have a greater potential in industrial applications. Apart from these cell wall polysaccharides, the different matrix polysaccharides including fucoidan, carageenan, alginate, and agar have been isolated and identified as texture-improving agents in cosmeceutical products.

25.3.2.1 Carageenan, Agar, Alginate, and Fucoidans

The structure of carageenan consists of additional sulfate groups with alternating (1-3)-β-D-galactopyranose and (1-4)-α-D-galactopyranose residues, while agar is composed of alternating D- and L-galactopyranose units. Both these compounds are well-known linear polysaccharides extracted from marine red algae (Kim et al., 2008).

Similar to the carageenan and agar in red algae, alginate is the main matrix polysaccharide of marine brown algae. This molecule is also a linear biopolymer of β-(1-4)-linked D-mannuronic acid and α-(1-4)-linked L-guluronoic acid (Smidsrod and Draget, 1996).

Even though alginate is the main matrix polysaccharide, marine brown algae are also composed of sulfated matrix polysaccharides, fucoidans, that are an industrially important polysaccharide and have received the recent attention of the global cosmeceutical industry. As reviewed by Li et al. (2008), the structure of fucoidans extracted from different marine organisms is highly complex and varies from species to species. Usually algal fucoidans are highly complicated, having heterogeneous and branched polysaccharide structures. However, the structure of fucoidans varies with sugar composition, sulfated content, position of the sulfate group, molecular weight, linkage mode, and sequence of the sugar residue. Many studies have shown that ordered fucoidans may contain a linear core structure that consists of (1-3)-α-L-fucose or alternating (1-3)-α-L-fucose and (1-4)-α-L-fucose or (1-2)-α-L-fucose. The sulfate groups of this complex structure are often attached to the carbon 2, 3, or 4 positions of the fucose residue. The presence of sulfate groups improves the biological properties of fucoidans, which increases their applications in the industrial sector.

25.3.2.2 Exopolysaccharides

More recently, exopolysaccharides from marine microorganisms have received greater attention due to the special properties that have been exhibited by them. As reviewed by Nicolaus et al. (2010), unlike many mesophilic producers that are producing pathogenic compounds, extremophilic marine bacteria produce nonpathogenic exopolysaccharides that have important industrial applications, especially in the food and cosmeceutical fields. These extremophilic marine bacteria secrete exopolysaccharides as a protective mechanism to extreme environmental conditions (Nichols et al., 2004, 2005a, 2010) or as an adhesive material to surfaces (Flemming and Wingender, 2001). These bacteria have been mainly isolated from harsh marine environments, such as intertidal areas, hydrothermal vents, or sea ice. Despite the marine ecosystem that they live in, the extremophilic marine bacteria are known to produce acidic polysaccharides that mainly consist of different combinations of neutral sugars, such as glucose, galactose, mannose, rhamnose, and fucose, along with glucuronic and galacturonic acids. Apart from these neutral sugars, some exopolysaccharides also consist of proteins, amino acids, N-acetyl glucosamine, or N-acetyl galactosamine and sulfate groups (Table 25.3). However, the amount of uronic acids present in exopolysaccharides has mainly contributed to their property of binding capacity. Therefore the uronic acid percentage present in exopolysaccharides is important for their industrial applications. Usually the exopolysaccharides extracted from the *Alteromonas macleodii* and *A. inferuns* that have been isolated from the deep-sea hydrothermal vents contain a high uronic acid percentage (Vincent et al., 1994; Raguenes et al., 1997b), whereas the sea ice bacteria, *Pseudoalteromonas*, have a lower uronic acid percentage (Nichols et al., 2004) than the deep-sea hydrothermal vent bacteria. In contrast to the above results, the study of Nichols et al.

TABLE 25.3
Composition of Exopolysaccharides from Different Marine Bacteria

Bacterial Species	Neutral Sugars (%)	Proteins (%)	Amino Acids (%)	Uronic Acid (%)	Sulfate (%)	Reference
Alteromonas macleodii sub sp. *Fijiensis* HYD 657	58	0.5	0	30	0.0	Cambon-Bonvita et al., 2002
Alteromonas macleodii sub sp. *Fijiensis*	42	4.0	NA	38	5.0	Cambon-Bonvita et al., 2002
Alteromonas macleodii sub sp. *Fijiensis*	38	4.0	2.3	42	5.0	Raguenes et al., 1996
Alteromonas macleodii	50	NA	3	40	NA	Vincent et al., 1994
Alteromonas inferuns	57	NA	0.7	42	NA	Raguenes et al., 1997b
Alteromonas inferuns	51	5.5	NA	37	4.0	Cambon-Bonvita et al., 2002
Alteromonas strain HYD 1644	52	0.0	NA	42	0.8	Cambon-Bonvita et al., 2002
Pseudoalteromonas						
CAM 025	74	2.0	NA	22	5.0	Nichols et al., 2004
CAM 036	50	3.0	NA	25	5.0	Nichols et al., 2004
Pseudoalteromonas strain HYD 721	51	8.0	NA	12	3.0	Cambon-Bonvita et al., 2002

Note: NA, not available.

(2005b) has revealed that there are some species of Antarctic marine bacteria that contain a high percentage of uronic acids. Therefore the uronic acid percentage in the exopolysaccharide varies with the species of bacteria rather than the type of ecosystem that they isolate.

Since the uronic acid extracted from marine bacteria contains an acidic carboxyl group that is ionizable at the sea water pH, the ionic strength of the solution has a great influence on the viscosity of the solutions in which the exopolysaccharide is dissolved (Vincent et al., 1994). Exopolysaccharide solutions with high viscosity can be applied as a thickening agent in cosmetic products. This type of unusual high molecular weight exopolysaccharide with high viscosity has been extracted from *A. macleodii* subsp. *fijiensis* grown in batch cultures, and these high-viscosity polysaccharides have potential for application in the commercial preparation of cosmeceutical products (Reguenes et al., 1996). Furthermore, the presence of sulfate as a functional group (Nichols et al., 2005b) and the ionic strength of the solvent of the exopolysaccharides (Vincent et al., 1994) also contribute to the viscosity-altering property of exopolysaccharide solutions. Usually the exopolysaccharides isolated from marine bacteria consist of around 5% sulfate (Table 25.3) in their structure, contributing to industrially important biological properties of exopolysaccharides (Raguensis et al., 1997; Nichols et al., 2004, 2005b). Moreover, the study of Nichols et al. (2005b) revealed that certain Antarctic marine bacteria produce sulfates up to 15% of the exopolysaccharide weight. Though sulfate is an important group for viscosity alteration, the exopolysaccharide from the bacterium strain HYD 657 (*A. macleodii* subsp. *fijiensis*), which is currently present in cosmetic products, has been shown to have important biological properties even without carrying any sulfate group in its structure (Canbon-Bonvita et al., 2002). This unusual exopolysaccharide possesses a different uronic acid residue called 3-*O*-(1-carboxyethyl)-D-glucuronic acid under a specific medium that leads to the biological properties of this exopolysaccharide being applied as a cosmeceutical (Dubreucq et al., 1996; Cambon-Bonvita et al., 2002). Hence these microbes from extreme environments have the ability to produce exopolysaccharides with a diversity of chemical and physical properties (Gueznuec, 2002). Moreover, chemical modifications of the complex structure of exopolysaccharides by depolymerization or sulfation have resulted in low molecular weight sulfated polysaccharide fractions with different biological activities (cited in Raguenes et al., 1997b). The recent study of Petit et al. (2006) has produced low molecular weight, size-defined exopolysaccharides with high reproducibility using free radical depolymerization with metallic catalysts. The authors have selected copper as the most suitable catalyst for the depolymerization of exopolysaccharides under this method for the production of cosmeceuticals, as a low concentration of copper can play an important antioxidant role by protecting skin against free radical-induced oxidative damage and UV radiation. Thus these low molecular weight exopolysaccharide fractions have the benefit of easy absorption through the skin; they can be applied as skin-protecting cosmeceuticals in the future.

However, the industrial application of such molecules needs large-scale production of exopolysaccharides in commercial operation systems. The bacteria isolated from marine ecosystems have shown the commencement of production of exopolysaccharides at the latter part of the exponential phase, which reaches the maximum level at the end of the stationary phase. The essential feature that has been observed for the production of exopolysaccharides is the presence of a glucose source in the bacteria culturing media. Since these diverse exopolysaccharides producing bacteria are isolated from extremely different ecosystems, such as hydrothermal vents and sea ice, the optimum conditions for exopolysaccharide production, such as temperature, differ highly with the bacterial species. The *Pseudoalteromonas* species, which have been isolated from marine ice, achieve their optimum exopolysaccharide production around 100 mg of exopolysaccharides per gram weight of cells (Nichols et al., 2004) at suboptimal temperatures, such as –2 or 10°C. Their exopolysaccharide yield at 20°C is minimum (Nichols et al., 2004, 2005a), reflecting that the high exopolysaccharide yield cannot be expected from temperatures above 20°C, which will be the optimum exopolysaccharide production temperature for the deep-sea hydrothermal vent *Alteromonas* sp. Their optimum exopolysaccharide production temperature varies from 25 to 35°C (Vincent et al., 1994; Raguenes et al., 1996, 1997b; Cambon-Bonvita et al., 2002), which suggests that the different optimum growth conditions needed by each bacterium should be provided essentially to obtain the exopolysaccharides with characteristic biological properties to be applied to them in the industrial context.

25.3.2.3 Chitin, Chitosan, and Their Derivatives

In the industrial scenario, chitin, chitosan, and their derivatives are the most important low-cost polysaccharides that can be found in nature, as they are present in sufficient amounts, as a result of the seafood processing industry.

An industrially important polysaccharide, chitin is the major structural component of the exoskeleton of invertebrates, including shrimp and crabs, and the cell walls of fungi. Hence this linear amino polysaccharide is the second most abundant biopolymer found in nature, after cellulose. The complex structure of chitin is composed of polysaccharide chains that are developed from chitin nanofibrils (Morganti and Morganti, 2008). These polysaccharide chains are named poly β(1→4)2-acetamido-2-deoxy-D-glucose, which is synthesized from the repeated units of 2-acetamido-2-deoxy-β-D-glucose linked by β(1→4) glycosidic bonds. The poly β(1→4)2-acetamido-2-deoxy-D-glucose structure of the chitin is deacetylated into different degrees by either acid or alkali treatment for the production of chitosan, which is known as (1→4)-2-amino-2-deoxy-D-glucose (Figure 25.4). Differing from the structure of chitin, chitosan has three types of reactive functional groups: an amino group at the C-2 position, and primary and secondary hydroxyl groups at the C-3 and C-6 positions, respectively (Furusaki et al., 1996). The chemical modifications of these three reactive functional groups have increased the use of chitosan in industrial applications. However, both chitin and chitosan have unusual multifunctional properties, enabling them to be applied in many industries, including the cosmeceutical field. Their high tensile strength, biodegradability (Simpson et al., 1994), biocompatibility, nonantigenecity, nontoxicity (Tharanathan and Kittur, 2003), antioxidant activity, and antimicrobial activity (Vidanarachchi et al., 2009) favor industrial applications, whereas low solubility at neutral pH due to high molecular weight and high viscosity limits the

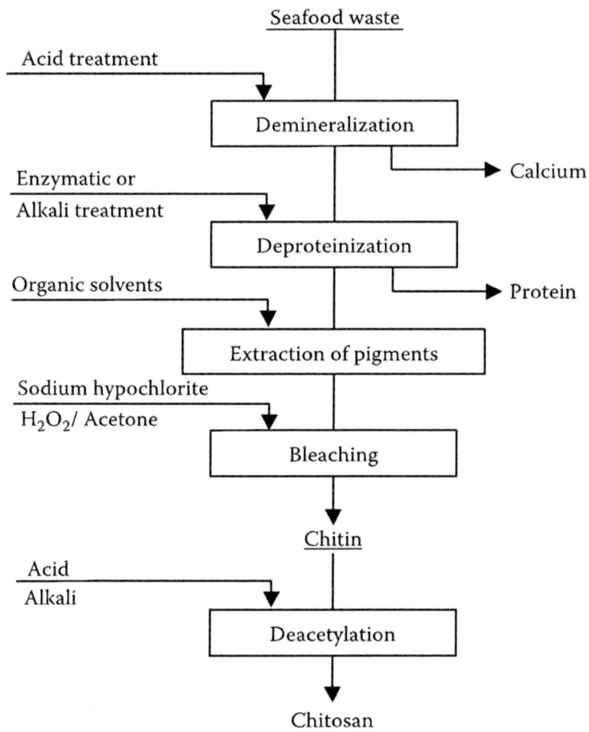

FIGURE 25.4 Flowchart of production of chitosan from seafood waste. (Adapted from Vidanarachchi, J. K., et al., in S. K. Kim (Ed.), *Chitin, Chitosan, Oligosaccharides and Their Derivatives: Biological Activities and Applications*, Taylor & Francis Group, Boca Raton, FL, 2009, 543–560.)

industrial applications of chitin and chitosan. Overcoming this limitation, chitooligisaccharides, which are soluble at neutral pH, have been produced from both chitin and chitosan by chemical hydrolysis (Rupley, 1964; Verum et al., 2001; Il'ina and Varlamov, 2004) or enzymatic hydrolysis (Izume et al., 1992; Yamasaki et al., 1993; Jeon and Kim, 2000a, 2000b). Moreover, during the process of chitooligosaccharide production, the uncontrolled hydrolysis of chitin or chitosan has resulted in a hydroxylate with glucosamines or N-acetyl glucosamines, which have been identified as important constituents of cosmeceutical products.

Apart from these chitin derivatives, the chitin nanofibrils have recently attracted the attention of the cosmeceutical industry as another important active ingredient delivering agent (Morganti et al., 2008).

25.3.3 Proteins or Peptides

Out of the active ingredients applied as cosmeceuticals, proteins or peptides are an important category. Among the different proteins present in the organisms, collagen is the major structural protein found in the skin and bones of all animals. The marine fish skin, bones, and fins that result from the by-products of the fish processing industry have been used widely for the extraction of collagen (Kim and Mendis, 2006), while the skin and bones of mammals, including cattle and pigs, have been widely used as raw materials for the collagen industry. Nevertheless, after the occurrence of bovine spongiform encephalopathy and foot and mouth disease outbreaks, the attention of the collagen industry has been directed toward the extraction of collagen from fish species. In the cosmeceutical industry collagen is the main marine-derived protein, which increases the moisturizing property of the final product.

However, the complex molecular structure of collagen is formed as triple helical protein chains that wrap around one another. Out of the four different types of collagen that have been identified, fish skin, bones, and scales mainly constitute type I collagen, which mainly consists of glycine, alanine, proline, and hydroxy proline. Usually glycine, which is the most abundant amino acid in collagen, is regularly spaced at every third residue throughout the α-chain of the collagen structure, allowing the three helical structures to pack tightly together (Bae et al., 2008). In addition, the amount of amino acids present in the collagen structure has great influence on its thermal stability (Kimura et al., 1988; Ogawa et al., 2004).

Though thermal stability has a wide industrial role, fish collagen with a denaturation temperature above 30°C is very limited. Therefore mammalian collagen is more stable than the collagen extracted from fish species. However, higher denaturation temperatures (34.1, 33.2, 32.2°C) have been reported from the collagen extracted from certain marine fish species, such as eagle ray, red stingray, and Yanti stingray (Bae et al., 2008). Similarly, Ogawa et al. (2004) have reported subtropical fish collagens with higher denaturation temperatures ranging from 34.7 to 35.7°C. The heat resistance ability of the skin, bone, and scale collagens of these fish species reflects the possibility of applying them in the cosmeceutical industry as a substitute for the risky collagen extracted from the mammals.

The complex structure of collagen is partially hydrolyzed either enzymatically or by heat treatment (Kim and Mendis, 2005) to produce gelatin, which consists of a heterogeneous mixture of high molecular weight, water-soluble proteins (Kim et al., 2008). During the hydrolyzing process, the hydrogen and covalent bonds that stabilize the triple helical structure of collagen are destroyed, resulting in the coiled confirmation of gelatin (Djabourov et al., 1993). Though the triple helical structure of collagen is absent after enzymatic or heat treatment, the denatured gelatins also have many industrial applications, including the cosmeceutical industry.

Even though both collagen and gelatin are important as moisturizing agents in skin preparations, peptides, which are short-chain sequences of amino acids, have a more specific role as cosmeceuticals. There are three main categories of cosmeceutical peptides: signal peptides, neurotransmitter-affecting peptides, and carrier peptides (Lupo and Cole, 2007).

25.3.4 Mycosporine-Like Amino Acids (MAAs)

Among the potential marine-derived cosmeceutical compounds, mycosporine-like amino acids have been identified as a multipurpose secondary metabolite (Oren and Gunde-Cimerman, 2007), which is an important source to absorb the energy of solar radiation. These are intracellular, low molecular weight (<400 Da), polar compounds that consist of cyclohexenone or cyclohexenimine chromophores with amino acid substituent (Singh et al., 2008). Most of these MAAs have a glycine subunit at the third carbon atom, while more complex secondary MAAs contain sulfate or glycosidic linkages in the third carbon position of the core MAA structure (Wu et al., 1997). Moreover, the presence of conjugated double bonds within the MAA molecular structure favored the photoprotective activity of them, which absorbs the energy of solar radiation (Karentz et al., 1991), and efficient dissipation of absorbed radiation as heat without producing reactive oxygen species (ROS) (Conde et al., 2000).

However, the synthesis of MAA begins with 3-dehydroquinate, which is an intermediate of the shikimate pathway. This molecule serves as the precursor for the synthesis of MAAs and 3-dehydroquinate is converted to deoxygadusol, which is the core of all MAAs via gadusol. In the presence of glycine, deoxygadusol is converted to the primary MAA: mycosporine-glycine (Singh et al., 2008). The biochemical conversions of this primary MAA during the late steps of the biosynthetic pathway have resulted in high diversity of secondary MAAs, which have been isolated from many marine organisms (Figure 25.5).

Though these MAAs have been isolated from a diverse group of marine organisms, the synthesis of them has been restricted to cyanobacteria, bacteria, and macro- and microalgae, whereas other

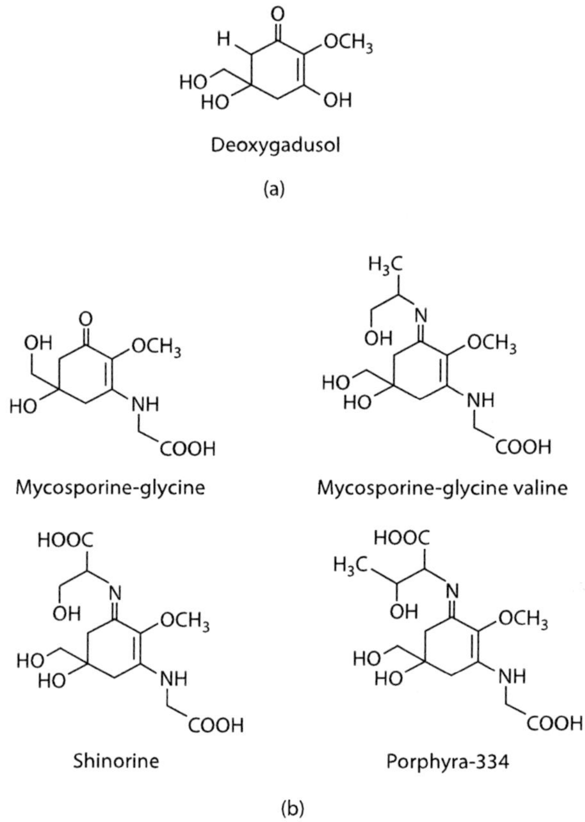

FIGURE 25.5 Structures of some of the cosmeceutically applied MAAs.

marine organisms acquire them through food chains. The recent study of Arbeola et al. (2010) has reported nondietary origin MAAs in sea anemones, and the authors suggested that these MAAs may have resulted from the biotransformation metabolism of dietary MAAs or their precursors. Furthermore, the genes encoding for enzymes of the shikimate pathway have been reported from the sea anemone, *Nematostella rectensis*, opening the gates for research on biosysnthesis of MAAs in marine animals (Starceve et al., 2008). Since the biosynthesis of MAAs from marine animals has yet to be discovered, extraction of these valuable compounds from marine algae, microalgae, and cyanobacteria for commercial purposes is more applicable.

MAAs producing cyanobacteria are abundant in hypersaline environments, and shinorine and porphyra-334 have been found to be the most dominant MAAs produced by them (Sinha et al., 2001, 2003).

Similar to cyanobacteria, several MAAs have been isolated from marine microalgal species. According to the recent review of Rastogi et al. (2010), MAAs predominantly occur in members of the Dinophyceae, Bacillariophyceae, and Prymnesiophyceae families. Llewellyn and Airs (2010) have also reported that the quantifiable amount of MAAs in microalgae of Dinophyceae have the highest amount of shinorine and palythenic acids. Among the MAA profile of microalgae, shinorine has been identified as the most common MAA. Apart from shinorine, several studies have isolated mycosporine-glycine, which is the most primary MAA, mycosporine-glycine valine, and porphyra-334 from different phytoplanktons (Moisan and Mitchell, 2001; Reimer et al., 2007). However, porphyra-334 is the predominant MAA found in macroalgae, having a considerable contribution of total MAA content from shinorine (Schmid et al., 2003; Zhang et al., 2005a, 2005b).

Usually MAAs in cyanobacteria, microalgae, and macroalgae are produced as photoprotective compounds responding to changes in the environment in which they live. Thus the identification of suitable environmental conditions for the maximum production of MAAs is essential for the commercial production of these valuable compounds. The UV radiation intensity and nutrient availability, especially the nitrogen content and salt concentration of the growth medium, have been identified as the factors that affect the production of photoprotective agents in these organisms (Oren and Gunde-Cimerman, 2007).

Even though different MAAs have been identified as important UV-absorbing compounds in Rhodophyceae, the occurrence of MAAs in Chlorophyceae and Phaeophyceae is limited (Rastogi et al., 2010). Thus these algae that lack MAAs may contain a different UV-absorbing or photoprotective compound, such as polyphenolic compounds having potential for use as cosmeceuticals.

25.3.5 Phenolic Compounds

The polyphenols found in marine algae, which are known as phlorotannins, have been identified as a natural photoprotective compound present in marine environments. This secondary metabolite is formed by polymerization of phoroglucinol (1,3,5-trihydroxy benzene) units and has only been found to exist within the group of brown algae (Amsler and Fairhead, 2005). In contrast, a recent study of Matanjun et al. (2008) reported that the high content of phenolic compounds described as phloroglucinol equivalents from two green seaweeds, *Caulerpa lentifera* and *C. racemosa*. However, many studies have isolated phlorotannins from brown algae (Ahn et al., 2007; Ye et al., 2009), which reflects the potential usage of brown algal phlorotannins in the industrial scenario, including the cosmeceutical industry.

In brown algae such as *Ecklonia bicyclis*, *E. cava*, and *E. kurome*, phlorotannins are accumulated in the outer corticle layer of the vegetative cells, which has the ability to quickly respond to marine environmental factors, such as salinity, nutrient, light availability, and ultraviolet radiation, especially stressful marine environmental conditions (Shibata et al., 2004). According to Koivikko et al. (2008), the nutrient enrichment of the algal growing environment has decreased the total phlorotanin content, reflecting that the phlorotannins are produced as a protective mechanism to environmental stresses. Furthermore, the same study revealed that there is a possibility of having a genetic relationship with the synthesis of phlorotannins in brown algae, as the phlorotannin profile changes with the brown algal population, and the observation of an unchanging phlorotannin profile with the changing nutrient enrichment.

However, the phlorotannin profiles have great influence on the antioxidant activity of the algal extract. Usually this phlorotannin profile contains phlorogucinol, eckol, and dieckol as major phlorotannins (Figure 25.6) (Ahn et al., 2007; Shibata et al., 2008), while some unidentified phlorotannin dimers, trimers (Ye et al., 2009), and tetramers (Shibata et al., 2008) are also reported. This pholotannin profile in the extract varies with the polarity of the solution that is dissolved (Ye et al., 2009). Moreover, the polarity of the solvent plays an important role in the extraction yield of phlorotannins from marine algae (Matanjun et al., 2008). Since the total phlorotannin yield and the phlorotannin profile have similar industrial importance, both these factors need to be considered in extracting phlorotannins for the commercial preparation of cosmeceuticals.

Similar to phlorotannins in brown algae, flavonoids that contribute to the total phenolic content of the algal extracts have been identified as antioxidant compounds in some green algae, such as *Enteromorpha compressa* and *Capsosiphon fulvescens* (Cho et al., 2010). Flavonoids are complex molecules with multiactive properties, which makes it possible to use them as an active ingredient in cosmeceuticals (Arct and Pytkowska, 2008). Therefore the presence of flavonoids in marine green algae and the observation of higher antioxidant properties from the chloroform fraction of

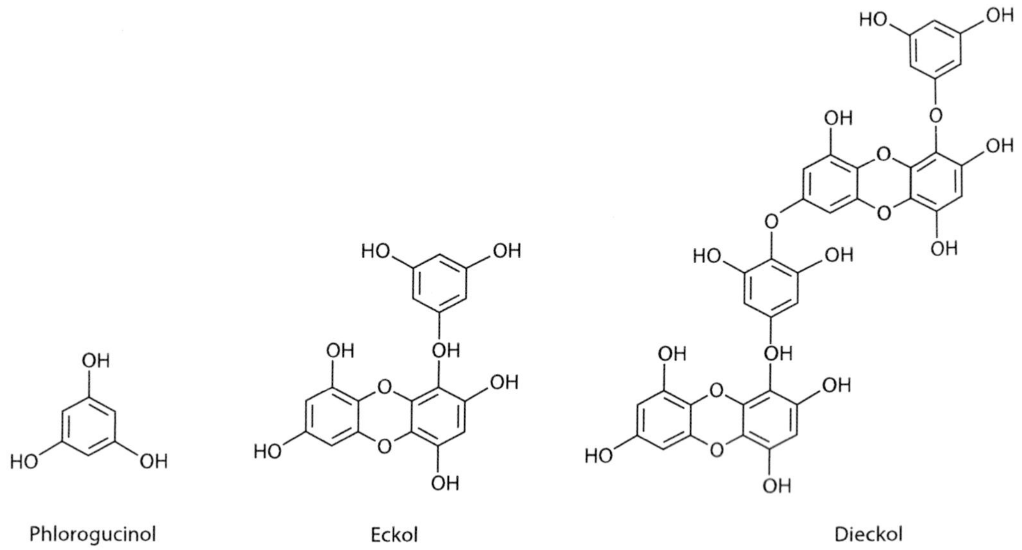

FIGURE 25.6 Structures of the major phlorotannins in algae.

their extract suggest that this would be an important natural antioxidant to be used in health care products, including cosmeceuticals.

25.3.6 Fatty Acids

Though many active substances found in cosmeceuticals are important for the prevention of possible damage to the skin, fatty acids have a different function as a cosmeceutical active ingredient. They have a wide role as a skin-nourishing agent that could be applied externally to get the maximum benefits. Of the different types of naturally occurring fatty acids, polyunsaturated fatty acids that contain fewer hydrogen atoms than the maximum possible number have received special attention due to their human health benefits. Among them, linoleic acid has been identified as a skin-nourishing cosmeceutical, as this fatty acid has a major role in the skin barrier function. Linoleic acid is part of the ceremide, which is the most important substance in the horny layer of the skin. As an important substance in the skin barrier, linoleic acid prevents transepidermal water loss, thereby improving skin moistness.

Since these polyunsaturated fatty acids, especially linoleic acid, are known to be produced within marine algae, considerable attention has been focused on the usage of marine algal polyunsaturated fatty acids as cosmeceuticals. Research has found C_{18} linoleic acid as the most dominant polyunsaturated fatty acid found in many marine algal extracts, including *Codium fernandezianum* (Goecke et al., 2010), *Chaetomorpha linum*, *Enteromorpha compressa* (Khotimchonko et al., 2002), and *Kappaphycus* sp. (Rajasulochana et al., 2010). Among these polyunsaturated fatty acid-producing algae, C_{18} polyunsaturated fatty acids can be found more abundantly in brown algae than in marine red algae. However, the fatty acid profile of these algal species has a relationship with the biogeographical distribution of them. This fatty acid profile varies according to the salinity, temperature, solar radiation, and nutrient availability present in the algal growing environment (Teoh et al., 2004).

Apart from skin-nourishing ability, the triglycerides present in the lipid profile of algal extracts have the potential to function as inhibitors in the photodegradation of collagen (Mansour et al., 2005).

25.3.7 Other Marine-Derived Compounds

Apart from the above-classified marine-derived compounds, several other compounds, such as vitamins, enzymes, and biopterin glucose, have been extracted from marine organisms that would have potential industrial applications as cosmeceuticals.

25.3.7.1 Vitamins

Vitamins have been identified as a widely used active ingredient in cosmeceutical preparations, especially vitamin E and vitamin C, which have skin-protecting antioxidant activities. Moreover, vitamin C acts as a cofactor for the enzymes involved in the synthesis of collagen, which is an important constituent in the skin. Since the human body is unable to synthesize both vitamin E and vitamin C, the major source of them is via diet. However, the optimum concentration of vitamin C in the skin is unable to be achieved even with massive oral doses, as the active transport of vitamin C across the gastrointestinal tract is limited (Burke, 2007). Therefore the topical application of vitamin C becomes important. The instability of the active form of vitamin C, L-ascorbic acid in cosmeceutical products has been a problem in developing the L-ascorbic acid included in cosmeceutical products. Therefore the stable esterified derivatives of L-ascorbic acid, such as magnesium ascorbyl phosphate or ascorbyl-6-palmitate (Farris, 2009), have been developed as a solution for cosmeceutical instability. Though the problem of instability is solved by esterifying, these esterified derivatives of L-ascorbic acid are not well absorbed through the skin (Darr et al., 1992). Even though very limited amounts of the esterified form of L-ascorbic acid are absorbed, they exhibit a low antioxidant activity (Kobayashi et al., 1998) due to their inability to be metabolized into the active form of vitamin C. Therefore as a solution to all these problems, Pinnell et al. (2001) have suggested delivering a high concentration of L-ascorbic acid, which is the active form of vitamin C, into the skin after stabilizing it by removing the ionic charge of the molecule by reducing the pH of the medium.

Similarly, instability of vitamin E in cosmeceutical products is a limitation in their application in the cosmeceutical industry. Hence esterified formulations of vitamin E have been used in commercial cosmeceutical preparations. However, α-tocopherol, which is the active form of vitamin E, is more biologically active than the esterified form, and its lypophilic structure favors the absorption of vitamin E through the skin (Burke, 2007). Since the concentration of topically applied vitamin E is high in the outermost layers of the stratum corneum, it can easily act against the oxidative stress of UV radiation.

A high antioxidant activity has been observed from vitamin E and vitamin C that have been extracted from the marine brown seaweed *Sargassum pallidum* (Ye et al., 2009). However, their extraction yields were very low, limiting the application of marine-derived vitamins in mass production.

25.3.7.2 Scytonemin and Biopterin Glucose

As the different marine organisms have different photoprotective mechanisms, cyanobacteria, which are a very primitive group of organisms, produce different pigments, such as scytonemin or biopterin glucose.

Scytonemin is a lipid-soluble pigment that can be found in the outer cuticle sheath of cyanobacteria as a photoprotective compound (Garcia-Pichel and Castenholz, 1991). The structure of scytonemin is composed of indolic and phenolic subunits linked by an olefinic carbon atom. This complexity in the structure of scytonemin enables this molecule to be involved in efficient UV absorption (Proteau et al., 1993).

Similarly, the UV-A-absorbing pigment glycoside of 2-amino-4-hydroxy-6-(1′,2′-dihydroxypropyl) pteridine or biopterin has been isolated from marine cyanobacterium *Oscillatoria* sp. (Matsunga et al., 1993). Since both these compounds are identified as photoprotective pigments in cyanobacteria, there is a possibility of applying them as sunscreen cosmetics.

25.3.7.3 Enzymes

The absorbing of UV radiation as well as the detoxifying of reactive oxygen species (ROSs) produced from UV radiation has similar importance in preventing photoaging. Extremozymes that have the ability to detoxify ROSs have been extracted from the deep-sea hydrothermal vent bacterium *Thermus thermophilus*, and currently these heat- and UV-stable enzymes have been incorporated into cosmeceutical products successfully (Linter et al., 2009).

Apart from the ROS detoxifying enzymes, the enzyme involved in the repair of UV-damaged deoxyribonucleic acid (DNA) has been isolated from the marine cyanobacterium *Anacyatis nodulans*. This photoreactivating enzyme, phytolase, consists of a single protein chain with two different intrinsic chromophores, which have been identified as 7,8-didemethyl-8-OH-5-deazariboflavin and FAD, in which the majority is present as FAD semiquinone (Eker et al., 1990). Since this enzyme is able to repair UV-damaged DNA, the preparation of cosmeceuticals with this enzyme is favorable for the prevention of photoaging by acting as an antiapoptosis agent.

25.3.7.4 Minerals

The antiapoptosis property as well as antioxidant and anti-inflammatory activities in the skin have been reported from certain minerals, enabling them to be used as cosmeceuticals. The main source of these marine minerals is Dead Sea mud, which consists of a high content of calcium, magnesium, sodium, and potassium (Ma'or et al., 2000). Different cosmeceutical preparations have been experimented upon using Dead Sea mud to investigate the possibility of nourishing the skin with these minerals and their bioavailability at the skin level (Abu-Jdayil and Mohameed, 2006) for application in the cosmeceutical industry.

25.3.7.5 Parabens

Even though the compounds that have direct cosmeceutical activity are more important as active ingredients, certain important additive compounds, such as preservatives, also play an important role in increasing the shelf life of cosmeceuticals. The chemically synthesized alkyl esters of 4-hydroxy benzoate are a widely used preservative in the cosmeceutical industry. This compound, which is called parabens, has been extracted from the marine bacteria of the *Microbulbifer* genus at a high yield (Peng et al., 2006). Therefore these marine bacteria can be used as a source for the mass production of preservatives in cosmeceutical products.

25.4 INDUSTRIAL APPLICATIONS OF MARINE COSMECEUTICALS

These marine-derived substances that are produced as secondary metabolites have different roles as cosmeceuticals. They can be applied as UV-absorbing agents, antioxidant compounds, antiphotoaging agents, moisturizers, active ingredient delivering systems, preservatives or colorants of cosmetic products, and texture-improving agents in the cosmeceutical industry.

25.4.1 Antiaging Cosmeceuticals

The development of novel antiaging cosmeceuticals has been favored by the discovery of biochemical and molecular pathways that lead to the aging of skin. These main pathways of skin aging include intrinsic skin aging and extrinsic skin aging, which manifest as dry skin, loss of elasticity and texture, thinning, appearance of spots, and finally wrinkles (Linter et al., 2009). Intrinsic skin aging is associated with the combined effects of telomere shortening and metabolic oxidative damage (Kosmadaki and Gilchrest, 2004), whereas extrinsic skin aging is due to environmental factors, including exposure to UV radiation (Mekta and Fitzpatrick, 2007). Exposure of skin to UV radiation may directly change the conformation of collagen molecules by removing water bonded to the hydrogen bonds of the collagen molecule (Sionkowska, 2006) or give rise to a cascade of pathways that ultimately result in degradation of collagen or elastosis, leading to photoaging of skin (Figure 25.7). The blockage of certain steps in this cascade by providing active ingredients externally has the possibility of preventing the photoaging of skin. The marine-derived substances that have the ability to act as these active ingredients have the potential to be applied as antiphotoaging cosmeceuticals.

25.4.2 Photoprotective Marine-Derived Cosmeceuticals

Both UV-B (280–325 nm) and solar UV-A radiation, which has a higher wavelength (320–390 nm), have detrimental effects for sun-exposed organisms, including humans (Lacroix et al., 1996; Takema and Imokawa, 1998; Mitchell, 2006). Absorption of these UV radiations by UV-absorbing compounds is the major step involved in the prevention of photoaging cascade (Figure 25.7). Topical application of UV-absorbing compounds such as MAAs, scytonemin, and biopterin glucose in human skin prevents the penetration of UV radiation into the deeper layers of skin.

According to the recent review of Rastogi et al. (2010), the UV-absorbing marine compound scytonemin has an ability to absorb UV radiation of a wide range of wavelengths, which includes both UV-B and UV-A solar radiation. Purified scytonemin has the highest absorption at 386 nm, which was higher than the maximum wavelength possible to be absorbed by MAAs, while absorbing the lower wavelengths, such as 252, 278, and 300 nm. Therefore incorporation of scytonemin into cosmeceutical products provides protection to the skin from both UV-A and UV-B radiation.

Though the maximum absorption wavelength is lower than that of scytonemin, MAAs are able to absorb a wide range of solar radiation, with the maximum absorption between 310 and 365 nm (Oren and Gunde-Cimerman, 2007). This range includes the wavelength of UV-A and UV-B radiation, reflecting their ability to provide protection for both types of radiation by including MAAs in cosmeceutical products. Usually MAAs with cyclohexenone-based structures are responsible for the absorption of UV-B radiation, while UV-A radiation is absorbed by the MAAs with cyclohexenimine-based structures (Karentz et al., 1991). Marine microalgae contain MAAs with both types of structures, which have the ability to absorb UV-A as well as UV-B radiation. 31.1% of the MAAs extracted from the phytoplankton *Nyctiphanes australis* contain mycosporine-glycine, which is mainly responsible for the absorption of UV-B radiation (Riemer et al., 2007). Similary, Moisan and Mitchell (2001) have also observed mycosporine-glycine from the MAAs extracted from *Phaeocystis antarctica*, which has a maximum absorption wavelength of 310 nm. Since the wavelength of UV-B radiation is lower than that of UV-A radiation, its penetrability in the skin is also lower than that of UV-A radiation. Thus UV-B radiation is completely absorbed in the epidermis, whereas UV-A radiation penetrates the epidermis and enters into the deeper layers of the skin, causing photoaging of the skin (Mitchell, 2006). Hence the importance of UV-A-absorbing MAAs in cosmeceutical production has received the attention of the cosmeceutical industry.

Apart from the UV-B-absorbing MAAs, marine microalgae also contribute to the UV-A-absorbing MAAs, such as shinorine, porphyria-334, palythine, and mycosporine-glycine valine (Riemer et al., 2007; Moisan and Mitchell, 2001). Among them shinorine is the predominant UV-A-absorbing MAA that can be found in microalgal MAA extracts, while porphyria-334 also contributes to a considerable amount (Riemer et al., 2007). Furthermore, these two MAAs are abundant in the marine red algae of the genus *Porphyra*. The ratio of porphyria-334 and shinorine in the MAA extract of *Porphyra umbilicalis* is 2:1, having a major contribution from porphyria-334. The pH, temperature, and polarity of porphyria-334 solvents have an influence on the UV absorption property of this molecule. High acidic solutions below pH 3 decrease the UV absorption with increasing acidity, whereas increments in the pH of the solvent above pH 3 increase the UV absorption by porphyria-334 (Zhang et al., 2005a, 2005b). However, the time period during which the compound can absorb UV radiation is essential for maintaining the efficacy of the cosmeceutical product. Zhang et al. (2005b) have observed a sharp decrease in UV absorption during the first 8 hours from porphyria-334 in pH 12 and pH 13 solutions, while the same level of UV absorption or a slight

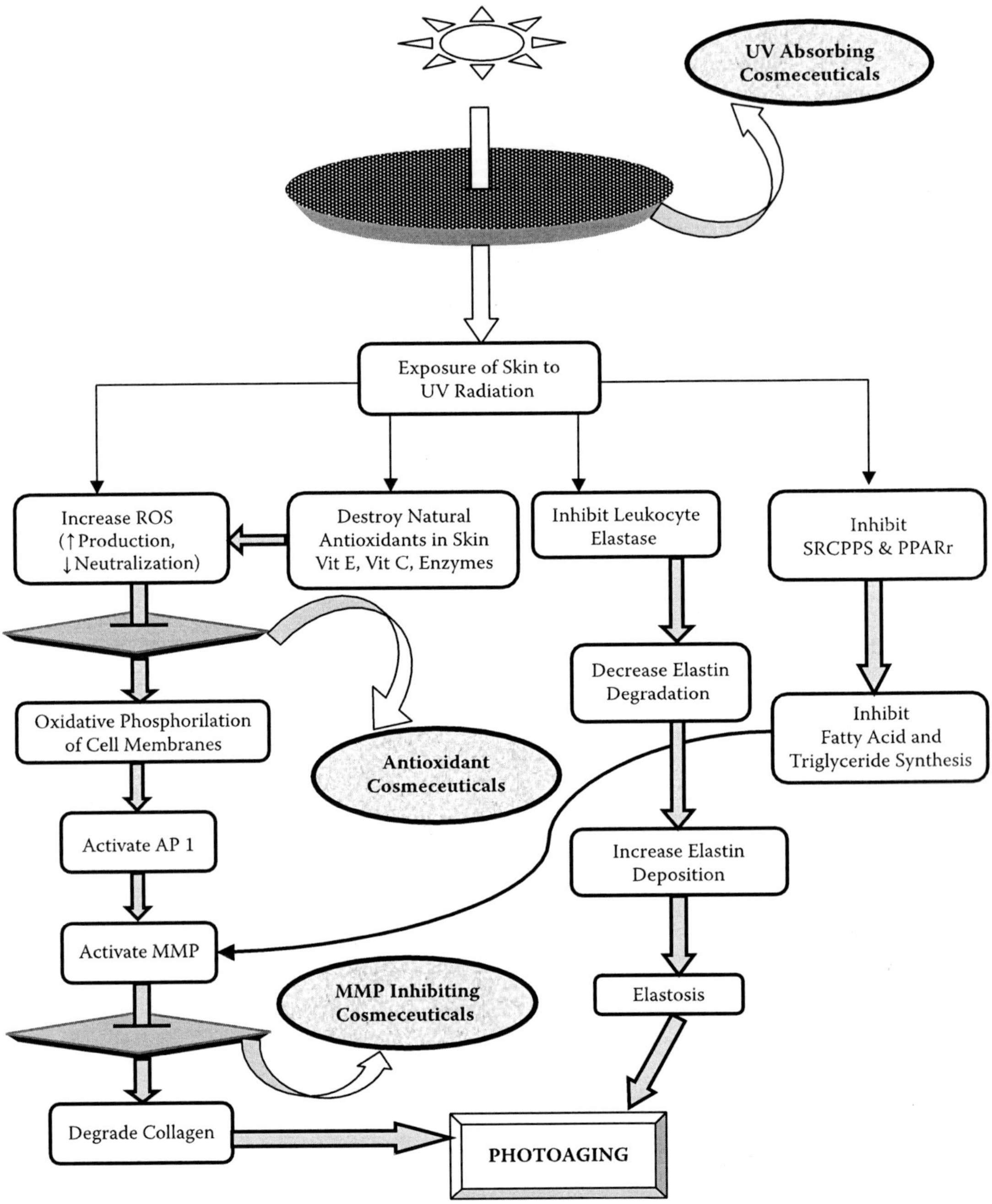

FIGURE 25.7 Schematic diagram of extrinsic skin aging and the cosmeceuticals involved in the inhibition of different steps.

decrease of UV absorption was reported from porphyria-334 in low basic or acidic solutions under room temperature for 80 days. Similarly, porphyria-334 showed the same level of UV absorption with increasing temperature up to 60°C in the aqueous solutions, while a rapid decrease has been exhibited at temperatures higher than 80°C (Zhang et al., 2005b). This was in accordance with the results of Sinha et al. (2000), which reported unaffected absorption properties of porphyria-334 and shinorine under heat treatment at 75°C for about 6 hours. Since the absorbed UV radiations are excreted as thermal energy, the heat stability of porphyria-334 is a greater advantage for them to be applied as cosmeceuticals.

Though aqueous solutions of porphyria-334 exhibit the same absorbance up to 60°C, the UV absorbance of porphyria-334 in ethanol or hexane is decreased even under the same conditions. Thus the type of solvent in which porphyria-334 is extracted plays an important role in the stability of the UV-absorbing compounds in cosmeceutical products. Furthermore, the polarity of the solvents affects the efficiency of UV absorption, as UV absorption is higher in polar solvents than in nonpolar inert solvents (Zhang et al., 2005a). Since porphyria-334 has more positive points to be applied as cosmeceuticals, the efficacy of porphyria-334- and shinorine-incorporated cosmeceuticals has been tested using both *in vitro* and *in vivo* human studies (Schmid et al., 2003, 2004a, 2004b). The study of Schmid et al. (2004b) demonstrates the efficient neutralization of UV-A effects from the cream produced by liposome encapsulated with 0.005% MAA, which was similar to the effects of cream with 1% synthetic UV-A filters and 4% UV-B filters. Four weeks' treatment with this cosmeceutical significantly improved the firmness and smoothness of human skin by 10 and 12%, respectively (Schmid et al., 2004b), while reducing the skin wrinkle depth by 20% compared to cosmeceutical untreated skin (Schmid et al., 2003).

Therefore marine-derived substances such as MAAs, which produce a photoprotective compound in marine organisms, have important applications in the cosmeceutical industry. Furthermore, certain other marine-derived compounds, such as biopterin glucoside, have the potential to be applied as sunscreening cosmeceuticals, as they can absorb UV-A radiation and block the first step of the premature skin photoaging pathway.

25.4.3 Antioxidant Activity of Marine-Derived Cosmeceuticals

Topically applied photoprotective compounds prevent absorption of UV radiation into sun-exposed skin. In the absence of these photoprotective compounds, UV radiation is absorbed into the chromophores in human skin, resulting in reactive oxygen species (ROS) such as superoxide anion and hydrogen peroxide (Mehta and Fitzpatrick, 2007). There are naturally occurring antioxidants such as vitamin E or vitamin C and enzyme systems that are responsible for the neutralizing of ROSs before damaging the skin. However, continuous exposure to UV radiation leads to thermal destruction of these enzymes, or antioxidant compounds naturally occur in human skin (Linter et al., 2009). Furthermore, the rate of neutralization of ROSs by these naturally occurring systems is insufficient to neutralize the ROSs produced due to continuous exposure to UV radiation, resulting in accumulation of ROSs, which causes the photoaging of skin.

Under these circumstances the provision of antioxidants externally may be useful for the neutralizing of ROSs produced in the skin. Various antioxidant compounds, including carotenoids, polyphenoles, vitamin E, vitamin C, fucoidans, MAAs, and exoenzymes, have been extracted from various marine organisms which have the potential to be used as antioxidant cosmeceutical compounds. These external antioxidants have different modes of actions, such as absorbing high energy of ROSs for neutralizing them, acting as electron donors, and acting as ROS neutralizing enzymes to prevent the ROS-induced photoaging of skin (Guerin et al., 2003; Ahn et al., 2007; Linter et al., 2009).

25.4.3.1 Energy-Absorbing Antioxidants

Carotenoids have been identified as important antioxidant compounds, as they can neutralize ROSs, thereby preventing ROS-induced skin damage. As reviewed by Guerin et al. (2003), carotenoids can absorb the excited energy of singlet oxygen onto their chains, leading to degradation of carotenoid molecules while preventing the possible damage caused by the singlet oxygen. Furthermore, carotenoids can act as radical scavengers under the conditions where singlet oxygen is not formed (Tsuchihashi et al., 1995; Murthy et al., 2005). According to Anggadiredja et al. (1997) and Shahab (2007), high antioxidant capacity has been reported from the carotenoids extracted from different algal species, including *Sargassum dentifolium*, *Laurencia papillosa*, *Jania corniculata*, *S. polycystum*, and *L. obtuse*.

However, some studies have revealed the absence of 2,2-diphenyl-1-picrylhydrasyl (DPPH) radical scavenging activity in carotenoids, such as β-carotene, zeaxanthin, and lutein, with the exception of fucoxanthin (Nomura et al., 1997). The presence of unusual double allenic carbon at the C-7′ position and two hydroxyl groups in the structure of fucoxanthin is suggested to be the reason for this special antioxidant property of the fucoxanthin molecule, which was not present in the other tested carotenoids (Yan et al., 1999; Sachindra et al., 2007). Therefore the structural differences among the pigments of the carotenoid group have resulted in variation in their antioxidant capacity. Similarly, the presence of an additional hydroxyl and ketone group on each β-ionine ring of the astaxanthin molecule has led to special biological properties of this molecule, including higher antioxidant capacity (Capelli and Cysoewski, 2006; Lyons and O'Brien, 2002). The higher antioxidant capacity of astaxanthin than of β-carotein, which has been observed by Kobayashi and Sakamoto (1999), is possibly due to the presence of those additional functional groups in the structure of the astaxanthin molecule. The comparison of this high antioxidant capacity of astaxanthin with those of other carotenoids has revealed that the antioxidant capacity of astaxanthin is approximately 10 times higher than that of carotenoids such as β-carotene, zeaxanthin, lutein, and canthaxanthin, and 100 times greater than α-tocopherol, which is a main active ingredient in current cosmeceutical products, reflecting the suitability of astaxanthin as an antioxidant cosmeceutical (Miki, 1991). Furthermore, their ability to prevent oxidative damage in UV-A-treated fibroblast cultures with 10 nM astaxanthin (O'Connor and O'Brien, 1998) and a high level of superoxide anion scavenging ability in human skin fibroblast cultures (Lyons and O'Brien, 2002) confirms their potential as cosmeceuticals in the protection of skin from UV-A-induced damage. Moreover, solving the problem of instability, which has been observed in the cosmeceutical β-carotene, the astaxanthin extracted from the microalgae *Haematococcus* is naturally present in the esterified form, increasing their solubility and stability in the cellular lipid environment, thereby favoring them to be applied as cosmeceutials (Lorenz and Cysewski, 2000).

Attempts have been made to increase the photostability and water solubility of β-carotene by producing β-carotene–humic acid complexes. Though the formation of this complex has increased the photostability and water solubility of β-carotene, its antioxidant capacity has been reduced, limiting its applicability as an antioxidant cosmeceutical (Martini et al., 2010). However, Darvin et al. (2010) increased the stability of β-carotene in cosmeceutical preparations with high antioxidant capacity in skin by incorporating β-carotene with other two antioxidants, vitamin E and vitamin C, making the carotenoids successful as cosmeceuticals.

25.4.3.2 Electron-Donating Antioxidant Compounds

Apart from the energy-absorbing antioxidant compounds such as carotenoids, marine organisms are known to produce electron-donating antioxidants. Polyphenols are electron-rich marine-derived compounds that act as electron donors to free radicals such as superoxide anion and hydrogen peroxide produced in the skin as a result of exposure to UV radiation. The phenoxyl radical produced after the electron donation is stabilized by several methods, including resonance delocalization of the unpaired electron to ortho or para positions of the phenyl ring, hydrogen bonding with an adjacent hydroxyl group, or dimerization of the phenoxyl radicals (Ahn et al., 2007). Shibata et al. (2008) have also suggested that the presence of phenolic hydroxyl groups in the structure of polyphenols may have an important role in their antioxidant activity.

High antioxidant capacity has been reported from phenolic compounds extracted from many macroalgae (Zubia et al., 2007; Matanjun et al., 2008; Ye et al., 2009; Cho et al., 2010). The study of Cho et al. (2010) has revealed that these antioxidant properties are mainly due to flavonoids, a main contributor of the extracted total phenolic content. They have observed a strong correlation between flavonoids and antioxidant capacity, which is not observed between the total phenoplic content and the antioxidant capacity, reflecting the importance of flavonoids as an antioxidant agent. However, the active compounds in the total phenolic extracts responsible for the antioxidant properties vary with the species of the marine organism. Matanjun et al. (2008) have suggested that the high superoxide radical scavenging ability of the phenolic mixture of the microalgae is mainly due to the phloroglucinol equivalents, which belong to the phlorotannins. Among the different phlorotannins found in the algal extracts, Ahn et al. (2007) have observed the highest superoxide radical scavenging activity from phloroglucinols, whereas eckol and dieckol are a trimer and hexamer of phloroglucinols, respectively. Such compounds have been found to possess higher scavenging activity toward the other ROSs. In contrast to these results, Shibata et al. (2008) have revealed a low superoxide radical scavenging ability from phloroglucinols and eckol. However, these authors have observed a very potent superoxide anion radical scavenging activity, which is 2 to 10 times more active than ascorbic acid or α-tocopherol from the phlorotannins. Similarly, Nakai et al. (2006) have observed high superoxide radical scavenging activity from a mixture of high molecular weight phlorotannins. Though the low molecular weight phlorotannins have the benefit of easy absorption through the skin after topical application of cosmeceuticals, those algal extracts of phlorotannins have greater potential to be applied as skin care cosmeceuticals due to their high antioxidant capacity.

In addition to the antioxidative benefits, the topical application of phlorotannins has exhibited antiphotocarcinogenic effects by activating the protective signaling cascade present in human skin (Hwang et al., 2006), which is important for the prevention of UV-induced skin cancers.

25.4.3.3 Antioxidant Enzymes

Apart from the naturally occurring antioxidants, skin has antioxidant enzymes such as catalase, superoxide dismutase (SOD), and peroxidases, which are involved in detoxifying the ROSs. The failure of the naturally occurring enzymes in the skin due to thermal or UV destruction ultimately results in accumulation of ROSs, causing harmful effects to skin. Artificial supplying of heat and UV-stable extremozymes with catalase, SOD, and glutathione peroxidase activity helps to improve skin condition. A solution containing these enzymes produced by the marine bacterium *Thermus thermophilus* has been shown efficient epidermis protection from UV radiation. Hence these enzymes are thermostable up to 80°C and are heat activated; their antioxidant activity will not be affected by UV-induced thermal stress. Therefore these enzymes have greater potential to be used as skin protective antioxidant enzymes in cosmeceutical products. Furthermore, topical application of cosmeceutical cream containing these enzymes has reduced transdermal water loss and has improved skin condition by water retention (Linter et al., 2009).

25.4.3.4 Undefined Antioxidant Activities

Despite the mode of action, the above-described compounds have similar importance as cosmeceuticals. However, there are certain other marine-derived antioxidant compounds, such as MAAs and fucoidans, whose mode of action is unclear due to their complex structure. High antioxidant capacity has been reported from fucoidans extracted from certain brown and red algae, including *Laminaria japonica* and *Fucus vesiculosus*. These algal extracts have exhibited high superoxide radical scavenging activity, which is important for the prevention of photodamage to the skin (So et al., 2007; Souza et al., 2007; Wang et al., 2008). Furthermore, Wang et al. (2008) and Souza et al. (2007) have observed a strong positive correlation between the sulfate content and the antioxidant capacity. Therefore the sulfate group present in the structure of the fucoidan molecule may have special involvement in the antioxidant activity of the fucoidan molecule.

Though the presence of a sulfate group has increased the antioxidant capacity of fucoidans, the low molecular weight, highly sulfated fractions of fucoidans have exhibited lower superoxide and hydroxyl radical scavenging activities than the low molecular weight, low sulfated fractions of fucoidans (Zhao et al., 2008). These authors suggested that these observations may have resulted from the prevention of reaction with oxygen radicals by sulfate groups that act as physical blocks. Therefore maintaining the appropriate amount of sulfate groups in the structure of low molecular weight fucoidans is essential in producing cosmeceuticals with easily absorbed skin-protecting antioxidants. Moreover, the structure of fucoidans has been phosphorylated and aminated to produce fucoidan derivatives with high antioxidant capacity, which have the potential to be applied as future cosmeceuticals.

25.4.3.5 Marine-Derived MMP Inhibiting Cosmeceuticals

Though naturally occurring antioxidant systems are present in the skin, inefficient neutralization of ROSs by these natural antioxidant systems or failure in the topically applied cosmeceutical antioxidants results in oxidative phosphorylation of cell surface receptors by ROSs. This leads to activation of transcription factor activation protein (AP-1), which simulates the matrix metalloproteinases (MMP 1) in the fibroblast and keratinocytes of skin. MMPs are matrix-degrading enzyme systems that ultimately cause the destruction of collagen in the skin. The inhibition of MMPs produced in the skin has the ability to solve the problem of photoaging by preventing UV-induced collagen destruction (Moon et al., 2009).

Various marine-derived MMP inhibitors such as triglycerides and fucoidans have been identified, and they have the potential to be applied as antiaging cosmeceuticals due to their skin health beneficial effects.

Triglycerides that result from the lipid synthesis in the epidermal cells are naturally occurring MMP 1 in human skin. However, chronic skin exposure to UV radiation reduces the lipid synthesis in the skin, thereby reducing the concentration of this natural MMP 1 inhibitor present in the skin (Kim et al., 2010). The same authors have revealed that the topical application of triglycerides is able to prevent UV-induced skin damage by inhibiting MMP expression. Therefore marine algal-derived triglycerides have the potential to be applied as an MMP inhibitor in skin care cosmeceuticals (Mansour et al., 2005).

Similarly, brown algal fucoidans also have been identified as an important MMP 1 inhibitor and can be included as an active ingredient in skin care cosmeceuticals. However, fucoidans prevent the expression of MMP 1 in a different way than triglycerides. Many studies have revealed the inhibition of the MMP 1 enzyme by fucoidans through preventing the expression of mRNA promoter activity in human skin fibroblasts (Moon et al., 2008a, 2008b, 2009) and in keratinocytes *in vitro* (Moon et al., 2008c; Lee and Ku, 2010). Furthermore, fucoidans prevent the expression of MMP 1 by inhibiting the extracellular signal-regulated kinase (ERK) pathway (Moon et al., 2008a; Lee and Ku et al., 2010). Despite the way of inhibition, the level of MMP inhibition is usually associated with the dose of fucoidans (Moon et al., 2008a, 2008b, 2008c, 2009; Lee and Ku et al., 2010). Though the increment in fucoidan concentration has resulted in increasing the level of MMP 1 inhibition, the level of MMP inhibition at 100 μl/ml concentration of fucoidans was not much higher than the other lower level of fucoidans, such as 10 or 50 μl/ml (Moon et al., 2008a, 2009). Moon et al. (2008a) have suggested that these observed results may be due to the slight toxication effects of fucoidans at a dose of 100 μl/ml concentration. Therefore incorporation of an appropriate dose of fucoidans to cosmeceutical products is essential to obtain the maximum health benefits from these cosmeceuticals.

Apart from the inhibition of MMP 1, fucoidan treatments have increased type I procollagen synthesis, which improves the health status of the skin (Moon et al., 2009). Therefore the important marine-derived compound fucoidan has multiple roles in the prevention of UV-induced photoaging of skin, increasing the potential application of fucoidans as antiphotoaging cosmeceuticals.

Obviously antiphotoaging cosmeceuticals have a greater contribution from marine-derived substances such as carotenoids, including β-carotene, astaxanthin, and fucoxanthin, phlorotannins, flavonoids, triglycerides, MAAs, scytonemin, biopterin glucoside, and fucoidans. These compounds have exhibited UV absorption, antioxidant, and MMP inhibition activities that involved the blockage of the skin photoaging pathway through collagen and elastin destruction.

25.4.3.6 Marine-Derived Antiapoptotic Cosmeceuticals

Similar to the prevention of degradation of extracellular structural components such as collagen and elastin, cosmeceuticals have a great role in the prevention of UV-induced cell apoptosis of the skin, thereby preventing skin photoaging. *In vitro* studies that have been conducted using Dead Sea minerals have exhibited an antiwrinkle property by reducing skin roughness (Ma'Or et al., 1997, 2000). Furthermore, Portugal-Cohen et al. (2009) have observed an efficient inhibition of UV-induced cytotoxicity by Dead Sea mineral-incorporated cosmeceuticals. They have suggested that this improved viability of the epidermal cells is due to the stimulation of the DNA repair process, which recovers the UV-damaged cells or decrement of the epidermal cell mutation rate.

Certain other enzymes are also reported to have the ability to prevent UV-induced cell apoptosis by repairing damaged DNA. According to the recent review of Rastogi et al. (2010), UV radiation is responsible for the formation of three major classes of DNA lesions, including cyclobutane pyrimidine dimers, which are responsible for skin cancers and cell death leading to the photoaging of skin. However, these pyrimidine dimers are efficiently removed by the cellular environment by the DNA repair enzyme phytolase (Rastogi et al., 2010). This enzyme is able to absorb energy from visible light, which is unable to be absorbed by the pyrimidine dimers, and this energy is presented to the pyrimidine dimers to get monomerized (Hearst, 1995). Therefore the external application of this enzyme favors the prevention of UV-induced cell death by repairing the damaged DNA. Stege et al. (2000) have found high DNA repairing ability from topically applied phytolase encapsulated into the liposomes that are derived from a marine cyanobacterium. The 7,8-didemethyl-8-OH-5-deazariboflavin chromophore (8HDF) of this enzyme absorbs energy from visible light and presents this energy to pyrimidine dimers either directly or through a FAD chromopohore to get monomerized (Figure 25.8) (Eker et al., 1990). Furthermore, the study of Stege et al. (2000) has revealed the reduction of UV-induced skin erythema through immunosuppression, which is due to the cyanobacterial phytolase. These beneficial effects have increased the applicability of cyanobacterial phytolase in the cosmetic and medicinal fields. Therefore marine-derived phytolase enzymes as well as Dead Sea minerals have greater potential in the development of novel antiphotoaging skin preparation, as they highly contribute to the repairing of DNA molecules in UV-damaged cells to prevent epidermal cell apoptosis.

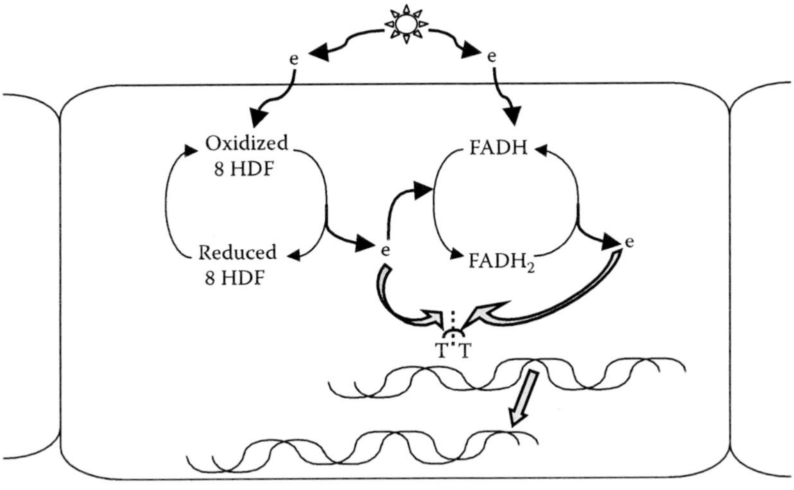

FIGURE 25.8 Schematic diagram of UV-damaged DNA repair. (Adapted from Eker, A. P. M., et al., *Journal of Biological Chemistry*, 265, 8009–8015a, 1990.)

25.4.3.7 Marine-Derived Anti-Inflammatory Cosmeceuticals

Apart from the above-described skin aging processes, the UV-induced inflammation that has been observed in the skin epidermis also highly contributes to the photoaging of skin. Exposure of skin keratinocytes to UV radiation leads to the expression of an intercellular adhesion molecule (ICAM-I) through a cascade of reactions. This expression of ICAM-I in the skin epidermis further initiates a series of events through interaction with lymphocytes in the skin, which ultimately results in inflammation of the skin epidermis. Since ICAM-I has a great role in the inflammatory process of the skin (Thibodeau, 2007), compounds that have the ability to reduce the expression of ICAM-I have greater potential to be applied as anti-inflammatory cosmeceuticals.

The recently isolated exopolysaccharide from the marine bacterium *Alteromonas macleodii* subspecies *Fijiensis* (strain HYD 657) has been identified as an important compound that involves the reduction of expression of ICAM-I. Alain and Atsushi (2006) have observed a 76% reduction of the expression of ICAM-I in keratinocytes treated with this important marine-derived exopolysaccharide, reflecting the importance of this compound as an anti-inflammatory cosmeceutical to reduce UV-induced inflammatory damage to the skin.

Furthermore, this exopolysaccharide is also responsible for the UV protection of skin defense cells called Langerhans cells and the proliferation of keratinocytes in the skin, which helps to renew skin. Due to these important biological properties, this exopolysaccharide has been incorporated into cosmetic formulations. Vacher and Fritsch (2009) have stated that the cosmetic composition of this exopolysaccharide needs to be between 0.001 and 5% to obtain the maximum health benefits from this cosmeceutical preparation. However, these exopolysaccharide-incorporated cosmeceuticals are currently present in the cosmeceutical market under the trade name Deepsane (Cambon-Bonvita et al., 2002).

25.4.4 Marine-Derived Skin Moisturizers

The maintenance of the moisture level in the skin also has similar importance as the prevention of photodamaging. The healthy stratum corneum is composed of approximately 30% water, in which 10% belongs to tightly held water. This tightly bound water closely depends on the presence of the intracellular natural moisturizing factor, which is composed of amino acids and their metabolites. Similar to the natural moisturizing factor, the length of the diffusion pathway as well as the function of the water barrier are also important for the maintenance of moisture content in the skin (Rawling and Harding, 2004). According to the review of Gao et al. (2008), the water barrier of the skin is an extracellular lipid membrane that is mainly (40%) composed of ceramide-I and the derivatives of ceramides. The lipids in this membrane are synthesized throughout the epidermis and later differentiated and contribute to the water barrier of the skin. However, the impairment of these naturally occurring systems fails to prevent water loss from the skin, leading to dry, scaly skin. Therefore the external application of lipid compounds that have the ability to occlude water loss or compounds that can produce bonds with water molecules may have the potential of mimicking the natural hydrating mechanisms of the skin. Therefore these molecules have the potential to be applied as skin moisturizing cosmeceuticals.

25.4.4.1 Water Barrier Mimicking Cosmeceuticals

Linoleic acid, which has been observed in marine algal extracts, is an important constituent in the water barrier layer of the skin. This fatty acid is a part of ceramide-I in the water barrier (Lautenschiager, 2003). Thus linoleic acid-containing cosmceuticals can improve the condition of the natural water barrier of the skin, thereby preventing excessive water loss through the skin.

Though some large molercular weight molecules are unable to incorporate into the natural water barrier of the skin, these compounds have the ability to prevent transepidermal water loss by acting as occlusive humectants that form a hydrophobic film on the skin surface. As a high molecular weight amino polysaccharide, chitosan has this type of mode of action on the skin surface; this compound can be applied as a skin moisturizer in cosmeceutical preparations. The efficacy of water-soluble chitosans as skin moisturizers has been tested by Chen and Hey (2000) by incorporating them into a moisture mask, and they have observed a superior water binding capacity from chitosans compared to that of methyl cellulose, which was used as the control. They also observed increasing water binding capacity of the moisture mask with increasing molecular weight of water-soluble chitosans, reflecting the suitability of high molecular weight chitosans in the occlusion of transepidermal water loss. Though these high moleculer weight chitosans are unable to penetrate the skin, they can act as skin moisturizing cosmeceuticals by their external application on the skin surface.

25.4.4.2 Water Molecule Binding Skin Moisturizing Cosmeceuticals

Similarly, certain other high molecular weight marine-derived compounds such as collagen have been identified as important skin moisturizers that act on the skin surface due to the limitations in the skin penetrance caused by their high molecular weight. However, their mode of action is different from that of chitosans. The externally applied collagen molecules form a layer on the skin surface and bind with the water molecules to increase skin humidity (Swatachek et al., 2002). Though the collagen molecules are known to act as skin moisturizers, Swatscheck et al. (2002) had a contradictory observation, such as low skin hydration effects from collagen extracted from certain marine sponges. However, these collagen molecules have shown an increase in the lipid content of the skin, indicating their suitability as a solution for dry skin that results from an insufficient sebum level. Therefore marine-derived collagen has an important skin moisturizing function by forming a layer on the skin surface.

In contrast to the high molecular weight collagen, certain other low molecular weight compounds such as glycerol and lactic acid have the ability to enter into the stratum corneum of the skin and increase the moisture content in the skin by binding with water molecules. Therefore glycerol has been identified as an important, most widely used moisturizing agent in cosmeceuticals. Research has found the presence of glycerol along with β-carotene extracted from the microalgae *Duniella*. These algal extracts have been incorporated into cosmetic preparations with minerals of Dead Sea mud (Ma'Or et al., 2000), and they have observed significant skin-hydrating effects, after 8 hours of application of this cosmeceutical. Since the glycerol molecules are known to be present in the algal extracts of *Duniella*, these glycerol molecules may have contributed to the skin hydration effects of this cosmeceutical. On the other hand, the Dead Sea minerals present in this preparation have beneficial effects on the natural moisturizing factor of the stratum corneum, and the synergistic effect of glycerol and Dead Sea minerals may have contributed to the increase of the skin-hydrating effect of the final cosmeceutical product. Apart from skin moisturizing effects, this cosmeceutical has been identified as a skin smoothener due to the antiwrinkle properties of Dead Sea minerals, reflecting the multiple roles of some of the marine-derived compounds in cosmeceuticals. The marine-derived substances with multiple properties are more important in the development of novel cosmeceuticals that have many beneficial effects on human skin health.

25.4.5 Marine-Derived Active Ingredient Delivering Cosmeceuticals

Even though the above-described active ingredients have many beneficial effects on human skin, the delivery of these active ingredients to the target site is a challenge. The efficacy of the cosmeceutical product has similar importance on both factors, the presence of the active ingredient and an adequate delivery system in the final product. Usually three main routes, including intracellular, intercellular, and via skin appendages, have been identified as the penetrance routes of the stratum corneum of the skin. The passage of the active ingredients through the above routes depends on the molar mass, molecular size, hydrophilic-lipophylic balance, distribution of polar and nonpolar groups in the molecule, and extent of the ionized state of the active ingredient (Forster et al., 2009).

However, poor skin-penetrating ability has been reported from certain active ingredients with high biological properties, limiting their applications as cosmeceuticals. As a solution for this limitation, active ingredients can be incorporated into the cosmeceutical vehicle, such as emulsions with high penetrability. Certain marine-derived substances have exhibited high skin-penetrating ability, making it possible to use them as cosmeceutical vehicles.

Ma'or et al. (2000) have used the biomass of microalgae *Duniella salina* as a cosmeceutical vehicle to deliver Dead Sea minerals into the skin. Though the skin penetration of this preparation is high, its success as a cosmeceutical vehicle is limited, as the algal biomass has shown a low biosorption of calcium and magnesium and a low release of minerals from the biomass at normal pH of the skin. Therefore biosorption property, active ingredient releasing ability at the target site, and the ability to maintain the efficacy of active ingredient throughout the storage period have similar importance as the high skin-penetrating ability of the cosmeceutical vehicle.

Among marine-derived cosmeceutical vehicles, chitosan and its derivatives have received greater attention due to the presence of these important biological properties. According to Anchisi et al. (2006), chitosan beads have been used to increase the storage time of less stable, volatile, essential oils extracted from *Mentha piperita*, which is a common active ingredient in current cosmeceuticals. According to this study, the cosmeceutical effect of this active ingredient persists for a 6-month storage time in chitosan, and chitosan molecules are able to release the active ingredient after application of them on the skin surface, reflecting the suitability of chitosan as a cosmeceutical vehicle. Furthermore, an efficient delivery of vitamin B_6 through the barrier of the stratum corneum has been reported after incorporating vitamin B_6 into the cosmeceutical vehicles chitosan and their derivatives (Orenti et al., 1999). However, those researchers observed higher vitamin B_6 delivering ability from the chitosan derivatives N-carboxyethyl chitosan and N-aminoethyl chitosan than from chitosan, reflecting the possibility of producing novel compounds from chitosan with higher biological properties, which increases their potential application as a cosmeceutical delivering agent.

Though chitosan has been identified as an important cosmeceutical vehicle, the development of global nantechnology has moved the attention of the cosmeceutical industry more toward chitin nanofibrils as a novel cosmeceutical vehicle. The electropositive structure of chitin nanofibrils enables them to entrap active ingredients of the cosmeceuticals, such as growth factors and endogenous and exogenous substances that need to be transported to the target sites by ionic bonding. After application of this prebound active ingredient of the chitin nanofibrillar polymer on the skin surface, the polymeric structure of the chitin nonofibrils is transformed into disaccharides or tetrasaccharides, which are smaller than the polymer structure (Figure 25.9). These disaccharides or tetrasaccharides are able to penetrate the stratum corneum and become localized in the epidermis or dermis to release the active substances of cosmeceutical preparation to the target sites, where they exert their function (Morganti and Muzzarelli, 2009).

Apart from the main function of chitin nanofibrils, acting as a cosmeceutical vehicle, high water binding capacity makes them suitable as skin moisturizers. The small molecular weight disaccharides and tetrasaccharides of chitin nanofibrils that penetrate the stratum corneum repolymerize enzymatically at the level of lipid lamellae and the extracellular substances of the dermis. These polymerized structures bind with many water molecules to improve skin hydration (Morganti and Muzzarelli, 2009). Furthermore, this chitin nanofibrillar structure neutralizes the ROSs produced in the skin (Morganti and Morganti, 2008), reflecting their importance as antiaging cosmeceuticals (Morganti et al., 2008). Therefore usage of marine-derived chitin nanofibrils as a cosmeceutical vehicle contributes to improving skin health by delivering active ingredients to the target sites, while neutralizing ROSs and hydrating the skin as additional benefits.

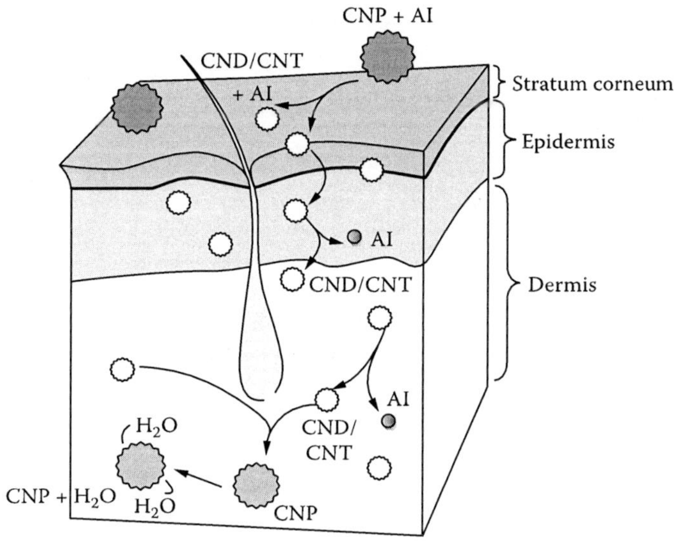

CNP - Chitin Nanofibrillar Polymers, CND - Disaccharides of Chitin Nanofibrils,
CNT - Tetrasaccharides of Chitin Nanofibrils, AI - Active Ingredient

FIGURE 25.9 Schematic diagram representing the cosmeceutical function of chitin nanofibrils on skin.

25.4.6 Other Important Marine-Derived Cosmeceuticals

Apart from the active ingredients that exert direct human health benefits, certain other compounds are essential for the maintenance of the efficiency of cosmeceutical products. These compounds have different important roles as thickening or gelling agents, colorants, and preservatives of cosmeceuticals.

25.4.6.1 Marine-Derived Thickening or Gelling Agents of Cosmeceuticals

Even though cosmeceutical thickening agents are unable to exert direct beneficial effects on human skin health, these compounds involve the maintenance of proper texture of the cosmeceutical product, which is needed to evenly distribute the active ingredients in the cosmeceutical product.

Marine-derived polysaccharides are mainly contributors to the cosmeceutical industry as thickening or gelling agents. The polysaccharides carageenans, alginates, and agar have been identified as important marine-derived thickening agents, which maintain the texture of the products in the cosmeceutical industry (Kim et al., 2008). Apart from the above polysaccharides, marine bacterial exopolysaccharides have been identified as novel thickening agents, which have the potential to be used as a cosmeceutical ingredient.

25.4.6.2 Marine-Derived Colorants of Cosmeceuticals

Coloration of cosmeceutical products increases their acceptance by cosmeceutical customers. However, there is an increasing demand for natural colorants, as such substances are known to be safe. Since marine algae have the ability to produce pigments with different colors, they are used as a source for the extraction of natural colorants for cosmeceutical products. Phycocyanin (blue pigment), β-carotene (yellow pigment), and astaxanthin (yellow to red pigment) that are extracted from *Spirulina*, *Duniella*, and *Haematococcus*, respectively, are industrially important colorants that have been applied widely in the current cosmeceutical industry (Dufosse et al., 2005).

25.4.6.3 Marine-Derived Preservatives of Cosmeceuticals

The preservation of cosmeceutical products increases their shelf life. Synthetic parabens are widely used in the cosmeceutical industry as a preservative. These synthetic parabens can be replaced by the recently isolated 4-hydroxy benzoate and its alkyl esters produced by marine bacteria, reflecting their potential in the cosmeceutical industry (Peng et al., 2006).

Though the parabens are incorporated into cosmeceuticals with the purpose of preservation, there are certain marine-derived active ingredients with antimicrobial properties other than their main biological activity. Dead Sea minerals have been identified as an important constituent in cosmeceuticals due to their antioxidant, anti-inflammatory, and antiapoptosis properties. However, their antimicrobial properties may have the property of preservation of the final cosmeceutical product, which increases the importance of Dead Sea minerals as cosmeceuticals (Ma'or et al., 2006).

25.5 CONCLUSIONS

In conclusion, it can be stated that the active ingredients derived from marine origin are widely used in the current cosmeceutical market, while some novel discoveries have increased their potential applications in the cosmeceutical industry. Since these compounds are natural, they are considered safe, and the demand for marine-derived active ingredients is increasing in the current cosmeceutical market. However, the development of different fields such as microbiology, biotechnology, and nanotechnology has increased their potential of applications in the cosmeceutical industry. Since these natural marine-derived compounds are known to have different biological properties such as UV absorption, photoprotective activity, MMP inhibition activity, anti-inflammatory activity, antiapoptotic activity, skin moisturizing ability, and active ingredient delivering ability, they have been identified as important active ingredients in cosmeceutical preparations. Furthermore, the synergistic effects of several marine-derived substances or multiple properties that have been observed from a single sunbstance increase the importance of them in the cosmeceutical industry. Therefore these active substances that produce secondary metabolites in the diverse group of marine organisms have a wide role in the current cosmeceutical industry as well as in the development of novel cosmeceuticals in the future.

REFERENCES

Abu-Jdayil, B., and Mohameed, H. A. 2006. A facial mask comprising Dead Sea mud. *Journal of Cosmetic Science* 57:441–454.

Ahn, G. N., Kim, K. N., Cha, S. H., et al. 2007. Antioxidant activities of phlorotannins purified from *Ecklonia cava* on free radical scavenging using ESR and H_2O_2-mediated DNA damage. *European Food Research and Technology* 226:71–79.

Alain, T., and Atsushi, T. 2006. The applications and functions of new exopolysaccharide 'Deepsane' from the deepest oceans. *Fragrance Journal* 34:61–68.

Althunibat, O. Y., Hashim, R. B., Taher, M., et al. 2009. *In vitro* antioxidant and antiproliferative activities of three Malaysian sea cucumber species. *European Journal of Scientific Research* 37:376–387.

Amsler, C. D., and Fairhead, V. A. 2005. Defensive and sensory chemical ecology of brown algae. *Advances in Botanical Research* 43:1–91.

Anchisi, C., Meloni, M. C., and Maccioni, A. M. 2006. Chitosan beads loaded with essential oils in cosmetic formulations. *Journal of Cosmetic Science* 57:205–214.

Anggadiredja, J., Andyani, R., Hayati, et al. 1997. Antioxidant activity of *Sargassum polycystum* (Phaeophyta) and *Laurencia obtuse* (Rhodophyta) from Seribu islands. *Journal of Applied Phycology* 9:477–479.

Arct, J., and Pytkowska, K. 2008. Flavonoids as components of biologically active cosmeceuticals. *Clinics in Dermatology* 26:347–357.

Bae, I., Osatomi, K., Yoshida, A., et al. 2008. Biochemical properties of acid-soluble collagens extracted from the skins of underutilized fishes. *Food Chemistry* 108:49–54.

Baharum, S. N., Beng, E. K., and Mokhtar, M. A. A. 2010. Marine microorganisms: potential application and challenges. *Journal of Biological Sciences* 10:555–564.

Ben-amotz, A. B. 1995. New mode of *Dunaliella* biotechnology: two phase growth for β-carotene production. *Journal of Applied Phycology* 7:65–68.

Ben-amotz, A., Shaish, A., and Avron, M. 1989. Mode of action of the massively accumulated β-carotene of *Dunaliella bardawil* in protecting the alga against damage by excess irradiation. *Plant Physiology* 91:1040–1043.

Borowitzka, M. A., Borowitzka, L. J., and Kessly, D. 1990. Effect of salinity increase on carotenoid accumulation in the green alga *Dunaliella salina. Journal of Applied Phycology* 2:111–119.

Burke, K. E. 2007. Interaction of vitamin C and E as better comeceuticals. *Dermatologic Therapy* 20:314–321.

Canbon-Bonvita, M. A., Raguenes, J. J., Vincent, P., et al. 2002. A novel polymer produced by a bacterium hydrothermal vent polychaete annelid. *Journal of Applied Microbiology* 93:310–315.

Capelli, B., and Cysewski, G. 2006. *Natural astaxanthin as king of the carotenoids.* Cyanotech Corporation Kailua Kona, HI.

Cho, M. L., Kang, I., Won, M. H., et al. 2010. The antioxidant properties of ethanol extracts and their solvent-partitioned fractions from various green sea weeds. *Journal of Medicinal Food* 13:1232–1239.

Chen, R. H., and Hey, R. S. 2000. Film-formation time, skin hydration effects, and physicochemical properties of moisture masks containing different water-soluble chitosans. *Journal of Cosmetic Science* 51:1–13.

Cifuentes, A. S., Gonzalez, M., Conejeros, M., et al. 1992. Growth and carotenogenesis in eight strains of *Dunaliella salina* Teodoresco from Chile. *Journal of Applied Phycology* 4:111–118.

Conde, F. R., Churio, M. S., and Previtali, C. M. 2000. The photoprotector mechanism of mycosporine-like amino acids. Exited-state properties and photostability of porphyra-pp3 in aqueous solution. *Journal of Photochemistry and Photobiology B: Biology* 56:139–144.

Darr, D., Combs, S., Dunsten, S., et al. 1992. Topical vitamin C protects porcine skin from ultraviolet radiation-induced damage. *British Journal of Dermatology* 127:247–253.

Darvin, M. E., Sterry, W., and Lademann, J. 2010. Resonance Raman spectroscopy as an effective tool for the determination of antioxidative stability of cosmetic formulations. *Journal of Biophotonics* 3:82–88.

Demirel, Z., Yilmaz-Koz, F. F., Yavasoglu, U. K., et al. 2009. Antimicrobial and antioxidant activity of brown algae from Aegean sea. *Journal of Serbian Chemical Society* 74:619–628.

Djabourov, M., Lechaire, J. P., and Gaill, F. 1993. Structure and rheology of gelatin and collagen gels. *Biorheology* 30:191–205.

Draelos, Z. D. 2009. Cosmeceuticals: undefined, unclassified and unregulated. *Clinics in Dermatology* 27:431–434.

Dubreucq, G., Domon, B., and Fournet, B. 1996. Structure determination of a novel uronic acid residue isolated from the exopolysaccharide produced by a bacterium originating from deep sea hydrothermal vents. *Carbohydrate Research* 290:175–181.

Dufosse, L., Galaup, P., Yaron, A., et al. 2005. Microorganisms and microalgae as souces of pigments for food use: a scientific oddity or an industrial reality? *Trends in Food Science and Technology* 16:389–406.

Eker, A. P. M., Kooiman, P., Hassels, J. K. C., et al. 1990. DNA photoreactivating enzyme from the cyanobacterium *Anacystis nidulans. Journal of Biological Chemistry* 265:8009–8015.

Fabregas, J., Otero, A. Maseda, A., et al. 2001. Two-stage culture for the production of astaxanthin from *Haematococcus pluvialis. Journal of Biotechnology* 26:65–71.

Farris, P. K. 2009. Cosmeceutical vitamins: vitamin C. In Z. Draelos (Ed.), *Procedures in cosmetic dermatology series: cosmeceuticals*, 51–56. Philadelphia: Elsevier.

Flemming, H. C., and Wingender, J. 2001. Relevance of microbial extracellular polymeric substances (EPSs). Part I. Structural and ecological aspects. *Water Science and Technology* 43:1–8.

Forster, M., Bolzinger, M. A., Fessi, H., et al. 2009. Topical delivery of cosmetics and drugs. *European Journal of Dermatology* 19:309–323.

Furusaki, E., Ueno, Y., Sakairi, N., et al. 1996. Facile preparation and inclusion ability of a chitosan derivative bearing carboxymethyl-β-cyclodextrin. *Carbohydrate Polymers* 9:29–34.

Gao, X. H., Zhang, L., Huachen, W., et al. 2008. Efficacy and safety of innovative cosmeceuticals. *Clinics in Dermatology* 26:367–374.

Garcia, F., Freile-Pelegrin, Y., and Robledo, D. 2007. Physiological characterization of *Dunaliella* sp. (Chlorophyta, Volvocales) from Yucatan, Mexico. *Bioresource Technology* 98:1359–1365.

Garcia-Pichel, F., and Castenholz, R. W. 1991. Characterization and biological implications of scytonemin, a cyanobacterial sheath pigment. *Journal of Phycology* 27:395–409.

Goecke, F., Hernández, V., Bittner, M., et al. 2010. Fatty acid composition of three species of *Codium* (Bryopsidales, Chlorophyta) in Chile. *Revista de Biología Marina y Oceanografía* 45:325–330.

Gomez, P. I., Barriga, A., Cifuentes, A. S., et al. 2003. Effect of salinity on the quantity and quality of carotenoids accumulated by *Dunaliella salina* (strain CONC-007) and *Dunaliella bardawil* (strain ATCC 30861) Chlorophyta. *Biological Research* 36:185–192.

Guerin, M., Huntley, M. L., and Olaizola, M. 2003. *Haematococcus astaxanthin*: applications for human health and nutrition. *Trends in Biotechnology* 21:210–216.

Gueznuec, J. 2002. Deep-sea hydrothermal vents: a new source of innovative bacterial exopolysaccharides of biotechnological interest? *Journal of Industrial Microbiology and Biotechnology* 29:204–208.

Hadi, M. R., Shariati, M., and Afsharzadeh, S. 2008. Microalgal biotechnology: carotenoid and glycerol production by the green algae *Dunaliella* isolated from the Gave-Khooni salt marsh, Iran. *Biotechnology and Bioprocess Engineering* 13:540–544.

Hearst, J. E. 1995. The structure of photolyase: using photon energy for DNA repair. *Science* 268:1858–1859.

Hejazi, M. A., Lamarliere, C., Rocha, J. M. S., et al. 2002. Selective extraction of carotenoid from the alga *Dunaliella salina* with retention of the viability. *Biotechnology and Bioengineering* 79:29–36.

Hwang, H., Chen, T., Nines, R. G., et al. 2006. Photochemoprevention of UVB-induced skin carcinogenesis in SKH-1 mice by brown algae polyphenols. *International Journal of Cancer* 119:2742–2749.

Il'ina, A. V., and Varlamov, V. P. 2004. Hydrolysis of chitosan in lactic acid. *Applied Biochemistry and Microbiology* 40:354–358.

Izume, M., Nagae, S., Kawagishi, H., et al. 1992. Action pattern of *Bacillus* sp. no. 7-M chitosanases on partially *N*-acetylated chitosan. *Bioscience, Biotechnology and Biochemistry* 56:448–453.

Jeon, Y. J., and Kim, S. K. 2000a. Production of chitooligosaccharides using ultrafiltration membrane reactor and their antibacterial activity. *Carbohydrate Polymers* 41:133–141.

Jeon, Y. J., and Kim, S. K. 2000b. Continuous production of chitooligosaccharides using a dual reactor system. *Process Biochemistry* 35:623–632.

Jimenez, C., and Niell, F. X. 1991. Growth of *Dunaliella viridis* Teodoresco: effect of salinity, temperature and nitrogen concentration. *Journal of Applied Phycology* 3:319–327.

Jin, E., and Melis, A. 2003. Microalgal biotechnology: carotenoid production by the green algae *Dunaliella salina. Biotechnology and Bioprocess Engineering* 8:331–337.

Kobayashi, M., and Sakamoto, Y. 1999. Singlet oxygen quenching ability of astaxanthin esters from the green alga *Haematococcus pluvialis. Biotechnology Letters* 21:265–269.

Karentz, D., McEuen, F. S., Land, M. C., et al. 1991. Survey of mycosporine-like amino acid compounds in Antarctic marine organisms: potential protection from ultraviolet exposure. *Marine Biology* 108:157–166.

Khotimchonko, S. V., Vaskovsky, V. E., and Titlyanova, T. V. 2002. Fatty acids of marine algae from the Pacific coast of North California. *Botanica Marina* 45:17–22.

Kim, E. J., Jin, X. J., Kim, Y. K., et al. 2010. UV decreases the synthesis of free fatty acids and triglycerides in the epidermis of human skin *in vitro* contributing to development of skin photoaging. *Journal of Dermatological Science* 57:19–26.

Kim, S. K., and Mendis, E. 2006. Bioactive compounds from marine processing byproducts—a review. *Food Research International* 39:383–393.

Kim, S. K., Ravichandran, Y. D., Khan, S. B., et al. 2008. Prospective of the cosmeceuticals derived from marine organisms. *Biotechnology and Bioprocess Engineering* 13:511–523.

Kimura, S., Zhu, X. P., Matsui, R., et al. 1988. Characterization of fish muscle type I collagen. *Journal of Food Science* 53:1315–1318.

Kleinegris, D. M. M., Janssen, M., Brandenburg, W. A., et al. 2010. The selectivity of milking of *Dunaliella salina. Marine Biotechnology* 12:14–23.

Kobayashi, S., Takaena, M., Kanke, M., et al. 1998. Post administration effect of magnesium-L-ascorbyl-phosphate on the development of UVB-induced cutaneous damage in mice. *Photochemistry and Photobiology* 67:669–675.

Koivikko, R., Eranen, J. K., Loponen, J., et al. 2008. Variation in phlorotannins among three populations of *Fucus vesiculosus* as revealed by HPLC and colorimetric quantification. *Journal of Chemical Ecology* 34:57–64.

Kosmadaki, M. G., and Gilchrest, B. A. 2004. The role of telomeres in skin aging/photoaging. *Micron* 35:155–159.

Lacroix, P. C., Michel, L., Moysan, A., et al. 1996. UVA-induced immune suppression in human skin: protective effect of vitamin E in human epidermal cells *in vitro. British Journal of Dermatology* 134:77–84.

Lautenschiager, H. 2003. Essential fatty acids—cosmetic from inside and outside. *Beauty Forum* 4:54–56.

Lee, J. H., Seo, Y. B., Jeong, S. Y., et al. 2007. Functional analysis of combinations in astaxanthin biosynthesis genes from *Paracoccus haeundaensis. Biotechnology and Bioprocessing Engineering* 12:312–317.

Lee, Y. H., and Ku, M. J. 2010. Effect of *Fucus evanescens* fucoidan on expression of matrix mettaloproteinase-1 promoter, mRNA, protein and signal pathway. *FASEB Journal* (Meeting Abstract Supplement) 484.3.

Li, B., Lu, F., Wei, X., et al. 2008. Fucoidan: structure and bioactivity. *Molecules* 13:1671–1695.

Linter, K., Chamberlin, C. M., Peschard, O., et al. 2009. Cosmeceuticals and active ingredients. *Clinics in Dermatology* 27:461–468.

Llewellyn, C. A., and Airs, R. L. 2010. Distribution and abundance of MAAs in 33 species of microalgae across 13 classes. *Marine Drugs* 8:1273–1291.

Lorenz, R. T., and Cysewski, G. R. 2000. Commercial potential for *Haematococcus* microalgae as a natural source of astaxanthin. *Trends in Biotechnology* 18:160–167.

Lyons, N. M., and O'Brien, N. M. 2002. Modulatory effects of an algal extract containing astaxanthin on UVA-irradiated cells in culture. *Journal of Dermatological Science* 30:73–84.

Lupo, M. P., and Cole, A. L. 2007. Cosmeceutical peptides. *Dermatologic Therapy* 20:343–349.

Mansour, M. P., Frampton, D. M. F., Nichols, P. D., et al. 2005. Lipid and fatty acid yield of nine stationary-phase microalgae: applications and unusual C_{24}-C_{28} polyunsaturated fatty acids. *Journal of Applied Phycology* 17:287–300.

Manela-Azulay, M., and Bagatin, E. 2009. Cosmeceutical vitamins. *Clinics in Dermatology* 27:469–474.

Ma'or, Z., Henis, Y., Alon, Y., et al. 2006. Antimicrobial properties of Dead Sea black mineral mud. *International Journal of Dermatology* 45:504–511.

Ma'or, Z., Simon, G. M., Yehuda, S., et al. 2000. Antiwrinkle and skin-moisturizing effects of a mineral-algal-botanical complex. *Journal of Cosmetic Science* 51:27–36.

Ma'or, Z., Yehuda, S., and Voss, W. 1997. Skin smoothing effects of Dead Sea minerals: comparative profilometric evaluation of skin surface. *International Journal of Cosmetic Science* 19:105–110.

Marin, N. M., Morales, F., Lodeiros, C., et al. 1998. Effect of nitrate concentration on growth and pigment synthesis of *Dunaliella salina* cultivated under low illumination and preadapted to different salinities. *Journal of Applied Phycology* 10:405–411.

Martini, S., D'Addario, C., Bonechi, C., et al. 2010. Increasing photostability and water-solubility of carotenoids: synthesis and characterization of β-carotene-humic acid complexes. *Journal of Photochemistry and Photobiology B: Biology* 101:355–361.

Matanjun, P., Mohamed, S., Mustapha, N. M., et al. 2008. Antioxidant activities and phenolics content of eight species of seaweeds from North Borneo. *Journal of Applied Phycology* 20:367–373.

Matsunga, T., Burgess, J. G., Yamada, N., et al. 1993. An ultraviolet (UV-A) absorbing biopterin glucoside from the marine planktonic cyanobacterium *Oscillatoria* sp. *Applied Microbiology and Biotechnology* 39:250–253.

Mehta, R. C., and Fitzpatrick, R. E. 2007. Endogenous growth factors as cosmeceuticals. *Dermatologic Therapy* 20:350–359.

Miki, W. 1991. Biological functions and activities of animal carotenoids. *Pure and Applied Chemistry* 63:141–146.

Mitchell, D. 2006. Revisiting the photochemistry of solar UVA in human skin. *PNAS* 103:13567–13568.

Moisan, T. A., and Mitchell, B. G. 2001. UV absorption by mycosporine-like amino acids in *Phaeocystis Antarctica* Karsten induced by photosynthetically available radiation. *Marine Biology* 138:217–227.

Moon, H. J., Lee, S. H. Ku, M. J., et al., 2009. Fucoidan inhibits UVB-induced MMP-1 promoter expression and down regulation of type I procollagen synthesis in human skin fibroblasts. *European Journal of Dermatology* 19:129–134.

Moon, H. J., Lee, S. R., Shim, S. N., et al. 2008a. Fucoidan inhibits UVB-induced MMP-1 expression in human skin fibroblasts. *Biological and Pharmaceutical Bulletin* 31:284–289.

Moon, H. J., Park, K. S., Ku, M. J., et al. 2008c. Effect of *Costaria costata* fucoidan on expression of matrix metalloproteinase-1 promoter, mRNA and protein. *Journal of Natural Products* 72:1731–1743.

Moon, H. J., Shim, S. N., Ku, M. J., et al. 2008b. Effect of fucoidan on mRNA and promoter activity of matrix mettaloproteinases (MMPs) in skin fibroblasts. *Journal of the Federation of American Societies for Experimental Biology* 31:284–289.

Morganti, P., Fabrizi, G., and Palombo, B., et al. 2008. Chitin nanofibrils: a new active cosmetic carrier. *Journal of Applied Cosmetology.* 26:113–128.

Morganti, P., and Morganti, G. M. 2008. Chitin nanofibrils for advanced cosmeceuticals. *Clinics in Dermatology* 26:334–340.

Morganti, P., and Muzzarelli, C. 2009. Spray-dried chitin nanofibrils, method for production and uses thereof. U.S. Patent 20090203642A1.

Murthy, K. N. C., Vanitha, A., Rajesh, M., et al. 2005. *In vitro* antioxidant activity of carotenoids from *Dunaliella salina*—a green microalga. *Life Sciences* 76:1381–1390.

Nakai, M., Kageyama, N., and Nakahara, K., et al. 2006. Phlorotannins as radical scavengers from the extract of *Sargassum ringgoldianum*. *Marine Biotechnology* 8:409–414.

Newburger, A. E. 2009. Cosmeceuticals: myths and misconceptions. *Clinics in Dermatology* 27:446–452.

Nicolaus, B., Kambourova, M., and Oner, E. T. 2010. Exopolysaccharides from extremophiles: from fundamentals to biotechnology. *Environmental Technology* 31:1145–1158.

Nichols, C. A. M., Garon, S., Bowman, J. P., et al. 2004. Production of exopolysaccharides by Antarctic marine bacterial isolates. *Journal of Applied Microbiology* 96:1057–1066.

Nichols, C. M., Bowman, J. P., and Guezennec, J. 2005a. Effect of incubation temperature on growth and production of exopolysaccharides by an Antarctic sea ice bacterium grow in batch culture. *Applied Environmental Microbiology* 71:3519–3523.

Nichols, C. M., Lardiere, S. G., Bowman, J. P., et al. 2005b. Chemical characterization of exopolysaccharides from Antarctic marine bacteria. *Microbial Ecology* 49:578–589.

Nomura, T., Kikuchi, M., Kubodera, A., et al. 1997. Proton-donative antioxidant activity of fucoxanthin with 1,1-diphenyl-2-picrylhydrazyl (DPPH). *Biochemistry and Molecular Biology International* 42:361–370.

O'Connor, I., and O'Brien, N. 1998. Modulation of UVA light-induced oxidative stress by β-carotene, lutein and astaxanthin in cultured fibroblasts. *Journal of Dermatological Science* 16:226–230.

Ogawa, M., Poetie. J. Moody, M. W., et al. 2004. Biochemical properties of bone and scale collagens isolated from the subtropical fish drum (*Pogonia cromis*) and sheepshead seabream (*Archosargus probatocephalus*). *Food Chemistry* 88:495–501.

Oren, A. (1998). Microbiology and biogeochemistry of halophilic microorganisms—an overview. In A. Oren (Ed.), *Microbiology and biogeochemistry of hypersaline environments*, 1–12. New York: CRC Press.

Oren, A., and Gunde-Cimerman, N. 2007. Mycosporines and mycosporine-like amino acids: UV protectants or multipurpose secondary metabolites? *FEMS Microbiology Letters* 269:1–10.

Orenti, I., Luppi, B., and Zecchi, V. 1999. Chitosana and its N-carboxyethyl and N-aminoethyl derivatives as vehicles for topical formulations. *Journal of Cosmestic Science* 50:307–313.

Patankar, M. S., Oehninger, S., Barnett, T., et al. 1993. A revised structure for fucoidan may explain some of its biological activities. *Journal of Biological Chemistry* 268:21770–21776.

Pedros-Alio, C. 2006. Marine microbial diversity: can it be determined? *Trends in Microbiology* 14:257–263.

Peng, X., Adachi, K., Chen, C., et al. 2006. Discovery of a marine bacterium producing 4-hydroxybenzoate and its alkyl esters, parabens. *Applied and Environmental Microbiology* 72:5556–5561.

Petit, A. C., Noiret, N., Guezennec, J., et al. 2006. Ultrasonic depolymerization of an exopolysaccharide produced by a bacterium isolated from a deep-sea hydrothermal vent polychaete annelid. *Ultrasonics Sonochemistry* 14:107–112.

Pinnell, S. R., Yang, H., Omar, M., et al. 2001. Topical L-ascorbic acid: percutaneous absorption studies. *Dermatologic Surgery* 27:137–142.

Pisal, D. S., and Lele, S. S. 2005. Carotenoid production from microalga, *Duneliella salina*. *Indian Journal of Biotechnology* 4:476–483.

Portugal-Cohen, M., Soroka, Y., Ma'or., Z., et al. 2009. Protective effects of a cream containing Dead Sea minerals against UVB-induced stress in human skin. *Experimental Dermatology* 18:781–788.

Proteau, P. J., Gerwick, W. H., Garcia-Pichel, F., et al. 1993. The structure of scytonemin, an ultraviolet sunscreen pigment from the sheaths of cyanobacteria. *Experientia* 49:825–829.

Raguenes, G., Christen, R., Guezennec, J., et al. 1997a. *Vibrio diabolicus* sp. nov., a new polysaccharide-secreting organism isolated from a deep-sea hydrothermal vent polychaete annelid, *Alvinella pompejana*. *International Journal of Systemic Bacteriology* 47:989–995.

Raguenes, G., Pignet, P., Gauthier, G., et al. 1996. Description of a new polymer-secreting bacterium from a deep-sea hydrothermal vent, *Alteromonas macleodii* subsp. *Fijiensis*, and preliminary characterization of the polymer. *Applied and Environmental Microbiology* 62:67–73.

Raguenes, G. H. C., Peres, A., Ruimy, R., et al. 1997b. *Alteromonas infernus* sp. noc., a new polysaccharide-producing bacterium isolated from a deep-sea hydrothermal vent. *Journal of Applied Microbiology* 82:422–430.

Rajasulochana, P., Krishnamoorthy, P., and Dhamotharan, R. 2010. Amino acids, fatty acids and minerals in *Kappaphycus* sps. *ARPN Journal of Agricultural and Biological Science* 5:1–12.

Rastogi, R. P., Richa., Kumar, A., et al. 2010. Molecular mechanisms of ultraviolet radiation-induced DNA damage and repair. *Journal of Nucleic Acids* 2010:1–32.

Rawling, A. V., and Harding, C. R. 2004. Moisturization and skin barrier function. *Dermatologic Therapy* 17:43–48.

Reed, R. E. 1962. The definition of "cosmeceutical." *Journal of Society of Cosmetic Chemist* 13:103–106.

Riemer, U., Lamare, M. D., and Peake, B. M. 2007. Temporal concentrations of sunscreen compounds (mycosporine-like amino acids) in phytoplankton and in the New Zealand krill, *Nyctiphanes australis* G.O. Sars. *Journal of Plankton Research* 29:1077–1086.

Rupley, J. A. 1964. The hydrolysis of chitin by concentrated hydrochloric acid, and preparation of low molecular-weight substrates for lysozyme. *Biochimica et Biophysica Acta* 83:245–248.

Sachindra, N. M., Sato, E., Maeda, H., et al. 2007. Radical scavenging and oxygen quenching activity of marine carotenoid fucoxanthin and its metabolites. *Journal of Agricultural and Food Chemistry* 55:8516–8522.

Salguero, A., Leon, R., Mariotti, A., et al. 2005. UV-A mediated induction of carotenoid accumulation in *Dunaliella bardawil* with retention of cell viability. *Applied Microbiology and Biotechnology* 66: 506–511.

Sankaran, J. K., and Mouly, V. S. 2007. Managing innovation in an emerging sector: the case of marine-based nutraceuticals. *R&D Management* 37:329–344.

Schmid, D., Schurch, C., Zulli, F., et al. 2003. Mycosporine-like amino acids: natural UV screening compounds from red algae to protect the skin against photoaging. *SÖFW-Journal* 129:1–5.

Schmid, D., Schurch, C., and Zulli, F. 2004a. UVA-screening compounds from red algae protect against photoageing. *Personal Care 1*:29–31.

Schmid, D., Cornelia, S., and Zulli, F. 2004b. UV-A sunscreening from red algae for protection against premature skin aging. *Cosmetics and Toiletries Manufacture Worldwide Yearbook*, 2004, 139–143.

Senevirathne, M., Kim, S., Siriwardena, N., et al. 2006. Antioxidant potential of *Ecklonia cava* on reactive oxygen species scavenging, metal chelating, reducing power and lipid peroxidation inhibition. *Food Science and Technology* 12:27–38.

Shanab, S. M. M. 2007. Antioxidant and antibiotic activities of some seaweeds (Egyptian isolates). *International Journal of Agriculture and Biology* 9:220–225.

Shibata, T., Ishimaru, K., Kawaguchi, S., et al. 2008. Antioxidant activities of phlorotannins isolated from Japanese Laminariaceae. *Journal of Applied Phycology* 20:705–711.

Shibata, T., Nagayama, K., Tanaka, R., et al. 2003. Inhibitory effects of brown algal phlorotannins on secretory phospholipase A_2s, lipoxygenases and cyclooxygenases. *Journal of Applied Phycology* 15:61–66.

Shibata, T., Kawaguchi, S., Hama, Y., et al. 2004. Local and chemical distribution of phlorotannins in brown algae. *Journal of Applied Phycology* 16:291–296.

Simpson, B. K., Gagne, N., and Simpson, M. V. 1994. Bioprocessing of chitin and chitosan. In A. M. Martin (Ed.), *Fisheries processing: biotechnological applications*, 155–173. London: Chapman & Hall.

Sinha, R. P., Klisch, M., Groniger, A., et al. 2000. Mycosporine-like amino acid in marine red alga *Gracilaria cornea*—effects of UV and heat. *Environmental and Experimental Botany* 43:33–43.

Sionkowska, A. 2006. Effects of solar radiation on collagen and chitosan films. *Journal of Photochemistry and Photobiology B: Biology* 82:9–15.

Smidsrod, O., and Draget, K. I. 1996. Chemistry and physical properties of alginates. *Carbohydrates in Europe* 14:6–13.

So, M. J., Kim, B. K., Choi, M. J., et al. 2007. Protective activity of fucoidan and alginic acid against free radical-induced oxidative stress under *in vitro* and cellular system. *Journal of Food Science and Nutrition* 12:191–196.

Souza, M. C. R. D., Marques, C. T., Dore, C. M. G., et al. 2007. Antioxidant activities of sulphated polysaccharides from brown and red seaweeds. *Journal of Applied Phycology* 19:153–160.

Stege, H., Rosa, L., Vink, A. A., et al. 2000. Enzyme plus light therapy to repair DNA damage in ultraviolet-B-irradiated human skin. *PNAS* 97:1790–1795.

Swatachek, D., Schatton, W., Kellermann, J., et al. 2002. Marine sponge collagen: isolation, characterization and effects on the skin parameters surface-pH, moisture and sebum. *European Journal of Pharmaceutics and Biopharmaceutics* 53:107–113.

Takema, Y., and Imokawa, G. 1998. The effects of UV-A and UV-B irradiation on the viscoelastic properties of hairless mouse skin *in vitro*. *Dermatology* 196:397–400.

Teoh, M. L., Chu, W. L., Merchant, H., et al. 2004. Influence of culture temperature on the growth, biochemical composition and fatty acid profiles of six Antarctic microalgae. *Journal of Applied Phycology* 16:421–430.

Tharanathan, R. N., and Kittur, F. S. 2003. Chitin—the undisrupted biomolecule of great potential. *Critical Reviews in Food Science and Nutrition* 43:61–87.

Thibodeau, A. 2007. Protecting the skin from environmental stresses with an exopolysaccharide formulation. In L. Rheins (Ed.), *Biotechnology in cosmetics: concepts, tools and techniques*, 267–284. Allured Publishing.

Tsuchihashi, H., Kigoshi, M., Iwatsuki, M., et al. 1995. Action of β-carotene as an antioxidant against lipid peroxidation. *Archives of Biochemistry and Biophysics* 323:137–147.

Vacher, A. M., and Fritsch, M. C. 2009. Cosmetic composition which includes at least one polysaccharide derived from bacteria of hydrothermal origin. U.S. Patent 0238782A1.

Verum, K. M., Ottøy, M. H., and Smidsrød, O. 2001. Acid hydrolysis of chitosans. *Carbohydrate Polymers* 46:89–98.

Vidanarachchi, J. K., Kurukulasuriya, M. S., and Kim, S. K. 2009. Chitin, chitosan, and their oligosaccharides in food industry. In S. K. Kim (Ed.), *Chitin, chitosan, oligosaccharides and their derivatives: biological activities and applications*, 543–560. Boca Raton, FL: Taylor and Francis Group.

Vincent, P., Pignet, P., Talmont, F., et al. 1994. Production and characterization of an exopolysaccharide excreted by a deep-sea hydrothermal vent bacterium isolated from the polychaete annelid *Alvinella pompejana*. *Applied and Environmental Microbiology* 60:4134–4141.

Wang, J., Zhang, Q., Zhang, Z., et al. 2008. Antioxidant activity of sulphated polysaccharide fractions extracted from *Laminaria japonica*. *International Journal of Biological Macromolecules* 42:127–132.

Yamasaki, Y., Hayashi, I., Ohta, Y., et al. 1993. Purification and mode of action of chitosanolytic enzyme from *Enterobacter* sp. G-1. *Bioscience, Biotechnology and Biochemistry* 56:1546–1551.

Yan, X., Chuda, Y., Suzuki, M., et al. 1999. Fucoxanthin as the major antioxidant in *Hijika fusiformis*, a common edible seaweed. *Bioscience, Biotechnology and Biochemistry* 63:605–607.

Ye, H., Zhou, C., Sun, Y., et al. 2009. Antioxidant activities *in vitro* of ethanol extract from brown seaweed *Sargassum palludum*. *European Food Research and Technology* 230:101–109.

Zhang, Z., Gao, X., Tashiro, Y., et al. 2005a. The isolation of porphyra-334 from marine algae and its UV-absorption behavior. *Chinese Journal of Oceanology and Limnology* 23:400–405.

Zhang, Z., Tashiro, Y., Matsukawa, S., et al. 2005b. Influence of pH and temperature on the ultraviolet-absorbing properties of porphyra-334. *Fisheries Science* 71:1382–1384.

Zhao, X., Xue, C. H., and Le, B. F. 2008. Study of antioxidant activities of sulphated polysaccharides from *Laminaria japonica*. *Journal of Applied Phycology* 20:431–436.

Zubia, M., Robledo, D., and Pelegrin, Y. F. 2007. Antioxidant activities in tropical marine macroalgae from the Yucatan, Mexico. *Journal of Applied Phycology* 19:449–458.

26 Aquatic and Marine Bioactive Antimicrobial Peptides and Biosurfactants for the Cosmeceutical Industry

Jen-Leih Wu and Jenn-Kan Lu

CONTENTS

26.1 ANTIMICROBIAL PEPTIDES FROM MARINE FISH AND INVERTEBRATES

Marine fish possess antimicrobial peptides as a part of their defense system, which are mainly present in the mucous layer, indicating that they eliminate pathogenic bacteria before they enter the skin barrier. However, marine invertebrates rely solely on innate immune mechanisms that include both humoral and cellular responses. Antimicrobial peptides (AMPs) are a major component of the innate immune defense system in marine invertebrates. They are defined as molecules less than 10 kDa in mass that show antimicrobial properties and provide an immediate and rapid response to invading microorganisms. Several AMPs have been isolated from marine fish, such as pleurocidins from winter flounder, *Pleuronectes americanus* (Walbaum), American plaice, *Hippoglossoides platessoides* (Fabricius), and Atlantic halibut, *Hippoglossus hippoglossus* (Cole et al., 1997; Douglas et al., 2001, 2003a, 2003b). Some examples are misgurin from loach *Misgurnus anguillicaudatus* (Cantor) (Park et al., 1997) and hagfish intestinal antimicrobial peptides (HFIAPs) from hagfish, *Eptatretus burgeri* (Girard) (Hwang et al., 1999); chrysophsins from red sea bream, *Chrysophrys major* (Iijima et al., 2003); piscidin or moronecidin from white bass, *Morone chrysops* (Rafinesque), and striped bass, *Morone saxatilis* (Walbaum) (Silphaduang and Noga, 2001; Lauth et al., 2002); grouper, *Epinephelus lanceolatus* (Lu et al., 2009); hepcidins (Shike et al., 2002; Douglas et al., 2003a); marine invertebrate species such as tunicates, *Styela clava* and *Halocynthia aurantium* (Lee et al., 1997, 2001); crustaceans, *Litopenaeus vannemei* (Destoumieux et al., 1997, 2009; Bachere et al., 2002) and *Penaeus monodon* (Chiou et al., 2005, 2007); chelicerates, *Tachyplesus tridentatus* and *Limulus polyphemus* (Nakamura et al., 1988; Miyata, et al., 1989); and mollusks (*Mytilis edulis* and *Mytilis galloprovincialii* (Charlet et al., 1996; Mitta et al., 2000) (Table 26.1).

Based on the nuclear magnetic resonance (NMR) structures of known peptides along with sequence analysis, AMPs are broadly classified into five groups. They are defined as molecules less than 10 kDa in mass that show antimicrobial properties and provide an immediate and rapid response to invading microorganisms. The major classes of antimicrobial peptides include (1) α-helices, (2) β-sheet and small proteins, (3) peptides with thio-ether rings, (4) peptides with an overrepresentation of one or two amino acids, (5) lipopeptides, and (6) macrocyclic cystine knot peptides.

The modes of action by which antimicrobial peptides kill bacteria are varied and include disrupting membranes, interfering with metabolism, and targeting cytoplasmic components. The initial contact between the peptide and the target organism would be electrostatic, as most bacterial surfaces are anionic. Their amino acid composition, amphipathicity, cationic charge and size allow them to attach to and insert into membrane bilayers to form pores by barrel stave, carpet, or toroidal pore mechanisms—or allow them to penetrate into the cell to bind intracellular molecules that are crucial to cell living. An intracellular binding model includes inhibition of cell wall synthesis, alteration of the cytoplasmic membrane, activation of autolysin, inhibition of DNA, RNA, and protein synthesis, and inhibition of certain enzymes. However, in many cases, the exact mechanism of killing is not known.

26.2 ANTIMICROBIAL PEPTIDES AS A NATURAL COSMETIC PRESERVATIVE

To keep products like cleansers, hand soaps, moisturizers, and toothpaste bacteria-free, parabens are used as a preservative in cosmetics. A recent study found that parabens can mimic natural hormones, including estrogen. Disruption of sex hormones may increase the risk of certain cancers. Parabens have been found in breast cancer tissue, although it is not known whether the parabens caused the cancer (Harvey and Everett, 2004).

Hydroquinone is banned in parts of Europe, but not in the United States, as a cancer risk (Taylor, 2006). It is still used in hair dye and skin fresheners in under 2% concentrations. Formaldehyde and acrylamide, both carcinogens and irritants, are also used in cosmetics. Allergies to cosmetic products can be due to chemicals such as man-made synthetics. This can lead to a skin rash where the product is applied. If an allergy to a cosmetic product is suspected, it is important to determine which ingredients may be causing the problem.

Nail salons are all at risk of causing diseases to individuals due to the services they provide and the products used. Nail products are considered cosmetics that are not tested by the U.S. Food and Drug Administration before release to the public. As with any product, allergic reactions can occur. For example, contact dermatitis is not uncommon, a skin rash characterized by redness, itching, and sometimes small oozing blisters. This reaction can be found on both the hands and any other body part touched by the nail, such as face, eyes, or neck. Chemicals used in conjunction with artificial nails can cause swelling and even separation of the natural nail from the nail bed. Blood-borne infections, such as HIV and hepatitis, are theoretically possible; however, none have been reported to date. Common nail problems are nail infections. Symptoms of infections include redness, pain, itching, discoloration, and pus in or around the nail area. Infections can be bacterial, fungal, or

TABLE 26.1
Natural Antimicrobial Peptides Isolated from Marine Organisms

Group/Species	AMP Name	Peptide Class[a]	References[b]
		Teleost Fish	
Pardachirus marmoratus	Pardaxin	_	Lazarovici et al. (1986)
Misgurnus anguillicaudatus	Misgurnin	_	Park et al. (1997)
Pseudopleuronectes americanus	Pleurocidin	_	Douglas et al. (2001), Cole et al. (1997)
Hippoglossoides platessoides	Pleurocidin	_	Patrzykat et al. (2003)
Morone chrysops/saxatilis	Moronecidin	_	Lauth et al. (2002)
Morone chrysops	Hepcidin	__	Shike et al. (2002)
Pleuronectes americanus	Hepcidin	___ __	Douglas et al. (2003a)
Salmo salar	Hepcidin	__	Douglas et al. (2003b)
Oncorhynchus mykiss	Hepcidin	__	EST (AF281354_1)
Paralichthys olivaceus	Hepcidin	___ __	EST (C23298.1, C23432.1)
Oryzias latipes	Hepcidin	__	EST (AU178966, AU179222, AU179314, AU179768, AU180044)
Oncorhynchus mykiss	Salmocidin	nd	Partial sequence (P81369)
Eptatretus burgeri	HFS-1	_	Hwang et al. (1999)
Epinephelus lanceolatus	Epinephecidin	_	Lu et al. (2009)
		Tunicates	
Styela clava	Clavanin	_	Lee et al. (2001)
Halocynthia aurantium	Dicynthaurin	_(2)	Lee et al. (2001)
		Chelicerates	
Tachyplesus tridentatus	Tachyplesin I	__	Nakamura et al. (1988)
Limulus polyphemus	Tachyplesin II	_2	Miyata et al. (1989)
		Crustaceans	
Carcinus maenas	6.5 kDa	nd	Schnapp et al. (1996)
Callinectes sapidus	Callinectin	nd	Knoo et al. (1999)
Penaeus vannemi	Penaeidins	Pro/__	Destoumieux et al. (1997)
Penaeus monodon	Monodocin	__	Chiou et al. (2007)
		Gastropods	
	Dolabella quriculara		Iijima et al. (2003)
		Mollusks	
Mytilis galloprovincialis	Mytilin	_4	Mitta et al. (2000)
Mytilis edulis	Mytimicin	_6	Charlet et al. (1996)

Source: Modified from Patrzykat, A., et al., *Antimicrob. Agents Chemother.*, 47, 2464–2470, 2003.

[a] Class designations are as follows:__, amphipathic alpha helix; _, structured; number refers to number of disulfide bonds; number in brackets refers to a dimer structure; nd, structure not determined.

[b] Published literature or database accession numbers for unpublished data.

viral. Bacteria, fungi, and viruses are everywhere, but due to the services provided and often unsanitary practices, there are many infections traced to nail salons. They are generally caused by bad practices: cutting cuticles, using equipment improperly so as to cause cuts or openings in the skin, working on clients with skin conditions, not washing hands between clients or using gloves, and using nondisposable equipment or supplies without sanitization.

Bacterial infections include acute or chronic paronychia (generally occurring in cuticles or skin folds) such as *Staphylococcus*. More serious is nontuberculous mycobacterial infections, which can cause severe skin and soft tissue infections. They are generally on the legs and caused by nail salon whirlpool footbaths that were insufficiently cleaned between users (skin and hair debris was later found behind inlet suction screens in a cluster of cases). Minor bacterial infections can be treated by soaking and draining pus; more serious ones, by topical ointments; and the most serious, by oral antibiotics. Fungal nail infections are called onychomycosis. They can be caused by yeasts such as *Candida* and dermatophytes such as ringworm. Fungal infections generally require oral anti-infective medicine. Viruses can transmit warts to users of nail salons. This is treated by removal of the wart by surgery, freezing, laser, or chemicals.

One of the primary requirements posed with respect to many drugs and cosmetics is their sterility and resistance to infections during application and storage. Hence the presence of preservatives also prevents the synthesis of the products of metabolism of microorganisms, which may be the cause of skin and mucosa irritation. The hitherto applied preservatives are mostly organic derivatives of organic acids, aldehydes, alcohols, and phenols, derivatives of guanidine, organic derivatives of mercury, and ammonia salts. Inorganic preservatives are above all boric acid, sodium sulfite, and iodate. Following the pro-ecological tendency, the world cosmetic industry tries to diverge from synthetic preservatives in favor of natural compounds. In particular, this pertains to biocosmetics, which by definition should be based on natural ingredients of plant or animal origin. The ideal preservatives used should exhibit various properties, e.g., be nontoxic, act in low concentrations, and have a wide spectrum of action, and they should not cause irritations or allergies. The preservatives currently used exhibit high antimicrobial activity. However, they have an adverse influence on the human body, for example, cause allergic reactions. Therefore there is a constant need for new preservatives. It seems it is worth considering the application of new compounds of peptide structure. Only a few peptides (from the group of lantibiotic compounds of peptide structure, of bacterial origin) have been used as preservatives. Some peptides (surfactin) have even been used as antimicrobial agents in masks to keep them for a long time (Figure 26.1).

Microbial infections constitute a serious problem in modern drug delivery. Despite considerable progress in the creation of new cosmetics and drugs of modern delivery systems, there are still various problems with preservation of these preparations.

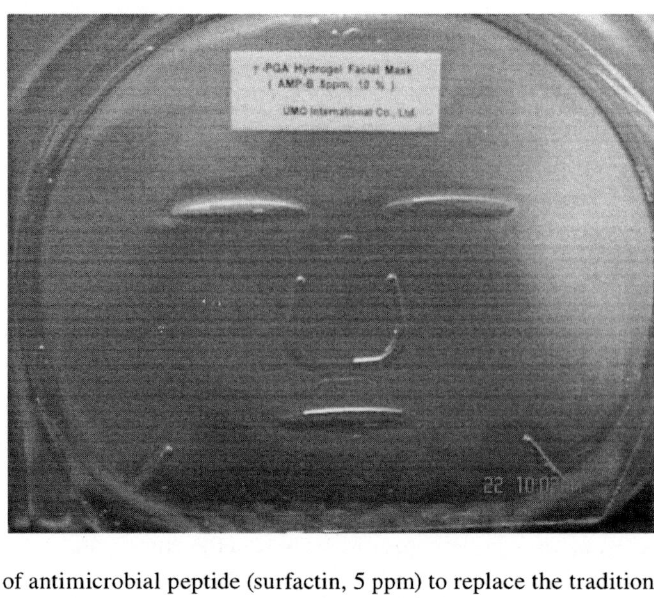

FIGURE 26.1 Use of antimicrobial peptide (surfactin, 5 ppm) to replace the traditional preservatives added to the mask, and the mask showing the appearance of nonperishability in 18 months.

TABLE 26.2
Inhibitory and Bactericidal Activity Assay of Marine Finfish, Shrimp, and Microbial AMP against Bacteria and Fungi Pathogens

	Fish AMP		Shrimp AMP		Microbial AMP MIC (μM)	
	MIC (μM)	MBC (μM)	MIC (μM)	MBC (μM)	ATCC21332	TH
Gram Positive						
Staphylococcus epidermidis	3.6	3.6			128.7	96.5
Propionibacterium acnes	3.6	3.6				
Aerococcus viridiam			20	100		
Gram Negative						
Escherichia coli	3.6	3.6	<1	<1	193	128.7
Vibrio alginolyticus	3.6	3.6	<1	<1	128.7	96.5
V. harveyi	22.5	27	<1	<1	128.7	96.5
Salmonella choleraesuis			2.4	4.8		
Aeromonas hydrophila	90	90			128.7	128.7
V. anguillarum	1.8	1.8			128.7	96.5
V. salmonicida	3.6	3.6			128.7	96.5
Fungi						
Aspergillus niger	45	—			193	128.7
Fusarium pisi			5	20		
F. oxysporum			10	20		
Chaetonium globosum	45	—				
P. citrinum	90	180				

Extending the durability of these products and preventing lesions caused by bacteria, fungi, and physicochemical factors is one of the most important issues in modern pharmacy. In recent years, a large number of natural antimicrobial peptides as well as several synthetic analogues have been tested with respect to their possible application as chemotherapeutic agents. Due to the low toxicity of peptide antibiotics, particularly ardent hopes are put on the possibility of the application of endogenous ones as biopharmaceuticals in modern chemotherapy of infectious diseases. Nevertheless, there are no reference data in the literature on attempts to use them as preservatives. There have been many recent attempts to find effective replacements for use against infections in human medicine. Marine antimicrobial peptides can be widely used as antimicrobial agents (for bacteria and fungi) in humans or throughout the animal kingdom due to their wide range of bacteriostatic and bactericidal effects as well as fungicidal activities (Table 26.2) (Chiou et al., 2005, 2007; Hu et al., 2006, Lu et al., 2010). Due to their bacteriostatic and bactericidal effects and fungicidal activities in a wide range of pathogen spectra, marine AMPs can serve medicinal purposes as a protein drug. Marine AMPs can be formulated in a pharmaceutical composition such as in the form of a solution, gel, emulgel, cream, ointment, lotion, transdermal system, injection fluid, suspension, or patch; be used as a bactericidal agent, fungicidal agent, preservative, or surface-active agent; and prevent infection of wounds for topical treatment in humans. The main trump card of the presented peptides is that they are natural compounds. The data presented here illustrate the potential of marine antibacterial peptides in the preservation of cosmetic preparations.

26.3 ANTIMICROBIAL PEPTIDE IN ACNE TREATMENT

Our skin is constantly challenged by microbes but is rarely infected. Cutaneous production of antimicrobial peptides (AMPs) is a primary system for protection, and expression of some AMPs further increases in response to microbial invasion.

Antimicrobial peptides such as cathelicidin and α-defensin protect the skin through two distinct pathways: (1) direct antimicrobial activity and (2) initiation of a host response resulting in cytokine release, inflammation, angiogenesis, and reepithelialization.

Staphylococci are the most abundant skin-colonizing bacteria and the most important causes of nosocomial infections and community-associated skin infections.

Acne is the most common skin disorder of adolescence and early adulthood, with a prevalence of approximately 85%. Generally mild to moderate cases are treated with topical anti-infectives or anti-inflammatories.

Marine antimicrobial peptides showed that marine AMPs had a strong inhibitory effect on *Propionibacterium acnes* and *Staphylococcus epidermidis*. The minimum inhibitory concentration (MIC) and minimum bactericidal concentration (MBC) values were the same (36 μM) for both bacterial species against *P. acnes* and *Staphylococcus epidermidis*, respectively (Table 26.2) (Chiou et al., 2007; Lu, 2009). Therefore these marine AMPs would be an interesting topic for further study and possibly for an alternative treatment for acne.

In response to *P. acnes* and inflammation the antimicrobial peptides have the ability to not only kill *P. acnes* but also significantly reduce its inflammatory potential. Antimicrobial peptides demonstrate great potential as a broad spectrum anti-infective and anti-inflammatory therapeutic for the treatment of dermatological conditions such as acne, acne rosacea, and atopic dermatitis. This molecule not only exhibits potent activity but maintains that activity in the presence of biological fluids (Lee et al., 2009). The antimicrobial activity is rapidly cidal, does not engender resistance, and is not cross-resistant with other mechanisms of action. AMP has the potential to provide significant clinical benefits.

26.4 MARINE BIOSURFACTANT

A surfactant is an amphiphilic agent with both lipophilic and hydrophilic structural moieties in its molecule. Surfactants are widely used for industrial, agricultural, food, cosmetic, and pharmaceutical applications. Most of these compounds are chemically synthesized and potentially cause environmental and toxic problems (Schramm et al., 2003). However, it is only in the past few decades that surface-active molecules of microbial origin, referred to as biosurfactants, have gained considerable interest (Desai and Banat, 1997; Healy et al., 1996).

Microbial surfactants constitute a diverse group of surface-active molecules and are known to occur in a variety of chemical structures. Based on the structure of their hydrophilic part, they are mainly classified into five categories: glycolipids, lipopeptides, fatty acids, polymer type, and particulate biosurfactants (Gautam and Tiagi, 2006).

Surface-active compounds produced by microorganisms are of two main types, those that reduce surface tension at the air-water interface (biosurfactants) and those that reduce the interfacial tension between immiscible liquids, or at the solid-liquid interface (bioemulsifiers). Biosurfactants usually exhibit emulsifying capacity, but bioemulsifiers do not necessarily reduce surface tension (Konishi et al., 2007; Langer et al., 2006; Freitas et al., 2009).

The features that make them commercially promising alternatives to chemically synthesized surfactants are their lower toxicity, higher biodegradability, and hence greater environmental compatibility, and better foaming properties (useful in mineral processing). Also, unlike most synthetic surfactants, many biosurfactants function effectively in extreme conditions such as pH, temperature, and salinity (Batista et al., 2006; Bento et al., 2005; Bhattacharyya et al., 2003; Bodour et al., 2003; Cameotra and Makkar, 2004; Chen et al., 2007; Christofi and Ivshina, 2002; Cohen and Exerowa, 2007).

TABLE 26.3
Marine Microorganisms and Some Types of Biosurfactants

Biosurfactant	Organisms	References
	Glycolipids	
Glucose lipids	*Alcanivorax borkumensis*	Abraham et al. (1998)
	Alcaligenes sp.	Poremba et al. (1991)
Trehalose lipids	*Arthrobacter* sp.	Schulz et al. (1991)
	Lipoproteins	
Ornithine lipids	*Myroides* sp. SM1	Maneerat et al. (2005a)
	Phospholipids and Fatty Acids	
Bile acids	*Myroides* sp. SM1	Maneerat et al. (2005b)
	Polymeric Biosurfactants	
Lipid-carbohydrate-protein	*Yarrowia lipolytica*	Zinjarde and Pant (2002)
	Pseudomonas nautica	Husain et al. (1997)
	Particulate Biosurfactants	
Whole cells	Variety of bacteria	Denger and Schink (1995)
		Gilewicz et al. (1997)
		Zinjarde and Pant (2002)

Source: Adapted from Maneerat et al., 2005.

Biosurfactants produced by certain species of bacteria and yeasts also exhibit effective antimicrobial, antifungal (Joshi et al., 2008; Rodrigues et al., 2004, 2006; Singh and Cameotra, 2004), antiviral, moisturizing, and antioxidant properties. However, some biosurfactants have demonstrated they are potentially useful surface-active agents for emulsion (Dhakephalkar et al., 2010; Amaral et al., 2009; Huang et al., 2010) in various industrial processes. Moreover, they can be used for their emulsification, foaming, water-binding capacity, phase dispersion, de-emulsification component, and wetting properties' effect on viscosity, and their product consistency can efficiently be utilized by the above cosmetics industry, with promising effects for antiaging and skin smoothing (Kosaric, 1992; Martinez-Checa et al., 2007). In pharmaceutics, biosurfactants can be used for gene delivery and recovery of intracellular products, and they can also serve as antimicrobial substances and emulsifying agents. Recently, biosurfactants received much attention in nanobiotechnology criteria (Koopmans and Aggeli, 2010).

The marine biosphere offers a wealth of flora and fauna, which represent a vast natural resource of important functional commercial grade products. Among the various bioactive compounds, biosurfactants (BSs)/bioemulsifiers (BEs) are attracting major interest and attention due to their structural and functional diversity. Marine biosurfactants produced by some marine microorganisms have been paid more attention, particularly for the bioremediation of the sea polluted by crude oil. Moreover, sponge-associated marine bacteria are the current focus of bioactive leads from the marine environment (Table 26.3).

Here, we discuss the role and applications of marine and aquatics biosurfactants, focusing on cosmeceutical perspectives.

26.5 BIOSURFACTANTS FOR COSMETICS

Cosmetic products using surfactants include insect repellents, antacids, bath products, acne pads, antidandruff products, contact lens solutions, hair colors and care products, deodorants, nail care, body massage accessories, lipsticks, eye shadows, mascaras, soap, tooth pastes and polishes, denture cleansers, adhesives, antiperspirants, lubricated condoms, baby products, foot care, mousses, antiseptics, shampoos, conditioners, shave and depilatory products, moisturizers, and health and beauty products (Schramm et al., 2003). All of these applications could be replaced with biosurfactants. Moreover, biosurfactants as emulsifiers, foaming agents, solubilizers, wetting agents, cleansers, antimicrobial agents, and mediators of enzyme action are produced in various forms such as creams, lotions, liquids, pastes, powders, sticks, gels, films, and sprays could be used and may be replaced by biosurfactants (Tuleva et al., 2002; Ueno et al., 2007).

Cosmetics with a petroleum base may contain cancer-causing impurities. Biosurfactants are not derived from petroleum and will not have those impurities. Biosurfactants have mild toxicity and are easily biodegradable compared with synthetic surfactants. They are low or anti-irritancy and are compatible with skin. Cosmetic ingredients can be contaminated with the cancer-causing impurity called 1,4-dioxane, which penetrates the skin. It is considered to be a carcinogen. Those impurities of petroleum-based ingredients are found in trace amounts in products such as facial moisturizers, anti-aging products, body lotion, eye creams, hair products (relaxers, dyes, bleaches, removers), baby soap, sunless tanning products, and body firming lotion. Cosmetic ingredients can contain the contaminant sodium laureth sulfate and ingredients that include the terms *PEG*, *xynol*, *ceteareth*, and *oleth*.

26.5.1 Marine Biosurfactant as Natural Antibiotics and Preservatives

The diverse structures of biosurfactants allow them to display versatile performance (Ajesh and Sreejith, 2009; Zhao et al., 2010). By its structure, a biosurfactant is supposed to exert its toxicity on the cell membrane permeability with a detergent-like effect. One useful property of many biosurfactants is their antimicrobial activity (Rahman and Ano, 2009).

Several biosurfactants have shown antimicrobial action against bacteria, fungi, algae, and viruses. The lipopeptide iturin from *B. subtilis* showed potent antifungal activity. Inactivation of an enveloped virus such as herpes and retrovirus was observed with 80 mM surfactin (Benson et al., 1976; Vollenbroich et al., 1997). Rhamnolipids inhibited the growth of harmful bloom algae species, *Heterosigma akashivo* and *Protocentrum dentatum*, at concentrations ranging from 0.4 to 10.0 mg/L. A rhamnolipid mixture obtained from *Pseudomonas aeruginosa* AT10 showed inhibitory activity against the bacteria *Escherichia coli*, *Micrococcus luteus*, *Alcaligenes faecalis* (32 mg/ml), *Serratia arcescens*, *Mycobacterium phlei* (16 mg/ml), and *Staphylococcus epidermidis* (8 mg/ml) and excellent antifungal properties against *Aspergillus niger* (16 mg/ml), *Chaetonium globosum*, *Enicillium crysogenum*, *Aureobasidium pullulans* (32 mg/ml), and the phytopathogenic *Botrytis cinerea* and *Rhizoctonia solani* (18 mg/ml) (Abalos et al., 2001). Sophorolipids and rhamnolipids were found to be effective antifungal agents against plant and seed pathogenic fungi. The mannosylerythritol mannosylerythritol lipid (MEL), a glycolipid surfactant from *Candida antartica*, has demonstrated antimicrobial activity particularly against Gram-positive bacteria (Borzeix and Frederique, 2003). Our study also indicated that marine biosurfactants (i.e., surfactin) are effective substances against all tested microorganisms (Lu, 2009). Das et al. (2009) have reported a biosurfactant produced by marine *B. circulans* that had a potent antimicrobial activity against Gram-positive and Gram-negative pathogenic and semipathogenic microbial strains, including MDR strains. The main trump card of surfactins is that they are natural compounds. The data presented here illustrate the potential of marine biosurfactants in preservation of cosmetic preparation.

26.5.2 Marine Biosurfactant as Emulsifier

Bioemulsifiers are classified according to their hydrophile-lipophile balance (HLB); those having a low HLB are strong lipophiles and are used as water-in-oil emulsifiers, whereas those having a high HLB are strong hydrophiles and are used as oil-in-water emulsifiers (Pruti and Cameotra, 1997).

TABLE 26.4
Important Types of Marine Microbial Emulsifiers

Bioemulsifier	Producing Microorganisms
BD4-emulsan	*Acinetobacter calcocaceticus* BD4
Trehalose-lipid-o-dialkyl monoglycerides-protein	*Pseudomonas fluorescens*
Polypeptide	*Acinetobacter* sp. A3
Emulsan	*A. calcoaceticus* RAG-1
	A. calcoaceticus A2
Biodispersan	*A. calcoaceticus* A2
Lipid-carbohydrate-protein complex	*Yarrowia lipolytica* NCIM 3589
Carbohydrate-protein complex	*Rhodotorula glutinis*
Protein-polysaccharide	*Streptomyces* sp. S1
HE39	*Halomonas* TG39
HE67	*Halomonas* TG67
Yansan	*Yarrowia lipolytica* IMUFRJ 50682

Source: Adapted from Nerurkar et al. 2009.

The marine biosphere offers a wealth of flora and fauna, which represent a vast natural resource of important functional commercial grade products (Table 26.4). Among the various bioactive compounds, biosurfactants (BSs)/bioemulsifiers (BEs) are attracting major interest and attention due to their structural and functional diversity. The discovery of potent BS/BE-producing marine microorganisms would enhance the use of environmental biodegradable surface-active molecules and hopefully reduce total dependence or number of new applications oriented toward the chemical synthetic surfactant industry. Marine microorganisms such as *Acinetobacter, Arthrobacter, Pseudomonas, Halomonas, Myroides, Corynebacteria, Bacillus,* and *Alteromonas* sp. have been studied for production of BSs/BEs and exopolysaccharides (EPSs).

We are currently testing the influence of pH, temperature, and biosurfactant type on emulsifier stability from diesel oil in water and the surfactin by *Bacillus subtilis* spp. on emulsifier stability from common cosmetic oils in water, and some data are depicted in Figure 26.2. The results showed that the surfactin has emulsion activity for diesel and common cosmetic synthetic oils when dissolved (4 ~ 40 μg/ml) in saline solution, and its emulsion activity remains half in 60°C (Lu, 2009).

26.5.3 Biosurfactant as a Foaming Agent and Skin Cleansing Cosmetic

There is a growing market for formulating proteins into a wide variety of products, including laundry detergents, bath soap, conditioning shampoo, and cream hand cleanser. Foams were formed by a bubbling technique. Excellent foaming properties of surfactin were shown by its high ability to form density and stabilize the foam at a concentration as low as 0.05 mg/ml, in comparison with major detergents such as sodium dodecyl sulfate (SDS), sodium laureth sulfate (SLS), and disodium lauryl sulfosuccinate (DSLSS), and bovine albumin (BSA), which are commonly used in household detergents such as washing up liquids, where it provides good detergency combined with an excellent wetting ability (Tamarkin et al., 2008). Furthermore, surfactin containing a lipidic chain length of 14 carbon atoms provides lipopeptides with the best foaming properties in terms of foam density and liquid stability. We are currently testing that marine and aquatic biosurfactants used to partially replace detergents not only retained excellent foaming properties but also exhibited enhanced emulsifying properties (Figure 26.3) (Lu, 2009). These results indicate that marine biosurfactants may be used in laundry and cosmetic products to fulfill market demand.

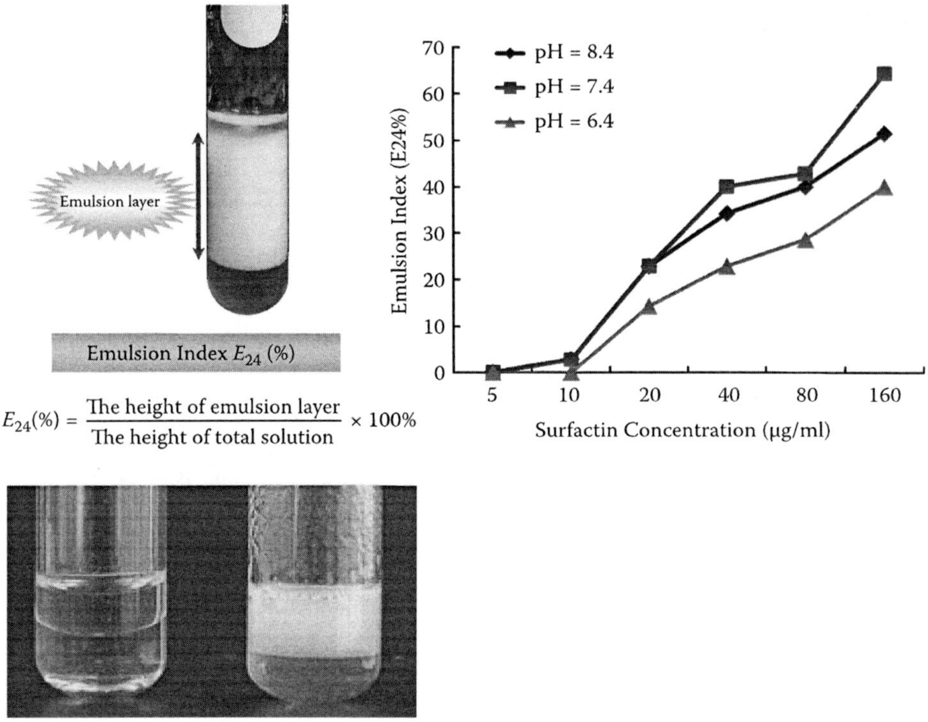

FIGURE 26.2 The emulsification activity of biosurfactant in diesel oil and common synthetic oils for cosmetics. (A) Surfactin from *Bacillus subtilis* spp. has diesel emulsion activity when dissolved in saline having various pH values. (B) Bioemulsifier (right-hand tube) stabilizing oils (such as dimethicone, isonomyl isononaoate, or myristyl myristate) and water emulsion.

Biosurfactants as detergents can advantageously be used to stabilize membrane proteins in the absence of a phospholipid bilayer in a manner that preserves the native conformation and permits the subsequent crystallization. Cleansing cosmetics containing biosurfactants exhibit excellent washability with extremely low skin irritation. Moreover, skin moisturizing and cleansing cosmetics have been prepared in oily gel with high stability (Razafindralambo et al., 1996, 1997).

26.5.4 Biosurfactant as Conditioning and Improved Coacervation

There is an increase in demand for hair care formulations offering improved conditioning. The market drivers behind this include demographic changes in age, ethnicity, and affluence in both

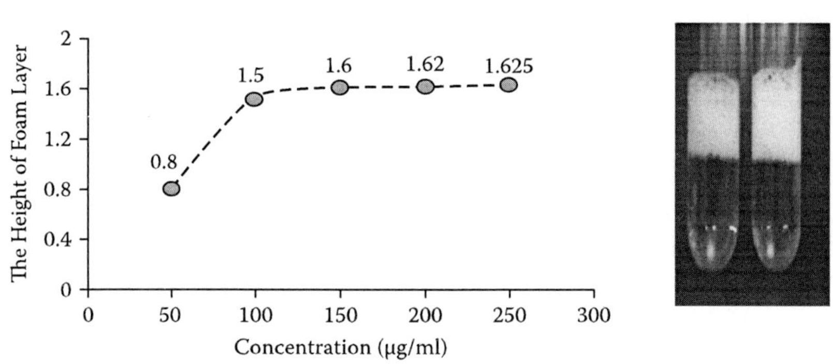

FIGURE 26.3 Foaming activity of various concentrations of biosynthetic surfactin addition.

developed and emerging economies. Further, there is an increase in the penetration and frequency of use of hair color, styling products, and shampoos. As a result, there is a greater need for products that maintain the healthy feel and appearance of hair. Formulators face the daunting task of developing hair care products that not only deliver enhanced conditioning, but also deliver it to a more diverse range of hair types.

To meet these requirements, formulators need ingredients that can improve the wet and dry properties of different hair types. Key improvements desired include slip, detangling, manageability, volume, luster, color retention, and repair. Formulators typically rely on the use of a wide variety of conditioning ingredients to address these needs. Conditioning ingredients are delivered to hair through a variety of ways, including conditioning shampoos, secondary conditioners, and hair treatments. The use of conditioning shampoos is an attractive option for many consumers, offering the convenience of cleansing and conditioning in a single step. The formulation of such products is challenging given the complexity of conflicting mechanisms at work; conditioning shampoos must be designed to remove and solubilize soil while simultaneously depositing conditioning ingredients.

While there are several proposed mechanisms for soil removal during the shampoo process, many agree that conditioning ingredients are likely deposited onto hair fibers through a mechanism referred to as coacervation. Coacervation theory suggests that anionic surfactants and cationic conditioning polymers can be formulated to produce a stable shampoo chassis. As a shampoo is diluted during use, destabilization occurs and a polymer-surfactant complex, also known as the coacervate, precipitates and deposits on the hair. During this process, small droplets of polymer are flocculated, entrapped in the coacervate, and deposited onto the surface of the hair (Kalantar et al., 2007). The amount of coacervate generated depends on the surfactant composition and the nature of cationic polymer, including molecular weight and cationic charge density.

We have tested the biosurfactant formulation with cationic polymers, which demonstrated higher levels of coacervate than equivalent formulations with benchmark polymer polyquaternium-10 (PQ-10; 1.06 meq/g) (Figure 26.4). Coacervates can effectively deposit on the hair surface, providing multiple benefits in the hair care reduction of combing friction of hair, mending of split ends, controlled delivery of insoluble actives, improved feel and appearance, etc. Gel-like coacervates provide more hair body, while liquid-like coacervates give a soft feel and volume enhancement of hair. Complex coacervate formation is crucially dependent on the molecular characteristics of the polymer and surfactant species (Lu, 2009).

26.5.5 Biosurfactant-Mediated Transdermal Drug Delivery

Transdermal drug delivery (TDD) is gaining prominence over other forms of drug delivery due to its potential advantages, including minimal trauma induction, noninvasiveness, reduction of the enzymatic degradation associated with oral delivery, increased patient compliance, and potential for continuous, controlled delivery (Brown, 1991). Consequently, in recent years, numerous transdermal products have been introduced into the market. The current U.S. market for transdermal patches is over $3 billion, and annual sales worldwide are estimated to be $31.5 billion by 2015. The worldwide transdermal drug delivery market is quite large, but only a small number of agents have FDA approval. The primary reason for such limited development of this market is the difficulty in permeating the stratum corneum layer of human skin, which is the outermost layer of skin. The stratum corneum is composed of keratin-filled, nonviable cells (corneocytes) embedded in a crystalline intercellular lipid domain. These intercellular lipids compose the continuous domain of the stratum corneum and provide its primary barrier properties. Stratum corneum lipids consist of an approximately equimolar mixture of free fatty acids, cholesterol, and ceramides.

Two main pathways for transdermal penetration are normally considered (although additional alternative pathways are proposed) (Cordell, 2002): either via the SC through the intercellular lipid lamellae or through cutaneous appendages, i.e., hair follicles and sweat glands (Corley, 2007). It is generally assumed that under normal circumstances, the predominant route is through the

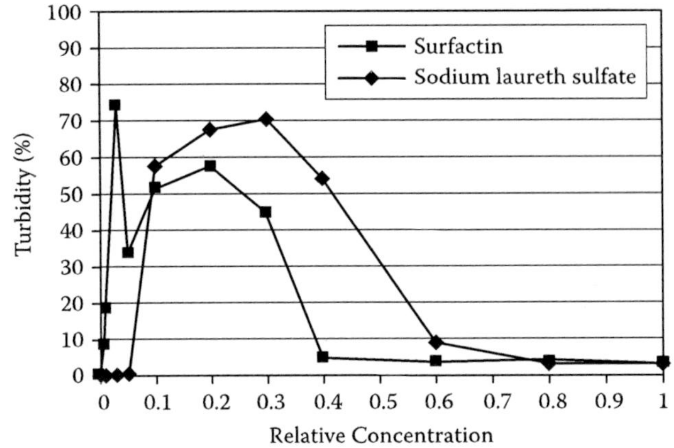

(a)

Shampoo 5% Shampoo

(b)

FIGURE 26.4 The range of coacervate. (a) Biosurfactant formulation with cationic polymers demonstrates higher levels of coacervate when compared to equivalent formulations with benchmark polymer polyquaternium-10 (PQ-10; 1.06 meq/g). (b) Shampoo containing surfactin, after dilution of a polymer-surfactant complex, also known as the coacervate, precipitates.

intercellular spaces in the SC. The diffusional path length is therefore much longer than the simple thickness of the stratum corneum (≈20 μm) and has been estimated to be as long as 500 μm.

The ideal penetration enhancers should have the following conditions:

1. Nonpharmacological effects, nontoxic, nonirritating, and nonallergic response
2. Immediately after the application
3. To restore normal skin barrier function after removing it
4. Do not cause the body of nutrients and moisture loss through the skin
5. Do not have physical and chemical effects with drugs and other materials
6. Colorless and odorless

Numerous methods have been suggested to overcome the skin barrier for transdermal drug delivery. Chemical penetration enhancement methods have received the most attention, where addition of various chemical agents, such as fatty acids, fatty esters, alcohols, terpenes, pyrrolidones, sulfoxides, and surfactants, has been tested to increase skin permeability. However, few have succeeded in delivering relevant agents at the appropriate flux levels without causing notable skin irritation or damage.

Liposome-like vesicles (such as ethosomes, noisomers, and transfersomes) have been developed and extensively investigated in transdermal delivery. Liposomes are phospholipid-based vesicles, and it is generally acknowledged that they are useful for local delivery to superficial skin layers. The delivery mechanisms may be a combination of a penetration enhancement effect of the surfactant in the skin and the increased flexibility it contributes to the vesicle structure. Transfersomes are composed of phospholipids and surfactants, which act as an "edge activator" conferring deformability on the vesicle.

Lipophilic derivatives of peptides can either increase the encapsulation efficiency into liposome or directly increase their delivery through skin. Cutaneous absorption of INF-a linked to various palmitoyl molecules was about 5 to 6 times greater than the parent peptides (Flodvari et al., 1999). The effectiveness of coupling a tetrapeptide linked with a short hydrocarbon chain can enhance transepidermal delivery of the protein inhibitor (Benson et al., 2008).

Membrane destabilization is a key mechanism of action of surfactin. A conformational change in the peptide chain facilitates hydrophobic interaction, which leads to penetration of the membrane by surfactin (Bernheimer and Avigad, 1970; Maget-Dana and Ptak, 1995; Carillo et al., 2003). Adequate penetration may occur at low concentrations ($\approx$ 20 μmol/L), and surfactin may be completely miscible with the phospholipids forming mixed micelles (Grau et al., 1999). We have recently tested the lipophilic biosurfactant (i.e., surfactin), which is capable of entrapping hydrophobic and hydrophilic drugs and increasing skin penetration of drugs (Figure 26.5A, B) and cosmetic moisturizers such as hyaluronic acid (HA) and γ-polyglutiamic acid (γ-PGA) (Figure 26.5C, D) (Lu, 2009). Biosurfactants can act as a local depot for sustained release of dermally active compounds. Biosurfactants can serve as a solubilizing matrix for hydrophilic and lipophilic types of drugs. The application of the surfactin approach to cutaneous delivery is relatively new and has merit on a scientific basis.

26.5.6 Other Applications of Biosurfactant in Cosmeceuticals

26.5.6.1 Biosurfactant as a Moisturizer

The term *moisturizer* does not necessarily imply that moisture or water is being added to the skin. Moisturizers are a key component of basic skin care, especially when there is alteration of the epidermal barrier function and reduced water content in the epidermis. Moisturizers are used to restore the barrier function of the epidermis, cover tiny fissures in the skin, provide a soothing protective

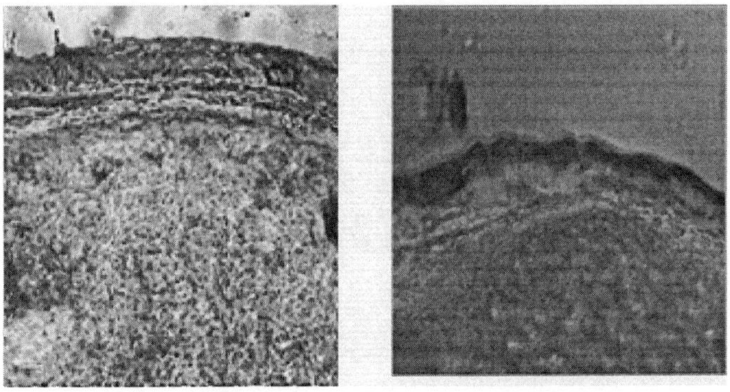

FIGURE 26.5A Evaluation of biosurfactant for transdermal delivery by confocal microscopy analysis. Increase in dye penetration from liposome to surfactin.

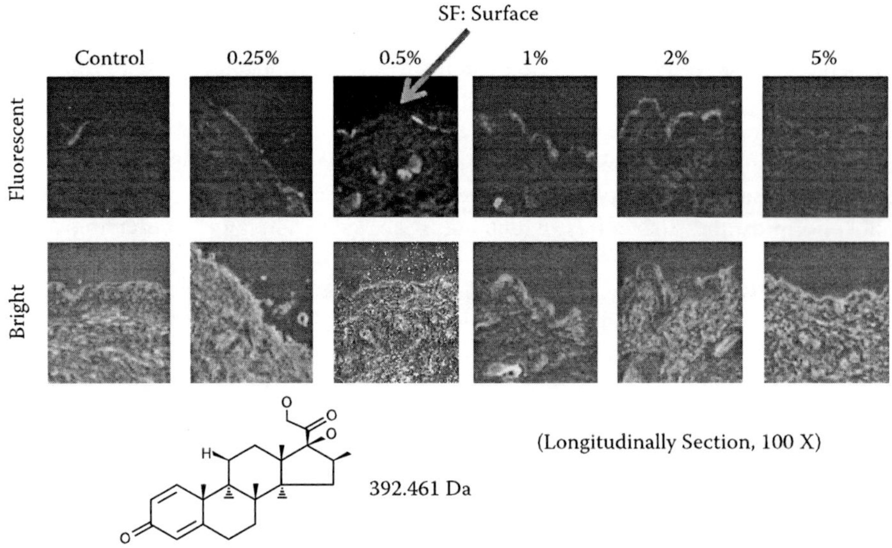

FIGURE 26.5B Evaluation of biosurfactant for transdermal delivery by confocal microscopy analysis. The enhancement effect of surfactin on the penetration of dexamethasone through mouse skin.

film, and increase the water content of the epidermis. They may thus slow evaporation and add to the skin's moisture, thereby maintaining hydration and improving the appearance and tactile properties of dry and aging skin (Ogawa et al., 1999).

The barrier function of the stratum corneum (SC), an important component in hydrating the skin, depends on the presence of unique intercellular lipids in the SC. In contrast, the water-holding capacity of the SC that keeps our skin surface smooth and soft is dependent on the amounts of intracellular lipids and water-soluble small molecules, such as amino acids of the SC as well as skin surface lipids.

Moisturizers are usually made in the form of an emulsion. An emulsion is a mixture of oil and water. This chemical is known as a surfactant and must have two parts, one that binds water and one

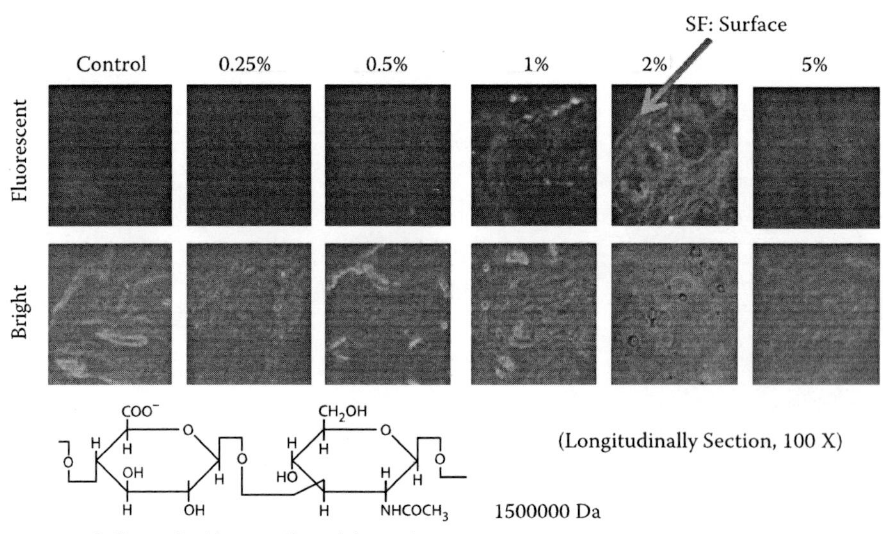

FIGURE 26.5C Evaluation of biosurfactant for transdermal delivery by confocal microscopy analysis. Penetration of HA through mouse skin was enhanced by surfactin.

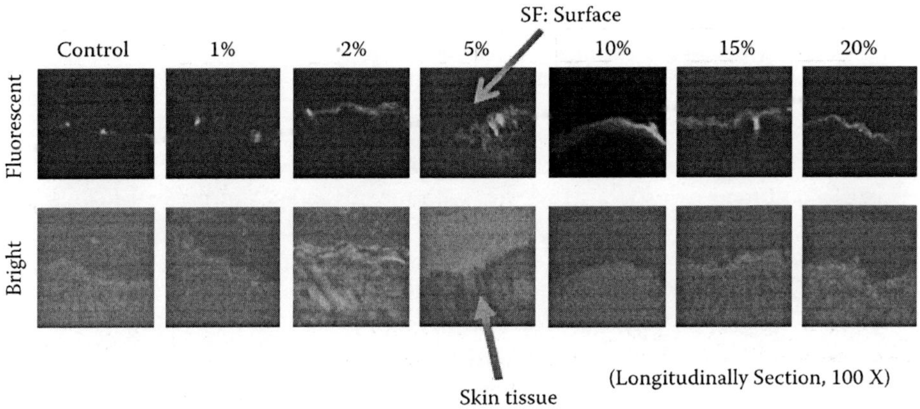

FIGURE 26.5D Evaluation of biosurfactant for transdermal delivery by confocal microscopy analysis. Penetration of γ-PGA through mouse skin was enhanced by surfactin.

that binds oil. When water is combined with an oil-based component and a surfactant, an emulsion is made when the components undergo a process that thoroughly mixes the three together.

26.5.6.2 Antiwrinkle, Antiaging, and Rejuvenation of Biosurfactants and Lipopeptides

The lipopeptides are amphiphatic and consist of a hydrophobic chain containing from about 8 to 24 carbon atoms and a peptide chain that is hydrophilic or has been rendered hydrophilic. They are applied, in particular, to emulsions of immiscible media and to the preparation of lyotropic liquid crystals. One of the applications for lipopeptides has been in antiwrinkle, antiaging, and rejunenation cosmetics (Guglielmo et al., 2003).

The use of sophorolipids in lactone form comprises a major part of diacetyl lactones as agents for stimulating skin dermal fibroblast cell metabolism and, more particularly, as agents for stimulating collagen neosynthesis, at a concentration of 0.01 ppm at 5% (p/p) of dry matter in formulation. This is applicable in cosmetology and dermatology. The purified lactone sophorolipid product is of importance in the formulation of dermis antiaging, repair, and restructuring products because of its effect on the stimulation of dermis cells. By encouraging the synthesis of new collagen fibers, purified lactone sophorolipids can be used both as a preventive measure against aging of the skin and in creams for the body, and in body milks, lotions, and gels for the skin (Borzeix and Frederique, 2003).

Lipopeptide (such as palmitoyl pentapeptide) is at least as effective against wrinkles as retinol but does not cause skin irritation, which is a common side effect of retinoids. It has been demonstrated that oligopeptides linked together and attached to a fatty acid enhance peptides' oil solubility and thus result in better skin penetration. It has been found that when added to the culture of fibroblasts (the key skin cells), palmitoyl pentapeptide stimulated the synthesis of the key constituents of the skin matrix: collagen, elastin, and glucosamnoglycans. How exactly palmitoyl pentapeptide did that remains unclear.

Furthermore, GHK peptide or glycine-histidine-lysine is believed to stimulate the feedback loop, triggering the synthesis of new collagen as well as other components of the skin matrix. When the key skin matrix-producing cells (fibroblasts) detect increased levels of GHK, they assume that the skin matrix is being lost at a higher rate and begin synthesizing it more vigorously. Thus, lipid-linked GHK (a version of GHK designed for better skin penetration) is intended to stimulate skin matrix replenishment via topical application, leading, presumably, to wrinkle reduction, skin firming, and other benefits (Montanari and Guglielmo, 2008).

26.6 CONCLUSIONS

26.6.1 Potential Applications of Marine Antimicrobial Peptides in the Cosmeceutical Industry

Chemically synthesized surface-active compounds are widely used in the pharmaceutical, cosmetic, petroleum, and food industries. However, with the advantages of biodegradability and production on renewable resource substrates, biosurfactants may eventually replace their chemically synthesized counterparts. So far, the use of biosurfactants has been limited to a few specialized applications because biosurfactants have been economically uncompetitive. There is a need to gain a greater understanding of the physiology, genetics, and biochemistry of biosurfactant-producing strains and to improve process technology to reduce production costs.

Biosurfactants (surfactin and other lipopeptides) are produced by marine microorganisms from renewable sources, are nontoxic and inexpensive to produce, and have excellent surface properties and biological activities suitable for cosmetic applications. These biosurfactants have good skin compatibility with low irritation and can potentially be applied as a topical dermatological product. The diversity of lipopeptide activities outlines their importance as a multifunctional cosmetic ingredient.

26.6.2 Potential Applications of Marine Biosurfactants in the Cosmeceutical Industry

Biosurfactants are nontoxic and inexpensive to produce and have excellent surface properties and biological activities suitable for cosmetic application. They can be used as green alternatives in a variety of applications, including bioremediation, pharmaceuticals, agricultural disease control, and cosmetics. Moreover, some biosurfactants have good skin compatibility with low irritation and can potentially be applied as a topical dermatological product.

Lipopeptides were used as emulsifiers and were claimed to have low skin irritation suitable for external skin preparations such as transparent cosmetics with a sequestering function. A low critical micelle concentration (CMC) property in biosurfactants is particularly suitable for topically applied dermatological products, and the low cytotoxicity toward mammalian cells confirmed the safety of these compounds.

REFERENCES

Abalos, A., A. Pinazo, M.R. Infante, M. Casals, F. Garcýa, and A. Manresa. 2001. Physicochemical and antimicrobial properties of new rhamnolipids produced by *Pseudomonas aeruginosa* AT10 from soybean oil refinery wastes. *Langmuir*, **17**: 1367–1371.

Abraham, W.-R., H. Meyer, and M. Yakimov. 1998. Novel glycine containing glucolipids from the alkane using bacterium *Alcanivorax borkumensis*. *Biochim. Biophys. Acta*, 1393: 57–62.

Ajesh, K., and K. Sreejith. 2009. Peptide antibiotics: an alternative and effective antimicrobial strategy to circumvent fungal infections. *Peptide*, 30: 999–1006.

Amaral, P., M. Colao, M.A. Coelho, G. Fontes, and M. Nele. 2009. Characterization of a bioemulsifier produced from glycerol and gloucose by *Yarrowia lipolytica*. *New Biotechnol.*, 25: s138.

Bachere, E., D. Destoumieux, and P. Bulet. 2000. Penaeidins, antimicrobial peptides of shrimp: a comparison with other effectors of innate immunity. *Aquaculture* 191: 71–88.

Batista, S.B., A.H. Mounteer, F.R. Amorim, and M.R. Totola. 2006. Isolation and characterization of biosurfactant/bioemulsifier-producing bacteria from petroleum contaminated sites. *Bioresour. Technol.*, 97: 868–875.

Bento, F.M., F.A.D.O. Camargo, B.C. Okeke, and W.T. Frankenberger. 2005. Diversity of biosurfactant producing microorganisms isolated from soils contaminated with diesel oil. *Microbiol. Res.*, 160: 249–255.

Bernheimer, A.W. and L.S. Avigad, 1970. Nature and properties of a cytological agent produced by *Bacillus subtilis*. *J. Gen. Microbiol.* 61: 361–369.

Besson, F., F. Peypoux, G. Michel, and L. Delcambe. 1976. Characterization of iturin A in antibiotics from various strains of *Bacillus subtilis*. *J. Antibiot.*, 29: 1043–1049.

Bhattacharyya, S., M. Ghosh, and D.K. Bhattacharyya. 2003. *Pseudomonas* strains as source of microbial surface active molecules. *J. Oleo Sci.*, 52: 221–224.

Bodour, A.A., K.P. Drees, and R.M. Maier. 2003. Distribution of biosurfactant-producing bacteria in undisturbed and contaminated arid southwestern soils. *Appl. Environ. Microbiol.*, 69: 3280–3287.

Borzeix, C., and K. Frederique. 2003. Use of sophorolipids comprising diacetyl lactones as agent for stimulating skin fibroblast metabolism. U.S. Patent 6596265.

Brown, M.J. 1991. Biosurfactants for cosmetic application. *Int. J. Cosmet. Sci.* 13: 61–64.

Cameotra, S.S., and R.S. Makkar. 2004. Recent applications of biosurfactants as biological and immunological molecules. *Curr. Opin. Microbiol.*, 7: 262–266.

Carillo, C., J.A. Teruel, F.A. Aranda, and A. Ortiz. 2003. Molecular mechanism of membrane permeabilization by the peptide antibiotic surfactin. *Biochim. Biophys. Acta.*, 1611: 91–97.

Charlet, M., S. Chernysh, H. Philippe, C. Hetrut, J. Hoffmann, and P. Bulet. 1996. Innate immunity. Isolation of several cysteine-rich antimicrobial peptides from the blood of a mollusc, *Mytilus edulis*. *J. Biol. Chem.*, 271: 21808–21813.

Chen, S.Y., Y.H. Wei., and J.S. Chang. 2007. Repeated pH-stat fed-bath fermentation for rhamnolipid production with indigenous *Pseudomonas aeruginosa* S2. *Applied Microbiol. Biotechnol.* 76(1): 67–74.

Chiou, T.T., J.K. Lu, J.L. Wu, T.T. Chen, C.F. Ko, and J.-C. Chen. 2007. Expression and characterisation of tiger shrimp *Penaeus monodon* penaeidin (mo-penaeidin) in various tissues, during early embryonic development and moulting stages. *Dev. Comp. Immunol.*, 31: 132–142.

Chiou, T.T., J.L. Wu, T.T. Chen, and J.K. Lu. 2005. Molecular cloning and characterization of cDNA of penaeidin-like antimicrobial peptide from tiger shrimp (*Penaeus monodon*) marine biotechnology. *Marine Biotechnol.*, 7: 119–127.

Christofi, N., & Ivshina, I.B. 2002. Microbial surfactants and their use in field studies of soil remediation. *J. Appl. Microbiol.* 93: 915–929.

Cohen, R., and D. Exerowa. 2007. Surface forces and properties of foam films from rhamnolipid biosurfactants. *Adv. Colloid Interfac.*, 134: 24–34.

Cole, A.M., Rabih, R.O. Darouiche, D. Legarda, N. Connell, and G. Dimond. 2000. Characterization of a fish antimicrobial peptide: gene expression, subcellular localization, and spectrum of activity. *Antimicrob. Agents Chemother.*, 44: 2039–2045.

Cole, A.M., P. Weis, and G. Diamond. 1997. Isolation and characterization of pleurocidin: an antimicrobial peptide in the skin secretions of winter flounder. *J. Biol. Chem.*, 272: 12008–12013.

Cordell, G.A. 2002. Natural products in drug discovery—creating a new vision. *Phytochem. Rev.,* 1: 261–273.

Corley, J.W. 2007. All that is good: naturals and their place in personal care. In *Naturals and organics in cosmetics: from R&D to the market place*, 7–12. Allured, Carol Stream, IL.

Das, P., A.K. Mukherjee, and R. Sen. 2009. Antiadhesive action of a marine microbial surfactant. *Colloids Surf B Biointerfaces*, 71: 183–186.

Denger, K., and B. Schink. 1995. New halo- and thermotolerant fermenting bacteria producing surface-active compounds. *Appl. Microbiol. Biotechnol.*, 44: 161–166.

Desai, J.D., and I.M. Banat. 1997. Microbial production of surfactants and their commercial potential. *Microbiol. Mol. Biol. Rev.*, 61: 47–64.

Destoumieux, D., P. Bulet, D. Loew, A. Van Dorsselaer, J. Rodriguez, and E. Bachère. 1997. Penaeidines, a new family of antimicrobial peptides isolated from the shrimp *Penaeus vannamei* (Decapoda). *J. Biol. Chem.*, 272: 28398–28406.

Dhakephalkar, P.K., A.G. Banpurkar, and B.A. Chopade. 2010. Biosurfactants, bioemulsifiers and exopolysaccharides from marine microorganisms. *Biotechnol. Adv.*, 10.1016/j.biotechadv.2010.02.006.

Douglas, S.E., J.W. Gallant, Z. Gong, and C. Hew. 2001. Cloning and developmental expression of a family of pleurocidin-like antimicrobial peptides from winter flounder, *Pleuronectes americanus* (Walbaum). *Dev. Comp. Immunol.* 25: 137–147.

Douglas, S.E., A. Patrzykat, J. Pytyck, and J.W. Gallant. 2003a. Identification, structure and differential expression of novel pleurocidins clustered on the genome of the winter flounder *Pseudopleuronectes americanus* (Walbaum). *Eur. J. Biochem.*, 270: 3720–3730.

Douglas, S.E., J.W. Gallant, R.S. Liebscher, A. Dacanay, and S.C. Tsoi. 2003b. Identification and expression analysis of hepcidin-like antimicrobial peptides in bony fish. *Dev. Comp. Immunol.*, 27: 589–601.

Freitas, F., V.D. Alves, M. Carvalheira, N. Costa, R. Oliveira, and M.A.M. Reis. 2009. Emulsifying behaviour and rheological properties of extracellular polysaccharide produced by *Pseudomonas oleovorans* grown on glycerol by product. *Carbohydr. Polym.*, 78: 549–556.

Gautam, K.K., and V.K. Tiagi. 2006. Microbial surfactants: a review. *J. Oleo Sci.*, 55: 155–166.

Gilewicz, M., T. Ni'matuzahroh, H. Nadalig, H. Budzinski, P. Doumenq, V. Michotey, and J.C. Bertrand. 1997. Isolation and characterization of a marine bacterium capable of utilizing 2-methylphenanthrene. *Appl. Microbiol. Biotechnol.*, 48: 528–533.

Grau, A., J.C. Gomez Fernandez, F. Peypoux, and A. Ortiz. 1999. A study on the interactions of surfactin with phospholipid vesicles. *Biochim. Biophys. Acta*, 1418: 307–319.

Guglielmo, M., and D. Montanari. 2003. Cosmetic preparation with anti-wrinkle action. World Patent 2003/000222. Gecomwert Anstalt, Vaduz.

Harvey, P.W., and D.J. Everett. 2004. Significance of the detection of esters of p-hydroxybenzoic acid (parabens) in human breast tumours. *J. Appl. Toxicol.*, 24 (1): 1–4.

Healy, M.G., C.M. Devine, and R. Murphy. 1996. Microbial production of biosurfactants. *Resour. Conserv. Recy.*, 18: 41–57.

Hu, S.Y., J.H. Huang, W.-T. Huang, Y.-H. Yeh, M.H.C. Chen, H.-Y. Gong, T.T. Chiou, T.H. Yang, T.T. Chen, J.K. Lu, and J.L. Wu. 2006. Structure and function of antimicrobial peptide penaeidin-5 from the black tiger shrimp *Penaeus monodon*. *Aquaculture* 260: 61–68.

Hwang, E.Y., J.K. Seo, C.H. Kim, H.J. Go, E.J. Kim, et al. 1999. Purification and characterization of a novel antimicrobial peptide from the skin of the hagfish *Eptatretus burgeri*. *Int. J. Food Sci. Nutr.*, 4: 28–32.

Huang, X.F., W. Guan, J. Liu, L.J. Lu, J. Cheng, and X.Q. Zhou, 2010. Characterization and polygenetic analysis of bioemulsifier-producing bacteria. *Bioresour. Technol.*, 101: 317–323.

Husain, D.R., M. Goutx, M. Acquaviva, M. Gilewicz, and J.-C. Bertrand. 1997. The effect of temperature on eicosane substrate uptake modes by a marine bacterium *Pseudomonas nautica* strain 617: relationship with the biochemical content of cells and supernatants. *World J. Microbiol. Biotechnol.*, 13: 587–590.

Iijima, R., J. Kisugi, and M. Yamazaki. 2003. A novel antimicrobial peptide from the sea hare *Dolabella auricularia*. *Dev. Comp. Immunol.* 27: 305–311.

Joshi, S., C. Bharucha, and A.J. Desai. 2008. Production of biosurfactant and antifungal compound by fermented food isolate *Bacillus subtilis* 20B. *Bioresour. Technol.*, 99: 4603–4608.

Kalantar T.H., Tucker C.J., Zalusky A.S., Boomgaard T.A., Wilson B.E., Ladika M., Jordan S.L., Li W.K., Zhang X., Goh C.G. 2007. High throughput workflow for coacervate formation and characterization in shampoo systems. *J Cosmet Sci.* 58(4): 375–83.

Khoo, L., D.W. Robinette, and E.J. Noga. 1999. Callinectin an antibacterial peptide from the blue crab *Callinectes sapidus*, hemocytes. *Mar. Biotechnol.* 1: 44–51.

Konishi, M., T. Morita, T. Fukuoka, T. Imura, K. Kakugawa, and D. Kitamoto. 2007. Production of different types of mannosylerythritol lipids as biosurfactants by the newly isolated yeast strains belonging to the genus *Pseudozyma*. *Appl. Microbiol. Biotechnol.*, 75: 521–531.

Koopmans, R.J., and A. Aggeli. 2010. Nanobiotechnology—quo vadis? *Curr. Opin. Microbiol.*, 13:1–8.

Kosaric, N. 1992. Biosurfactants in industry. *Pure Appl. Chem.*, 64: 1731–1737.

Langer, O., O. Palme, V. Wray, H. Tokuda, and S. Lang. 2006. Production and modification of bioactive biosurfactants. *Process Biochem.*, 41: 2138–2145.

Lauth, X., H. Shike, J.C. Burns, M.E. Westerman, V.E. Ostland, J.M. Carlberg, J.C. Van Olst, V. Nizet, S.W. Taylor, C. Shimizu, and P. Bulet. 2002. Discovery and characterization of two isoforms of moronecidin, a novel antimicrobial peptide from hybrid striped bass. *J. Biol. Chem.* 277: 5030–5039.

Lazarovici, P., N.I. Primor, and L.M. Loew. 1986. Purification and pore-forming activity of two hydrophobic polypeptides from the secretion of the Red Sea Moses sole (*Pardachirus marmoratus*). *J. Biol. Chem.* 261: 16704–16713.

Lee, D.-Y., C.-M. Huang, T. Nakatsuji, D. Thiboutot, S.-A. Kang, M. Monestier, and R.L. Gallo. 2009. Histone H4 is a major component of the antimicrobial action of human sebocytes. *J. Invest. Dermatol.* 129(10): 2489–2496.

Lu, J.K. 2009. Peptide having antimicrobial and/or biosurfactant property. World Patent 2009/044279. UMO, Tokyo, Japan.

Lu, J.K. Z.H. Lai., 2009. An antimicrobial peptide from giant grouper (*Epinephelus lanceotatus*). Republic of China patent I 316944, YangSheng, Inc. Taiwan.

Lu, J.K., J.L. Wu, and T.T. Chiou. 2010. Antimicrobial peptide isolated from *Penaeus monodon*. U.S. Patent 7670836 B2. Acdemia Sinica, Taiwan.

Maget-Dana, R., and M. Ptak. 1995. Interactions of surfactin with membrane models. *Biophys. J.* 68: 1937–1973.

Maneerat, S. 2005. Biosurfactants from marine microorganisms. *Songklanakarin J. Sci. Technol.* 27(6): 1263–1272.

Maneerat, S., T. Bamba, K. Harada, A. Kobayashi, H. Yamada, and F. Kawai. 2005a. A novel crude oil emulsifier excreted in the culture supernatant of a marine bacterium, *Myroides* sp. strain SM1. *Appl. Microbiol. Biotechnol.* (online).

Maneerat, S., T. Nitoda, H. Kanzaki, and F. Kawai. 2005b. Bile acids are new products of a marine bacterium, *Myroides* sp. strain SM1. *Appl. Microbiol. Biotechnol.*, 67: 679–683.

Martinez-Checa, F., F.L. Toledo, K.E. Mabrouki, E. Quesada, and C. Calvo. 2007. Characteristics of bioemulsier V2-7 synthesized in culture media added of hydrocarbons: chemical composition, emulsifying activity and rheological properties. *Bioresour. Technol.*, 98: 3130–3135.

Mitta, G., F. Vandenbulcke, and P. Roch. 2000. Original involvement of antimicrobial peptides in mussel innate immunity. *FEBS Lett.* 486: 185–190.

Miyata, T., F. Tokunaga, T. Yoneya, K. Yoshikawa, S. Iwanaga, M. Niwa, T. Takao, and Y. Shimonishi. 1989. Antimicrobial peptides, isolated from horseshoe crab hemocytes, tachyplesin II, and polyphemusins I and II: chemical structures and biological activity. *J Biochem.* 106 (4): 663–668.

Montanari, D., and M. Guglielmo. 2008. Cosmetic composition for the treatment and/or prevention of skin stretch marks. World Patent 2008/080443. Labo Cosprophar Ag, Basel.

Nakamura, T., H. Furunaka, T. Miyata, F. Tokunaga, and T. Muta. 1988. Tachyplesin, a class of antimicrobial peptide from the hemocytes of the horseshoe crab (*Tachypleus tridentatus*): isolation and chemical structure. *J. Biol. Chem.*, 263: 16709–16713.

Nerurkar A.S., K.S. Hingurao, and H.G. Suthar 2009. Bioemulsifier from marine microorganisms. *J. Sci. Indus. Res.* 68: 273–277.

Ogawa, Y., H. Kawahara, and N. Yagi. 1999. Synthesis of a novel lipopeptides with Art was here, but was deleted.-melanocyte-stimulating hormone peptide ligand and its effect on liposome stability. *Lipids*, 34: 387–394.

Park, C.B., J.H. Lee, I.Y. Park, M.S. Kim, and S.C. Kim. 1997. A novel antimicrobial peptide from the loach, *Misgurnus anguillicaudatus. FEBS Lett.*, 411: 173–178.

Patrzykat, A., J.W. Gallant, J.-K. Seo, J. Pytyck, and S.E. Douglas. 2003. Novel antimicrobial peptides derived from flatfish genes. *Antimicrob. Agents Chemother.* 47: 2464–2470.

Poremba, K., W. Gunkel, S. Lang, and F. Wagner. 1991. Marine biosurfactants. III. Toxicity testing with marine microorganisms and comparison with synthetic surfactants. *Z. Naturforsch.*, 46c: 210–216.

Pruti, V., and S.S. Cameotra. 1997. Production of a biosurfactant exhibiting excellent emulsification and surface active properties by *Serratia marcescens. World J. Microbiol. Biotechnol.*, 13: 133–135.

Rahman, M.S., and T. Ano. 2009. Production characteristics of lipopeptide antibiotics in biofilm fermentation of *Bacillus subtilis. J. Environ. Sci.*, 21: s36–s39.

Razafindralambo, H., M. Paquot, and A. Baniel. 1996. Foaming properties of surfactin, a lipopeptides biosurfactant from *Bacillus subtilis. J. Am. Oil Chem. Soc.* 73: 149–151.

Razafindralambo, H., M. Paquot, and A. Baniel. 1997. Foaming properties of a natural cyclic lipoheptapeptide belonging to a special class of amphiphilic molecules. *Food Hydrocol.*, 11: 59–62.

Rodrigues, L., I.M. Banat, J. Teixeira, and R. Oliveira. 2006a. Biosurfactants: potential applications in medicine. *J. Antimicrob. Chemother.*, 57: 609–618.

Rodrigues, L., H.C.V. Mei, J. Teixeira, and R. Oliveira. 2004. Influence of biosurfactants from probiotic bacteria on formation of biofilms on voice prostheses. *Appl. Environ. Microbiol.*, 70: 4408–4410.

Schramm, L.L., E.N. Stasiuk, and D.G. Marangoni. 2003. Surfactants and their applications. *Ann. Rep. Program Chem. Sec.*, 99: 3–48.

Schulz, D., A. Passeri, M. Schmidt, S. Lang, F. Wagner, V. Wray, and W. Gunkel. 1991. Marine biosurfactants. I. Screening for biosurfactants among crude oil degrading marine microorganisms from the North Sea. *Z. Naturforsch.*, 46c: 197–203.

Schnapp, D., Reid, C.J., and A. Harris, 1998. Localization and expression of human defenesin-1 in the pancreas and kidney. *J. Pathol.* 186: 99–103.

Shike, H., X. Lauth, M.E. Westerman, V.E. Ostland, J.M. Carlberg, et al. 2002. Bass hepcidin is a novel antimicrobial peptide induced by bacterial challenge. *Eur. J. Biochem.*, 269: 2232–2237.

Silphaduang, U., and E.J. Noga. 2001. Peptide antibiotics in mast cells of fish. *Nature*, 414: 268–269.

Singh, A., J.D. Hamme, and O.P. Ward. 2007. Surfactants in microbiology and biotechnology. Part 2. Application aspects. *Biotechnol. Adv.*, 25: 99–121.

Tamarkin, D., D. Friedman, and M. Eini. 2008. Oleaginous pharmaceutical and cosmetic foam. U.S. Patent 0063 607. Foamix Ltd., Miami.

Taylor S. 2006. Government Printing Office, Federal Register, Aug. 29, 2006; Vol. 71:pp. 51146–51155.

Tuleva, B.K., G.R. Ivanov, and N.E. Christova. 2002. Biosurfactant production by a new *Pseudomonas putida* strain. *Z. Naturforsch. C*, 57: 356–360.

Ueno, Y., N. Hirashima, Y. Inoh, T. Furuno, and M. Nakanishi. 2007. Characterization of biosurfactant-containing liposomes and their efficiency for gene transfection. *Biol. Pharm. Bull.*, 30: 169–172.

Vollenbroich, D., M. ÖZel, J. Vater, R.M. Kamp, and G. Pauli. 1997. Mechanism of inactivation of enveloped viruses by the biosurfactant surfactin from *Bacillus subtilis*. 25: 289–297.

Zhao, Z., Q. Wang, K. Wang, K. Brian, C. Liu, and Y. Cou. 2010. Study of the antifungal activity of *Bacillus vallismortis* ZZ 185 *in vitro* and identification of its antifungal components. *Bioresour. Technol.*, 101: 292–297.

Zinjarde, S.S., and A. Pant. 2002. Emulsifier from a tropical marine yeast, *Yarrowia lipolytica* NCIM 3589. *J. Basic Microbiol.*, 42: 67–73.

27 Industrial Prospects of the Cosmeceuticals Derived from Marine Mucin

Se-Kwon Kim and Fatih Karadeniz

CONTENTS

27.1 INTRODUCTION

Mucins are of much interest currently among pharmaceutical and cosmeceutical researchers. The potential of natural sources for novel bioactive agents as well as biomarkers of common diseases has turned the attention of researchers to elucidating natural products of high bioactivity with fewer side effects. In this context, marine-based sources such as animals, plants, bacteria, and fungi have been thoroughly studied and considered as alternatives to find new therapeutic agents (Haefner, 2003; Hill and Fenical, 2010). Until now, several cosmeceutical compounds have been extracted and characterized from marine sources (Kim et al., 2008). As a reported cosmeceutical agent, mucin has also started to be isolated in different forms with higher bioactivity than marine mollusks (Kimura et al., 2003).

Mucins belong to a family of large, extracellular, high molecular weight O-glycosylated proteins that are formed by tandem repeats of amino acid chains rich in cysteine, serine, and threonine coated with oligosaccharide side chain (Sheehan et al., 1991; Gerken, 1993). Mucins' gel-like structures make them able to serve in lubrication, cell signaling, and most importantly, forming chemical barriers in mainly epithelial and connective tissues, as well as several tissues with mucosa layers (Linden et al., 2008). *In vivo*, mucins are found to be in secreted and membrane-bound forms. Secreted mucins have been initially identified as produced by epithelia with the ability to form viscoelastic gels that are responsible for the elasticity and liquid state of mucus, while membrane-bound mucins are reported to have roles in cell-cell and cell-extracellular matrix signaling. In addition, membrane-bound mucins have been studied for their possible biological role in cancer development (McGuckin et al., 1995; Ajioka et al., 1996).

The tandem amino acid repeats with linked sugar chains of mucins make them able to hold significant amounts of water while protecting the compound to undergo proteolysis. Therefore mucins can easily keep their mucosal state, enabling their utilization in terms of protective or drug delivery agent (Ahuja et al., 1997). Reported antibacterial effects of some mucins and their derivatives increase the ability of promoting mucins as an important cosmeceutical lead for upcoming discoveries of novel products (Slomiany et al., 1996).

In addition to mentioned properties of human mucins, several other mucins have been isolated from marine sources. It has been suggested that these mucins are of higher efficiency in terms of skin protection and moisturizing. In this regard, further detailed researches are to be carried out to improve mucins' ability to be used in the cosmeceutical area.

27.2 STRUCTURE OF MUCIN

Mucins are unique in terms of biophysical and biochemical properties due to their large size and densely glycosylated regions. They are quite large glycoproteins with molecular weights ranging from 0.5 to 20 MDa. These extracellular, highly glycosylated proteins are found in secreted and membrane-bound forms that share many common features. About 50 to 80% of mucin molecular mass consists of highly glycosylated parts, largely N-acetylglucosamine, N-acetylgalactosamine, N-acetylneuraminic acid, fucose, and galactose (Offner and Troxler, 2000). The high hydrophobicity of the carbohydrate part is the integrated part of the membrane-bound mucins. The carbohydrate chains are linked to the protein core of the mucins by O-linked glycosylation of hydroxyl side chains of threonine and serine, which primarily make up the protein core (Perez-Vilar and Hill, 1999). The rest of the molecular mass (20 to 50%) is formed by the protein core. The protein structure of mucin includes distinct regions. Extensive tandem repeats of serine, threonine, and proline comprise the main glycosylation region through their hydroxyl side chains, which enables the O-linked glycosylation of carbohydrate chains of mucin. Other regions are amino and carboxyl ends of the protein core and have high amounts of cysteine. Large amounts of cysteine in these regions are suggested to be responsible for globular formation and the subsequent polymerization of forming multimers from dimers via disulfide bond formation (Carlstedt et al., 1983).

Mucins are a family of proteins that consist of several members, as all animals, including marine ones, produce mucin and mucin-like glycoproteins. Although mucins share common features, mucins from different origins show distinct properties, as the content of carbohydrate chain and protein core and the fashion of O-glycans uniquely vary and result in various efficiencies in terms of biophysical properties (Park, 2007).

Complex O-glycosylation and the large size of the molecule make the characterization of mucins in a biochemical manner difficult. However, as interest in mucin is growing with time, mucins from several organisms have been elucidated and tested for their unique properties as therapeutical agents to date.

27.3 SOURCES OF MUCIN

Among other compounds of extracellular matrix and mucosa, mucins exert their effects while they are introduced externally. Hence the production of these compounds in related industries on a large-scale basis exhibits great advantage as pharmaceuticals, cosmetics, and food additives. However, in case of mucin, dietary supply is on a basic level, and mass production has not been introduced yet. Therefore mucin is considered to be extracted and purified from raw materials.

As the prime component of mucus layers in animals, mucin is found abundantly in various organisms. Recently, human mucins were thoroughly identified in both gene expression and glycoprotein secretion levels (Moniaux et al., 2001; Thornton et al., 2008). The close relation between the mucin gene and cancer development argues for the detailed identification and revealing of the mechanism of mucin function and production (Wakatsuki et al., 2008). In this context, the biochemistry of mucins has been studied heavily in recent years, which helps researchers to study the unique properties and therapeutic actions of mucins from different sources.

Mucins are usually isolated from mammalian saliva and intestinal epithelia. Several mucin genes in humans have been identified and studied in detail (Moniaux et al., 2001). However, characterization of human mucin is aimed to identify its role in diseases rather than its therapeutic activity. Various diseases are reported to be in close relation with the mucosa layer and mucin precisely. In order to treat mucin-related diseases or find novel bioactive mucins, several organisms are of great interest for mucin isolation, such as slugs and mollusks, which are rich in mucus layers and consist of high amounts of unique mucins.

27.4 MARINE MUCIN

With marine species comprising approximately one-half of the total global biodiversity, the sea offers an enormous resource for bioactive compounds to be extracted and purified for therapeutic purposes (Aneiros and Garateix, 2004). Furthermore, various kinds of substances have been acquired from marine organisms on the grounds that they are living in a quite oppressing, competitive, and aggressive environment—very diverse in many aspects from the terrestrial environment, a situation that enables the production of unique and potent active molecules. In this regard, marine resources are considered to be important on account of the need for isolation of novel mucin-like substances to aid their potential pharmaceutical utilization.

In addition to terrestrial slugs, marine species comprise numerous organisms that produce various mucins to obtain protective barriers and lubrication (Urai et al., 2009; Kimura et al., 2004; Ohta et al., 2009). Until now, mollusks have been the main source of extracellular matrix and mucosa-based substances because they exhibit quite large amounts of mucus layers to protect themselves in the marine environment. Recent studies have shown that jellyfish have the potential to be a significant resource of mucin and mucin-like glycoproteins (Masuda et al., 2007). Several researchers have isolated and characterized distinctive mucin components from various jellyfish species. On the other hand, for therapeutic purposes, some species of squid are reported to produce bioactive mucins and offer the ability to extract large amounts of mucin with a designed protocol (Kimura, 2005). High amounts of mucin can be extracted from marine mollusks in addition to the diversity of components that have already been identified. Therefore mucin from marine resources ought to be investigated in further detail, in consideration of the importance of mucin in therapeutic applications.

27.5 MARINE MUCINS AS COSMECEUTICALS

As mentioned earlier, marine organisms are an extremely heterogeneous group in the ocean and are excellent reservoirs for identifying and extracting biologically active substances with a potential to act as cosmeceuticals (Kim et al., 2008). Therefore with an increasing interest to investigate the potential biological activities of mucin, marine resources are on the stage for procuring diverse mucin-like glycoproteins.

Mucins and mucin-like glycoproteins are considered to have unique physical properties due to their high sugar coating and a protein core that varies in different tissues or organisms. The repeating structure of carbohydrate chains of mucin, as well as tandem repeating amino acid chains, promote mucin as a useful compound in various areas of interest. However, as a secretory large protein, *in vitro* utilization of mucin is limited. Novel marine mucin-like glycoproteins are purified and characterized in order to present new applications for mucin rather than *in vitro* utilization. On the other hand, as the interest in mucin is growing, a way to produce mucin that can be utilized in addition to its application may be exerted. Additionally, effects of mucin have been suggested to be widened or improved by using a composition with several components, such as gum, honey, disinfectants, gelatin, preservatives, and so on (Momoh et al., 2008; Adikwu and Alozie, 2007).

As reported by studies to date, mucin from different organisms possesses numerous effects and uses, such as retaining moisture of mucosa layers and epithelia, lubrication, tissue protection, substitute for original mucus-based layers like saliva, drug delivery carrier for effective absorption, and a barrier for protection against bacteria, viruses, and infectious components like tissue particles (Svensson and Arnebrant, 2010). Considering these effects, it can be easily suggested that mucin holds quite good potential to be a highly effective cosmeceutical. Cosmeceutical use of marine mucin, however, has not been studied widely yet. Nonetheless, there are reports that present valuable information regarding how significant is further research on mucin efficiency and its applications in the pharmaceutical and cosmetic industries.

In this context, recently several products, including snail mucin, have been made available on the market that claim to be effective cosmeceutical agents with moisturizing, wound healing, and antibacterial effects. Wound healing properties of snail mucin have been recently reported (Adikwu and Ikejiuba, 2005). Snail mucin application significantly reduced the wound size on the fourth day of application in comparison to nontreated wounds. Moreover, administration of snail mucin with honey exerted a higher wound healing effect than that of mucin alone. In addition to wound healing effect, snail mucin showed an antibacterial effect against both gram-positive and gram-negative bacterial strains (Iguchi et al., 1982). Recent studies also reported mucin as an enhancer for insulin absorption and a potential agent for drug delivery (Adikwu, 2005). In light of detailed studies on the above-mentioned properties of snail mucin, marine equivalents or derivatives of mucin-like glycoproteins have also attracted the attention of researchers in order to develop novel products of complementary treatment and food additive industries.

A novel mucin, qniumucin, has been isolated from several species of jellyfish and its structure has been characterized (Masuda et al., 2007). This mucin is reported to be similar to human-type mucin MUC5AC, which also consists of eight residues. The study has shown that this mucin has a repeating tandem of Val-Val(Ile)-Glu-Thr(-GalNAc)-Thr(-GalNAc)-Ala-Ala-Pro. Interestingly, it has been exhibited that qniumucin comprises some different sugars in its glycosylation parts, along the threonine-N-acetyl-D-galactosamine linkage. Unlike vertebrate mucins, this marine mucin does not include sialic acid. In accordance with these unique properties, the potential of this mucin to be used as a therapeutic agent is also exhibited. Mucin from moon jelly (*Aurelia aurita*) was investigated for hygroscopic and moisturizing properties in comparison with a strong moisturizing agent, hyaluronic acid, which is being used widely in cosmetics. The results clearly showed that this marine mucin from jellyfish exhibits three times more moisturizing and hygroscopic activity than hyaluronic acid (Ushida et al., 2008). Furthermore, chemical and biophysical properties of this marine mucin lead the way of utilization and application of this compound as a protective biomaterial with lubrication and moisturizing effects, as well as a starting material of large-scale industrial synthesis of bioactive mucin-like glycoproteins. In addition, mucin extraction yield from the jellyfish was quite high and paves the way for further studies and use in the cosmeceutical and food industries.

Moreover, recently a marine mucin has been isolated from the nidamental gland mucilage of a marine mollusk, *Todarodes pacificus*, at the factory production level by solubilization with a mild alkali followed by precipitation with alcohol containing NaCl (Aso and Kimura, 2006). The protocol that has been used for the isolation of this marine mucin is considered to strengthen the possibility of extraction of large-scale mucins to be used by the cosmeceutical industry. Similar to qniumucin, this mucin also does not include sialic acid and shows unique properties that are different than vertebrate mucins. The protein core of this mucin consists of threonine, proline, and isoleucine, and has a protein:sugar:sulfate weight ratio of 16.4:80.3:3.3. It has been reported that the aqueous solution is very viscous and transparent. According to the reported properties of this mucin from squid, it is highly expected that marine mucin from squid is a new and efficient cosmeceutical with high moisturizing and skin-protecting functions due to its similarity in structure but diversity in consistency and content (Aso et al., 2003).

27.6 OTHER INDUSTRIAL APPLICATIONS OF MUCIN

Concentration of the secreted mucin is the key aspect of the physical state of mucus, and its dependence on environmental factors such as ionic strength and pH is reported to be involved in several diseases (Lee et al., 2005). Mucus, with mucin as its main component, serves as a physical barrier to extracellular toxins and mainly bacteria. On the other hand, some bacteria possess adhesins that bind to mucin and let the bacteria reside in the mucus. This concept is reported to be the cause of several bacterial infections (Dharmani et al., 2008). For example, *Helicobacter pylori* residing in the mucosa layer of the stomach is the main cause of ulcers (Slomiany and Slomiany, 1991). Therefore keeping physiochemical properties as a barrier to bacteria offers a promising way to keep mucus free of bacterial infection. Several studies have shown that mucins from different tissues and organisms are able to suppress bacterial adhesion (Shi et al., 2000). Correspondingly, snail mucus mucin is reported to possess strong antibacterial activity against both gram-positive bacteria, *Bacillus subtilis* and *Staphylococcus aureus*, and gram-negative bacteria, *Escherichia coli* and *Pseudomonas aeruginosa* (Iguchi et al., 1982). Interestingly, mucins have the ability of not only showing antibacterial activity but also acting as a protective surfactant to cover biomaterials in order to suppress immunological response (Sandberg et al., 2009). Moreover, the lubricative activity of mucin leads the way to produce oral surface covering substitutes for saliva (Berg et al., 2004). Additionally, these properties of mucin are to be investigated and developed to prevent postsurgical adhesions, present ease of clinical usage of stents, and protective treatment for contact lenses (Svensson and Arnebrant, 2010). Overall, mucins possess numerous biological effects that attract future studies in order to develop industrial approaches toward the large-scale production of mucins in the areas of pharmaceutical, cosmeceutical, and food industries.

In addition to being a physical barrier, mucus layers are the first step of nutrient and drug diffusion. Further, optimizing the mucus adhesion of nutrients and drugs is gaining more and more interest for improved absorption of nutrients and more importantly, directed drug delivery. In this context, the gel-forming ability of mucins that protect the molecule from the acidic proteolysis promotes great potential for designing pH-specific drug delivery. Besides, among bioadhesive drug delivery methods, mucoadhesive drug delivery systems are gaining importance in the pharmaceutical field due to their therapeutic advantage in controlling the rate and amount of drug release (Shaikh et al., 2011). This activity of mucin in drug delivery promotes mucin studies to be a promising novel area in the medicine industry, especially in drug delivery applications.

27.7 CONCLUSION

Marine mucin studies have attracted quite a lot of interest recently. Several mucin-like glycoproteins have been isolated and characterized from jellyfish, squid, marine worms, and so on. Novel marine mucins have been reported to have unique properties and contents that are different from those of terrestrial mucin. Already studied properties of terrestrial mucin have shown the significance of mucin in wound healing, moisturizing, and skin protection from bacteria, viruses, and inflammatory infections. However, little has been studied about the possible cosmeceutical effect of purified marine mucin. Nonetheless, the reported data about the moisturizing and skin protective effects of marine mucin are quite promising, and further studies into detailed effects of marine mucin in terms of cosmeceutical efficiency are expected to produce important results by way of wide applications of marine mucin.

ACKNOWLEDGMENTS

This work was supported by a grant from the Marine Bioprocess Research Center of the Marine Bio 21 Center funded by the Ministry of Land, Transport and Maritime, Republic of Korea.

REFERENCES

Adikwu, MU. 2005. Evaluation of snail mucin motifs as rectal absorption enhancer for insulin in non-diabetic rat models. *Biological and Pharmaceutical Bulletin* 28 (9):1801–1804.

Adikwu, MU, and BU Alozie. 2007. Application of snail mucin dispersed in detarium gum gel in wound healing. *Sci Res Essay* 2:195–198.

Adikwu, MU, and CC Ikejiuba. 2005. Some physicochemical and wound healing properties of snail mucin. *Bollettino Chimico Farmaceutico* 145 (2):5.

Ahuja, A, RK Khar, and J Ali. 1997. Mucoadhesive drug delivery systems. *Drug Development and Industrial Pharmacy* 23 (5):489–515.

Ajioka, Y, LJ Allison, and JR Jass. 1996. Significance of MUC1 and MUC2 mucin expression in colorectal cancer. *Journal of Clinical Pathology* 49 (7):560.

Aneiros, A, and A Garateix. 2004. Bioactive peptides from marine sources: Pharmacological properties and isolation procedures. *Journal of Chromatography B* 803 (1):41–53.

Aso, Y, and S Kimura. 2006. Development and application of marine mucin for cosmetics. *Fragrance Journal* 34 (3):14.

Aso, Y, S Kimura, and T Miyata. 2003. New marine cosmetic materials: Sharkskin collagen and marine mucin. *Fragrance Journal* 31 (5):104–108.

Berg, H, L Lindh, and T Arnebrant. 2004. Intraoral lubrication of PRP-1, statherin and mucin as studied by AFM. *Biofouling* 20 (1):65–70.

Carlstedt, I, H Lindgren, and JK Sheehan. 1983. The macromolecular structure of human cervical-mucus glycoproteins. Studies on fragments obtained after reduction of disulphide bridges and after subsequent trypsin digestion. *Biochemistry Journal* 213 (2):427–35.

Dharmani, P, V Srivastava, V Kissoon-Singh, and K Chadee. 2008. Role of intestinal mucins in innate host defense mechanisms against pathogens. *Journal of Innate Immunity* 1 (2):123–135.

Gerken, TA 1993. Biophysical approaches to salivary mucin structure, conformation and dynamics. *Critical Reviews in Oral Biology and Medicine* 4 (3):261.

Haefner, B. 2003. Drugs from the deep: Marine natural products as drug candidates. *Drug Discovery Today* 8 (12):536–544.

Hill, RT, and W Fenical. 2010. Pharmaceuticals from marine natural products: Surge or ebb? *Current Opinion in Biotechnology* 21 (6):777–779.

Iguchi, SM, T Aikawa, and JJ Matsumoto. 1982. Antibacterial activity of snail mucus mucin. *Comparative Biochemistry and Physiology Part A: Physiology* 72 (3):571–574.

Kim, SK, Y Ravichandran, S Khan, and Y Kim. 2008. Prospective of the cosmeceuticals derived from marine organisms. *Biotechnology and Bioprocess Engineering* 13 (5):511–523.

Kimura, S 2005. Studies on utilization of marine mucin. *Bio Industry* 22 (5):74–79.

Kimura, S, T Gohda, and Y Sakurai. 2003. Characterization of nidamental mucin from Japanese common squid *Todarodes pacificus*. *Journal of Tokyo University of Fisheries* 89:7–14.

Kimura, S, Y Higuchi, M Aminaka, JR Bower, and Y Sakurai. 2004. Chemical properties of egg mass mucin complexes of the ommastrephid squid *Todarodes pacificus*. *Journal of Molluscan Studies* 70 (2):117.

Lee, SH, M Müller, K Rezwan, and ND Spencer. 2005. Porcine gastric mucin (PGM) at the water/poly(dimethylsiloxane) (PDMS) interface: Influence of pH and ionic strength on its conformation, adsorption, and aqueous lubrication properties. *Langmuir* 21 (18):8344–8353.

Linden, SK, P Sutton, NG Karlsson, V Korolik, and MA McGuckin. 2008. Mucins in the mucosal barrier to infection. *Mucosal Immunology* 1 (3):183–197.

Masuda, A, T Baba, N Dohmae, M Yamamura, H Wada, and K Ushida. 2007. Mucin (qniumucin), a glycoprotein from jellyfish, and determination of its main chain structure. *Journal of Natural Products* 70 (7):1089–1092.

McGuckin, MA, MD Walsh, BG Hohn, BG Ward, and RG Wright. 1995. Prognostic significance of MUC1 epithelial mucin expression in breast cancer. *Human Pathology* 26 (4):432–439.

Momoh, MA, MU Adikwu, and SO Eraga. 2008. The effect of cast films of bovine mucin and prosopis gum containing a Cicatrin® powder on wound healing. *Dhaka University Journal of Pharmaceutical Sciences* 7 (1):7–10.

Moniaux, N, F Escande, N Porchet, JP Aubert, and SK Batra. 2001. Structural organization and classification of the human mucin genes. *Frontiers in Biosciences* 6 (1):1192–1206.

Offner, GD, and RF Troxler. 2000. Heterogeneity of high-molecular-weight human salivary mucins. *Advances in Dental Research* 14 (1):69–75.

Ohta, N, M Sato, K Ushida, et al., 2009. Jellyfish mucin may have potential disease-modifying effects on osteoarthritis. *BMC Biotechnology* 9 (1):98.

Park, M. 2007. Viscosity and wettability of animal mucin solutions and human saliva. *Oral Diseases* 13 (2):181–186.

Perez-Vilar, J, and RL Hill. 1999. The structure and assembly of secreted mucins. *Journal of Biological Chemistry* 274 (45):31751–31754.

Sandberg, T, MK Ott, J Carlsson, A Feiler, and KD Caldwell. 2009. Potential use of mucins as biomaterial coatings. II. Mucin coatings affect the conformation and neutrophil-activating properties of adsorbed host proteins—Toward a mucosal mimic. *Journal of Biomedical Materials Research Part A* 91A (3):773–785.

Shaikh, R, TR Singh, MJ Garland, AD Woolfson, and RF Donnelly. 2011. Mucoadhesive drug delivery systems. *Journal of Pharmacy and Bioallied Sciences* 3 (1):89.

Sheehan, JK, DJ Thornton, M Somerville, and I Carlstedt. 1991. Mucin structure. The structure and heterogeneity of respiratory mucus glycoproteins. *American Review of Respiratory Disease* 144 (3 Pt 2):S4.

Shi, L, R Ardehali, KD Caldwell, and P Valint. 2000. Mucin coating on polymeric material surfaces to suppress bacterial adhesion. *Colloids and Surfaces B: Biointerfaces* 17 (4):229–239.

Slomiany, BL, VL Murty, J Piotrowski, and A Slomiany. 1996. Salivary mucins in oral mucosal defense. *General Pharmacology: The Vascular System* 27 (5):761–771.

Slomiany, BL, and A Slomiany. 1991. Role of mucus in gastric mucosal protection. *Journal of Physiological Pharmacology* 42 (2):147–61.

Svensson, O, and T Arnebrant. 2010. Mucin layers and multilayers—Physicochemical properties and applications. *Current Opinion in Colloid and Interface Science* 15 (6):395–405.

Thornton, DJ, K Rousseau, and MA McGuckin. 2008. Structure and function of the polymeric mucins in airways mucus. *Annual Review of Physiology* 70:459–486.

Urai, M, T Nakamura, J Uzawa, et al., 2009. Structural analysis of O-glycans of mucin from jellyfish (*Aurelia aurita*) containing 2-aminoethylphosphonate. *Carbohydrate Research* 344 (16):2182–2187.

Ushida, K, A Masuda, N Dohmae, H Furukawa, and A Miyawaki. 2008. Novel mucin glycoprotein and use thereof. EP Patent 1,947,111.

Wakatsuki, K, Y Yamada, M Narikiyo, et al., 2008. Clinicopathological and prognostic significance of mucin phenotype in gastric cancer. *Journal of Surgical Oncology* 98 (2):124–129.

Index

B

D

G

H

M

T